Introduction to Differential and Algebraic Topology

T0180343

Kluwer Texts in the Mathematical Sciences

VOLUME 9

A Graduate-Level Book Series

The titles published in this series are listed at the end of this volume.

Introduction to Differential and Algebraic Topology

by

Yuri G. Borisovich

Voronezh State University,
Voronezh, Russia

Nikolai M. Bliznyakov

Voronezh State University,
Voronezh, Russia

Tatyana N. Fomenko

Moscow Institute of Steel and Alloys,
Moscow, Russia

and

Yakov A. Izrailevich

Voronezh State University,
Voronezh, Russia

KLUWER ACADEMIC PUBLISHERS
DORDRECHT / BOSTON / LONDON

A C.I.P. Catalogue record for this book is available from the Library of Congress.

ISBN 978-90-481-4558-4

Published by Kluwer Academic Publishers,
P.O. Box 17, 3300 AA Dordrecht, The Netherlands.

Kluwer Academic Publishers incorporates
the publishing programmes of
D. Reidel, Martinus Nijhoff, Dr W. Junk and MTP Press.

Sold and distributed in the U.S.A. and Canada
by Kluwer Academic Publishers,
101 Philip Drive, Norwell, MA 02061, U.S.A.

In all other countries, sold and distributed
by Kluwer Academic Publishers Group,
P.O. Box 322, 3300 AH Dordrecht, The Netherlands.

Printed on acid-free paper

This is a completely revised second edition of the original Russian work with
the same title, Nauka Moscow © 1980.

TABLE OF CONTENTS

PREFACE

Topology as a subject, in our opinion, plays a central role in university education. It is not really possible to design courses in differential geometry, mathematical analysis, differential equations, mechanics, functional analysis that correspond to the temporary state of these disciplines without involving topological concepts. Therefore, it is essential to acquaint students with topological research methods already in the first university courses.

This textbook is one possible version of an introductory course in topology and elements of differential geometry, and it absolutely reflects both the authors' personal preferences and experience as lecturers and researchers. It deals with those areas of topology and geometry that are most closely related to fundamental courses in general mathematics. The educational material leaves a lecturer a free choice in designing his own course or his own seminar.

We draw attention to a number of particularities in our book. The first chapter, according to the authors' intention, should acquaint readers with topological problems and concepts which arise from problems in geometry, analysis, and physics. Here, general topology (Ch. 2) is presented by introducing constructions, for example, related to the concept of quotient spaces, much earlier than various other notions of general topology thus making it possible for students to study important examples of manifolds (two–dimensional surfaces, projective spaces, orbit spaces, etc.) as topological spaces, immediately. Later, smooth structures are defined on them. The theory of two–dimensional surfaces is not confined to one section but is distributed over Ch. 1, Ch. 2, and Ch. 3 depending on the presentation of the basic ideas of topology in this book. The concepts of category and functor are introduced in the homotopy theory section (Ch. 3) and the idea of the algebraization of topological problems is explained as well. The functorial approach helps us to explain homotopy and homology theories uniformly and to complete the description of various homology theories with the Steenrod–Eilenberg axiomatics, compensating, to some extent, for the absence of a proof of the invariance of the simplicial homology theory in this textbook. Computational techniques in homotopy are limited to the calculation of the fundamental groups of the circle and closed surfaces. The equality $\pi_n(S^n) \cong \mathbf{Z}$, $n \geq 2$, (and a number of other equalities), however, are given without proof and serves as a basis for the introduction of the degree of a mapping of spheres and the characteristic of a vector field (with a deduction of the Brouwer fixed point theorem and a proof of fundamental theorem of algebra). It seems to us that a number of versions of homology theory (singular, simplicial, cellular) should be studied at a very early stage, since the reader can come across any of them in even the simplest applications. The versions mentioned above are explained in Ch. 5, where the concept of cohomology is also introduced. In the section on homology groups (Ch. 5) the tools are extended to exact sequences. In particular, the groups $H_k(S^n, \mathbf{Z})$, $H_k(\mathbf{R}P^n, \mathbf{Z})$, and $H_k(\mathbf{C}P^n, \mathbf{Z})$ are computed, and the Brouwer, Lefshetz, and Hopf fixed point theorems are proved; the theory of the degree of a mapping and the rotation of a vector field, which are important for applications are also developed. In spite of having everything prepared for a further development of the technique involved, we deliberately leave the subject at this stage in line with the idea of a textbook.

The concepts of a smooth manifold, smooth structure, and tangent bundle (Ch. 4) have been elaborated as precisely as possible, the general concept of a locally trivial fibre bundle is introduced, and the theory of coverings is presented. Special attention has been paid to the relations of the subject with mechanics, dynamical systems, and Morse theory.

A reader should take into account that in some of the sections (§ 4, Ch. 1; § 5, Ch. 2; § 9, Ch. 4) elements of complex variable function theory are applied. Therefore, while skipping through for the first time he can neglect these sections without any loss in understanding the following material. The exercises in the text of a section often replace simple arguments or proofs and are meant to stimulate the reader's reflection. Supplementary material is presented in smaller type. We indicate the end of the proofs with the sign □, and if it is necessary to separate an example from the following text, the sign ◇ is used.

Note that this textbook is based on lectures delivered by Yu. G. Borisovich to students of the Mathematics Department of Voronezh University. However, the text composed by N. M. Bliznyakov and T. N. Fomenko has been essentially revised by adding suplementary material by the lecturer together with N. M. Bliznyakov (Ch. 4), Ya. A. Israilevich (Ch. 4, 5), and T. N. Fomenko (Ch. 2, 3, 5). All the textual drawings are by T. N. Fomenko, while the cover and chapter title illustrations are by A. T. Fomenko (Moscow University), whom the authors would like to thank sincerely.

To conclude, we should like to express our sincere gratitude to a number of mathematicians: D. A. Anosov, A. V. Chernavsky, A. T. Fomenko, D. B. Fuks, S. V. Matveev, A. S. Mishchenko, S. P. Novikov, M. M. Postnikov, E. G. Skliarenko, Yu. P. Solovyov, and others, whose advice and critical remarks during different stages of work enabled improving the book. We also are grateful to a group of young staff members and post–graduates at the department of algebra and topological methods in analysis of Voronezh University for their helpful discussions and remarks, and are especially grateful to G. N. Borisovich for her support and technical help.

The authors

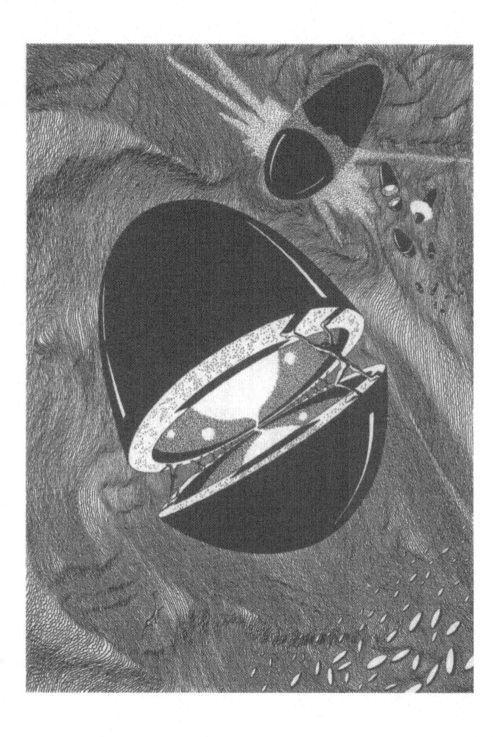

The main purpose of this chapter is to prepare the reader for the systematic study of topology as it is presented in the subsequent chapters. Here we give a popular review of some of the problems, of which the solution led to the formation of topology as a branch of mathematics and to its intensive development at present. The sources of the notions of topological spaces and manifolds are also discussed.

An object depicted in the foreground, a sphere cut along the circumference of a large circle, is one of the main examples of objects, of which various properties are analyzed in practically all parts of topology. The philosophical meaning of this composition is transparent. The mind of an inquisitive human being is always directed towards the kernel of things, cognition of essence; its persistence succeeds; but how small and helpless he seems to be in the Great and Eternal Cosmos.

CHAPTER I

FIRST NOTIONS OF TOPOLOGY

The purpose of this chapter is to prepare reader for the systematic study of topology as it is presented in the subsequent chapters. Here we review those problems whose study has led to the formation of topology as a mathematical discipline and its present extensive development. We also included some applications in modern physics.

§ 1. What is topology?

> Quant à moi, toute les voies diverses
> où je m'étais engagé sucessivement
> me conduisaient à l'Analysis Sitûs.[*]
>
> H. Poincaré

1. Topology as a science was, as it is generally believed, formed through the works of the great French mathematician Henri Poincaré at the end of the 19–th century. The first observations of a topological nature go back to L. Euler and C. F. Gauss. The beginning of topology research may be dated from the works of B. Riemann in the middle of the 19–th century. In his investigations of function theory, he developed new methods based on geometric representations. He also made an attempt to formulate the concept of a multi–dimensional manifold and to introduce higher orders of connectedness. These notions were specified by E. Betti (1871) but it was H. Poincaré who, proceeding from requirements of function theory and differential equations, introduced a number of most important topological concepts, developed a profound theory, and applied it to research in different fields of mathematics and

---

[*] As for me, all the various journeys, on which one by one I found myself engaged, were leading me to Analysis Situs (Position analysis).

mechanics. His ideas and the problems raised by him have had a considerable influence in the development of topology and its applications up to now.

Poincaré defined Analysis situs (as topology was called at that time) as follows: "L'Analysis Sitûs est la science qui nous fait connaître les propriétiés qualitatives des figures géometriques non seulment dans l'espace ordinaire, mais dans l'espace à plus de trois dimensions.

L'Analysis Sitûs à trois dimensions est pour nous une connaissance presque intuitive, L'Analysis Sitûs à plus de trois dimensions présente au contraire de difficultés enormes; il faut pour tenter de les surmonter être bien persuadé de l'extrême importance de cette science.

Si cette importance n'est pas comprise de tout le monde, c'est que tout le monde n'y a pas suffisamment réfléchi"*) [66, vol 3].

In order to understand what is meant by the qualitative properties of geometrical figures, imagine a sphere in the form of a rubber baloon that can be shrunk and stretched in any manner without tearing it or "gluing" any two distinct points together. These transformations of a sphere are called *homeomorphisms*, and different figures obtained as a result of homeomorphisms are said to be mutually *homeomorphic*.

Clearly, we can discuss homeomorphisms and qualitative properties of other figures as well. Such qualitative properties are usually called *topological properties*. In the example above, one topological property of a sphere, i.e. its one–wholeness (or connectedness), is obvious. Its more subtle properties are revealed if an attempt to establish homeomorphism of a sphere with a ball (or a solid sphere) is made. It is easy to conclude that such a homeomorphism is impossible. However, in order to prove this it is necessary to indicate different topological properties for a sphere and a ball. One of these properties is the "contractibility" of a ball into one of its points by means of a "smooth" changing (deformation), the contraction along its radii towards centre, and the "non–contractibility" of a sphere into any of its points. It is useful to pay attention to the topological difference between a volleyball bladder and an inner tube. These intuitive ideas need to be strictly founded.

---

*)Analysis situs is a science which lets us learn the qualitative properties of geometric figures not only in ordinary space, but also in spaces of more than three dimensions. Analysis situs in three dimensions we understand almost intuitively. Analysis situs in more than three dimensions presents, on the contrary, enormous difficulties, and to attempt to surmount them, one should be persuaded of the extreme importance of this science. If this importance is not understood by everyone, it is because everyone has not sufficiently reflected upon it.

Exercise 1°.   Proceeding from the evident assumptions, verify that: *a*) an annulus, i.e. a disc with a hole in it, is not homeomorphic to a disc; *b*) the number of "holes" in a geometrical figure is a topological property.

The research by Poincaré started one of the branches in topology, viz. combinatorial or algebraic topology. The essence of this method is to associate to geometrical figures by a rule common to all figures, certain algebraic objects (i.e. groups, rings, etc.) in such a way that the certain relations between figures correspond to the algebraic relations between objects. The study of the properties of those algebraic objects then sheds light on the properties of geometrical figures. The algebraic objects constructed by Poincaré are in fact the homology groups and the fundamental group.

The development of algebraic topology method led to interaction with the ideas of set–theoretic topology (G. Cantor, the end of the 19–th century; F. Hausdorff, the first decade of the 20–th century). Indeed, the study of qualitative properties of sets in spaces of arbitrary dimension later took the form of the idea of a topological space, i.e. a fundamental concept that has penetrated all of mathematics. It is not only related to the investigation of geometrical figures in finite–dimensional spaces, but due to the development of the theory of functions of a real variable and functional analysis, i.e. to the construction of function spaces, which are, as a rule, infinite–dimensional.

"The first sufficiently general definitions of a topological space occurred in the works of M. R. Fréchet, F. Riesz and F. Hausdorff. The complete definition of a topological space was formulated by K. Kuratowski and P. S. Alexandrov" [41].

Topological spaces and their continuous mappings, and the study of their general properties form one of the branches of topology known as "general topology".

Algebraic and set–theoretic directions in topology were combined in the work of L. E. J. Brouwer in his study of the notion of the dimension of a space (1908–1912). Later, it was considerably developed by J. W. Alexander, S. Lefschetz, P. S. Alexandrov, P. S. Uryson, H. Hopf, L. A. Lyusternik, L. G. Shnirelman, M. Morse, A. N. Tikhonov, L. S. Pontryagin, A. N. Kolmogorov, E. Čech, et al.

It is impossible to describe precisely the results (and even to formulate the problems) without an initial introduction to the elements of general and algebraic topology. Here, we only give an idea of some of the problems that have stimulated topological research.

If S^1 is a circle in the Euclidean plane $\mathbf{R}^2$, then a set $\mathbf{R}^2 \backslash S^1$ is decomposed into two mutually complementary open sets, viz. the interior A and the exterior B with respect to S^1. The circle S^1 serves as a boundary between A and B. Can a continuous simple path be drawn between arbitrary points $a \in A$ and $b \in B$ in such a way that it does not intersect the boundary S^1? (A *simple continuous path* is a homeomorphic mapping of the interval $[0, 1]$ of the real line into the plane). The answer is negative. In fact, if $\rho(x, y)$ is the Euclidean distance between points x, y of the plane $\mathbf{R}^2$, and $\gamma(t)$ is such a path, $0 \leq t \leq 1$, $\gamma(0) = a$, $\gamma(1) = b$, then the function $f(t) = \rho(\gamma(t), 0)$, where 0 is the centre of the circle, is continuous, and $f(0) < r$, $f(1) > r$, where r is the radius of the circle S^1. According to property of continuous functions $f(t)$ takes the value r at some point t_0, and, consequently, $\gamma(t_0) \in S^1$.

Now we replace the circle S^1 with a homeomorphic image Γ (such a curve is said to be *simple closed*). The question arises, whether the set $\mathbf{R}^2 \backslash \Gamma$ can be partitioned into non–intersecting open sets so that the curve Γ remains the boundary of each of them. The answer is positive (Jordan's theorem), but the proof involves subtle topological concepts. Thus, the curve Γ just as the circle S^1 bounds two open sets.

The problem gets even more complicated, if instead of a simple closed curve we deal with a homeomorphic image of an n–dimensional sphere lying in $(n + 1)$–dimensional Euclidean space. The generalization of Jordan's theorem for this case was presented by L. E. J. Brouwer in 1911–1913. A more extensive generalization of this result led to duality theorems (J. W. Alexander, L. S. Pontryagin, P. S. Alexandrov, et al.) which determined the development of algebraic topology for a long time thereafter.

Another important problem is the generalization of the concept of dimension. Dimension of an Euclidean space is well known as an algebraic concept; but is it a topological property, i.e. will homeomorphic Euclidean spaces be of the same dimension? The positive answer was given by A. Lebesgue (1911).

Concerning geometrical figures that lie in Euclidean spaces, to begin their study one needs to formulate also for them the concept of *dimension*. Poincaré was the first to propose such a definition. The dimension of an empty set is equal to -1. Now, by induction, if we already know the dimension up to $n - 1$, then the dimension n of a set A ($\dim A = n$) signifies that it can be partitioned into parts as small as wanted by a set of dimension $n - 1$, but it cannot be done by a set of dimension $n - 2$. These ideas were elaborated by L. E. J. Brouwer, K. Menger, P. S. Uryson, P. S. Alexandrov et al.

Another important direction in topology which is closely related to applications, is fixed–point theory. Even in algebra and the elements of analysis, we meet the question whether there exist solutions of equations of the form

$$(1) \qquad\qquad f(x) = 0,$$

where $f(x)$ is a polynomial or a more complicated function. Equation (1) is equivalent to the equation

$$(2) \qquad\qquad f(x) + x = x$$

or (writing $F(x) = f(x) + x$) to the equation

$$(3) \qquad\qquad F(x) = x$$

The solutions of equation (3) are called the *fixed–points of the mapping F*. If equation (1) deals with vectors, i.e. if it represents a system of equations in n ($n > 1$) unknowns, then the corresponding equation (3) refers to vectors, and, consequently, the fixed–points are in the multi–dimensional Euclidean space $\mathbf{R}^n$.

An extremely important problem is to find sufficiently general and efficient features that indicate the existence of fixed–points. L. E. J. Brouwer obtained a remarkable result with most extensive applications in modern research. It is surprisingly simple to formulate: any continuous mapping of a convex, bounded closed set into itself has a fixed–point. Convex sets can be considered both in three–dimensional and multi–dimensional Euclidean spaces. For example, a continuous mapping into itself of a closed (i.e. considered together with its boundary) disc in a plane, or a ball in a space necessarily has a fixed–point.

Exercise 2°. Show that the analogue of the Brouwer theorem for an annulus does not hold.

The Brouwer theorem was further developed by H. Hopf, S. Lefschetz et al. It was also generalized for mappings of function spaces (O. D. Kellog, G. D. Birkhoff, Yu. P. Schauder, J. Leray) which extended its applications. It should be noted that even Poincaré was interested in theorems of existence of fixed–points when reducing certain celestial problems to them.

2. Note that the problems discussed above are far from representing a complete set of topological problems. We now present other examples. B. Riemann was the first to introduce the notion of an *n–dimensional manifold* as a space, where points have n numerical coordinates defined at least on sufficiently small parts of the space. In contemporary mathematics topological and smooth manifolds are distinguished. This fact is connected with certain possibilities of compatible systems of coordinates which are given on separate parts of the manifold. Two parts of a manifold can intersect, and then the intersections possess different systems of coordinates; moreover, each system of coordinates can be expressed into the other by a continuous or smooth (differentiable) mapping. In the first case, the manifold is called topological, and in the second it is smooth.

As a generalization of the notion of surface in three–dimensional Euclidean space, the concept of manifold has embraced a number of geometrical objects that appear in classical mechanics, differential equations, and surface theory. Poincaré gave the final shape to the concept of a manifold and developed the basics of analysis on such spaces.

Later, these concepts were elaborated in smooth manifold theory (G. de Rham, L. S. Pontryagin, H. Whitney et al.). Using algebraic topology methods to such spaces there were associated new algebraic objects: "cohomology rings of exterior differential forms". Smooth manifolds themselves were "organized" into a "ring of internal homologies" (V. A. Rokhlin, the beginning of fifties). Algebraic objects of another type, i.e. *homotopic groups* π_n, $n > 1$, of a topological space were introduced by W. Hurewicz in the thirties. These were the extensive generalization of the notion of the *fundamental group* π_1, which was introduced by H. Poincaré. The groups π_n are "the most important invariants in topology" [62, p. 25]. The problem to calculate these invariants by means of geometrical methods was very important in topology (L. S. Pontryagin, G. F. Freudenthal, V. A. Rokhlin, thirties – early fifties). In the late twenties the *homology groups* H_n, $n \geq 0$, of topological spaces were studied (the formal–algebraic definition was given by E. Noether). The "dual" theory, with respect to homology, "cohomology" theory appeared (A. N. Kolmogorov, J. W. Alexander, middle of the thirties). The accumulation of different algebraic objects in topology gave a start to the emergence and development of so–called "homological algebra". In the thirties and forties of this century the theory of *fibre spaces* (fibrations) emerged from differential geometry and developed in an independent direction. A fibre space can be considered as a continuous family of spaces, i.e. fibres which are homeomorphic to each other and are "labeled" by the points from another space which is the base of the fibration. A simplest example is the family of normal or tangent planes of a two–dimensional surface in the Euclidean space (the base of fibration). However, in general, the base of a fibration can be more complicated. The problem of classifying fibrations and constructing its invariants ("characteristic classes") was solved by L. S. Pontryagin, H. Whitney, E. Stiefel, S. S. Chern).

3. In the post–war period, algebraic topology has been essentially restructured. At the beginning of fifties many results of algebraic topology had accumulated. Thus, problem arose to work out a common point of view for such a variety of facts, and to construct new common methods. This reconstruction of topology was influenced by the French topological school (J. Leray, R. Thom, H. Cartan, J. P. Serre et al.).

Since the fifties, the development of topology has reached a high level in many directions. An active part in this has been taken by Russian mathematicians. Following the review [62] we shall touch on some of the more important directions.

In this period, most of the attention is still paid to smooth manifolds and fibre spaces, and their mappings. The following directions have been intensively developed: homotopy theory, cohomological operations, spectral sequences of fibrations; theory of characteristic classes and cobordisms; geometrical and homotopical structures of smooth manifolds; theory of categories and functors; K–theory and the index theory of elliptic operators that is closely related to K–theory; theory of fibrations, representation theory of finite and compact groups; calculation of the cohomology of Lie algebras of vector fields and interactions with "general topology".

A number of most important achievements is linked with the names of such mathematicians as J. Leray, J. P. Serre, R. Bott, F. Hirzebruch, R. Thom, J. F. Adams, J. Milnor, S. P. Novikov, and M. M. Postnikov.

In particular, developments in algebraic topology led to the solution of a number of major problems, viz. J. Leray has constructed a fundamental algebraic method for computing homological groups with the help of spectral sequences; M. M. Postnikov has completely solved the problem of defining a homotopical type of spaces and investigated the number of "smooth structures" in the given topological manifold (i.e. ways of turning it into a smooth manifold; note that J. Milnor obtained an efficient result about the existence of twenty eight smooth structures on the seven–dimensional sphere S^7); S. P. Novikov has proved that the Pontryagin's classes are "topologically invariant". The problem of triangulation of smooth manifolds, i.e. the possibility to divide them up into *simplices* which fit well together, was considered. Figures that are composed of simplices (the so–called polyhedrons) were already considered by Poincaré when constructing homology theory. Recall that a simplex is the convex hull of linearly independent points in the Euclidean space, or the homeomorphic image of it. These points are called the vertices of a simplex, and the number of them reduced by one is the dimension of a simplex. Each subset of the set of vertices of a simplex also defines a simplex and these form the boundary of the initial simplex. Saying that simplices fit well to form a polyhedron means that they can intersect only along a common boundary. One of the principal questions in classical topology is whether two homeomorphic polyhedrons of dimension n allow "identical", from the combinatorial point of view, triangulations. This so–called "main problem of combinatorial topology", was solved for dimension $n \leq 3$ by E. E. Moise; for the higher dimensions the solution, in general, is negative (M. Friedman and S. Donaldson managed to solve it for the case of $n = 4$ only in eighties). S. P. Novikov, R. Kirby, and L. Siebenmann have investigated the possible number of non–equivalent triangulations of a manifold.

4. We point out some more of the important achievements in topology itself and some applications: the classification of multi–dimensional ($n \geq 5$) manifolds (S. P. Novikov–C. T. C. Wall—W. Browder); classifying and computing the number of linearly independent vector fields over spheres (J. F. Adams); the proof by A. V. Chernavski of the local contractibility of the group of continuous homeomorphisms of a topological closed compact manifold M^n (or a space $\mathbf{R}^n$); the proof by a number of topologists, A. S. Mishchenko, Yu. P. Solovyov, G. G. Kasparov, at the conjecture by S. P. Novikov concerning "highest signatures", for the case of arbitrary manifolds and for discrete subgroups of the Lie groups; the topological solution of a problem on computing the index of elliptic operators (M. F. Atiyah, I. M. Singer); general solution of a problem about the existence of minimal surfaces in spectral bordism classes by A. T. Fomenko, etc. (We have presented a list of important results in topology at present, which is far from complete.) Details on the developments in topology may be found in the "History of Soviet mathematics" [41] cited above, and also in the fundamental review [62]. An important feature of the contemporary level of development of topology should be noted, viz. the most extensive penetration of topological methods into other fields of modern mathematics such as variational calculus and geometry as a whole, topology of Lie groups and homogeneous spaces, topology of complex and algebraic manifolds, qualitative (topological) theory of dynamical systems and fibrations, topological methods in Hamiltonian mechanics, topology of elliptic and hyperbolic equations with partial derivatives. The penetration of topological methods into classical analysis evolved into a new mathematical theory, the so–called theory of singularities, which V. I. Arnold (one of the creators of this theory, as is R. Thom) characterized as an "immense generalization of investigating functions by their maxima and minima" [10]. Since the sixties, "global analysis", i.e., the topology of infinite–dimensional manifolds and their mappings, has been developed. It is interlocked with the variational calculus in the large and classical non–linear functional analysis (or to be more exact, with that topological part which is related to the theory of Morse–Smale critical values, and the topology of Fredholm mappings).

Topology has become a powerful instrument in mathematical research, and its language has gained universal importance.

A remarkable fact is that in the seventies and eighties of this century there appeared a complex of applications of topology in modern physics, a fact that is important not only for physics, but also for topology itself. "In many cases, it is impossible to comprehend the essence of real physical phenomena without topological notions ... Topology has acquired a number of perfect applications in the most various problems for defining qualitative, constant properties for different mathematical and physical objects... [62, p. 6–7].

Of course, the opposite effect should also be noted, i.e., the influence of physical objects on topology itself. S. P. Novikov has widely propagated the theory of mutual relations of topology and physics, and participates actively in the development of this branch. He considers that the most important ideas in topology of the last ten years have come from the outer world, and is certain that in the twenty first century topology will become an indespensible instrument of applications to the analysis, and everyone will need to be acquainted with it.

§ 2. Generalization of the concepts of space and function

1. Metric space. We have already mentioned that in topology an esentially wider idea of space is considered than that of an Euclidean. To start with we shall consider the notion of a metric space (which is less general than that of the topological space) because of its greater simplicity and because of the wide use of this notion in modern mathematics.

In Euclidean space $\mathbf{R}^3$, for each pair of its points $x = (\xi_1, \xi_2, \xi_3)$, $y = (\eta_1, \eta_2, \eta_3)$ the following distance is defined

$$\rho(x, y) = \left(\sum_{i=1}^{3} (\xi_i - \eta_i)^2 \right)^{1/2}.$$

When studying $\mathbf{R}^3$ the following properties of the distance are used:

I. $\rho(x, y) \geq 0$ for any x, y.

II. $\rho(x, y) = 0$ iff $x = y$.

III. $\rho(x, y) = \rho(y, x)$.

IV. $\rho(x, y) \leq \rho(x, z) + \rho(z, y)$ for any $x, y, z \in \mathbf{R}^3$ (the *triangle inequality*).

There also can exist other real functions of a pair of points x, y on $\mathbf{R}^3$ which satisfy properties I–IV.

Exercise 1°. Let (ξ_1, ξ_2, ξ_3), (η_1, η_2, η_3) be the coordinates of the points $x, y \in \mathbf{R}^3$. Show that the function $\rho(x, y) = \max_{1 \leq i \leq 3} |x_i - \eta_i|$ satisfies properties I–IV.

Functions like this can also exist on other kind of sets.

Exercise 2°. Let X be an arbitrary set. Set $\rho(x, y) = 0$ if x and y are coincident elements in X, and $\rho(x, y) = 1$ otherwise. Show that such a function ρ satisfies properties I–IV.

The functions ρ from both exercises are naturally called the *distances between the elements* of the corresponding sets.

To introduce a general concept of the distance recall the definition of the product of two sets. If X and Y are two sets then their product $X \times Y$ is the set consisting of all ordered pairs (x, y), where $x \in X$, $y \in Y$. In particular, the product $X \times X$ is also defined.

Definition 1. A set X together with a mapping $\rho : X \times X \to \mathbf{R}^1$, associating to each pair $(x, y) \in X \times Y$ a real number $\rho(x, y)$ and satisfying properties I—IV is called a *metric space* and is denoted by (X, ρ).

The mapping ρ is called the *distance* or *metric on the space* X. The elements of the set X are usually called the points.

Any set can be turned into a metric space by providing it with the metric described in exercise 2. This metric space is called *discrete*. However, this way of "metrization" is not effective (or interesting).

Example 1. Let $X \subset \mathbf{R}^3$ be a subset of Euclidean space. A distance in $\mathbf{R}^3$ can simultaneously serve as a distance in X. In this case, the metric on X is obtained by restricting the metric on $\mathbf{R}^3$.

If (X, ρ) is a metric space and $Y \subset X$ a subset, then $(Y, \bar{\rho})$ is also a metric space, where $\bar{\rho} : Y \times Y \to \mathbf{R}^1$ is the restriction to the mapping ρ on the subset $Y \times Y$.

It is said that the metric on Y is *induced* by (inherited from) the metric on X, and Y is called a *subspace of the metric space* X.

A number of examples of metric spaces naturally arise from the problems of analysis.

Example 2. Consider the set of all continuous functions in the interval $[0, 1]$. It is often denoted by $C_{[0,1]}$. If $x(t)$, $y(t)$ are two continuous functions from $C_{[0,1]}$, then set

$$(1) \qquad \rho(x, y) = \max_{t \in [0,1]} |x(t) - y(t)|.$$

Exercise 3°. Verify that function (1) is a metric.

The set $C_{[0,1]}$ with the metric described above is called the *space of continuous functions* and is important in analysis.

Exercise 4°. Let A be an arbitrary set; and let X be the set of bounded real functions on A. If $f : A \to \mathbf{R}^1$, $g : A \to \mathbf{R}^1$ are arbitrary elements from X, then we set

$$\rho(f, g) = \sup_{t \in A} |f(t) - g(t)|.$$

Show that ρ is a metric on X.

Exercise 5°. Let ρ be a prime number. If $n > 0$ is an integer and, when decomposed into prime factors, contains a power p^α, then we put $v_p(n) = \alpha$; if p does not divide n $v_p(n) = 0$. Now extend the function v_p from the set of positive integers to the set $\mathbf{Q} \backslash \{0\}$ of nonzero rational numbers using the formula $v_p(\pm r/s) = v_p(r) - v_p(s)$. Set

$$\rho(x, y) = \rho^{-v_p(x-y)}, \quad x \neq y$$
$$\rho(x, x) = 0$$

for arbitrary x, y from $\mathbf{Q}$. Show that the function $\rho(x, y)$ is well defined and is a metric on $\mathbf{Q}$ (the *p–adic distance*).

2. Convergent sequences and continuous mappings.
The notions that generalize these initial concepts of mathematical analysis are naturally introduced for a metric space (X, ρ).

A mapping $n \mapsto x_n$ of the set of natural numbers into a metric space (X, ρ) is called a *sequence of points* of this space and is denoted by $\{x_n\}$. It is said that the sequence $\{x_n\}$ converges to a point a (has a limit a), if for any $\epsilon > 0$ there exists a natural number $n_0(\epsilon)$ such that $\rho(x_n, a) < \epsilon$ for all $n \geq n_0(\epsilon)$.

This fact is often written as

$$x_n \xrightarrow{\rho} a \text{ (or just } x_n \to a).$$

Exercise 6°. Let $\{x_n = \left(\xi_1^n, \xi_2^n, \xi_3^n\right)\}$ be a sequence of points of three–dimensional Euclidean space; let ρ be the Euclidean metric. Prove that $x_n \xrightarrow{\rho} a$ iff $\xi_i^n \to \xi_i^0$ $(i = 1, 2, 3)$ as $n \to \infty$, where $a = (\xi_1^0, \xi_2^0, \xi_3^0)$.

Considering the sequence of continuous functions $x_n(t)$, $0 \leq t \leq 1$, as a sequence in the metric space $C_{[0,1]}$, we can speak of the convergence of this sequence to the element $x_0 = x_0(t) : x_n \xrightarrow{\rho} x_0$. Such a convergence is often said to be *uniform* on the interval $[0, 1]$.

Exercise 7°. Show that the sequence of functions $x_n(t) = n^2 t e^{-nt}$ on the interval $[0, 1]$ converges pointwise to the zero function (i.e. for each fixed $t \in [0, 1]$), but it does not converge uniformly on the interval $[0, 1]$.

We now define a concept of continuous mapping of a metric space (X, ρ_1) into a metric space (Y, ρ_2).

Definition 2. Let $f : X \to Y$ be a mapping of a set X into a set Y. If for any point $x_0 \in X$ and any sequence $x_n \xrightarrow{\rho_1} x_0$ in X, the sequence of images in Y converges to $f(x_0) : f(x_n) \xrightarrow{\rho_2} f(x_0)$, then the mapping f is said to be a *continuous mapping of the metric space* (X, ρ_1) into the metric space (Y, ρ_2).

The definition, evidently, is a generalization of the concept of a continuous numerical function; it covers a wide class of mappings of geometrical figures in Euclidean spaces.

If the continuity property defined above, is considered at a certain point x_0, then the definition of a *continuous mapping at a point* x_0 is obtained.

Exercise 8°. Let S^2 be a sphere in the Euclidean space $\mathbf{R}^3$ with its centre at the origin of coordinates. Setting $f(x) = -x$ (central symmetry) prove that f is continuous.

Exercise 9°. Give an example of continuous mapping of a plane square into itself that has fixed points only on the boundary.

Evidently, an equivalent definition of continuous mapping of metric spaces can also be given in terms of ϵ, δ:

A *mapping* $f : X \to Y$ *is continuous* if for any $x_0 \in X$ and any $\epsilon > 0$ there exists a $\delta = \delta(\epsilon, x_0) > 0$ such that $\rho_2(f(x), f(x_0)) < \epsilon$ as soon as $\rho_1(x, x_0) < \delta$.

If, in this definition, δ is not dependent on the choice of the point x_0, then the mapping f is said to be *uniformly continuous*.

Exercise 10°. Let $f : \mathbf{R}^1 \to \mathbf{R}^1$ be a continuous function. Prove that the mapping $F : C_{[0,1]} \to C_{[0,1]}$, where $F(x)(t) = f(x(t))$, is continuous.

Recall that a mapping of sets $f : X \to Y$ is said to be *surjective* if each element from Y is the image of a certain element from X; that is *injective* if different elements of X are mapped to different elements of Y; and that it is *bijective* if the mapping is both surjective and injective at the same time.

Now we can define a homeomorphism of metric spaces.

Definition 3. A mapping $f : X \to Y$ of metric spaces is called a *homeomorphism*, and the spaces X, Y are called *homeomorphic* if 1) f is bijective, 2) f is continuous, 3) the inverse mapping f^{-1} is continuous.

This definition makes more precise the rough idea of homeomorphic figures as it was discussed intuitively in § 1. Thus, the notion of topological properties of figures now obtain a firmer ground: the *topological properties of metric spaces* are those which persist (are invariant) under homeomorphisms. Homeomorphic spaces are said to be *topologically equivalent*.

Exercise 11°. Prove that 1) an annulus in $\mathbf{R}^2$ is homeomorphic to a cylinder in $\mathbf{R}^3$; 2) an annulus without boundary (the interior of an annulus) is homeomorphic to $\mathbf{R}^2$ without one point, and to S^2 without two different points.

Exercise 12°. Prove that the mapping of the half–interval $[0, 1)$ into the circle in the complex plane defined by the function $z = e^{2\pi i t}$, $0 \leq t < 1$, is not a homeomorphism (the metric on the complex plane is given by the formula $\rho(z_1, z_2) = |z_1 - z_2|$).

Exercise 13°. Prove that 1) a closed ball and a closed cube in $\mathbf{R}^3$ are homeomorphic; 2) the sphere S^2 with a deleted point N (i.e. the space $S^2 \backslash N$, where N is the North pole of the sphere) is homeomorphic to the plane $\mathbf{R}^2$. (Apply the stereographic projection for 2).)

If a mapping $f : X \rightarrow Y$ is a homeomorphism on its image $f(X)$ which is considered as a subspace in Y, then f is called the *imbedding of the space X into Y*.

§ 3. From metric space to topological space (visual material)

1. The "gluing" method. We now discuss a more general concept of space than a metric space, i.e. the concept of a topological space, and give an initial conception of such spaces. First of all, we describe a way of constructing new spaces, which immediately deviates from metric spaces.

Let (X, ρ) be some metric space (for simplicity, one can think of X as a certain subset of the Euclidean space $\mathbf{R}^2$). Let X be divided into non–intersecting subsets A_α:

$$X = \bigcup_\alpha A_\alpha; \quad A_\alpha \cap A_\beta = \emptyset, \quad \text{if } \alpha \neq \beta.$$

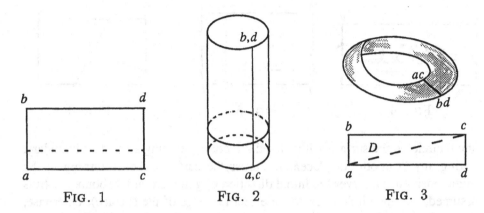

FIG. 1   FIG. 2   FIG. 3

If all the points from X that fall into any A_α are called equivalent and then "glued" (identified) to one point[*] a_α, then a new set $Y = \bigcup_\alpha a_\alpha$ is obtained. This is called the *quotient set* with respect to the given equivalence. Note, that Y is not a subset of X; therefore, the metric ρ, generally speaking, has nothing to do with the "space" Y.

By means of gluing we can obtain a number of well–known surfaces in Euclidean space. We now consider some of them. Let X be a rectangle (Fig. 1). If we "glue" only those points of teh sides ab and cd that lie on a common horizontal, then we get a quotient set which can be identified with a cylinder (Fig. 2). If we "glue" those points that are on the sides ab and cd and symmetric with respect to the centre 0 of the rectangle, then we obtain the "Möbius strip" (Fig. 3).

A Möbius strip can be made of a sheet of paper by gluing opposite sides in the appropriate way. This model visually presents a number of properties of the Möbius strip.

The Möbius strip possesses many remarkable properties: it has one edge (the closed line $adbca$) and, contrary to a cylinder, it has one side, thus it can be painted in one colour by continuous movement without passing over the edge (these properties are easily seen on a paper model). The Möbius strip is a non–orientable surface. Recall that a surface is orientable if any sufficiently

---

[*]Strictly speaking, each set A_α of equivalent points from X is considered as one element of the new set.

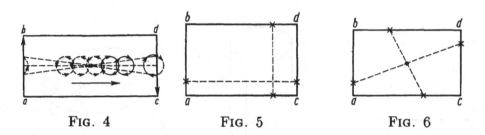

FIG. 4   FIG. 5   FIG. 6

small circle on the surface with a fixed direction for going around its boundary, during any "smooth" displacement across the surface when returning to the initial position, preserves the initial direction of going around its boundary (it is assumed, that the circle does not intersect the edge of the surface); otherwise, the surface is non–orientable. The non–orientability of the Möbius strip is obvious from Fig. 4.

If the sides ab and cd of the strip $abcd$ are glued together point by point which lie on the same horizontal, and, at the same time, the sides bd and cd are glued together point by point which lie on the same vertical, then a surface called a *torus* is obtained (Fig. 5).

If the sides ab and cd, as well as the sides bd and ac, are glued together point by point which are symmetric with respect to the centre 0 (Fig. 6), then the quotient set cannot be represented as a figure in three–dimensional Euclidean space. Or to be more exact, such an attempt to glue equivalent points would lead to a surface that would pierce itself without selfintersecting. This surface can be placed in $\mathbf{R}^3$ only by tearing it into parts in an appropriate manner, but then this would violate our "continuity" principle of gluing (i.e., the points which are close to the equivalent points, remain the same close points after gluing). The obtained quotient set is called the *projective plane* and is denoted by $\mathbf{RP}^2$.

Note that the rectangle $abdc$ is homeomorphic to the disc with the boundary $abdc$, and the projective plane can be described as a disc (Fig. 7) whose diametrically opposite boundary points are glued together, or, finally, it can be described as a hemisphere which has each pair of diametrically opposite boundary points glued into one point (Fig. 8).

Thus, the forming of a quotient set in the first three cases considered, leads to the figures which are again in the Euclidean space $\mathbf{R}^3$, while for the last

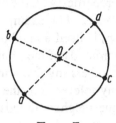

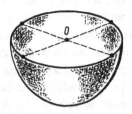

FIG. 7   FIG. 8

case it gives a new object which cannot be represented in $\mathbf{R}^3$.

Exercise 1°. Verify that a cylinder, torus and sphere are orientable surfaces, and that a projective plane is non–orientable.

Exercise 2°. By a suitable gluing operation (forming quotients), obtain a circle from an interval, a sphere from a disc, a circle from $\mathbf{R}^1$, and a torus from $\mathbf{R}^2$.

2. On the concept of topological space. We can now explain how the idea of a topological space arises. It was mentioned above that it is not always possible to place a quotient set in a metric space in a natural way, and therefore a metric can not be induced on it. One of the functions of a metric is to characterize how close are two points, and in the definition of continuous mapping, a metric plays this role (compare with § 2). One can geometrize the notion of proximity by considering (open) balls

$$D_r(x_0) = \{x \in X \; : \; \rho(x, x_0) < r\}, \quad r > 0,$$

with centre at the point x_0 and radius r. Thus, a point x is ϵ–near to the point x_0, if $x \in D_\epsilon(x_0)$.

It is easy to verify that the continuity of a mapping $f : X \to Y$ of two metric spaces can be characterized in the following equivalent way: let $x_0 \in X$ be an arbitrary (fixed) point, and let $y_0 = f(x_0)$ be an element from Y. Then for a ball $D_\epsilon(y_0)$ there always exists a ball $D_\delta(x_0)$ such that $f(D_\delta(x_0)) \subset D_\epsilon(y_0)$.

We may say that the continuity property of a mapping signifies the preservation of the nearness of points. The concept of proximity allows us to formulate exactly the concept of neighbourhood of a point: a part Ω of a metric space is a neighbourhood of a point x_0 in Ω, if each point, which is sufficiently near to x_0, belongs to Ω. Thus, neighbourhood structures arise in metric spaces.

"However, the spaces so defined possess many properties which can be formulated without reference to the concept of distance. For example, every subset which contains a neighbourhood of a point x_0 is itself a neighbourhood of the point x_0, and the intersection of two neighbourhoods of the point is also a neighbourhood of x_0. These properties and some others involve a multitude of consequences which can be deduced without any reference to the concept of "distance" which originally enabled us to define neighbourhoods. Thus, we obtain statements without even mentioning magnitude, distance, etc." [16, p. 12]

If a distance is not introduced in a set X, then the concept of proximity has no exact meaning and the definition of neighbourhoods given above is inappropriate. However, the inverse process proves to be effective, i.e. to each element $x_0 \in X$ is associated a certain collection of subsets $\{\Omega(x_0)\}$ in X in such a way that the main properties (axioms) of neighbourhoods hold. This collection is then called a system of neighbourhoods and the elements from a neighbourhood $\Omega(x_0)$ are then said to be Ω–near to x_0. In this case a set X is said to be endowed with a *topological structure*, or a *topology*. It is called a *topological space*, and the elements of X are called points.

"Once topological structures have been defined, the concept of continuity can be made precise. Intuitively, a function is continuous at a point if its value changes as little as we please, whenever the argument remains sufficiently near to the point considered. We can see that the concept of continuity will have an exact meaning whenever the spaces of arguments and the space of values of the function are topological spaces" [16, p. 14].

Thus, replacing balls in the definition of a continuous mapping by neighbourhoods we obtain the concept of a continuous mapping, and then the concept of a homeomorphism of topological spaces. Homeomorphic topological spaces are called *topologically equivalent*.

Example. Let C be the complex plane. The extended plane of complex numbers $\tilde{C} = C \cup \infty$ is a topological space. The spherical neighbourhoods of points $z \in C$, and the neighbourhoods of the point ∞ of the form

$$D_r(\infty) = \{z \in C : |z| > r\} \cup \infty,$$

and also the subsets containing them, define a topological structure on $\tilde{C}$.

Exercise 3°. Determine the homeomorphism of the extended plane of complex numbers and the sphere S^2 so that the north pole N is the image of the point ∞, and the south pole the image of the point 0.

Hint: Apply the stereographic projection to the equatorial plane C, $u=\frac{x+iy}{1-z}$, $(x,y,z) \in S^2 \setminus N$.

For a metric space, the topological structure of a quotient set arise naturally from a topolgical structure in a metric space by gluing neighbourhoods together. Thus, a quotient set becomes a topological space (*quotient space*).

3. Gluing two–dimensional surfaces together. Now we investigate in more detail the factor sets that are obtained by gluing plane figures together. Consider a polygon Π in the plane $\mathbf{R}^2$ and tkae the induced metric from $\mathbf{R}^2$. Obviously, the spherical neighbourhoods of a point $x \in \Pi$ consist of intersections of open discs centred at x with Π. Thus, sufficiently small spherical neighbourhoods of the point x are open discs iff x does not lie on the boundary of the polygon, and are sectors of an open disc (together with boundary radii), if x lies on the boundary (Fig. 9).

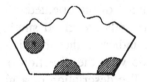

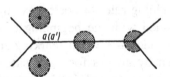

FIG. 9   FIG. 10

Let there be two polygons Π and Π'. Mark their two sides a and a'. We can glue Π and Π along these sides by specifying a homeomorphism $\alpha : a \to a'$ and declaring that the image and preimage are equivalent. The topology of the quotient space $(\Pi \cup \Pi')/R$ according to this equivalence, consists of open discs for internal points $x \in \Pi$, $x' \in \Pi'$, of sectors glued together for equivalent points $x \in a$, $x' \in a'$, and of sets containing the neighbourhoods mentioned. Fig. 10 illustrates the case when the identification is carried out by joining polygons along the equal sides a and a'.

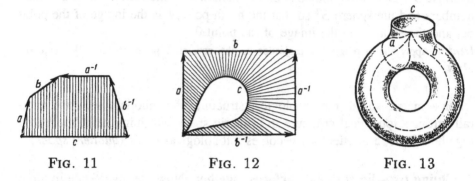

FIG. 11   FIG. 12   FIG. 13

Similarly, two sides of one polygon can be glued together (see the examples from Subsection 1).

Exercise 4°. Describe the topologies of a cylinder, torus, Möbius strip and projective plane.

Exercise 5°. Verify that the examples of quotient spaces (Subsection 1) are homeomorphic to their realizations in the Euclidean space $\mathbf{R}^3$. Verify that the different models (Fig. 6, 7, 8) of the projective plane are homeomorphic.

Now consider again the gluing of surfaces. Let us glue together the sides of the pentagon (see, Fig. 11) that are denoted by the same letters. The arrows denote the gluing rule for corresponding sides (the beginning of the oriented interval is glued with the beginning of the other, and the ends are glued similarly). The index -1 when marking some of the sides, indicates that the direction given by arrows for these sides does not coincide with the direction of the orientation (which is clockwise) of the polygon. The description of a gluing scheme can be obtained by writing in consequtive order the designations of sides into a "word" by going around the polygon clockwise. For instance, if we start at the side a, then the gluing scheme is $aba^{-1}b^{-1}c$. This scheme characterizes gluing, because it determines the sides in the polygon which are glued, and the gluing rule itself. It is not difficult to see that this quotient set can also be obtained in another topologically equivalent way (Fig. 12). Here the factor space is presented by torus with a cut along the curve c (Fig. 13, where the dotted lines denote the lines glued aa^{-1} and bb^{-1}). A torus with a hole is called a *handle*.

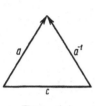

FIG. 14

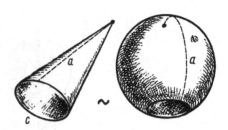

FIG. 15

FIG. 16

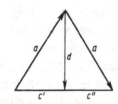

FIG. 17

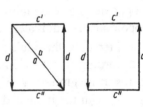

FIG. 18

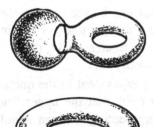

FIG. 19

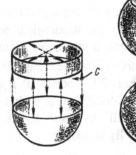

FIG. 20

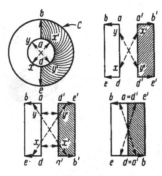

FIG. 21

Consider the gluing of two adjacent sides of a triangle. If the orientations are opposite, i.e. the gluing scheme is $aa^{-1}c$ (Fig. 14), then the quotient space is topologically equivalent to a sphere with a hole (Fig. 15).

Consider now the gluing of two adjacent sides with the same orientation, i.e. according to the scheme aac (Fig. 16). This triangle can be represented as a result of gluing two right triangles together along the common height d (Fig. 17) with the indicated orientation. Let us change the order in which they are glued: first, identify the hypotenuses a, and then the legs d (Fig. 18). Thus we obtain a Möbius strip (compare with Fig. 3); moreover, the latter quotient space is homeomorphic to the original one (Fig. 16).

Now, having cut a disc out of the sphere S^2, we can glue either a handle or a Möbius strip along the available side c; the latter can be seen as the circle S^1 (the boundary of the cut out disc). In the first case, a torus is obtained (Fig. 19) (verify the topological equivalence of the figures in the drawing), and in the second a projective plane $\mathbf{RP}^2$. Let us verify this.

The projective plane (see, Fig. 8) is topologically equivalent to the quotient space from Fig. 20. Indeed, it remains to show only, that the upper "cap" (Fig. 20) is a Möbius strip with the edge c. Representing it as an annulus with diametrically opposite points of the internal circle being identified, we can topologically transform it into a Möbius strip (Fig. 21).

Further constructions can be developed in two ways: 1) by cutting p discs out of the sphere and gluing p handles in the holes; 2) by cutting out q discs and gluing in Möbius strips. Thus, two series of surfaces are obtained

$$(1) \qquad M_0, M_1, M_2, \dots, M_p, \dots; \quad N_1, N_2, \dots, N_q, \dots$$

(clearly, M_0 and N_0 are the sphere S^2).

Now we discuss the properties of these surfaces. First of all, it is easy to see that these series are obtained from a finite number of convex polygons by gluing together their sides and the subsequent topological representations. Such spaces are said to be *finitely triangulable*, and the partition of a space into "curvilinear" polygons is called a *triangulation**)*. The surfaces M_p, N_q are connected, meaning that they consist of one whole "piece", i.e. cannot be split into two non–intersecting groups of polygons. This follows from the fact that any two vertices of triangulation polygons are joined by a continuous path made of their sides. The surfaces considered have no edge as any boundary side of a polygon is glued to another (and only one) side. Hence, each point of this surface has a neighbourhood which is homeomorphic to an open disc; these spaces are called *two–dimensional manifolds*.

Finitely–triangulable, connected, two–dimensional manifolds are called *closed surfaces*. If we did not glue all the pairs of sides of the polygons, but left some sides free, then we would obtain a *non–closed surface* (or a *surface with a boundary*). A point of the edge has a neighbourhood homeomorphic to a semidisc. An example can be a sphere S^2 with several holes in it.

Note also that the surfaces M_p are orientable, and they can be placed in $\mathbf{R}^3$ as two–sided surfaces without self–intersections. On the contrary, the surfaces N_q are non–orientable (they are called one–sided, analogously to the Möbius strip), and cannot be embedded in $\mathbf{R}^3$ without self–intersections (but it is possible in $\mathbf{R}^4$!).

In Ch. 2, it is shown that any closed surface is homeomorphic to a surface of the type M_p or N_q (the numbers p, q are called the *genus of the surface*). The surfaces M_p and N_q, $q \geq 1$, are never homeomorphic, since the orientability of a surface is a topological property. Two different M_p and M_p' surfaces (or N_q, N_q' surfaces) cannot be homeomorphic (see the following section). Thus, the list (1) gives a complete topological classification of closed surfaces. If we glue in p handles and $q \geq 1$ Möbius strips (having cut $p + q$ holes) to a

---

*)The exact definition of a triangulation of a surface can be found in Ch. 2.

sphere then the surface obtained is topologically equivalent to the sphere with $2p + q$ Möbius strips glued in.

Exercise 6°. Glue a cylinder along its edges to a sphere with two holes. Prove, that the surface obtained is homeomorphic to a sphere with a glued handle, i.e. to a torus.

Exercise 7°. Show that an annulus and a Möbius strip can be obtained from a disc by gluing its boundary to two sides of a rectangle.

Exercise 8°. Prove the equivalence of the following definitions of RP^2 to those given above: 1) diametrically opposite pairs of points are identified in S^2; 2) the edge of a Möbius strip is contracted to one point; 3) the edge of a Möbius strip is glued to a disc by a certain homeomorphism of boundary circles.

Exercise 9°. Define RP^1 by identifying diametrically opposite points of the circle S^1. Show, that 1) RP^1 is homeomorphic to the circle S^1; 2) $RP^1 \subset RP^2$; 3) there exists a neighbourhood of RP^1 in RP^2 which is homeomorphic to a Möbius strip.

Exercise 10°. Prove the equivalence of the following definitions of the surface N_2 (the *Klein bottle*): 1) rectangle (Fig.22) with the sides glued according to the scheme $aba^{-1}b$; 2) an annulus with the boundary circles glued together and their orientations reversed (such a gluing can be represented in the following way: "reflect" the internal circle over any diameter and, after that, glue together the points of the internal and external circle, which are on the same radius; in Fig. 22, x, y are the points to be glued); 3) two Möbius strips glued together along the edge; 4) an annulus to each edge of whose a Möbius strip is glued.

A topological space homeomorphic to a convex polygon is called a *topological polygon*. Accordingly, we say that the images of vertices (sides) are the *vertices* (*edges*) of the topological polygon. Without any loss of generality, we can assume that the triangulation of a surface consists of topological polygons which are joined by the edges (in order to achieve this, the convex polygons whose sides are identified to obtain the surface should be divided a priori in-

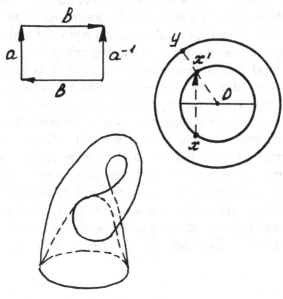

FIG. 22

to sufficiently small polygons, say triangles). Below we deal only with such triangulations.

For a triangulated surface Π we define a number $\chi(\Pi) = e - k + f$, where e is a number of vertices, k a number of edges, f a number of triangulation polygons, known as the *Euler characteristic of the surface* Π. It possesses the remarkable property that it does not depend on triangulation, i.e. it is a topological invariant of the surface.

Exercise 11°. Verify that the Euler characteristic of the sphere S^2 is equal to 2, of a torus it is 0, of a disc 1, of a handle −1, and of a Möbius strip 0.

It is easy to prove the topological invariance of the Euler characteristic $\chi(S^2)$ using the Jordan theorem[*], which states that any simple closed curve, i.e. a curve homeomorphic to a circle, divides a sphere or a plane into two non–intersecting regions and is the boundary of them.

Thus, consider a triangulation of S^2. It can be achieved in stages by fixing a vertex (∗) and drawing one edge after another; we draw the first edge from

---

[*] The proof of the Jordan theorem is quite long, so we skip it.

the vertex (∗) to a new vertex, and then take care that each subsequent edge should start at the vertex of an edge already drawn. At every step we count the number obtained of vertices e, edges k, and regions f which are bounded by a closed simple curve of edges. At the beginning, we set $e = 1$, $k = 0$, $f = 1$ (the vertex (∗) and its complementary region). It is easy to notice that the number $e - k + f$ does not change if we add a new edge. Indeed, if an edge comes to a new vertex then no new regions appear, and the numbers e and k increase by 1. If a new edge joins two of the original vertices, then it will close a certain path of edges, and a new region will appear (according to the Jordan theorem); so k and f will both increase by 1, and e will remain the same. By drawing the last edge we shall completely restore the triangulation, and then $e - k + f = \chi(S^2)$; originally, $e - k + f = 2$. Consequently, $\chi(S^2) = 2$.

If Π_1, Π_2 are two surfaces with boundaries l_1, l_2 homeomorphic to S^1, then these surfaces can be glued along their boundaries by a homeomorphism $\alpha : l_1 \rightarrow l_2$. Let $\Pi_1 \cup_\alpha \Pi_2$ denote the factor space obtained. We prove the formula

$$(2) \qquad \chi(\Pi_1 \cup_\alpha \Pi_2) = \chi(\Pi_1) + \chi(\Pi_2).$$

Let us triangulate Π_1 and Π_2 so that on the boundaries l_1, l_2 there arise homeomorphic triangulations (the triangulation of S^1 consists of l vertices and the same number of edges). After gluing, the numbers of vertices, edges and polygons are equal to $e_1 + e_2 - l$, $k_1 + k_2 - l$, $f_1 + f_2$, respectively. Formula (2) follows from the equality

$$(e_1 + e)^2 - l) - (k_1 + k_2 - l) + (f_1 + f_2) =$$
$$= (e_1 - k_1 + f_1) + (e_2 - k_2 + f_2).$$

Formula (2) is sometimes convenient for calculating the Euler characteristic.

Let $_pS^2$ be a sphere with p holes. If we glue p discs back in, then we obtain S^2. Formula (2) gives us the equality $\chi(S^2) = \chi(_pS^2) + p$, and hence $\chi(_pS^2) = 2 - p$.

The surface M_p can be obtained by gluing $_pS^2$ to p handles, each having Euler characteristic equal to (-1). From (2) we obtain that $\chi(M_p) = 2 - 2p$. Analogously, $\chi(N_q) = 2 - q$, as the Euler characteristic of a Möbius strip is equal to zero. Since $\chi(M_{p_1}) = \chi(M_{p_2})$ only if $p_1 = p_2$, and $\chi(N_{q_1}) = \chi(N_{q_2})$

only if $q_1 = q_2$, the surfaces M_{p_1} and M_{p_2} cannot be homeomorphic when $p_1 \neq p_2$ because of the topological invariance of the Euler characteristic; the same can be said about the surfaces N_{q_1}, N_{q_2}, when $q_1 \neq q_2$.

Other interesting applications of the Euler characteristic can be found in the theory of convex polyhedra. Imagine the surface of a convex polyhedron which is obtained by gluing a finite number of convex polygons (its faces) by identical mappings of the edges glued. We immediately obtain the Euler formula for a convex polygon:

$$\alpha_0 - \alpha_1 + \alpha_2 = 2,$$

where α_0 is a number of vertices, α_1 a number of edges, α_2 a number of faces of the polygon. In fact, the left–hand side is the Euler characteristic of the polyhedron surface which is, clearly, homeomorphic to S^2.

If m faces meet at each vertex, and each face is a convex n–gon then it is said that the type of that polyhedron is $\{n, m\}$. If the n–gons are regular then the polyhedron is said to be regular. Having the type $\{n, m\}$ we can calculate α_0, α_1, α_2. Indeed, m edges meet in each vertex, therefore $\alpha_0 m = 2\alpha_1$; there are n edges in each face, so $n\alpha_2 = 2\alpha_1$ (each edge joins two vertices and two faces). Thus,

$$\frac{\alpha_0}{m^{-1}} = \frac{\alpha_1}{2^{-1}} = \frac{\alpha_2}{n^{-1}} = \frac{\alpha_0 - \alpha_1 + \alpha_2}{m^{-1} - 2^{-1} + n^{-1}} =$$
$$= \frac{2}{m^{-1} - 2^{-1} + n^{-1}} = \frac{4mn}{2n + 2m - mn},$$

and from this the values of α_0, α_1, and α_2 can be computed. The natural condition for α_0, α_1, and α_2 to be positive leads to the inequality for positive integers n, m:

$$2n + 2m - nm > 0, \quad \text{hence } (n - 2)(m - 2) < 4.$$

It is easy to see that there are only five solutions:

(3) $\{3, 3\}, \{4, 3\}, \{3, 4\}, \{5, 3\}, \{3, 5\}.$

Five kinds of regular polyhedra are known from elementary geometry: tetrahedron, cube, octahedron, dodechedron, and icosahedron (Fig. 23), whose types coincide with (3).

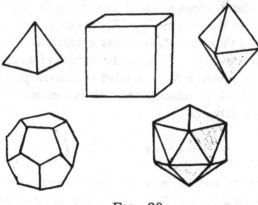

FIG. 23

Thus, a complete classification of polyhedra of the type $\{n, m\}$ has been given.

§ 4. The concept of Riemann surface

One of the paths which lead to the basic topological ideas, is related to the study of algebraic functions and their integrals. It was discovered by Riemann as far back as the middle of the last century.

Consider an algebraic equation

$$(1) \qquad a_o(z)w^n + a_1(z)w^{n-1} + \ldots + a_n(z) = 0, \quad a_0(z) \neq 0,$$

with complex coefficients which are polynomials of a complex variable z; its roots are functions $w = w(z)$ of z and analytic under certain conditions. For example, if all the roots of equation (1) differ in the point z_0, then in the neighbourhood of the point z_0, there exist n solution functions $w_i(z)$, $i = 1, \ldots, n$, that analytically depend on z.

An analytic function $w = w(z)$ satisfying equation (1) is called an *algebraic function*. Equation (1) determines several branches $w_i(z)$ of algebraic functions, the number of which, generally speaking, varies, and which change into one another as z varies. Therefore, a multi–valued function $w(z)$ defined by equation (1), and its branches $w_i(z)$ are considered. Riemann proposed the idea of replacing the z–plane C by such a surface on which the function $w(z)$ is single–valued, and its branches $w_i(z)$ are the values of $w(z)$ on separate parts of the surface (these surfaces are called *Riemann surfaces*).

It is not complicated to construct such a surface. Consider the extended plane $\tilde{C} = C \cup \infty$ of complex numbers (the z–sphere) and the Cartesian product $\tilde{C} \times \tilde{C}$ consisting of ordered pairs (z, w). Neighbourhoods in $\tilde{C} \times \tilde{C}$ are naturally defined as the Cartesian products of neighbourhoods (and all the sets containing them). Then an algebraic equation (1) determines a subset on $\tilde{C} \times \tilde{C}$, i.e. the graph of the multi–valued algebraic function $w(z)$ over the complex plane C consisting of those pairs $(z, w) \in C \times C$ which satisfy equation (1). This is the Riemann surface Π of the multi–valued algebraic function $w(z)$; indeed, the projection $\Pi \to \tilde{C}$ which is given by the rule

$$(2) \qquad\qquad (z, w) \mapsto w,$$

determines a single–valued function on the Riemann surface which has the values of all the branches of the multi–valued function. An interesting question arises about constructing the surface Π and about the distribution of the branches of the function w on it. To study the questions mentioned, it is very useful to extend the graph Π by adding certain "infinitely distant" points from $\tilde{C} \times \tilde{C}$; the obtained extension $\tilde{\Pi}$ is said to be a compact Riemann surface.

The simplest multi–valued algebraic function is related to the equation of the second degree

$$(3) \qquad\qquad w^2 + a_1(z)w + a_2(z) = 0.$$

The change of variables $v^2 = 2w + a_1$ reduces this equation to a simpler form $v^2 - p(z) = 0$, where $p(z)$ is a polynomial. Therefore, instead of equation (3) we consider the equation

$$(4) \qquad\qquad w^2 - p(z) = 0$$

Let $p(z) = z$. Then for the algebraic equation $w^2 - z = 0$ a Riemann surface, i.e. the graph Π_1 in $\tilde{C} \times \tilde{C}$ on which the function w is single–valued, is determined. Having added the point (∞, ∞) to Π_1, we obtain the "extension" $\tilde{\Pi}_1$ which is a compact Riemann surface.

Now we show that $\tilde{\Pi}_1$ is homeomorphic to $\tilde{C}$, i.e. to the two–dimensional sphere S^2. Indeed, the mapping (2)

$$w = w(t), \text{ where } t = (z, w) \in \Pi_1,$$

with the inverse mapping $t = (w^2, w)$ defines, as it is easy to see, a homeomorphism $\widetilde{\Pi}_1$ on the w–sphere S^2. Let us present another construction of a Riemann surface, the one that is usually used in the theory of functions of a complex variable. Equation (4) determines the two–valued algebraic function $w = \sqrt{z}$. If $z = re^{i\phi}$ then its two values $w_1 = \sqrt{r}\,e^{i\phi/2}$ and $w_2 = -\sqrt{r}\,e^{i\phi/2}$ differ by signs and change into each other when the point z moves around the point $z = 0$ along a closed path. In

order to prevent passing of the branch w_1 to the branch w_2, we make a cut in the z–sphere along the positive real half axis (Fig. 24). This cut joins the points 0 and ∞. Two edges, viz. (+), the upper, and (−), the lower, adjoin on the cut. Consider the union (non–intersecting) of two sheets (replicas) I and II of the cut z–sphere. Call the sheet I the carrier of the branch w_1,

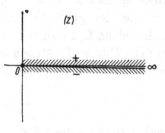

Fig. 24

and the sheet II the carrier of the branch w_2 (assuming that on each sheet I, II, $w_i = \infty$ for $z = \infty$). The function w on a two–sheet surface $I \cup II$ is single–valued. In order to detect the effect of passing from the branch w_1

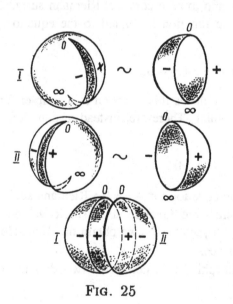

Fig. 25

to the branch w_2, we glue the (−) edge of sheet I with the (+) edge of sheet II, and the (+) edge of sheet I with the (−) edge of sheet II. Thus we obtain a quotient space Π_1' which is the two–sheeted Riemann surface of the function $w = \sqrt{z}$. It is easy to notice that Π_1' is homeomorphic to the sphere S^2; Fig. 25 shows the sheets I and II glued together after their preliminary topological transformation into a hemisphere by moving their edges apart, and that results the sphere S^2. Although Π_1' does not lie in $\mathbf{R}^3$ (the sheets I and II pierce each other, see the gluing scheme, Fig. 26), nevertheless it gives a good visual demonstration of interconnection of the branches w_1 and w_2. It can be immediately checked that the mapping $w : \Pi_1' \to \tilde{\mathbf{C}}$ given by the multi–valued function $w = \sqrt{z}$, is also a homeomorphism on the w–sphere S^2. Thus, $\tilde{\Pi}_1$ and Π_1' are homeomorphic to each other and to the sphere S^2.

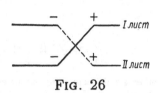

FIG. 26

Let us define a projection $\tilde{\Pi}_1 \to \tilde{\mathbf{C}}$ by the formula $z(t) = z$ and identify $\tilde{\mathbf{C}}$ with S^2. Then we have two diagrams

$$\Pi_1 \xrightarrow{\ w = w(t)\ } S^2(\omega - \text{sphere}) \qquad \Pi_1 \xrightarrow{\ t = \left(w^2, w\right)\ } S^2(\omega - \text{sphere})$$

$$z = z(t) \searrow \quad \swarrow z = w^2 \qquad\qquad z = z(t) \searrow \quad \swarrow z = w^2$$

$$S^2 \qquad\qquad\qquad\qquad S^2$$

$$(z - \text{sphere}) \qquad\qquad\qquad (z - \text{sphere})$$

These diagrams are commutative, i.e. the composition of two mappings (in the direction of arrows) is equal to the third mapping which completes the triangle. The horizontal mappings in the diagrams are inverse to each other.

The mapping $S^2 \xrightarrow{z=w^2} S^2$ is called a *two–sheeted (ramified) covering* of the sphere S^2 with the branch points $z = 0$ and $z = \infty$ (verify that the circuit around point $z = \infty$ also leads to a change of the branch).

This covering explains the idea of substitutions used in analysis for the rationalizing of integrands in familiar integrals $\int R(z, \sqrt{z})dz$, where $R(z, w)$ is a rational function of two variables z, w. Let R be given on the domain $\mathbf{C} \times \mathbf{C}$. Consider the integral

(5)
$$\int_{z_0}^{z} R(z, \sqrt{z})dz = \int_{\gamma} R(z, \sqrt{z})dz,$$

which is regarded as a curvilinear integral along certain path γ in the z–plane $\mathbf{C}$ which does not intersect the cut $\overline{0\infty}$ and joins the points z_0, z, where $\sqrt{z}$ is one of the branches of the multi–valued algebraic function $w=\sqrt{z}$. The point $t=(z,\sqrt{z})$ lies in the Riemann surface Π_1 and runs through the path $\tilde{\gamma} \subset \Pi_1$, while z runs over the path $\gamma \subset C$. The integrand in (5) is given on the curve $\tilde{\gamma}$: $R(z,\sqrt{z})=R(t)$, $t\in\tilde{\gamma}$. With the help of the homeomorphism of the w–sphere S^2 $\tilde{\Pi}_1$: $t=(w^2,w)$, we find a path $\bar{\gamma}$ on the w–sphere S^2 homeomorphic to $\tilde{\gamma}$. The relation between z–sphere and w–sphere is specified by mapping $z=w^2$ and makes it possible to transform the integral (5):

$$\int_{\gamma} R(z,\sqrt{z})\,dz = \int_{\tilde{\gamma}} 2R(w^2,w)w\,dw = \int_{w_0}^{w_1} 2R(w^2,w)w\,dw,$$

where w_0 and w_1 are the beginning and the end of the path $\bar{\gamma}$ on the w–plane C. The latter integral is an integral of a rational function.

Now let $p(z) = a_0 z^2 + a_1 z + a_2$, where $a_0, a_1, a_2 \in C$, $a_1^2 - 4a_0 a_2 \neq 0$, $a_0 \neq 0$. Denoting the roots of the polynomial $p(z)$ by r_1 and r_2, $r_1 \neq r_2$, we obtain the algebraic function

$$(6) \qquad w = \sqrt{a_0(z - r_1)(z - r_2)}.$$

Obviously, it is also two–valued. Considerations, similar to those presented previously, shows that one branch passes to another while going around each of the points r_1 and r_2; a loop around both points (along a closed path surrounding the points r_1 and r_2) as well as a loop around the point ∞ does not change the value of the branch. Consequently, the Riemann surface Π_2' of the function considered can be obtained from two replicas of the z–sphere which are cut along the interval $\overline{r_1 r_2}$; moreover, the edges of sheets I and II are glued together as they were in the first example. Note that Π_2' contains two infinitely distant points ∞_1 and ∞_2 that lie on the sheets I and II and are not branch points. Evidently, the space is still topologically equivalent to the sphere. We again have a two–sheeted covering of the sphere S^2 with two branch points $z = r_1$, $z = r_2$.

Consider the graph Π_2 of the multi–valued algebraic function $w = w(z)$ on the complex plane C for the algebraic equation

$$(6') \qquad w^2 - a_0 \cdot (z - r_1)(z - r_2) = 0$$

It is useful to note that if from C we delete the points r_1 and r_2, then the remaining part of the graph Π_2 on $C\backslash\{r_1, r_2\}$ (denote it by $\widehat{\Pi}_2$) is homeomorphic to the part of the graph Π_1 on $C\backslash\{0\}$ (denote it by $\widehat{\Pi}_1$). This homeomorphism, as is easy to check, is given by the mapping $\Phi : (z, w) \mapsto (v, \tau)$, where

$$\tau = \frac{z - r_1}{z - r_2}, \quad v = \frac{1}{\sqrt{a_0}} \cdot \frac{w}{z - r_2},$$

and it transforms equation (6′) into the equation $v^2 - \tau = 0$, i.e. into the Riemann surface Π_1 considered above. However, if we have obtained the extension $\widetilde{\Pi}_1$ in a simple and natural way, it is more complicated to get the extension $\widetilde{\Pi}_2$, therefore we do not consider it. Nevertheless, we have its homeomorphic image Π_2' constructed above.

Thus, we get a commutative diagram

(7)

(where $t = (\tau, v)$, and not all mappings are homeomorphisms).

If the integral

$$\int_{z_0}^{z} R(z, \sqrt{a_0(z-r_1)(z-r_2)})dz = \int_{z_0}^{z} R(z, w(z))dz,$$

is taken along a path γ which is in $C\backslash\{r_1, r_2\}$, then the horizontal mappings of diagram (7) enable us to transform the integral above into an integral on the v–sphere S^2:

$$\int_{v_0}^{v} \widetilde{R}(v^2, v)dv,$$

where $\widetilde{R}$ is a rational function. This accounts for the rationalization of the integrand by the formal Euler substitution

$$v = \sqrt{\tau} = \sqrt{\frac{z-r_1}{z-r_2}}.$$

We will arrive at an essentially new result if we consider a polynomial $p(z)$ of the third degree. Thus, consider an algebraic function of the form

$$(8) \qquad w = \sqrt{a_0(z - r_1)(z - r_2)(z - r_3)},$$

where r_1, r_2, r_3 are pairwise different. The function w has two branches, but now they are "joined" in a more complicated way. A loop around the point r_1 leads to a change of a branch of the function w, while loops around the other two points preserve the branches of the function w; a loop around three points, as well as loop around ∞, alters a branch. In order to "forbid" these changeovers it is sufficient to make the cuts, $\overline{r_1 r_2}$ and $\overline{r_3 \infty}$, on the z–sphere. Then each branch of the function w is one–valued on such a sheet with the cuts. To transform one branch into another in the required way, we glue replicas I and II along the cuts $\overline{r_1 r_2}$ and $\overline{r_3 \infty}$, respectively, the edges being glued together as before. The topological space Π_3' thus obtained is evidently the Riemann surface of function (8). A considerable difference between the surfaces Π_3' and Π_2' is that Π_3' is topologically equivalent to a sphere with a handle (Fig. 27, where the cuts are first expanded into "holes", then the tubes are stretched and glued together along the edges in the required way). The natural mapping $\Pi_3' \to \widetilde{C}$ is a two–sheeted covering S^2 with the branch points r_1, r_2, r_3, ∞.

For the function $w = \sqrt{a_0(z - r_1)(z - r_2)(z - r_3)(z - r_4)}$, where r_1, r_2, r_3, r_4 are pairwise different, we obtain a Riemann surface Π_4' homeomorphic to Π_3'. This follows from the single–valued branches being separated by two cuts $\overline{r_1 r_2}$ and $\overline{r_3 r_4}$, and the point r_4 acting as the point ∞ of the previous example (the latter point in the case Π_3' not being a branch point).

Note that integration of rational functions on the surfaces Π_3, Π_4 leads to the theory of elliptic integrals.

It is also not complicated to investigate the case of an algebraic function

$$(9) \qquad w = \sqrt{a_0(z - r_1)\ldots(z - r_n)},$$

where the r_i are pairwise different. Make $n/2$ cuts $\overline{r_1 r_2},\ldots,\overline{r_{n-1} r_n}$, if n is even, and $(n+1)/2$ cuts $\overline{r_1 r_2},\ldots,\overline{r_{n-2} r_{n-1}}, \overline{r_n \infty}$, if n is odd. Having taken two replicas of the z–sphere with such cuts, we glue them together along the corresponding cuts. Constructions similar to those presented in Fig. 27, give a sphere with $\left(\frac{n}{2} - 1\right) = \frac{n-2}{2}$ or with $\frac{n+1}{2} - 1 = \frac{n-1}{2}$ handles. This is the

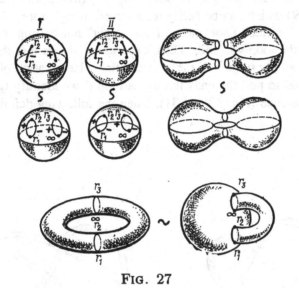

FIG. 27

Riemann surface of function (9). The number of handles p (the genus of a surface) is related to the number of branch points V of the Riemann surface by the equality $V = 2(p + 1)$.

Thus, the multi–valued algebraic function determined by equation (3) has a Riemann surface which is topologically equivalent to a sphere with handles. This statement is valid for any multi–valued algebraic function.

Exercise 1°. Construct the Riemann surface of the algebraic function $w^n - z = 0$, where $n > 2$ is an integer, and verify that it is n–sheeted and topologicallly equivalent to a sphere.

Investigation of non–algebraic analytic functions $w(z)$ in the z–plane, which satisfy the equation $f(z, w) = 0$ with a non–algebraic analytic function f, also leads to Riemann surfaces on which analytic functions are single–valued.

Exercise 2°. Consider the logarithmic function determined by the equation $e^w - z = 0$, and construct its Riemann surface.

§ 5. Something about knots

Intuitively, the concept of knot seems to be not complicated. The simplest examples of knots are the "simple" knot (Fig. 28) (trefoil knot) and the "figure eight" (Fig. 29) which can be easily represented using a rope. Any attempt to transform the "simple" knot into a "figure eight" not pulling the ends of the rope through the loop will fail. Thus, experiment shows that these knots are different, and so the subject of the mathematical classification of knots arises.

We can forbid to pull the ends through a loop if we identify (glue together) the ends of the rope (Fig. 30 and 31). Then the following definition becomes natural.

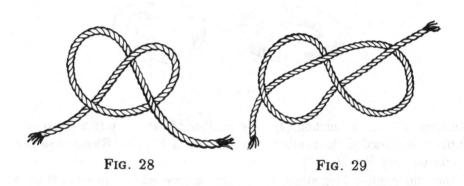

FIG. 28   FIG. 29

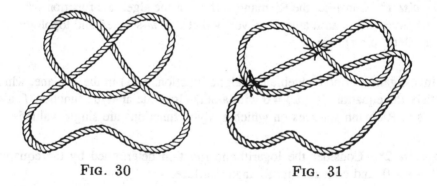

FIG. 30   FIG. 31

Definition 1. A homeomorphic image of the circle S^1 in $\mathbf{R}^3$ is called a *knot*.

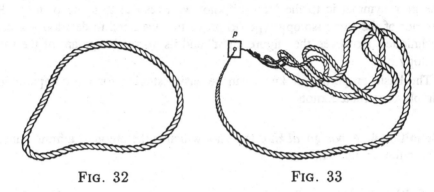

FIG. 32   FIG. 33

Examples. *a*) the trivial knot (or unknot) (Fig. 32);   *b*) the "simple" knot or "cloverleaf", or "trefoil" (Fig. 30); the "figure eight" or a "bretzel" knot (Fig. 31).

Note that by the definition all knots are homeomorphic. Therefore, to classify the knots, we classify imbeddings (homeomorphisms) with a help of which the circle can be imbedded in $\mathbf{R}^3$.

Definition 2. *The knots K_1 and K_2 are equivalent* if there exists a homeomorphism of $\mathbf{R}^3$ onto itself which maps K_1 into K_2.

A more precise classification of knots is based on the concept of isotopy of the space $\mathbf{R}^3$. A continuous mapping $H : [0, 1] \times \mathbf{R}^3 \rightarrow \mathbf{R}^3$ is called an *isotopy* if for each $t \in [0, 1]$, the mapping H maps $\mathbf{R}^3$ homeomorphically into itself; while, for $t = 0$ it is the identity mapping. Thus, an isotopy is a family of homeomorphisms of the space $\mathbf{R}^3$ which depends on a parameter t and which changes continously as t increases, starting with the identity when $t = 0$.

Definition 3. Two knots K_1 and K_2 are of the same *isotopy type* if there exists an isotopy $H(t, x)$ of the space $\mathbf{R}^3$, $t \in [0, 1]$, $x \in \mathbf{R}^3$, such that $H(1, K_1) = K_2$.

Exercise 1°. Show that the belonging to the same isotopy class is an equivalence relation.

There are examples of knots which are equivalent in the sense of definition 2, but of different isotopy types. Thus, a "trefoil" knot and its mirror image, i.e. the knot symmetric to the "trefoil" knot with respect to some plane in $\mathbf{R}^3$, are not of the same isotopy type (to prove this we need to develop a special technique). However, the "figure eight" and its mirror image are of the same isotopy type.

The basic properties of knots can be easily studied for the comparatively simply constructed knots.

Definition 4. A *polygonal knot* is a knot which is the union of a finite number of rectilinear intervals.

Definition 5. A knot which is equivalent to a polygonal knot, is said to be *tame*, and a knot which is not equivalent to a polygonal one is called *wild*.

Examples. The trivial, "trefoil" and "figure eight" knots are tame. Fig. 33 demonstrates a wild knot. The number of loops in this knot increases indefinitely, whereas their size decreases indefinitely when approaching point p. It is of interest to note, that if the number of loops is finite, then the knot would be equivalent to the trivial knot.

The classification of knots is closely related to the properties of the spaces which are complementary to knots. For instance, if a topological invariant of the complements of knots K_1 and K_2 differs, then the knot K_1 is not equivalent to K_2 (and not isotopic as well). A useful topological invariant is the fundamental group of the knot complement (see, Ch. 3). Note also that the set of all equivalence classes of knots (or isotopic equivalence) can be provided with an algebraic structure. The idea about this structure can be presented in the following way: by the *composition (product)* $K_1 * K_2$ of two knots K_1 and K_2 we understand the result of tying them one after the other. The order in which they are tied does not matter, or, more exactly, the knot $K_1 * K_2$ is equivalent to the knot $K_2 * K_1$. Thus, we define the composition of equivalence classes of knots which is commutative and associative. The equivalence class of a trivial knot plays a role of the identity element. However, an attempt to solve the equation $K * X = 1$ (i.e. to untie K by tying up the knot X) fails with exception of the case $K = 1$. Consequently, the equivalence classes of knots form only a semigroup, and not a group.

§ 6. On some topological applications in physics

In physics of condensed matter different ordered structures in substances are considered. Topological representations necessarily appear in the investigations of stability of the defects which arise in these structures. Some of these defects are as follows. In crystals one has the so–called dislocations (violations of the order of a crystal structure); in liquid crystals there are disclinations (violations of the continuity in the field of directions of molecules); in the superfluid liquids He^3, He^4 and ferromagnets, and other stable geometric configurations there can occur vortices. The topological basis of these phenomena will be discussed in this section.

1. The concept of a (vector) field and a singular point of it (or a singular curve) are the first to occur in the mathematical description of defects. By a vector field on a surface $U \subset \mathbf{R}^3$ in three–dimensional Euclidean space we usually understand a mapping $F : U \to \mathbf{R}^3 \backslash 0$, which associates to every point $x \in U$ a vector $F(x) \in \mathbf{R}^3$, $F(x) \neq 0$ (we identify points x with their radius–vectors $\overrightarrow{x}$). If for every point $x \in U$ we draw a vector $\overrightarrow{F(x)}$ starting in that point, then we get a geometric picture (Fig. 34) called a vector field on U.

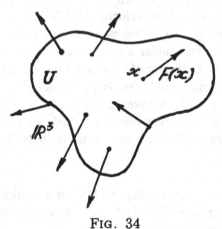

In addition, it is desirable to require a continuous relation between the vector $\overrightarrow{F(x)}$ and the point x (i.e. continuity of the mapping $F : U \to \mathbf{R}^3 \backslash 0$).

However, this requirement not always is satisfied for all the points from U, viz. in separate points (called "singular") or on certain curves $\gamma \subset U$ ("singular curves") the field $\overrightarrow{F}$ can be discontinuous or even indeterminate; in particu-

FIG. 34

lar, the zero value $F(x) = 0$ is also considered as indeterminate (i.e. the direction of the vector is undefined). In such cases one speaks of a singularity of the vector field.

For instance, if a domain U is occupied by a ferromagnet, then one can define in its points a magnetic moment, i.e., a vector $\overrightarrow{M}$ (even when the magnetic field is absent), if the temperature for the given object is lower than the critical one. The vector field $\overrightarrow{F}(x) = \overrightarrow{M}(x)$ on the domain U may have singular points and singular curves. The simplest singular point is of a field which is directed along the radius–vectors $\overrightarrow{x}$: $\overrightarrow{M}(x) = M(x) \cdot \frac{\overrightarrow{x}}{|\overrightarrow{x}|}$, $0 \in U$, $M(x) \neq 0$ is a numerical function on U; the field here is not determined in the point $x = 0$; so $x = 0$ is the singular point called the "hedgehog".

If the vector $\overrightarrow{M}(x)$ in all of the points where it is determined, is parallel to a subspace $\mathbf{R}^2 \subset \mathbf{R}^3$, then $\overrightarrow{M}(x) \in \mathbf{R}^2$, ot to a plane Πx^0 which passes through the point $x^0 \in U$ and is parallel to $\mathbf{R}^2$. There is the vector field $\overrightarrow{M}$: $V_{x^0} \to \mathbf{R}^2$ given on the intersection $V_{x^0} = U \cap \Pi x^0$ and it can have a singular point x_*^0 in the plane domain V_{x^0}. If the plane Πx^0 is moved in a parallel way (by changing x^0), then these singular points x_*^0 can merge into a singular curve l of the vector field $\overrightarrow{M}$ on U (Fig. 35). Such a singular curve l is called a "vortex" of the field $\overrightarrow{M}$. Vortices naturally occur in continuous media, viz. fluids and gas, when the particles from the medium describe circular movements around some curve l. Usually, by the term vorticity (or vortex) we understand in fact this circular movement of the medium; the curve l is then called the "axis" of vorticity.

Here, the role of the field $\overrightarrow{M}$ is played by the velocity field $\overrightarrow{v}(x)$ of the points x of the medium; l is a singular curve of the velocity field $\overrightarrow{v}(x)$.

A remarkable fact is the discovery and studies of the vorticity movements in the superfluid helium He^4. The liquid He^4 under a temperature T which is close to the absolute zero, behaves as a mixture of two components, viz. "normal" (density ρ_s, velocity $\overrightarrow{v_n}$) and "superfluid" (density ρ_n, velocity $\overrightarrow{v_s}$); $\rho_s + \rho_n = \rho$ is the total density. When $T = 0$ we have $\rho_n = 0$, $\rho_s = \rho$. It is the superfluid component of He^4 that forms vortices which are stable in spite of the viscous friction forces. Vortices in fluids and gases do not possess this stability.

2. In order to explain topological effects in the stability and laws of evolution of vortices in a ferromagnet and in the superfluid He^4, it is necessary to introduce some of the simplest topological invariants.

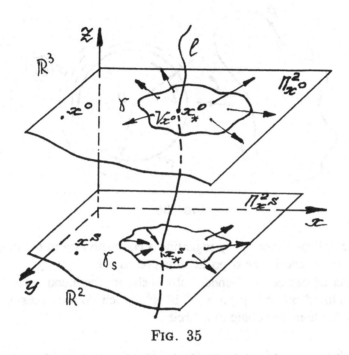

FIG. 35

Let us fix a two–dimensional plane $\Pi^2 \subset \mathbf{R}^3$ and a path $\gamma \subset \Pi^2$ which is homeomorphic to a circle and is provided with an orientation (a fixed direction of the loop). Let there be given a continuous mapping $f : \gamma \to S^1 \subset \mathbf{R}^2$ into the circle of radius 1 with centre at the origin and also provided with an orientation. While the point $p \in \gamma$ runs through the path $\widetilde{p_0 p}$ starting at the point $p_0 \in \gamma$ in the direction of the orientation of γ, the point $q = f(p)$ on S^1 runs through a path on S^1 either in the direction of the orientation, or opposite to it (depending on the position of the point p), counting from the point $q = f(p_0)$ (see, Fig. 36).

While the point p makes one full turn along the path γ, the point $q = f(p)$, as a result, makes k_+ full turns around S^1 in the direction of orientation of S^1, and k_- turns in the opposite direction. The degree of the mapping f of the oriented curve γ to the oriented curve S^1 is the difference $k_+ - k_-$ and is denoted $\deg f$. This degree can be also defined as a sum of complete

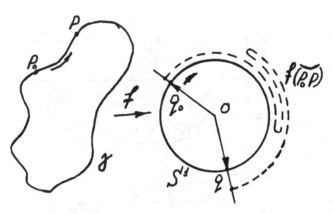

FIG. 36

turns of the radius–vector $\vec{q}$ by regarding a complete turn in the direction of orientation as +1 and in the opposite direction as −1.

The notion of degree is extended, obviously, to continuous mappings f : $\gamma \rightarrow \Gamma$ of closed oriented paths γ, Γ in $\mathbf{R}^3$, which do not necessarily lie in planes, but are homeomorphic to a circle.

3. The degree, $\deg f$, is a topological invariant (or to be more exact, a homotopy invariant), i.e., it remains the same if the mapping f is continuously changed (changed homotopically). This signifies that if f is included into a family of mappings f_λ : $\gamma \rightarrow \Gamma$ that depends on a numerical parameter λ, $0 \le \lambda \le 1$, in such a way, that $f_1 = f$, and the point $q = f_\lambda(p)$ continuously depends on the variables $(\lambda, p) \in [0, 1] \times \gamma$, then for every member f_λ of the family, the degree $\deg f_\lambda$ is constant for all λ from the interval $[0, 1]$.

Thus, changing the mapping f continuously (without sharp jumps and discontinuities) one cannot decrease the total number of positive and negative loops of the image $f(\gamma)$ which are described on Γ, i.e. the number $k_+ - k_-$.

Another important property of the degree $\deg f$ is the following: if γ, γ_1 and γ_2 are paths homeomorphic to a circle, with fixed orientations, and f : $\gamma \rightarrow \gamma_1$, $g : \gamma_1 \rightarrow \gamma_2$ are continuous mappings with degrees $\deg f$ and $\deg g$, respectively, then the degree $\deg(gf)$, of the composition $gf : \gamma \rightarrow \gamma_2$ is equal to the product of the degrees f and g, i.e. $\deg(gf) = (\deg g)(\deg f)$.

This property can be verified by calculating the total number of loops of the image $(gf(\gamma))$ which are described on γ_2, by means of analogous sums for the images $f(\gamma)$ and $g(\gamma_1)$.

Let associate to each mapping $f : \gamma \to \Gamma$ the class $[f]$ of all mappings $f' : \gamma \to \Gamma$ homotopic to it. It is not difficult to see, using obvious arguments, that the homotopy relation of mappings is an equivalence relation. Consequently, a set of all continuous mappings from γ into Γ is divided into a set of non–intersecting equivalence classes $\{[f]\}$, the so–called homotopic classes. For all mappings f' from the class $[f]$ deg $f' = $ deg f, therefore to every class we can assosciate the integer $[f] = $ deg f. The fact that this correspondence is bijective is non–trivial (see, Ch. 3, § 3). Therefore, the integer $n = $ deg f is a homotopy invariant of the class of mappings $[f]$. It is convenient to label this class by the index n, i.e. $[f]_n$. The class $[f]_0$ consists of all mappings which are homotopic to a fixed mapping $f_0 : \gamma \to q_*$ of the path γ into a fixed point $q_* \in \Gamma$.

4. Now let us return to the magnetic vorticity in a ferromagnet (Fig. 35) and to the vorticity in the superfluid He^4.

In a plane P_{x^0}, take an oriented closed path γ which has no singular points, but inside of which there is a unique singular point x_0. Let this path be homeomorphic to a circle, and run around the singular point x_0 once; let $S^1 \subset \mathbf{R}^2$ be the oriented circle of radius 1 with its centre at 0. A vector field $\overrightarrow{M}$ considered only in the points $x \in \gamma$, determines a mapping $f : \gamma \to S^1$, defined by

$$(1) \qquad \overrightarrow{f(x)} = \frac{\overrightarrow{M}(x)}{|\overrightarrow{M}(x)|}, \quad x \in \gamma.$$

The degree, deg f, of f is called the rotation of a vector field $\overrightarrow{M}$ on γ is defined by the mapping (1); we denote this rotation by $\kappa(\overrightarrow{M}; \gamma)$.

Now let us continuously move the plane Π_{x^0} parallelly to itself to a new position Π_{x^1}, and let the contour γ go through the intermediate positions γ_s (on the corresponding plane Π_{x^s}) to the contour $\gamma_1 \subset \Pi_{x^1}$, $0 \le s \le 1$; moreover, $\gamma_{s=0} = \gamma$, $\gamma_{s=1} = \gamma_1$. It is assumed that all contours γ_s enclose a singular curve l, that all the γ_s are homeomorphic to the initial one: $\alpha(s) : \gamma \to \gamma_s$, that they are oriented equally, and that the homeomorphism $\alpha(s)$ preserves orientation (i.e. deg $\alpha(s) = +1$) and depends continuously on the parameter

s; and $\alpha(0) = 1_\gamma : \gamma \to \gamma$ is the identity mapping of γ into itself. The vector field $\vec{M}(x)$ considered on the contour γ_s can be "transferred" by the homeomorphism $\alpha(s) : \gamma \to \gamma_s$ to the contour γ, according to the formula $\vec{F_s}(x) = \vec{M}(\alpha(s)(x))$, $x \in \gamma$. Evidently, the family $\vec{F_s}(x)$ on γ continuously depends on (x, s) and, therefore, it is a homotopy of vector fields on γ between $\vec{F_1}(x) = \vec{M}(\alpha(1)(x))$, $F_0(x) = \vec{M}(x)$. By the degree properties, we have

$$\kappa(\vec{M}, \gamma) = \kappa(\vec{F_1}, \gamma) = (\deg \alpha(1))\kappa(\vec{M}, \gamma_1) = \kappa(\vec{M}, \gamma_1),$$

where the latter rotation is the rotation of the field $\vec{M}$ on γ_1 in the plane Π_{x^1}.

This general rotation of the field $\vec{M}$ in an arbitrary plane Π_{x^1} is called the topological index $\kappa(l)$ of the vortex l in a ferromagnet (or the topological charge of vortex). The conditions required above are fulfilled, say, for a ferromagnet which has an "easy plane of magnetization" $\mathbf{R}^2$ (physical terminology). In this case the stability condition of vorticity in a ferromagnet (under homotopies of the magnetization field $\vec{M}$ in the plane $\mathbf{R}^2$) is the difference between the degree $\deg f$ and zero, or, what ammounts to the same, between the rotation $\kappa(\vec{M}, \gamma)$ and zero.

Consider the superfluid component of He^4. Under temperatures which are close to absolute zero, the superfluid part of the liquid occuring in He^4 ("superfluid condensate") is described in terms of quantum mechanics by a complex–valued wave function $\Psi(x)$, $x \in \mathbf{R}^3$. We have $\Psi(x) = |\Psi(x)|e^{i\Phi(x)}$, where $|\Psi|$ is the absolute value of the complex number Ψ, and Φ is the phase of the wave function. In addition, if a superfluid condensate is in an equilibrium state $(\vec{v_s} = 0)$, then $|\Psi|$ is constant (not dependent on x), Φ is an indeterminate constant; if a superfluid condensate is in a non–equilibrium state $(\vec{v_s}$ not identically zero), then $|\Psi|$ remains a constant, but $\Phi = \Phi(x)$ becomes a function of the point $x \in \mathbf{R}^3$; moreover, the velocity field of the superfluid component is calculated by the formula $\vec{v_s}(x) = \frac{\hbar}{m}\mathrm{grad}\,\Phi$, where $\hbar$ is the Planck constant and m is the atom mass of He^4.

Let U be a domain in $\mathbf{R}^3$ occupied by the superfluid condensate. For any point $x \in U$, where $\vec{v_s}(x) \neq 0$, the phase $\Phi(x)$ is defined; it corresponds to a ray in $\mathbf{R}^3$ which leaves the point zero at an angle $\Phi(x)$ taken counter clockwise with respect to some fixed direction. For a point $x \in U$, where $\vec{v_s}(x) = 0$, the angle $\Phi(x)$ is not determined. Thus, on the domain U we have a vector field of unit length vectors $\vec{d}(x)$, $d : U \to S^1$, which are directed along the rays by the corresponding values of the phase $\Phi(x)$, with the exception of singular points in which the phase $\Phi(x)$ is undefined. If we select a closed oriented path γ in U which has no singular points, then the mapping $d : \gamma \to S^1$ has its degree, $\deg d$, defined (given a fixed orientation on S^1).

Consider now a vortex in the superfluid He^4 with vorticity curve l; the value of the phase Φ is not defined on l, as it consists of singular points of the field $\overrightarrow{d}(x)$. In the capacity of γ, we take an oriented closed path which loop around the vorticity curve l once. Analogously to the case of vorticity in ferromagnet, the degree, $\deg d$, of the mapping $d : \gamma \to S^1$ is called the topological index (or topological charge) of the vortex. The stability of a vortex in the superfluid He^4 is characterized again by the difference between the topological index of vorticity and zero.

5. From the physical point of view, the stability of vortices in a ferromagnet and superfluid He^4 signifies that they are observed under natural conditions despite external disturbances, which are impossible to get rid of completely. Mathematically, this stability, as will be shown below, takes the form of the difference between the topological index of vorticity and zero, and, thus, it is related to the homotopy properties of the circle S^1.

If a magnetic field $\overrightarrow{M}(x)$ or a superfluid velocity $\overrightarrow{v_s}(x)$ corresponding the phase $\Phi(x)$ is changed sufficiently little uniformly for all x, then the new values $\overrightarrow{M_1}(x) = \overrightarrow{M}(x) + \Delta\overrightarrow{M}(x)$, $(\overrightarrow{v_1})_s(x) = \overrightarrow{v_s}(x) + \Delta\overrightarrow{v_s}(x)$ (with phase $\Phi_1(x) = \Phi(x) + \Delta\Phi(x)$) under sufficiently small increases, will be homotopic to the initial ones. The homotopies involved can be written, for instance, in the form $\overrightarrow{M_\lambda}(x) = \overrightarrow{M}(x) + \lambda\Delta\overrightarrow{M}(x) \in \mathbf{R}^2$, $\Phi_\lambda(x) = \Phi(x) + \lambda\Delta\Phi(x)$, $0 \leq \lambda \leq 1$. These homotopies yield homotopies of the vector fields $\overrightarrow{f(x)}$, $\overrightarrow{d(x)}$, i.e. yield mappings $f_\lambda : \gamma \to S^1$, $d_\lambda : \gamma \to S^1$ (f_λ is defined by formula (1)), which depend continuously on the parameter λ and coincide with $f_0 = f$, $d_0 = d$, when $\lambda = 0$, and with f_1, d_1, i.e. with the corresponding fields $\overrightarrow{M_1}(x)$, $(\overrightarrow{v_1})_s(x)$, when $\lambda = 1$.

Hence, $\deg f_\lambda$, $\deg d_\lambda$ do not alter under the changes in λ, i.e. $\deg f_0 = \deg f_1$, $\deg d_0 = \deg d_1$. Thus, slight physical disturbances do not change the degree on the contour γ. If the topological index of vorticity, $\deg f$ or $\deg d$, differs from zero then after a small disturbance it would still be the case that $\deg f_1 \neq 0$ or $\deg d_1 \neq 0$ on the curve γ. Hence, the central statement that the curve γ encloses a singular curve of the altered vector field $\overrightarrow{M_1}(x)$ or $(\overrightarrow{v_1})_s(x)$ follows. Indeed, choosing the curve γ as a circle (which lies in the plane Π_{x^0}) and assuming that the closed disc $D \subset \Pi_{x^0}$ bounded by that curve has no singular point of the field $\overrightarrow{M_1}$ or $(\overrightarrow{v_1})_s$ (i.e. has no trace of intersection between the singular curve of vorticity and D), we can obtain an extension $\widetilde{f_1} : D \to S^1$ $(\widetilde{d_1} : D \to S^1)$ of the mapping $f_1(d_1)$. Define a deformation of the disc D with respect to its centre a, e.g. $\phi_\lambda(x) = a + \lambda(x-a)$, $0 \le \lambda \le 1$. This deformation yields a homotopy $\widetilde{f}1,\lambda : \gamma \to S^1$ according to the formula $\widetilde{f}_{1,\lambda}(x) = \widetilde{f_1}(\phi_\lambda(x))$, and, similarly, a homotopy $\widetilde{d}_{1,\lambda}$. If $\lambda = 1$, then $\widetilde{f}_{1,1}(x) = f_1(x)$; if $\lambda = 0$, then $\widetilde{f}_{1,0}(x) = \widetilde{f_1}(a)$ is a constant mapping. We have shown, thus, that the mapping $f_1 : \gamma \to S^1$ is homotopic to a constant if only it is extended to D (and, analogously, for $d_1 : \gamma \to S^1$). It is evident, that the degree $\deg \widetilde{f}_{1,0}$ $(\deg \widetilde{d}_{1,0})$ of a constant mapping is equal to zero, and from the invariance property of degree under homotopy it follows that $\deg f_1 = 0$ (and, analogously, $\deg d_1 = 0$); this contradicts the previously obtained conclusion that $\deg f_1$ $(\deg d_1)$ differs from zero. Thus, we have that a singular point of the field $\overrightarrow{M_1}(x)$ or $\overrightarrow{d_1}(x)$ exists inside the disc D; moving the disc D along the vorticity curve l we obtain a family of singular points which merge into a vorticity curve l_1 of the fields $\overrightarrow{M_1}(x)$ or $(\overrightarrow{v_1})_s(x)$.

Therefore, slight physical disturbances of the magnetic field in a ferromagnet (in the plane $\mathbf{R}^2$) or of the velocity field of the superfluid component of He^4, cannot kill a vortex of a non–zero topological index; this, really, is the fact discussed above concerning the stability of vortices. There is, of course, a stronger statement of the "topological stability", viz. a vortex persists under any homotopy $\overrightarrow{M_\lambda}(x)$, $(\overrightarrow{v_\lambda})_s(x)$ not necessarily a small one, but such that it yields continuous homotopies $f_\lambda : \gamma \to S^1$, $d_\lambda : \gamma \to S^1$.

As for vorticities of topological index 0, they are topologically unstable (i.e. can be destroyed during a homotopy).

6. There is a clear physical reason for the appearance of the circle S^1 in the analysis of vortices in subsection 5. According to a well–known principle in physics, stable states of matter observed in experiments, correspond to the local minimal values of energy. The energy is calculated with respect to the given field (vector, tensor, etc.), which describes the state of matter. For example, the energy of a ferromagnet is defined by the vector field of the magnetic moment $\overrightarrow{M}(x)$, and that of the helium He^4, by the wave function $\Psi(x)$. The so–called "degenerate states" of matter are characterized by the non–uniqueness of the field for which the energy takes the value of a local minimum. For instance, the energy of a ferromagnet can take a minimal value when the vectors $\overrightarrow{M}(x)$ are all orthogonal to the certain crystal axis (i.e., they lie in a two–dimensional plane $\mathbf{R}^2$); in addition, they have a fixed absolute value $|\overrightarrow{M}(x)|$ and can have any direction (this is exactly the case of an "easy plane" of magnetization). A set of all such vectors $\overrightarrow{M}(x)$ forms a circle S^1 of radius $|\overrightarrow{M}(x)| = $ const and is called the domain of degenerate states (with respect to the parameter of magnetization) of a ferromagnet. If the energy of a ferromagnet depends only on the absolute value of magnetization $|\overrightarrow{M}| = $ const, then the domain of degeneracy is the two–dimensional sphere S^2; this case corresponds to amorphous (isotrop) matter.

The case of the superfluid helium He^4 differs from that of a ferromagnet, and it is related to quantum mechanics: for a superfluid condensate which is in equilibrium, the phase Φ of the wave function $\Psi = |\Psi| \exp(i\Phi)$ remains arbitrary, the absolute value $|\Psi|$ stays constant, and the energy does not depend on Φ; thus, we get degeneracy with respect to the phase Φ, and the domain of degenerate states, i.e. the set of all possible values of the function Ψ in the complex plane, is the circle S^1 of radius $|\Psi|$.

Topological indices of vortices as considered above, are defined by mappings $f, d : \gamma \rightarrow S^1$ of closed curves into the domains of degeneracy of a ferromagnet (with an easy plane of magnetization) and the superfluid He^4. Thus, the existence of topologically stable vortices in these matters is related to the topological properties of the domain of degeneracy, which in this case is the circle S^1.

7. In physics, however, one can find domains of degeneracy different from S^1. The case of isotopic ferromagnet already illustrates the possibility of a two–dimensional domain of degeneracy, viz. the sphere S^2. A conclusion that topologically stable vortices are absent in the situation described can be drawn easily. If D is an arbitrary disc with a boundary γ (a circle), then any mapping $f : \gamma \rightarrow S^2$ is homotopic to a constant mapping $f_0 : \gamma \rightarrow \overrightarrow{c} \in S^2$. Such a homotopy, for instance, is realized by moving the points of the curve $f(\gamma)$ on S^2 to a fixed point $\overrightarrow{c}$ along the meridians which run from $(-\overrightarrow{c})$ to $\overrightarrow{c}$ (for simplicity, assume that $(-\overrightarrow{c})$ does not lie on the image $f(\gamma)$). Hence, $\deg f = \deg f_0 = 0$, and the disc D does not necessarily intersect with a curve of topologically stable vorticity. This fact is implication of topological properties of the domain of degeneracy, i.e. the sphere S^2.

Because of the same properties there exist simpler topological singularities for an isotropic ferromagnet, viz. isolated points of the vector field $\overrightarrow{M}(x)$. An example of a singular point of the "hedgehog" type was discussed in subsection 1. In order to construct the topological index of a singular point it is necessary to define the mapping f (see (1), subsection 4) on the sphere $S^2_\epsilon(x^*)$ of a sufficiently small radius ϵ with centre at the singular point x^*, taking values in the unit sphere S^2, i.e. $f : S^2_\epsilon(x^*) \to S^2$. For this kind of a mapping we generalize the notion of the degree of mapping which is denoted by $\deg f$, as above. The construction of the degree is much more complicated compared with that of the mapping of a circle into a circle. First of all, the spheres should be oriented by orienting their tangent planes in such a way that the orientations in points that are near are the same. In the second place, it is necessary to define the algebraic number of layers of the image $f(S^2)$, which lie on the sphere S^2; in addition, a layer gets number (+1) if its orientation is the same as that of S^2, and (−1), otherwise. To give a more precise meaning to these words in the case of continuous mappings would be a rather long job (the reasons for this are discussed in Ch. 3, § 4; Ch. 4, § 5; Ch. 5, § 6); in the case of the differentiable mappings, the $\deg f$ can be defined with a help of methods of differential geometry (see, e.g., [64]). In this section we use not a precise but the obvious description of $\deg f$ given above. The properties of the degree of the mappings of circles (subsection 3) hold also for the degree of mappings of spheres. Thus, for a ferromagnet with a magnetic vector field $\vec{M}(x)$ and a singular point x^* the integer $\deg f$ is defined; it is called the topological index of the singular (isolated) point x^* and denoted by $\kappa(x^*)$. This number does not depend on the radius $\epsilon > 0$ of the sphere $S^2_\epsilon(x^*)$, if ϵ is small enough. In physics the number $\kappa(x^*)$ is called the topological charge of the singular point x^*.

If we have a singular point of the "hedgehog" type (subsection 1), then $\kappa(x^*) = +1$. The topological index $\kappa(x^*)$ plays the same role in the investigation of singular points as the topological index $\kappa(l)$ when studying vorticities. A singular point is topologically stable if $\kappa(x^*) \neq 0$; such a point can be physically observed and existence of such a point is preserved under homotopies of the magnetic field; on the other hand, if $\kappa(x^*) = 0$ then the singular point x^* can be eliminated by a suitable homotopy of the magnetic field, i.e. it is topologically unstable.

In the beginning of the seventies of this century, the domains of degeneracy of superfluid phases for helium He^4 had been studied. There were found two corresponding phases, called A and B; the more complicated and interesting appeared to be the phase A. The domain of degeneracy of the phase A is characterized by a set of four vectors $(\vec{e_1}, \vec{e_2}, \vec{e_3}, \vec{v})$ from $\mathbf{R}^3$, where the first three $(\vec{e_1}, \vec{e_2}, \vec{e_3})$ are all possible orthonormalized frames with a fixed orientation, and $\vec{v}$ is an arbitrary unit length vector (the vectors $\vec{e_i}$, $i = 1, 2, 3$,

$\overrightarrow{v}$ have definite physical meaning [86] which is not considered here). The set of frames $(\overrightarrow{e_1}, \overrightarrow{e_2}, \overrightarrow{e_3})$ can be identified with the group $SO(3)$, i.e., with the group of orthogonal 3×3 matrices or the group of rotations of a solid body. The set of vectors $\overrightarrow{v}$ is the unit sphere S^2. Consequently, the domain of degeneracy of the A–phase is the Cartesian product $S^2 \times SO(3)$ (its dimension equals to 5). Under definite physical conditions the vector v is fixed and then the domain of degenerate states is reduced to the group $SO(3)$. In this situation the degree of mapping does not work and a more complicated part of the aparatus of homotopic topology should be applied, i.e., the concept of a fundamental group. Corresponding analysis shows that there exist two classes of vortices: one class contains topologically stable vorticities, the other one's are topologically unstable. One can observe experimentally not only stable but also non–stable vorticities by placing the A–phase into a rotating vessel. The observed new phenomena such as a vortex with an end, a rotational flow without a singular curve, vortices on a surface, allow theoretical explanation based on topological ideas.

8. As it was noted above, in order to investigate more complicated domains of degeneracy than the sphere S^2 or the circle S^1, we have to apply a more complex part of the apparatus of the homotopy theory. To start, note that the concept of homotopy and homotopical classes is naturally extended to mappings $f : X \rightarrow Y$ of arbitrary sets which lie in Euclidean (or even metrical and topological) spaces; the family of all such mappings again is divided into homotopic classes $\{[f]\}$, i.e. equivalence classes; this collection of equivalence classes is denoted by $\pi[X, Y]$. However, the description of these classes becomes more complicated, and it is one of important problems in homotopic topology.

Most often are considered mappings $f : S^n \rightarrow Y$ of the n–dimensional unit sphere into Y, $n \geq 1$; their homotopy classes $\pi[S^n, Y]$ are called the n–dimensional homotopy classes.

Let $[f]$ be a homotopy class from $\pi[S^n, Y]$. The mappings of the class $[f]$ are said to be n–dimensional spheroids (or "loops", for $n = 1$). It is useful to fix points $x_0 \in S^n$, $y_0 \in Y$ and to restrict the class of spheroids $f : S^n \rightarrow Y$ by the condition $f(x_0) = y_0$, i.e. the spheroids at the distinguished point y_0. Their images can be seen as strongly distorted spheres which are attached to the point y_0. For a homotopy f_λ, $0 \leq \lambda \leq 1$, of the spheroid f we also introduce the condition $f_\lambda(x_0) = y_0$, $0 \leq \lambda \leq 1$; this means that although the image $f_\lambda(S^n)$ moves in Y as λ varies, but it remains attached to the fixed point y_0. Then the corresponding homotopy classes of spheroids ("loops", for $n = 1$) in the point y_0 are denoted by $\pi_n(Y, y_0)$.

An important property of the classes $\pi_n(Y, y_0)$ is that a notion of product can be introduced in them, and so $\pi_n(Y, y_0)$ becomes a group (commutative for $n > 1$). It is easy to describe the formation of products for $n = 1$. Let S^1 be oriented. If $f : S^1 \to Y$ is a loop of the class $[f]$, then by moving the point x along S^1 starting at the point x_0 in the direction of orientation, the point $y = f(x)$ draws a path in Y with beginning and end at the point y_0 (a "loop" in the point y_0). Let f, g be two loops in the point y_0 from the classes $[f], [g] \in \pi_1(Y, y_0)$; then one can consider the "composite" loop in the point y_0 composed of the first loop f followed by the loop g while this "double" loop corresponds to one complete traversing of the circle S^1 by the point x (Fig. 37).

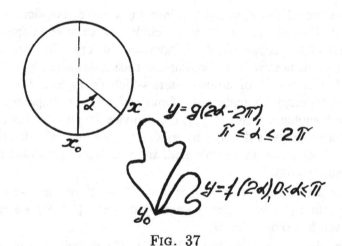

FIG. 37

It is this double loop that defines the class from $\pi_1(Y, y_0)$ which, by definition, is equal to the product $[f] \circ [g]$ of the classes $[f]$ and $[g]$ (in the order indicated). The loop $e : S^1 \to y_0 \in Y$ (constant mapping) defines the class $[e]$ which is the identity in $\pi_1(Y, y_0)$: $[e] \circ [f] = [f] \circ [e] = [f]$ for any $[f] \in \pi_1(Y, y_0)$. Any loop from the class $[e]$ is homotopic to the constant mapping e. The inverse element $[f]^{-1}$ is defined by the loop f which is traversed in the opposite direction, i.e. $y = f(2\pi - \alpha)$, $0 \leq \alpha \leq 2\pi$. It is easy to check that the axioms for groups hold.

It is more complicated to describe the multiplication of homotopy classes in $\pi_n(Y, y_0)$, when $n > 1$, and we refer the reader to Ch. 3, where the elementary information on the groups π_1 and π_n, $n > 1$, is presented in detail. The group $\pi_1(Y, y_0)$ is called the fundamental group of the space Y at the point y_0. For path–connected spaces (in which any two points can be joined by a path), the group $\pi_1(Y, y_0)$ does not depend on the chosen point y_0 (i.e. $\pi_1(Y, y_0)$ and $\pi_1(Y, y_1)$ are isomorphic for any points y_0, y_1). In fact, given a loop $f \in \pi_1(Y, y_1)$, it can be "transfered" into $\pi_1(Y, y_0)$, where it is associated to the loop $\overrightarrow{f}$ composed of three parts $\widehat{y_0 y_1}$, f, and $\widehat{y_1 y_0}$ which are traversed in the indicated order, where $\widehat{y_0 y_1}$ is a path joining the points y_0 and y_1, and $\widehat{y_1 y_0}$ is the inverse path. This rule defines an isomorphism of the groups $\pi_1(Y, y_1)$, $\pi_1(Y, y_0)$.

The isomorphism mentioned allows us to identify groups $\pi_1(Y, y_0)$, $\pi_1(Y, y_1)$ for path–connected spaces; they are denoted by the symbol $\pi_1(Y)$.

Examples. If $Y = S^1$ then $\pi_1(S^1)$ is a commutative group isomorphic to the group of integers $\mathbf{Z}$ with repsect to addition; this fact is written as the equality $\pi_1(S^1) = \mathbf{Z}$. A similar statement holds for $Y = S^n$, $n > 1$: $\pi_n(S^n) = \mathbf{Z}$. The generating element γ_n of the group $\pi_n(S^n)$, $n > 1$, is represented by the identity $1_{S^n} : S^n \to S^n$; therefore, an arbitrary element $[f] \in \pi_n(S^n)$ is of the form $[f] = k\gamma_n$. The integer k, by definition, is the degree of the mapping $f : S^n \to S^n$ and it is denoted by $\deg f$. In the special case $f : S^1 \to S^1$, the number k has the following geometric interpretation : the loop f is homotopic to the loop which is obtained by repeating the loop $1_{S^1} : S^1 \to S^1$ k times in the positive direction, if $k > 0$, and in the opposite direction $|k|$ times, if $k < 0$; in the case when $k = 0$ the loop f is homotopic to the constant loop e.

Thus, in the case considered we have $[f] = (\deg f)\gamma_1$ for $f : S^1 \to S^1$, and, more generally, $[f] = (\deg f)\gamma_n$, if $f : S^n \to S^n$, $n > 1$. Thereby a one-to-one correspondence between the homotopic class $[f]$ and the degree $\deg f$ of a mapping f is defined. In particular, if $Y = S^2$, we obtain the definition of $\deg f$ for the case of mappings $f : S^2 \to S^2$, which is needed when considering singular points of a ferromagnet. If $Y \neq S^n$, then, generally speaking, such a connection may not exist, or a generalization of the notion of degree, $\deg f$, may be required; this depends on the algebraic construction of the homotopic group $\pi_n(Y)$. For instance, when classifying vortices of the superfluid phase A of the helium He^4, it is necessary to consider the mappings, i.e. loops $f : S^1 \to SO(3)$, and to have their homotopy classification. Since $\pi_1(SO(3)) = \mathbf{Z}_2$, we have, as above, an expression for a homotopy class $[f] = k\gamma_*$, where γ_* is the generating class in $\pi_1(SO(3))$, and $k \in \mathbf{Z}_2$ is a residue modulo 2 with values either 0 or 1. Defining $\deg_2 f$, i.e. the degree modulo 2, by the equality $\deg_2 f = k$, we come up to a generalization of the integer valued degree. In addition, the main properties of degree remain (as an exercise, verify it!), and we obtain two types of vortices: topologically stable ones, with a topological index of vorticity 1, and topologically unstable ones, with the index 0.

9. Singular points and singular curves (vortices) also appear in another class of matter, viz. in the so-called liquid crystals that have been studied intensively during the last twenty years. Under this term, there are known a number of matterials in which a definite "order structure" is observed, and this order is intermediate between the "order structure" of the common liquids and that of solid crystals.

"This outbreak of interest was caused by many reasons. First of all, liquid crystals have speeded up a technical revolution in devices for the visual representation of information (displays)... Second, the liquid crystal state is characteristic for any biologically active system including the human body... Third, and this is the most important from our point of view, the physics of liquid crystals appeared to be unusually complicated" [65, p. 21–22].

The simplest such structure appears in the "nematic" liquid crystals (or shortly, "nematics"), which consist of elongated (rod–like) molecules. The molecules of a nematic have only one (elongated) symmetry axis. The nematics have the characteristic orientation order of symmetry axes of molecules, when the axes of the molecules that are near, are almost parallel. By setting in each point x of a nematic the direction ("director") determined by the axis of the molecule that passes through x, we obtain a "direction field". The state of a nematic is determined by the director field similarly to the manner in which the field of magnetization of a ferromagnet determines the state of the latter. In order to define the director field analytically, it is convenient to associate to each direction a unit vector $\vec{d}(x)$ in $\mathbf{R}^3$ which is parallel to the director in the point x; thus, on the domain $U \subset \mathbf{R}^3$ occupied by a nematic there is a vector field $\vec{d}(x)$. The director field is defined by the vector field $\vec{d}(x)$; however, it should be distinguished from the vector field $\vec{d}(x)$, as the vectors $\pm\vec{d}(x)$ define the same direction. The ends of the vector $\pm\vec{d}(x)$ on

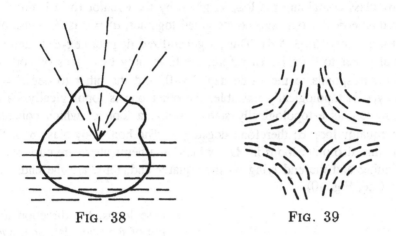

FIG. 38   FIG. 39

the unit sphere in $\mathbf{R}^3$ define a pair of central–symmetric points which can be seen as a point $\overrightarrow{d}\,(x)$ of the projective space $\mathbf{R}P^2$; this point is obtained from the two–dimensional sphere S^2 by gluing (identifying) diametrically opposite points together; recall that $\mathbf{R}P^2$ can also be obtained from the hemisphere by gluing together diametrically opposite points on the equator (see, § 3). Thus, the director field is completely characterized by a mapping $\overrightarrow{\widehat{d}}\ : U \to \mathbf{R}P^2$ of the domain U into the projective space $\mathbf{R}P^2$. It is $\mathbf{R}P^2$ that is the domain of degenerate states of a nematic, as there are no specific restrictions for the directions of the axes of molecules (contrary to a number of other types of liquid crystals).

The requirement of continuity for the mapping $\widehat{d}$ on the whole domain U is natural, but not always possible, and in the domain U there occur (as in the case of a vector field) singular points (and singular curves) in which (and on which) the mapping is undetermined or discontinuous. Singular curves can be observed optically in the form of fine threads in the nematic; hence the term "nematic" (from the greek word "thread"). The Fig. 38, 39 present two pictures of director fields in nematics for the plane field $\widehat{d}(x)$ (i.e. $\widehat{d}(x) \in \mathbf{R}^2$); a vortex is here represented by the central point.

The topological classification of vortices is handled in this case according to the previous scenario. Take a circle S^1 which encloses a singular curve, and consider the director field on it, which is defined by a mapping $\widehat{d} : S^1 \rightarrowtail$ $\mathbf{R}P^2$. This gives a homotopy class $[\widehat{d}] \in \pi(\mathbf{R}P^2)$. The structure of the group $\pi_1(\mathbf{R}P^2)$ is known: $\pi_1(\mathbf{R}P^2) = \mathbf{Z}_2$, where the generator $a \in \mathbf{Z}_2$, is the homotopy class containing the loop α given by the equator (of a hemispher) with the diametrically opposite points glued together, $a^2 = 0$ is the class of a constant loop (see, Ch. 3, § 4). Thus, a generalized degree, $\deg_2 \widehat{d}$, is defined; it is equal either to 0 or 1. Therefore, we have only two types of vortices: one type corresponds to the value $\deg_2 \widehat{d} = 0$, and the other to $\deg_2 \widehat{d} = 1$; the first type is topologically unstable, the other one is topologically stable. Fig. 38 represents a topologically stable vortex, a loop of which coincides with the equator–loop α; therefore, $\deg_2 \widehat{d} = 1$. The homotopy class α of this loop contains also other loops in $\mathbf{R}P^2$ which on a hemisphere are curves with the beginning and the end lying on the equator and, thus, are the ends of a diameter (see, Fig. 40).

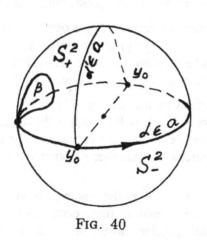

Fig. 40

For these loops, the direction $\widehat{d}(x)$ goes out of the plane $\mathbf{R}^2$, and, nvertheless, thay are not homotopic to a constant loop in $\mathbf{R}P^2$. The product of two such loops is zero, i.e., in the class of a constant loop in $\mathbf{R}P^2$. All the loops with coinciding initial and end points on the hemisphere also belong to this class. For instance, the vortex in Fig. 39 is characterized by a loop from the class 0. All the conclusions noted can be seen geometrically assuming that the distinguished point $y_0 \in Y = \mathbf{R}P^2$ with respect to which the group $\pi_1(Y, y_0)$ is computed, is given by the ends of a diameter on the equator.

Consequently, in view of the topological difference of the domains of degeneracy in the case of an isotropic ferromagnet and a nematic (S^2 and $\mathbf{R}P^2$, respectively), we draw different physical conlusions: non–observability of the vortices for the first case, and observability for the second case (we talk about vortices with index $\deg \widehat{d}_2 = 1$). The experiments support the theory. For that matter, in physics, topologically stable vortices are called "vortices of strength 1/2", stressing by this, that the directions $\widehat{d}(x)$ when traversing the loop, changes by an angle $1/2(2\pi)$, i.e. by an angle π. If the direction $\widehat{d}(x)$ changes by angle $N \cdot (2\pi)$, where N is integer, then the vortex is said to be of "strength N". In our classification a vortex of strength N has a topological index $\deg_2 \widehat{d} = 0$, and it is topologically unstable (an example of such a vortex of strength $N = -1$ is presented in Fig. 39).

Experiments show strong diffusion of the vortex curves of strength $N = \pm 1$ and it is interpreted as the "vorticity flowing out into a third dimension"; the latter means that the director near the disclination curve turns and reorients itself along this curve, and, thus, the singular curve of the vorticity ceases to exist. Topology predicts the non–existence of disclinations of integer strengths N. Indeed, the discovery of the effect of the vorticity flowing out into the third dimension, together with the investigation of defects in matter with more complicated domains of degeneracy (like He^4) led the Russian physicists G. E. Volovik, V. P. Mineev, and French physicists G. Toulouse and M. Clement (in 1976) to the necessity of a topological description of defects by using homotopic topology; it appears that multiplicative properties of the group π_1 determine the ways in which the vorticities can combine fusion of vorticities (with corresponding addition of topological indices), or a decomposition of a vortex into several ones with preservation of the total topological index (charge); these are important laws in the physics of condensed matter.

A nematic, as well as a ferromagnet, can have point defects, i.e. isolated singular points in the direction field of the director $\widehat{d}(x)$. If x^* is such a point, then, in order to define the topological index (charge) of it, it is necessary to surround x^* by a sphere $S^2_\epsilon(x^*)$ (as in the case of ferromagnet) of a sufficiently small radius ϵ and not containing other singular points, and then to consider the mapping $\widetilde{d} : S^2_\epsilon(x^*) \rightarrow \mathbf{R}P^2$ and its homotopy class $[\widetilde{d}] \in \pi_2(\mathbf{R}P^2, y_0)$, where $y_0 = \widetilde{d}(x_0)$, x_0 is the distinguished point in $S^2_\epsilon(x^*)$. Since $\pi_2(\mathbf{R}P^2) = \mathbf{Z}$ is a free abelian group (with the generator γ_2 induced by the mapping of gluing diametrically opposite points together: $S^2 \rightarrow \mathbf{R}P^2$), $[\widetilde{d}] = k\gamma_2$, where $k \in \mathbf{Z}$, and this number is called an integer index, $\deg \widetilde{d}$. This degree does not depend on $\epsilon > 0$, if $\epsilon \rightarrow 0$, and is called the topological index (or a charge) $\kappa(x^*)$ of a singular point x^* (as in the case a ferromagnetic, it can be defined more precisely: $\kappa(x^*) = k$, where $[f] = k\gamma_2$, γ_2 is a generator of a free group $\pi_2(S^2) = \mathbf{Z}$). There is topological stability of singular points with $\kappa(x^*) \neq 0$; however, point defects with $|\kappa(x^*)| > 1$ are not observed experimentally (point defects are observed naturally if a nematic is enclosed into a capillary cylinder and the director on its boundary is orthogonal to the wall of the capillary cylinder). More generally, the restricting surfaces of matter can induce new classes of defects since they can change the topology of a domain of degeneracy just as in the case of other classes of nematics, for instance, the biaxial classes, the defects of which remind us the defects of the superfluid A–phase of He^3. Note that domain of degeneracy of the latter $SO(3)$, is homeomorphic to $\mathbf{R}P^3$; therefore, (see Ch. 3, § 4) $\pi_2(SO(3)) = \pi_2(\mathbf{R}P^3) = 0$, and hence it follows that topologically stable point singularities do not occur in He^3.

REVIEW OF THE RECOMMENDED LITERATURE

The first chapter has touched upon the subjects from many parts of topology. Bibliographical indications will be given accordingly and added to after each corresponding chapter. Here we present sources for an initial acquaintance with the subject, as well as books which systematically present (at one level or another) a course in topology.

A systematic introduction to the basic concepts in topology for beginners – [14, 19, 18].

An elementary approach to various questions is presented in [34, 35, 38, 19, 18].

The proof of the Jordan theorem for a closed simple curve – [1,6].

Initial information about metric spaces and their mappings – [46, 51].

The visual material that illustrates the notion of the topological space (§ 3) can be supplemented by the books [14, 34, 35, 38, 88, 19, 18]; in particular, the classification of two–dimensional surfaces is covered by [14, 47], and the representation of the fiber spaces is presented in [14].

For initial information about the Riemann surfaces, see [77], and for applications to elliptic integrals – [80].

Initial information on knot theory is in [14, 47]; a systematic presentation of knot theory is in [20].

As systematic courses for beginners on topology, we indicate the books [47, 78], for topology and differential geometry [60, 64], and also the series of books [71–73].

An extensive presentation of the process of development of the ideas, methods and results in modern topology is given in the fundamental work [62]; the first two chapters and the beginning of the third can one be used for initial synthetic studies of topology.

For a historical survey of the development of topology in the Soviet Union, see also [41].

For applications of topology in the investigation of critical functions on smooth manifolds (Morse theory, Lyusternik–Shnirel'man theory), see [55, 70, 29, 30].

Applications of topology in the theory of singularities can be found in the survey [10] aimed at a wide circle of readers, and also in the specialized monographs [12, 13]; the role of topology in the problem of minimal surfaces is described in the monographs [21, 30] (see, also [18]).

Applications of topology in physics of condensed states of the matter (some of them are described in § 6) can be found in the surveys [86, 37, 75, 65]. A popular account of the topology of spaces in elementary particle theory in modern physics is given in [44]. And, finally, the inverse influence of ideas of theoretical physics in modern topology of manifolds can be found in the specialized monograph [32]; the introduction and the first chapter of the book deals with the results by S. Donaldson and M. Friedman (1981–1982) concerning the classification of four–dimensional manifolds, which are obtained by investigating the space of solutions of the Yang–Mills equations of theoretical physics. These results lead, in particular, to the fact that the classical space $\mathbf{R}^4$ can have non–standard smooth structures (so–called "exotic" ones), and even an uncountable number of them C. Taubes, 1987). We note also the fundamental monograph [76] dedicated to topological methods in quantum field theory and the theory of condensed matter.

The recent book [89] contains a lot of interesting visual material explaining topological ideas, and it can serve as an excellent introduction to algebraic topology, topology of manifolds, symplectic topology, and mechanics.

As it has been mentioned above, the concept of metric space is insufficient for the development of many mathematical problems. In the XX century, a more general concept of space, the concept of topological space, appeared and developed in mathematics. Nowadays, this concept has become universal in that the "structure" of topological space is quite ubiquitous and substantial and usually precedes the introduction of other structures. The language of topological spaces has become standard in other branches of mathematics which are related to the concept of space. This chapter is devoted to the theory of topological spaces and their continuous mappings

The abstract composition of this drawing rises associations of deformation, interpenetration, and the complicated structure of a topological space. We are surrounded by an intricate, changeable, and unsteady world, which is, at the same time, beautiful in its harmony. What is stable, and what is transient in this world? General topology, from its specific point of view, helps to look for properties of spaces, to see and understand their "individuality" and interaction.

CHAPTER II

GENERAL TOPOLOGY

As was mentioned above, the concept of a metric space is not sufficient for the development of a number of important mathematical problems. In the twentieth century, a more general concept of space arouse and has been developed in mathematics: the concept of a topological space. By now, this concept has become universal since the "structure" of a topological space being basic and profound, usually precedes the introduction of other geometric structures. The terminology of the theory of topological spaces is generally accepted in all fields of mathematics that are related to the concept of space. This chapter is devoted to the the theory of topological spaces and their continuous mappings.

§ 1. Topological spaces and continuous mappings

1. Definition of a topological space. Let X be an arbitray set and $\tau = \{U\}$ a family of its subsets which possesses the following properties:

   (1) $\emptyset, X \in \tau$;
   (2) the union of any family of sets from τ belongs to τ;
   (3) the intersection of any finite number of sets from τ belongs to τ.

Such a family of sets τ is called a *topology* on X. A set X with a topology τ defined on it, is called a *topological space* and denoted by (X, τ); the subsets from the family τ are said to be *open* (in the space (X, τ)).

Instead of (X, τ) we can simply write X, if it does not cause any misunderstandings.

Example 1. X is the number line $\mathbf{R}^1$. A topology on $\mathbf{R}^1$ can be given by the following collection of subsets: the empty set $\emptyset$, all possible intervals and their union $U = \bigcup_\alpha (a_\alpha, b_\alpha)$ (verify!).

Example 2. $X = \mathbf{R}^2$. We call a set open in $\mathbf{R}^2$ if together with each of its points, it contains a sufficiently small open disc with the centre at that point, and also the empty set. It is easy to verify that the system of all open sets in $\mathbf{R}^2$ forms a topology.

Example 3. X is an arbitrary set. The family $\tau_0 = \{\emptyset, X\}$ gives a topology on X (verify!).

Example 4. X is an arbitrary set, $\tau_1 = \{\text{all possible subsets from } X\}$. The family τ is a topology on X (verify!).

The topology τ_1 is said to be *maximal* or *discrete*, and the topology τ_0 *minimal* or *trivial*. Thus, we can introduce different topologies, e.g. the trivial and the discrete one, on the same set.

The concept of open set in the topological space (X, τ) is closely related to the dual notion of *closed set*: a set with an open complement. So, if $U \in \tau$, then $X \backslash U$ is closed, and, conversily: if F is closed, then $X \backslash F$ is open.

Exercise $1°$. Verify that the following sets are closed: an interval $[a, b]$ in $\mathbf{R}^1$ with the topology from example 1, a closed disc in $\mathbf{R}^2$ with the topology from example 2.

In virtue of the duality of the set–theoretic operations, the family $\{F\}$ of all closed sets from a topological space (X, τ) satisfies the following properties:
   (1) $X, \emptyset \in \{F\}$;
   (2) the intersection of any family of sets from $\{F\}$ belongs to $\{F\}$;
   (3) the union of any finite number of sets from $\{F\}$ belongs to $\{F\}$.
These properties completely characterize the closed sets of the topological space (X, τ), and thus the topology τ itself (as the sets from τ are the complements of the closed sets); these properties can also be used as the axioms of a topological space. Thus, the topology on X can be given by indicating the family $\{F\}$ of subsets X which satisfies properties (1)–(3); in this case, the family $\{X \backslash F\}$ will be a topology on X.

The different topologies on the same set form a partially ordered set.

Definition 1. It is said that a topology τ on X is *weaker* (*coarser*) than a topology τ' on X ($\tau \prec \tau'$) if from $U \in \tau$ it follows that $U \subset \tau'$, i.e. $\tau \subset \tau'$. In this case the topology τ' is *stronger* (*finer*) than the topology τ.

Note that for any topology τ we have $\tau_0 \prec \tau \prec \tau_1$. It is clear, that there also exist incomparable topologies. The topologies τ' and τ'' are uncomparable if each of them contains at least some sets which do not belong to the other one.

Consider now how to construct a topology. First we present an important definition.

Definition 2. A family $B = \{V\}$ of open sets from topological space (X, τ) is called a *base of a topology* τ if for any open set $U \in \tau$ and any point $x \in U$ there can be found a set $V \in B$ such that $x \in V$ and $V \subset U$.

Consequently, any non–empty open set of a topological space (X, τ) can be represented as the union of open sets from a base for the topology τ (this property characterizes a base and is often used as the definition of a base). In particular, X is equal to the union of all the sets from the base (a system of subsets from X the union of which is equal to X, is said to be a *covering* of X). If $\{V_\alpha\}$ is some covering of X, then the question arises: under what conditions can a topology on X be constructed so that the family $\{V_\alpha\}$ is a base for this topology?

Theorem 1. (A criterion for a base) *Let* $X = \bigcup_\alpha V_\alpha$. *A covering* $B = \{V_\alpha\}$ *is a base for a certain topology iff for each* V_α, *each* V_β *from B and each* $x \in V_\alpha \cap V_\beta$ *there exists* $V_\gamma \in B$ *such that* $x \in V_\gamma \subset V_\alpha \cap V_\beta$

Proof. If $B = \{V_\alpha\}$ is a base for a topology then $V_\alpha \cap V_\beta$ is an open set, and, by the definition of a base, for every $x \in V_\alpha \cap V_\beta$ there exists $V_\gamma : x \in V_\gamma \in V_\alpha \cap V_\beta$.
Conversely: if $B = \{V_\alpha\}$ satisfies the condition of the theorem, then the sets $U = \cup V_\alpha$ (all possible unions) and the empty set $\emptyset$ form, as it is easy to check, a topology on X for which $B = \{V_\alpha\}$ is a base. $\qquad\square$

Note that in the proof we indicated a method of constructing a topology if the family B satisfying the condition of the theorem, is determined.

Is it possible to construct a topology on the set X by its arbitrary covering $\{S_\alpha\}$? The following theorem gives an answer to this question.

Theorem 2. it A covering $\{S_\alpha\}$ naturally generates a topology on X: more precisely, the collection of sets $\{V = \bigcap_{\alpha \in K} S_\alpha\}$, where K is an arbitrary finite subset from $\{\alpha\}$, is a base of that topology.

Proof. Verify that the collection $\{V\}$ satisfies the criterion of a base. Indeed, we can take $V_\gamma = V_\alpha \cap U_\beta$ for arbitrary V_α, V_β. Evidently, $V_\gamma \in \{V\}$, therefore, the criterion of a base is fulfilled. $\square$

Thus, a covering $\{S_\alpha\}$ of a set X determines topology on X, for which the open sets are all possible unions $\cup(\cap_{\alpha \in K} S_\alpha)$ and the empty set.

Definition 3. A family $\{S_\alpha\}$ is a *subbase* for the topology which it generates.

Example 5. Let $X = \mathbf{R}^1$. Sets of the form $S_\alpha = \{x : x < \alpha\}$, $\alpha \in \mathbf{R}^1$, and $S_\beta = \{x : x > \beta\}$, $\beta \in \mathbf{R}^1$, form a subbase for the topology of the number line $\mathbf{R}^1$, indicated in example 1.

Example 6. Let $X = \mathbf{R}^n$ be an n–dimensional vector space. As a base of the topology on $\mathbf{R}^n$ we can take a system of sets $B = \{V_{\alpha,\beta}\}$, where $V_{\alpha,\beta} = \{x \in \mathbf{R}^n : a_i < \xi_i < b_i, i = 1, \dots, n\}$, ξ_i is the coordinate of the vector $x = (\xi_1, \xi_2, \dots, \xi_n)$; $\alpha = (a_1, \dots, a_n)$, $\beta = (b_1, \dots, b_n)$ are arbitrary vectors in $\mathbf{R}^n$, and $a_i < b_i$.
These sets $V_{\alpha,\beta}$ are called *open parallelepipeda* in $\mathbf{R}^n$.

Exercise 2°. Prove that the system of parallelepipeda described in example 6 forms a base for the topology $\mathbf{R}^n$. Make sure that the topology defined by this base, for $n = 1, 2$, coincides with the topologies on $\mathbf{R}^1$, $\mathbf{R}^2$ given in examples 1, 2.

In future, if the topology on $\mathbf{R}^n$ is not indicated, then we will assume that $\mathbf{R}^n$ has the topology with the base given in example 6.
It is natural for a topological space to select a base with the least possible number of elements. For example, in $\mathbf{R}^1$, the sets $V = (t_1, t_2)$, where t_1, t_2 are rational numbers, form a base consisting of a countable number of elements.
Similarly, we can select a base for the topology on $\mathbf{R}^n$ consisting of the countable set of parallelepipeda with rational vertices, i.e. of the form

$$V_{r_1, r_2} = \{x : r_1^i < \xi_i < r_2^i, \ i = 1, \dots, n\},$$

where r_1, r_2 are rational vectors in $\mathbf{R}^n$.

2. Neighbourhoods. Let (X, τ) be a topological space, and $x \in X$ an arbitrary point.

Definition 4. By a *neighbourhood of a point* $x \in X$ we understand a subset $\Omega(x) \subset X$ which satisfies the conditions: (1) $x \in \Omega(x)$, (2) there exists a $U \in \tau$ such that $x \in U \subset \Omega(x)$.

Consider the collection of all neighbourhoods of a given point x. It possesses the following properties:

(1) the union of any collection of neighbourhoods of a point x is a neighbourhood of x;

(2) the intersection of a finite number of neighbourhoods of a point x is a neighbourhood of x;

(3) any set containing some neighbourhood of a point x is a neighbourhood of x.

Theorem 3. *A subset A ($A \neq \emptyset$) of a topological space (X, τ) is open iff it contains some neighbourhood of each of its points.*

Proof. Let A be open, $x \in A$. It is clear that A is a neighbourhood of x; so A contains a neighbourhood of any of its points. Suppose that for each $x \in A$ there exists a neighbourhood of the point x, lying wholly in A. By the definition of a neighbourhood, it contains some open set U_x, $x \in U_x$. Consider the union $\bigcup_{x \in A} U_x$ of such sets. It is open and coincides with A. Indeed, $A \subset \bigcup_{x \in A} U_x$, since any point of the set A belongs to $\bigcup_{x \in A} U_x$. On the other hand, for every x we have $U_x \subset A$, i.e. $\bigcup_{x \in A} U_x \subset A$. Therefore, $A = \bigcup_{x \in A} U_x$, and thus, A is open. $\qquad\square$

Neighbourhoods are used for separating points from each other.

Definition 5. A topological space (X, τ) is called *Hausdorff* if for any two different points x, y in it, there can be found neighbourhoods $U(x)$, $U(y)$ of these points such that $U(x) \cap U(y) = \emptyset$.

A topological space (X, τ) with the trivial topology is not Hausdorff if it contains more than one point (verify!).

The properties of the neighbourhood of a point considered above (now declared to be axioms), is a basis for the following definition of a topological space.

Definition 6. A *topological space* is a set X with a given non–empty system of subsets $\{\Omega_\alpha(x)\}$, for every x of it, which are called the *neighbourhoods of a point* x, satisfying the following properties:

(1) x belongs to each of its neighbourhoods $\Omega_\alpha(x)$;

(2) if a set $U \subset X$ contains some $\Omega_\alpha(x)$, then U is also a neighbourhood of x;

(3) for any neighbourhoods $\Omega_{\alpha_1}(x)$, $\Omega_{\alpha_2}(x)$ of the point x, their intersection $\Omega_{\alpha_1}(x) \cap \Omega_{\alpha_2}(x)$ is also a neighbourhood of the point x;

(4) for any neighbourhood $\Omega_\alpha(x)$ of the point x there exists a neighbourhood $\Omega_{\alpha_1}(x) \subset \Omega_\alpha(x)$ such that it is a neighbourhood of each of its points.

Exercise $3°$. Show that the sets that are neighbourhoods of each of their points and the empty set $\emptyset$ form a topology on X.

3. Continuous mapping. Homeomorphism. Now we discuss the definition of a continuous mapping of topological spaces.

Let (X, τ), (Y, σ) be two topological spaces with topologies τ, σ, respectively. Let $f : X \to Y$ be a mapping of sets.

Definition 7. The mapping f is said to be a *continuous mapping* of topological spaces if the full preimage $f^{-1}(V)$ of any open set V of the space (Y, σ) is an open set of the space (X, τ).

Exercise $4°$. Formulate the definition of a continuous mapping in terms of bases of the two topologies.

Exercise $5°$. Show that a continuous numerical function $y = f(x)$ $(-\infty < x < +\infty)$ determines a continuous mapping $f : \mathbf{R}^1 \to \mathbf{R}^1$.

Exercise $6°$. Prove, that $f : X \to Y$ is continuous iff the preimage $f^{-1}(F)$ is closed in X for any closed set F in Y.

If $f : X \to Y$, $g : Y \to Z$ are mappings of topological spaces then the superposition $gf : X \to Z$ is naturally defined by the rule $(gf) : x \mapsto g(f(x))$.

Theorem 4. *If f, g are continuous then gf is also continuous.*

The proof follows easily from the remark:

$$(gf)^{-1}(W) = f^{-1}(g^{-1}(W)),$$

where $W \subset Z$ is an arbitrary set.

Definition 8. Two topological spaces (X, τ), (Y, τ) are said to be *homeomorphic* if there exists a mapping $f : X \to Y$ satisfying the conditions: 1) $f : X \to Y$ is a bijective mapping; 2) f is continuous; 3) f^{-1} is continuous.

Note that in the case of metric spaces this definition precisely reproduces the definition of a homeomorphism of metric spaces, introduced in Ch. 1, § 2; cf also § 2 below.

Definition 9. A mapping $f : X \to Y$ is called *open* (*closed*) if the image of each open (closed) set in X is open (closed) in Y.

Exercise 7°. Prove that a mapping $f : X \to Y$ is a homeomorphism iff the mapping $f^{-1} : Y \to X$ is defined, and the mappings f and f^{-1} are both open and closed simultaneously.

Thus, a homeomorphism transforms open sets into open sets, and closed sets into closed sets.

Associating each open set U of the space X with its image $f(U)$ under a homeomorphism $f : X \to Y$ establishes a bijective correspondence between the topologies of the spaces X and Y. Therefore, any property of the space X formulated in terms of the topology on this space, is also valid for the space Y which is homeomorphic to X, and it is similarly formulated in terms of the topology on Y. Thus, homeomorphic spaces X and Y possess identical properties and, from this point of view, are undistinguishable.

The properties of topological spaces that are preserved under homeomorphisms are called *topological properties*[*]. In this connection, note that the

---

[*] While investigating topological properties, homeomorpic spaces X and Y often are not distinguished.

main task of topology (and still partially unsolved) is to work out an effective method of distinguishing between non–homeomorphic spaces.

Exercise 8°. Show that a homeomorphism defines a correspondence between bases and subbases of homeomorphic spaces.

Exercise 9°. Show that the relation of a homeomorphism is an equivalence relation.

Exercise 10°. Show that the interval $(-1, +1)$ of the number axis is homeomorphic to the whole number axis; give a formula for this homeomoprhism.

Exercise 11°. Show that a closed interval and an open interval on the number axis are not homeomorphic.

There exists a quite useful extension of the notion of homeomorphism, viz., a *local homeomorphism*. This is a continuous mapping $f : X \rightarrow Y$ such that for any pair of points x, y, $y = f(x)$, there can be found neighbourhoods $U(x)$, $V(y)$ for which $f : U(x) \rightarrow V(x)$ is a homeomorphism.

Exercise 12°. Verify that the mapping $\mathbf{R}^1 \backslash \{0\} \rightarrow \mathbf{R}^1 \backslash \{0\}$ given by the formula $y = x^2$, is a local homeomorphism.

4. A subspace of a topological space. As we can see from the above, the subsets of metric and topological spaces are often considered as original objects. In addition, a subset Y of a metric space X naturally inherits a metric from X. We now define the concept of the inherited topology on a subset Y, when X is a topological space.

Let (X, τ) be a topological space, $Y \subset X$ a subset in X. Consider the following system of subsets of the set Y:

$$\tau_Y = \{V : V = U \cap Y, \quad U \in \tau\}.$$

Theorem 5. *The system τ_Y is a topology on Y.*

The proof is left to the reader (it is obvious).

The topology τ_Y is called the *induced* or *inherited topology* from X. The space (Y, τ_Y) is called a *subspace* of the space (X, τ).

Subsets of topological spaces are considered, as a rule, with the induced topology.

If $f : X \to Z$ is a continuous mapping of topological spaces (X, τ), (Z, σ), and Y is a subspace of X, then one can consider the mapping $f : Y \to Z$ called the *restriction* of f to Y and denoted by $f|_Y$.

Theorem 6. *The mapping* $f_Y : Y \to Z$ *is continuous.*

Proof. Let $W \in \sigma$. Then $(f|_Y)^{-1}(W) = f^{-1}(W) \cap Y$. Since $f^{-1}(W) \in \tau$, we have $(f|_Y)^{-1}(W) \in \tau_Y$. $\qquad\square$

Exercise 13°. Show that an open set in a subspace Y of a topological space X is not necessarily open in X. Analyze the examples for $X = \mathbf{R}^1$, $\mathbf{R}^2$, $\mathbf{R}^3$. There is an analogous question for closed sets in Y. Prove first that any closed set F_Y in Y is of the form $F_Y = F \cap Y$, where F is a closed set in Y.

Exercise 14°. Let $A, B \subset X$ be closed sets of a topological space X; and let $X = A \cup B$. Then the mapping $f : X \to Y$ is continuous iff $f|_A : A \to Y$, $f|_B : B \to Y$ are continuous.

We now introduce another useful concept. A mapping $i : Y \to X$ of the topological spaces Y, X is called an *imbedding* of Y into X if 1) i is continuous, 2) $i : Y \to i(Y)$ is a homeomorphism, where $i(Y) \subset X$ is considered as a subspace of X.

Imbeddings are useful when we need to "distinguish" a subspace $Y \subset X$ from its surrounding space X and consider it separately. The connection with X is preserved by the natural mapping $Y \to X$ which associates an element of Y with the same element from X and is an imbedding.

§ 2. Topology and continuous mappings of metric spaces. The spaces $\mathbf{R}^n$, S^{n-1} and D^n

1. Topology in a metric space. Let (X, ρ) be some metric space with a metric ρ. We can construct a topology on X in a natural way. Consider all possible

possible sets $D_\epsilon(x) = \{y : \rho(y, x) < \epsilon\}$, where $x \in X$, $\epsilon > 0$. A set $D_\epsilon(x)$ is called an *open ball* of radius ϵ with centre at the point x.

The family $\{D_\epsilon(x)\}$ of all open balls forms a covering of the metric space for which the criterion of a base holds (theorem 1, § 1). Indeed, let $D_{\epsilon_1}(x_1)$ and $D_{\epsilon_2}(x_2)$ be two open balls with a non–empty intersection. Let $y \in D_{\epsilon_1}(x_1) \cap D_{\epsilon_2}(x_2)$ and $\delta = \min\{\epsilon_1 - \rho(y, x_1),\ \epsilon_2 - \rho(y, x_2)\}$, and let $z \in D_\delta(y)$. Then

$$\rho(z, x_1) \leq \rho(z, y) + \rho(y, x_1) < \delta + \rho(y, x_1) \leq \epsilon_1,$$
$$\rho(z, x_2) \leq \rho(z, y) + \rho(y, x_2) < \delta + \rho(y, x_2) \leq \epsilon_2.$$

Therefore, $z \in D_{\epsilon_1}(x_1) \cap D_{\epsilon_2}(x_2)$, from which $D_\delta(y) \subset D_{\epsilon_1}(x_1) \cap D_{\epsilon_2}(x_2)$. Thus, the conditions of theorem 1 are fulfilled.

Definition 1. The topology τ_ρ determined by the base consisting of all open balls in the metric space (X, ρ) is called the topology *induced by the metric* ρ, or the *metric topology*.

Thus, the open sets of the topology τ_ρ are all the possible unions of open balls of the metric space (X, ρ) and the empty set $\emptyset$.

Theorem 1. *The constructed topology τ_ρ is Hausdorff.*

Proof. Let $x \neq y$. Then $\rho(x, y) = \alpha > 0$ (according to the properties of a metric). Take $\epsilon = \alpha/3$, and consider $D_\epsilon(x)$, $D_\epsilon(y)$. It can be easily seen that $D_\epsilon(x) \cap D_\epsilon(y) = \emptyset$. In fact, by assuming the opposite, for a point $z \in D_\epsilon(x) \cap D_\epsilon(y)$ we would have

$$\alpha = \rho(x, y) \leq \rho(x, z) + \rho(z, y) < \frac{\alpha}{3} + \frac{\alpha}{3} = \frac{2\alpha}{3},$$

which is impossible.   □

Another, equivalent definition of open sets in a metric space can be given.

Definition 2. A set $U \neq \emptyset$ is *open* if for any point $x \in U$ there can be found an open ball $D_\delta(x)$ with centre at x lying wholly in U.

Note that precisely in this way, we defined the topology on $\mathbf{R}^2$ (§ 1), and, consequently, the latter coincides with the topology τ_ρ which is generated by the Euclidean metric ρ on the plane $\mathbf{R}^2$. Checking the equivalence of the two definitions of open sets is left to the reader.

Consider a mapping $f : X \to Y$ of a metric space (X, ρ_1) into a metric space (Y, ρ_2). We can present two definitions of the continuity of the mapping f, viz., as a mapping of metric spaces and as a mapping topological spaces as well. These two definitions are equivalent, and so the following theorem holds.

Theorem 2. *A mapping $f : X \to Y$ of a metric space (X, ρ_1) to a metric space (Y, ρ_2) is continuous (for the topologies induced by the metrics) iff for every $x_0 \in X$ and every sequence $\{x_n\}$ in X which converges to x_0, the sequence $\{f(x_n)\}$ converges to $f(x_0)$ in Y.*

Proof. Let $f : X \to Y$ be a continuous mapping for the topologies on X, Y which are induced by the metrics; and let $x_n \xrightarrow{\rho_1} x_0$. We then show that $f(x_n) \xrightarrow{\rho_2} f(x_0)$. The latter means that for any $\epsilon > 0$ there can be found a natural number $N = N(\epsilon, x_0)$ such that $\rho_2(f(x_n), f(x_0)) < \epsilon$ when $n > N$.

Consider an open ball $D_\epsilon(f(x_0))$ in Y; denote it by V_ϵ. Its preimage $f^{-1}(V_\epsilon)$ is an open set in X by the continuity of f; in addition, $x_0 \in f^{-1}(V_\epsilon)$. The point x_0 belongs to $f^{-1}(V_\epsilon)$ together with some ball $D_\delta(x_0)$ of radius δ. There exists a number $N(N = N(\epsilon, x_0))$ such that x_n belongs to $D_\delta(x_0)$ (and to $f^{-1}(V_\epsilon)$) for $n > N$. But then $f(x_n) \in V_\epsilon$ (i.e. $\rho_2(f(x_n), f(x_0) < \epsilon)$ for $n > N$. Therefore, the mapping f is continuous as a mapping of metric spaces.

Let the condition $f(x_n) \xrightarrow{\rho_2} f(x_0)$ be satisfied for any sequence $\{x_n\}$ which is convergent to some point x_0 in the space X. We show that in this case, the preimage of any open set is open. Let V be an open set in Y, $U = f^{-1}(V)$. We can show that U is open in the space X by using definition 2 of an open set. Let $x \in f^{-1}(V)$. Then it is sufficient to find $\epsilon > 0$ such that $D_\epsilon(x) \subset f^{-1}(V)$. Let us assume that such ϵ does not exist. Then there exist sequences $\{\epsilon_n\}$, $\{x_n\}$ such that $\epsilon_n \to 0$, $x_n \in D_{\epsilon_n}(x)$, but $x_n \notin f^{-1}(V)$. Therefore, $x_n \xrightarrow{\rho_1} x$, which implies $f(x_n) \xrightarrow{\rho_2} f(x)$. By noticing that $f(x)$ belongs to V together with a certain open ball, we can conclude that $f(x_n) \in V$ and $x_n \in f^{-1}(V)$ starting at some number, and this contradicts the assumption. Thus, the mapping f is continuous in topologies on the spaces X, Y which are induced by the metrics. $\square$

2. The space $\mathbf{R}^n$. We shall consider an important example of a metric space, viz. the *Euclidean space*

$$\mathbf{R}^n = \{(\xi_1, \dots, \xi_n), \quad -\infty < \xi_i < +\infty, \quad i = 1, \dots, n\},$$

consisting of all ordered sets (called *points* or *vectors*) of n real numbers; the numbers ξ_i are called the *coordinates* of a point, or vector.

A metric (the Euclidean metric) on $\mathbf{R}^n$ ($n \geq 1$) is introduced analogously to the metric on $\mathbf{R}^3$:

$$(1) \qquad \rho(x, y) = \left(\sum_{i=1}^{n} (\xi_i - \eta_i)^2 \right)^{1/2},$$

where $x = (\xi_1, \dots, \xi_n)$, $y = (\eta_1, \dots, \eta_n)$ are two arbitrary vectors from $\mathbf{R}^n$.

Let us verify that this is a metric. Obviously, properties (1), (2), (3) of a metric (see, Ch. 1, § 2) are fulfilled. To check property (4), it is required to prove the inequality

$$\left(\sum_{i=1}^{n} (\xi_i - \eta_i)^2 \right)^{1/2} \leq \left(\sum_{i=1}^{n} (\xi_i - \zeta_i)^2 \right)^{1/2} + \left(\sum_{i=1}^{n} (\zeta_i - \eta_i)^2 \right)^{1/2}$$

for arbitrary real numbers ξ_i, η_i, ζ_i, $i = 1, \dots, n$. The proof is divided into two lemmas.

Lemma 1. (The Cauchy–Bunyakovski inequality) *For any real numbers ξ_i, η_i, $i = 1, \dots, n$, the following inequality holds*

$$\sum_{i=1}^{n} (\xi_i \eta_i) \leq \left(\sum_{i=1}^{n} \xi_i^2 \right)^{1/2} \left(\sum_{i=1}^{n} \eta_i^2 \right)^{1/2}.$$

Proof. For an arbitrary real number λ we have $\sum_{i=1}^{n} (\xi_i + \lambda \eta_i)^2 \geq 0$, from which $\sum_{i=1}^{n} \xi_i^2 + 2\lambda \sum_{i=1}^{n} \xi_i \eta_i + \lambda^2 \sum_{i=1}^{n} \eta_i^2 \geq 0$. Consider the left–hand side of the inequality as a polynomial in λ. It cannot have two different real roots, therefore, the discriminant is non–positive and we get the inequality

$$\left(\sum_{i=1}^{n} \xi_i \eta_i \right)^2 \leq \sum_{i=1}^{n} \xi_i^2 \sum_{i=1}^{n} \eta_i^2,$$

which implies the required one. □

Lemma 2. (**The Minkowski inequality**) *For arbitrary real numbers* ξ_i, η_i, $i = 1, \ldots, n$ *the following inequality holds*

$$\left(\sum_{i=1}^{n} (\xi_i + \eta_i)^2 \right)^{1/2} \leq \left(\sum_{i=1}^{n} \xi_i^2 \right)^{1/2} + \left(\sum_{i=1}^{n} \eta_i^2 \right)^{1/2}.$$

Proof. We shall use the Cauchy–Bunyakovski inequality:

$$\sum_{i=1}^{n} (\xi_i + \eta_i)^2 = \sum_{i=1}^{n} (\xi_i^2 + 2\xi_i \eta_i + \eta_i^2) \leq$$

$$\leq \sum_{i=1}^{n} \xi_i^2 + 2 \left(\sum_{i=1}^{n} \xi_i^2 \right)^{1/2} \left(\sum_{i=1}^{n} \eta_i^n \right)^{1/2} + \sum_{i=1}^{n} \eta_i^2 =$$

$$= \left[\left(\sum_{i=1}^{n} \xi_i^2 \right)^{1/2} + \left(\sum_{i=1}^{n} \eta_i^2 \right)^{1/2} \right]^2.$$

By taking the square root of both sides of the inequality, we obtain the required inequality. □

We can now complete the verification of property (4) of a metric. By using the Minkowski inequality, we obtain

$$\left(\sum_{i=1}^{n} (\xi_i - \eta_i)^2 \right)^{1/2} = \left(\sum_{i=1}^{n} [(\xi_i - \zeta_i) + (\zeta_i - \eta_i)]^2 \right)^{1/2} \leq$$

$$\leq \left(\sum_{i=1}^{n} (\xi_i - \zeta_i)^2 \right)^{1/2} + \left(\sum_{i=1}^{n} (\zeta_i - \eta_i)^2 \right)^{1/2}.$$

Thus, ρ is a metric on $\mathbf{R}^n$. □

It is not difficult to see that the metric topology τ_ρ on $\mathbf{R}^n$ induced by the Euclidean metric ρ, coincides with the topology on $\mathbf{R}^n$, the base of which was given in § 1, example 6.

Let $x_0 = (\xi_1^0, \ldots, \xi_n^0)$ be the centre of the ball $D_r(x_0)$, and $x = (\xi_1, \ldots \ldots, \xi_n)$ an arbitrary point in it. Then the coordinates of the point x satisfy the inequality

$$(2) \qquad |\xi_1 - \xi_1^0|^2 + \ldots + |\xi_n - \xi_n^0|^2 < r^2.$$

A ball $D_r(x_0)$ in $\mathbf{R}^n$ is often denoted by $D_r^n(x_0)$ and called an *open n–disc*. By the *closed ball* (*closed n–disc*) $\bar{D}_r^n(x_0)$ we understand the set of points x with its coordinates satisfying the inequality

$$(3) \qquad |\xi_1 - \xi_1^0|^2 + \ldots + |\xi_n - \xi_n^0|^2 \leq r^2;$$

The $(n-1)$–dimensional sphere $S_r^{n-1}(x_0)$ of radius r with centre at the point x_0 is defined by the equality

$$(4) \qquad |\xi_1 - \xi_1^0|^2 + \ldots + |\xi_n - \xi_n^0|^2 = r^2.$$

The sphere $S_r^{n-1}(x_0)$ is the *boundary of the disc* $\bar{D}_r^n(x_0)$ or $D_r^n(x_0)$.
There are other metrics on $\mathbf{R}^n$. For instance,

$$(5) \qquad \rho(x, y) = \max_{i=1,\ldots,n} \{|\xi_i - \eta_i|\}.$$

Exercise 1°. Describe a ball in $\mathbf{R}^n$ under the metric (5). Show that the Euclidean metric and the metric (5) induce identical topologies.

Consider the complex n–dimensional space $\mathbf{C}^n$:

$$\mathbf{C}^n = \{z : z = (z_1, \ldots, z_n), \ z_k = x_k + iy_k, \ y_k \in \mathbf{R}^1, \ k = 1, \ldots, n\}.$$

A metric on it is introduced similarly to that in the real case:

$$\rho(z', z'') = (|z_1' - z_1''|^2 + \ldots + |z_n' - z_n''|^2)^{1/2},$$

where $z' = (z_1', \dots, z_n')$, $z'' = (z_1'', \dots, z_n'')$ are elements from $\mathbf{C}^n$. The topology on $\mathbf{C}^n$ induced by this metric, can also be given by the metric

$$\rho'(z', z'') = \max_{k=1,\dots,n} |z_k' - z_k''|.$$

We now formulate a condition on continuity of mappings of Euclidean spaces. A mapping $f : \mathbf{R}^n \to \mathbf{R}^m$ associates to every point $(\xi_1, \dots, \xi_n)$ a certain point $(\eta_1, \dots, \eta_m)$, so that we can write

(6)
$$\eta_1 = f_1(\xi_1, \dots, \xi_n),$$
$$\dots\dots\dots\dots\dots$$
$$\eta_m = f_m(\xi_1, \dots, \xi_n),$$

where f_i, $i = 1, \dots, m$ is a numerical function of n variables. This function determines the mapping $f_i : \mathbf{R}^n \to \mathbf{R}^1$ by the rule

(7)
$$\eta_i = f_i(\xi_1, \dots, \xi_n).$$

It is obvious, that the continuity of the mapping f_i is equivalent to the continuity of the numerical function $f_i(\xi_1, \dots, \xi_n)$, as it is defined in analysis.

We call mapping (7) the i-th component of the mapping f. The mapping f is defined by specifying all its components f_i, $i = 1, \dots, m$.

Theorem 3. *A mapping $f : \mathbf{R}^n \to \mathbf{R}^m$ is continuous iff each of its components $f_i : \mathbf{R}^n \to \mathbf{R}^1$, $i = 2, \dots, m$, is continuous.*

Proof. This follows from the remark that $f(x^k) \to f(x^0)$, $k \to \infty$, is equivalent to $f_i(x^k) \to f_i(x^0)$, $k \to \infty$, for $i = 1, \dots, m$.   □

3. The disc D^m is homeomorphic to $\mathbf{R}^m$. Consider some subsets of $\mathbf{R}^n$, $n \geq 2$. Let S^{n-1}, D^n be the sphere and the open n–disc of unit radius and centre at the point $(0, \dots, 0)$, respectively. Denote by S_+^{n-1} the part of the sphere, where $\xi_n > 0$ (the north hemisphere). We prove that the disc D^{n-1} is homeomorphic to the hemisphere S_+^{n-1}.

It can be assumed that the space $\mathbf{R}^{n-1}$ coincides with the subspace of the points $(\xi_1, \dots, \xi_{n-1}, 0)$ of the space $\mathbf{R}^n$, if the points $(\xi_1, \dots, \xi_{n-1})$ and

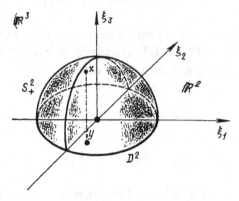

FIG. 41

$(\xi_1, \dots, \xi_{n-1}, 0)$ are identified. Then D^{n-1} and S_+^{n-1} lie in $\mathbf{R}^n$, and are given in the following form:

$$S_+^{n-1} = \{(\xi_1, \dots, \xi_n) : \xi_1^2 + \dots + \xi_n^2 = 1, \ \xi_n > 0\},$$
$$D^{n-1} = \{(\xi_1, \dots, \xi_n) : \xi_1^2 + \dots + \xi_{n-1}^2 < 1, \ \xi_n = 0\}.$$

In the case of $\mathbf{R}^3$, we have the following situation: S_+^2 is the upper half of the sphere without the equator, D^2 is the interior of the unit circle in $\mathbf{R}^2$ (Fig. 41),

$$x = (\xi_1, \xi_2, \xi_3) \in S_+^2, \quad y = (\xi_1, \xi_2, 0) \in D^2.$$

The projection $(\xi_1, \xi_2, \xi_3) \mapsto (\xi_1, \xi_2, 0)$ is a homeomorphism of the hemisphere S_+^2 and the disc D^2.

In the case of $\mathbf{R}^n$, we proceed similarly: the projection

$$f : (\xi_1, \dots, \xi_{n-1}, \xi_n) \mapsto (\xi_1, \dots, \xi_{n-1}, 0)$$

determines a continuous bijective mapping of S_+^{n-1} onto D^{n-1} (verify!). Consider the inverse mapping. It is easy to verify that it has the form

(8)
$$f^{-1} : (\xi_1, \ldots, \xi_{n-1}, 0) \mapsto$$
$$\mapsto (\xi_1, \ldots, \xi_{n-1}, (1 - \xi_1^2 - \ldots - \xi_{n-1}^2)^{1/2}$$

and is continuous. Thus, we have constructed a homeomorphism of the disc D^{n-1} and hemisphere S_+^{n-1}. Call the set of points of the sphere S^{n-1} which satisfy the inequality $\xi_n \geq 0$, the *closed hemisphere* $\bar{S}_+^{n-1}$. It is clear that $\bar{S}_+^{n-1} = S_+^{n-1} \cup S^{n-2}$. It is natural to call the sphere S^{n-2} the *boundary of the hemisphere* $\bar{S}_+^{n-1}$ (or S_+^{n-1}). Note that S^{n-2} is at the same time the boundary of the disc $\bar{D}^{n-1}$ (or D^{n-1}). It is easy to notice, that homeomorphism (8) is also well defined on S^{n-2} and that $f^{-1}|_{s^{n-2}} = 1_{s^{n-2}}$. Thus, $\bar{D}^{n-1}$ is homeomorphic to $\bar{S}_+^{n-1}$.

Now we establish another important homeomorphism.

Theorem 4. *The disc D^m is homeomorphic to the space $\mathbf{R}^m$, $m \geq 1$.*

Proof. Set $m = n - 1$ and use the previous construction. Transfer the space $\mathbf{R}^{n-1} \subset \mathbf{R}^n$, $n \geq 2$, so that its origin of the coordinates concides with the point $(0, \ldots, 0, 1) \in \mathbf{R}^n$, the north pole of the sphere S^{n-1}. The points in the new plane have the form $(\xi_1, \xi_2, \ldots, \xi_{n-1}, 1)$. Through every point $x = (\xi_1, \ldots, \xi_n) \in S_+^{n-1}$ we draw $\eta_1 = t\xi_1$, $\eta_2 = t\xi_2, \ldots$, the half line $\eta_n = t\xi_n$, $t \geq 0$. It intersects the constructed plane at a unique point corresponding to the value $t(x) = 1/\xi_n$. By associating this intersection point to the point x, we obtain a mapping $\Phi : S_+^{n-1} \to \mathbf{R}^{n-1}$ which is given by the rule

$$(\xi_1, \ldots, \xi_n) \mapsto (\xi_1/\xi_n, \ldots, \xi_{n-1}/\xi_n, 1).$$

This mapping, as it is easy to check, is a homeomorphism. The superposition of homeomorphisms

$$\Phi f^{-1} : D^{n-1} \to \mathbf{R}^{n-1}, \quad n \geq 2,$$

is the required homeomorphism. $\square$

Exercise 2°. Formulate a continuity criterion for a mapping $f : \mathbf{C}^n \to \mathbf{C}^m$ of complex spaces.

Exercise 3°. Prove that $\mathbf{C}^n$ is homeomorphic to $\mathbf{R}^{2n}$.

Exercise 4°. Prove that the balls in the space $\mathbf{R}^n$ which are defined by using metrics (1), (5), are homeomorphic.

Exercise 5°. Prove the continuity of the functions

$$f(\xi_1, \xi_2) = (\xi_1^2 + \xi_2^2),$$
$$f(\xi_1, \dots, \xi_n) = (\xi_1^2 + \dots + \xi_n^2)^{1/2}.$$

Exercise 6°. Determine discs and a sphere in the space $\mathbf{C}^n$ by conditions (2)–(4) and denote them by $D_{\mathbf{C},r}^n$, $\bar{D}_{\mathbf{C},r}^n$, $S_{\mathbf{C},r}^{n-1}$, respectively. Prove that they are homeomorphic to D_r^{2n}, $\bar{D}_r^{2n}$, S_r^{2n-1}, respectively.

Exercise 7°. Prove that discs of any radius are homeomorphic in $\mathbf{R}^n$; prove the same for spheres.

§ 3. Quotient space and quotient topology

1. The definition of a quotient topology. We will give a precise definition of a topology in a quotient space, i.e. a quotient topology, and from the new point of view, we analyze the examples from Ch. 1, § 3.

Let a relation $x \sim y$ between some elements $x, y \in X$ be defined on an abstract set X. This relation is called an *equivalence* if the following properties are fulfilled: (1) $x \sim x$ for any $x \in X$ (reflexivity); (2) if $x \sim y$, then $y \sim x$ (symmetry); (3) if $x \sim y$ and $y \sim z$, then $x \sim z$ (transitivity).

The set X can be decomposed into non–intersecting classes of mutually equivalent elements, called *equivalence classes*.

The set $\{D_\alpha\}$ of all equivalence classes we denote by X/R, where R denotes the equivalence in X.

Definition. The set X/R is called the *quotient set* of the set X with respect to the equivalence relation R.

Let (X, τ) be a topological space, and let an equivalence relation R be defined on the set X. Then we may introduce a natural topology on the quotient set X/R in the following way: we say that a subset $V \subset \{D_\alpha\}$ consisting of elements D_α is open iff the union $\cup D_\alpha$ of the sets D_α as the subsets of X is open in the space (X, τ); the empty set is also an open set. This family of open subsets in X/R is a topology and denoted by τ_R.

Exercise 1°. Verify that τ_R is a topology on X/R.

The topology τ_R is called the *quotient topology*; and it is usually taken into account when speaking of a quotient space.

The reasons for defining the topology τ_R in this way become clearer if the mapping $\pi : X \to X/R$ associating an equivalence class D_x to any of its elements $x \in X$ is considered. This mapping is called the *projection* of the space X on the quotient space X/R. It is easy to see that the set $V \subset X/R$ is open iff the set $\pi^{-1}(V)$ is open in X. Thus, the projection π is continuous as a mapping from (X, τ) into $(X/R, \tau_R)$. (Note that from this the principle of the continuity of "gluing" mentioned in Ch. 1, § 3, follows.)

Of course, there may also exist other topologies on the set X/R, for which the projection π is continuous. The following theorem characterizes the topology τ_R.

Theorem 1. *The topology τ_R is the strongest among all topologies on X/R under which the mapping π is continuous.*

Proof. If $\{W\}$ is a topology on X/R for which the mapping π is continuous, then for any $W \in \{W\}$ the set $\pi^{-1}(W)$ is open in X. Therefore, W is open in the quotient space X/R, i.e. $W \in \tau_R$. This means that the topology $\{W\}$ is weaker than the topology τ_R. $\qquad\square$

Exercise 2°. Let $X = [0, 1] \subset \mathbf{R}^1$. We define the equivalence: $x \overset{R}{\sim} y \Leftrightarrow x - y$ is rational. Show that the quotient space X/R is not Hausdorff.

2. Examples of quotient spaces. Consider the examples of Ch. 1, § 3. If X is a rectangle $abcd$ and the equivalence relation R is defined so that $x \sim x$ for every $x \in X$, and $x \sim y$ iff $x \in ab$, $y \in cd$ and x, y lie on the same horizontal in X, then X/R is a topological space homeomorphic to the cylinder (see, Fig. 1, 2).

Indeed, a base for the topology of the cylinder is formed by two-dimensional "discs", i.e. the intersections of the balls in $\mathbf{R}^3$ with the cylinder (Fig. 42). If

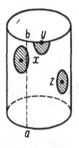

FIG. 42

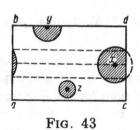

FIG. 43

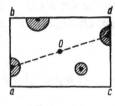

FIG. 44

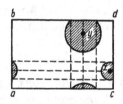

FIG. 45

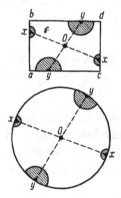

FIG. 46

we cut the cylinder along the line ab and unroll it into a rectangle, then the "discs" turn into a base for the topology of the rectangle, and, in addition, the "discs" intersecting the line ab will be cut into "segments" which complement each other to discs and lie on the opposite sides of the rectangle. It is evident from this that it is necessary to glue complementary "segments" together along the line of the cut in order to obtain a base for the topology on X/R (Fig. 43). Now we can easily verify that by associating equivalent points of the rectangle to the point into which they have been "glued" together, we obtain a homeomorphism of the considered quotient space X/R with the cylinder.

The topology of a Möbius strip can be investigated precisely in the same way (see the following example of the "gluing" in Ch. 1,§ 3). Fig. 44 presents some open sets of a Möbius strip. Here the segments are "glued" together by the points which are symmetric relatively to the centre and lie on the boundaries ab, cd.

In the third example of "gluing", the corresponding quotient space is homeomorphic to the torus, and the elements of a base for its topology are represented in Fig. 45. Here, the corresponding segments are not only glued together by identifying the vertical boundaries ab and cd, but also the horizontal boundaries ac and bd.

Finally, the last example gives a projective plane; the elements of a base for its topology are represented in Fig. 46. Here, the segments are glued together by identifying the diametrically opposite points of their boundaries both on the vertical and horizontal boundaries of the rectangle.

We present another useful example of how to form a quotient space. Let $Y \subset X$ be a subspace of a topological space X. We say that all the points in Y are mutually equivalent, and all the points $x \in X \backslash Y$ are equivalent only to themselves. The quotient space, with respect to this equivalence, is denoted by X/Y, and the projection $\pi : X \to X/Y$ is called the *contraction of the set* Y to a point. For example, $S^1 = I/\{0, 1\}$ is the quotient space of the interval $I = [0, 1]$ with respect to the set of end points.

3. Mappings of quotient spaces. Let X, X' be the topological spaces, and R, R' be the equivalences on them. Consider a mapping $f : X \to X'$. We say that the mapping f *preserves equivalence* if it follows from $x \overset{R}{\sim} y$ that $f(x) \overset{R'}{\sim} f(y)$. For this kind of mappings it is natural to define the mapping $f : X/R \to X'/R'$ of the quotient spaces in the following way: let D_α be the equivalence class in X, $x \in D_\alpha$ any element; let D'_β be the equivalence class in X' containing the point $f(x)$. Then $\widehat{f}(x)(D_\alpha) = D'_\beta$.

Exercise 3°. Show that the mapping $\widehat{f}$ is well defined. The mapping $\widehat{f}$ is called the *quotient mapping*.

Theorem 2. *If a continuous mapping* $f : X \to X'$ *preserves equivalence, then the corresponding quotient mapping* $\widehat{f} : X/R \to X'/R'$ *is continuous.*

Proof. Denote the projections of the spaces X, X' on the corresponding quotient spaces by π, π', respectively. The diagram

$$
\begin{array}{ccc}
X & \xrightarrow{\ f\ } & X' \\[4pt]
{\scriptstyle \pi}\downarrow & & {\scriptstyle \pi'}\downarrow \\[4pt]
X/R & \xrightarrow{\ \widehat{f}\ } & X'/R'
\end{array}
$$

is commutative, i.e. for each $x \in X$ we have $(\widehat{f}\pi)(x) = (\pi'f)(x)$. If the set V is open in X'/R', then $(\pi'f)^{-1}(V)$ is open in X, because $\pi'f$ is continuous. But $(\widehat{f}\pi)^{-1}(V) = (\pi'f)^{-1}(V)$, and, thus, the set $(\widehat{f}\pi)^{-1}(V)$ is open in X. Since $(\widehat{f}\pi)^{-1}(V) = \pi^{-1}(\widehat{f}^{-1}(V))$, the set $\widehat{f}^{-1}(V)$ is open in X/R (by the definition of topology on a quotient space), and so $\widehat{f}$ is continuous. $\square$

We now formulate a test for showing that quotient spaces are homeomorphic.

Theorem 3. *If* $f : X \to X'$ *is a homeomorphism and the mappings* f, f^{-1} *preserve equivalence, then the quotient mapping* $\widehat{f} : X/R \to X'/R'$ *is a homeomorphism.*

Indeed, the mapping f^{-1}, in this case, determines the quotient mapping $\widehat{f}^{-1} = (\widehat{f})^{-1}$ (verify!), and we can apply theorem 2 both to $\widehat{f}$ and $(\widehat{f})^{-1}$. $\square$

We add three more "models" to those of the projective plane $\mathbf{R}P^2$ listed in Ch. 1, § 3. The first is obtained from the sphere $X = S^2$ by gluing together diametrically opposite points (Fig. 47). The second model consists of lines in $\mathbf{R}^3$ that pass through zero ($x \overset{R}{\sim} y \Leftrightarrow x, y$ lie on one of such lines, and $x \neq 0$, $y \neq 0$) (Fig. 47).

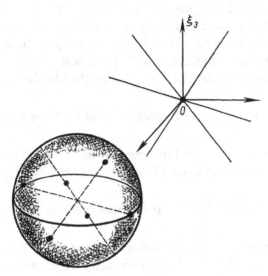

FIG. 47

Exercise 4°. Describe the topology of the spaces obtained as the topology of the quotient spaces S^2/R and $(\mathbf{R}^3\backslash 0)/R$, respectively.

A third model of $\mathbf{R}P^2$ is the following. Consider in $\mathbf{R}^3$ an arbitrary plane P which does not pass trough the origin of the coordinates. Fix a point a on P. According to the second model of $\mathbf{R}P^2$ just considered, this space consists of the lines in $\mathbf{R}^3$ that pass through the origin. Now we associate each of these lines to the point where it intersects the plane P (if it does intersect P), or associate to this line the line in P passing through the point a and parallel to the given one (if it does not intersect P). The line obtained on the plane P is symbo-

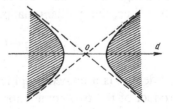

FIG. 48

lically identified with the point at infinity where these parallels meet.

Thus, we obtained a one–to–one correspondence between $\mathbf{R}P^2$ (the second model) and the plane P with the points at infinity added to it, one point for each (two–sided) direction (i.e. for each line passing through the origin) in P. In the set obtained, the plane P with the ususal topology is considered; a *neighbourhood of a point at infinity* corresponding to some direction d on P is defined as a part of the plane P (the shaded part on fig. 48), which is bounded by an arbitrary hyperbola with the axis d. The set of all points at infinity that are added to the plane P is called the *absolute* or *line at infinity*.

Exercise 5°. Prove that all these realizations of $\mathbf{R}P^2$ are homeomorphic.

Consider a closed disc $\bar{D}^n$ and its boundary S^{n-1}. We identify all points of the boundary. The quotient space obtained is denoted by $\bar{D}^n/S^{n-1}$.

Theorem 4. *The space $\bar{D}^n/S^{n-1}$ is homeomorphic to the sphere S^n.*

Proof. In § 2, subsection 3, it was shown that the disc $\bar{D}^n$ is homeomorphic to the closed hemisphere $\bar{S}^n_+$. This homeomorphism is the identity on the common boundary (S^{n-1}) of these sets; therefore, the equivalence relation from $\bar{D}^n$ induces one on $\bar{S}^n_+$, and, according to theorem 3, $\bar{D}^n/S^{n-1}$ is homeomorphic to $\bar{S}^n_+/S^{n-1}$.

Now we show that $\bar{S}^n_+/S^{n-1}$ is homeomorphic to S^n. The inclusion $\bar{S}^n_+ \to S^n$ is natural. We denote the south pole $(0,0,\ldots,0,-1)$ of the sphere S^n by $*$. Then there exists a continuous surjective mapping $\phi : \bar{S}^n_+ \to S^n$ such that $\phi(S^{n-1}) = *$, and $\phi|_{S^n_+} : S^n_+ \to S^n\backslash\{*\}$ is a homeomorphism. It can be constructed, for instance, as follows: if $x \in \bar{S}^n_+$ and $x \neq N$ (N is the north pole), then through the points 0, N, x we draw a two–dimensional plane which intersects S^n along a circumference (meridian). Moving x along the meridian through the arc that is twice the arc xN, we obtain a point $\phi(x)$; set $\phi(N) = N$. Thus, the quotient mapping is defined

$$\hat{\phi} : \bar{S}^n_+/S^{n-1} \to S^n/\{*\} = S^n,$$

which, evidently, is a homeomorphism.

The product of the two homeomorphisms

$$\bar{D}^n/S^{n-1} \to \bar{S}^n_+/S^{n-1}, \quad \bar{S}^n_+/S^{n-1} \to S^n$$

is the desired homeomorphism. □

§ 4. Classification of surfaces

1. Surfaces and their triangulation. Let us return to the study of closed surfaces. The definitions of a topological space, quotient space and homeomorphism of topological spaces given above, and the examples considered provides us with a strong basis for a proof of the theorem mentioned in Ch. 1, § 3. It states that any closed surface is topologically equivalent to one of the surfaces of the form M_p or N_q, i.e. to a sphere with p handles or q Möbius strips glued to it. We shall make the corresponding concepts more accurate and give the proof of the theorem mentioned.

A topological space X in which each point has a neighbourhood homeomorphic to an open disc we call a *two–dimensional manifold*. It is more convenient to study these surfaces by dividing them into pieces which are topologically equivalent to triangles on the two–dimensional Euclidean plane. We make this representation more precise.

Definition 1. A pair (T, ϕ), where T is a subspace in X, and $\phi : \triangle \rightarrow T$ is a homeomorphism of a certain triangle $\triangle \subset \mathbf{R}^2$ onto T, is called a *topological triangle* in X.

If the homeomorphism $\phi : \triangle \rightarrow T$ is fixed (when this cannot be a cause of misunderstanding), then the subspace $T \subset X$ will be called, for short, a topological triangle. The images of the vertices, and the sides of the triangle $\triangle$ (together with the restriction of the homeomorphism ϕ) are called the *vertices*, and the *edges* of the topological triangle, respectively. For uniformity, it is also convenient to call the sides of the triangle $\triangle$ its edges.

We now define an orientation of a triangle. Different ordered triples of points can be formed from the vertices of $\triangle$. We define that two triples are equivalent if one can be obtained from the other by a cyclic permutation. It is clear, that we get two equivalence classes. A *triangle* $\triangle$ is *oriented*, if one of these equivalence classes is given. A *topological triangle* (T, ϕ) *is oriented*, if the triangle $\triangle$ is oriented. Obviously, the orientation of the triangle $\triangle$ amounts to the problem of determining a direction of going around" its vertices (clockwise or counterclockwise). This direction of going around the vertices determines, via the homeomorphism ϕ, a direction of going around the vertices of the topological triangle, i.e. an orientation induced by the homeomorphism ϕ. An orientation of a triangle, obviously, determines orientations of its edges (as ordered pairs of its vertices).

Note for the future that an orientation of the n–gon and its edges for $n > 3$ is defined in a similar way (by defining a direction for passing through its vertices).

Definition 2. A finite set $K = \{(T_i, \phi_i)\}_{i=1}^{k}$ of topological triangles in X which satisfies the properties: (1) $X = \bigcup_{i=1}^{k} T_i$; (2) the intersection $T_i \cap T_j$ of any pair of triangles from K is either empty or coincides with a common vertex or common edge, is called a *triangulation* of the two–dimensional manifold.

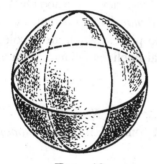

A manifold which has a triangulation is said to be *triangulable*. If any two vertices of the triangles from K can be joined by a path composed of edges, then we call X connected.

Fig. 49 presents an example of triangulation of the sphere S^2 consisting of eight triangles.

FIG. 49

Definition 3. We call a connected, triangulable, two–dimensional manifold a *closed surface*.

Note that the examples of closed surfaces which can be triangulated into topological polygons, considered in Ch. 1, § 3, are examples of closed surfaces in the sense of definition 3 (to verify this, it is sufficient to triangulate polygons).

Exercise 1°. Construct triangulations of the torus and the projective plane. Verify that they are closed surfaces.

Topological properties of a closed surface are defined by constructing a triangulation of it. In order to investigate the latter, it is convenient to consider its schematic representation in the plane. Besides, we can assume that the plane triangles Δ_i which are the preimages of the triangles $T_i \subset K$, belong to the same plane and do not intersect.

We will describe this representation. Let (T_i, ϕ_i), (T_j, ϕ_j) be two triangles from K, and $T_i \cap T_j = a$ a common edge; let $a_i = \phi_i^{-1}(a)$, $a_j = \phi_j^{-1}(a)$ be the corresponding edges in Δ_i, Δ_j. Then the gluing homeomorphism is defined

$$\phi_{ij} = \phi_j^{-1}|_a \phi_i|_{a_i} : a_i \to a_j.$$

Thus, the triangulation K can be associated to a system $\Delta = (\{\Delta_i\}_{i=1}^k, \{\phi_{ij}\})$ of triangles of the plane together with the homeomorphisms ϕ_{ij} for corresponding pairs of the edges. We declare that in $\bigcup_{i=1}^k \Delta_i$, the points corresponding to each other under the homeomorphisms ϕ_{ij} are equivalent. We will denote this equivalence by R.

Lemma. *The quotient space* $\left(\bigcup_{i=1}^k\right)/R$ *is homeomorphic to the surface* X.

Proof. The homeomorphisms $\phi_i : \Delta_i \to T_i$ naturally determine a surjective mapping $\Phi : \bigcup_{i=1}^k \Delta_i \to X$, moreover, the preimage $\Phi^{-1}(x)$ for any $x \in X$ is precisely an R–equivalence class. The quotient mapping $\widehat{\Phi} : \left(\bigcup_{i=1}^k \Delta)_i\right)/R \to X$ is a continuous mapping with respect to theorem 2, § 3. Obviously, it is bijective, and the mapping $\widehat{\Phi}^{-1}$ inverse to it is continuous. □

2. Development of a surface. In future, we shall need some systems similar to the system Δ which schematically represents a triangulation K of the surface X, but such that along with triangles they may contain n–gons ($n > 3$).

Definition 4. A system $Q = (\{Q_i\})$, $\{\phi_{ij}\}$, where $\{Q_i\}$ is a finite set of non–intersecting plane polygons, and $\{\phi_{ij}\}$ is a finite set of gluing homeomorphisms of pairs of edges of polygons from the set $\{Q_i\}$, each edge being glued together with precisely one (other) edge, is called a *development*. Gluing the edges of the same polygon together is allowed.

In particular, the system $\Delta = (\{\Delta_i\}, \{\phi_{ij}\})$ mentioned above is a development; it is said that Δ is the development of a surface X together with a triangulation K.

Note that if the location of a polygon Q_i on the plane is altered by a homeomorphism α_i, then, naturally, new homeomorphisms $\{\alpha_i \phi_i \alpha_1^{-1}\}$ that glue its edges are defined. Later we shall not distinguish these homeomorphisms from the homeomorphisms $\{\phi_{ij}\}$.

For an arbitrary development Q, consider the quotient space $\widehat{Q}$ of the union $\bigcup_i Q_i$ with respect to the equivalence R determined by the homeomorphisms $\{\phi_{ij}\}$, $\widehat{Q} = \left(\bigcup_i Q_i\right)/R$. We call $\widehat{Q}$ the quotient space of the development Q. Evidently, the quotient space of a development is a two–dimensional manifold; it admits a triangulation generated by a sufficiently small triangulation of polygons Q_i. Thus, if the quotient space $\widehat{Q}$ is connected, then it is a closed surface (below we consider only such $\widehat{Q}$). In such a case we call Q a *development of the space* $\widehat{Q}$.

The quotient mapping induces a decomposition of the surface $\widehat{Q}$ into images of polygons, images of edges (decomposition edges), images of vertices (decomposition vertices); this decomposition, generally speaking, is not a triangulation.

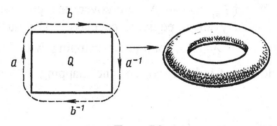

FIG. 50

Fig. 50 shows a development of the torus represented by a polygon. The arrows and labels of its edges indicate the gluing rule for obtaining the torus.

Hereafter, we shall orient development polygons by fixing an orientation for each of them. The orientations of polygons give corresponding orientations of the edges. Under the gluing homeomorphism $\phi_{ij} : a_i \rightarrow a_j$ of two edges, the edge a_j acquires an orientation induced (from the orientation of the edge a_i) by the homeomorphism ϕ_{ij}, which, in general, can differ from the orientation of the edge a_j.

A development A is said to be *orientable* if for a given orientation of all its polygons (for example, the going counterclockwise around its vertices), the homeomorphisms of gluing edges induce the reverse orientation in the image edge. Otherwise, (i.e. if at least for one edge the orientation coincides with the induced one) the development is said to be non–orientable.

A surface X is said to be *orientable* (*non–orientable*) depending upon whether its development is orientable (non–orientable).

3. The classification of developments

Definition 5. Two developments Q and Q' are called *equivalent* if their quotient spaces are homeomorphic.

Let us now introduce some elementary operations on a development, which transform it into an equivalent one.

Subdivision. Suppose there is an n–gon Q_i ($n > 3$) in a development Q. Let us draw a diagonal d which divides Q_i into two polygons Q_i' and Q_i'', move these polygons Q_i' and Q_i'' apart, and construct a new development $\tilde{Q}$ from Q by replacing the polygon Q_i by the two polygons Q_i' and Q_i''. In addition, we shall connect the two new edges d' and d'', which are the replicas of the diagonal d, by the natural identity homeomorphism, and keep all the homeomorphisms of the old edges. The development $\tilde{Q}$ is called a *subdivision of the development* Q; obviously, Q and $\tilde{Q}$ are equivalent.

Enlargement. This operation is inverse to the subdivision. Two polygons Q_i' and Q_i'' of the development Q are glued together into one polygon Q_i by using one of the gluing homeomorphisms involving, say, the edges d' and d''. The homeomorphisms involving the remaining edges of Q_i' and Q_i'' induce homeomorphisms for the edges of the polygon Q.

Convolution. Suppose that two adjacent edges with opposite orientations in the polygon Q_i of the development Q are glued together. By "gluing" these edges we obtain a development $\tilde{Q}$ which, instead of Q_i, contains a polygon in which the number of vertices is two less than that of Q_i, and the number of homeomorphisms of the development $\tilde{Q}$ is one less than that of Q (Fig. 51).

We stress that the operations described preserve the development equivalence class (verify!).

For convenience, further on, we shall describe each development by a set of

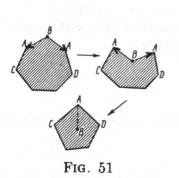

FIG. 51

special symbolic words according to the following rule. Let $Q = (\{Q_\Pi, \{\phi_{ij}\})$ be a development; fix an orientation for each polygon of the development (to be definite, we assume that all the polygons of the development are counterclockwise oriented). Denote the edges of polygons of the development Q by letters in such a way that the edges which are glued together, would be denoted by the same letters, and the non–glued ones by different letters.

The gluing rules for the edges given by the homeomorphisms ϕ_{ij} we shall indicate on the figures with the help of arrows thus determining the directions of the edges glued; the directions are given in such a way that the beginning of one edge is glued with the beginning of the other edge, and similarly the ends are glued together (besides, the direction for one of the edges from the pair glued together, can be given arbitrary; the direction for the other edge is then determined uniquely by the corresponding gluing homeomorphism ϕ_{ij}). Thus, we orient all the edges of polygons, which are glued with other edges. Because of this, it can turn out that the orientation of some edges does not coincide with the orientation of its polygon. We add an index -1 to the letter label for these edges. As in Ch. 1, § 3, we also write down in sequence the labels of the edges of a polygon Q_i into a word $w(Q_i)$ by passing through the edges in a given direction. The word $w(Q_i)$ characterizes the scheme of "gluing" a polygon Q_i in the development Q, and a set of such words for all polygons of the development Q characterizes the development Q.

Two basic types of developments are distinguished.

Definition 6. The development consisting of one polygon which is determined by a word of the form aa^{-1} or

$$a_1 b_1 a_1^{-1} b_1^{-1} a_2 b_2 a_2^{-1} b_2^{-1} \ldots a_m b_m a_m^{-1} b_m^{-1}, \quad m > 0,$$

is called a *type I canonical development*.

Definition 7. The development consisting of one polygon with a word of the form $a_1a_1a_2a_2\ldots a_ma_m$, $m > 0$, is called the *type II canonical development*.

We now formulate the basic·result.

Theorem 1. *Any development is equivalent to a type I or II canonical development depending on whether it is orientable or not.*

Proof. At first we shall make two remarks. To begin with, it is easy to see that with the help of enlargement it is always possible to transform a development corresponding to a triangulation K of a surface X into a development consisting of one polygon. Therefore, in future, we shall consider only such developments. Secondly, if a development with word different from aa^{-1} has a combination of the form aa^{-1}, then we can get rid of it by convolution of the edges a and a^{-1} around their common vertex A. The word of the new development is obtained from the original by crossing out the combinations aa^{-1}.

Finally, we come either to a word of two letters (aa^{-1} or aa), or to a word of no less than four letters long, and in which combinations of the form aa^{-1} are missing (recall that the surface is closed). Thus, as the words aa, aa^{-1} describe a canonical development, only the last case should be investigated further.

Let us divide this analysis into a number of steps:

(1) The obtained development Q' can be transformed into a development for which all vertices are equivalent, i.e. are glued under taking the quotient. In fact, assume that Q' contains non–equivalent vertices. Then in Q', there exists an edge a with non–equivalent ends A, B. Let b be another edge with ends at vertices B and C. Connect A and C by a diagonal d. In this case the edge b' which has to be glued with the edge b, is outside the triangle ABC. Otherwise, either $b = a$ or $b = a^{-1}$ what contradicts either the non–equivalence of the vertices A and B or the assumption of absence of combinations of the form aa^{-1}. Now apply the operation of subdivision along the diagonal d, and later use the operation of enlargement with respect to the edge b (we glue it to the edge b'). In the obtained development P', the set of vertices which are equivalent to A increases by one, and that equivalent to B reduces by one (Fig. 52). If, in addition, combinations of the form aa^{-1} appeared in the word of the development P', then we would eliminate them by convolution. It should be noted, that the last reconstruction cannot change the difference

between the set of vertices, which are equivalent to B, and the set of those equivalent to A (verify!).

Furthermore, if there still remain some vertices that are not equivalent to A, then we repeat the whole procedure until we obtain a development with the needed property.

Thus, in future, we can assume that all the vertices in the development considered are equivalent and that there are no combinations of the form aa^{-1} in it.

(2) We show now that two identical letters in the word of a development can always be placed together. Assume that the letters a and a are not next to each other. Then in a polygon we draw a diagonal d that connects the initial points of the edges a and a. Subdivide by d, and then glue by a. The new word has no letter a, but the combination dd appears, which is what is needed (Fig. 53). (It is easy to verify that the results of the first step are preserved.)

In precisely the same manner, we proceed with other identical letters situated apart from each other.

In addition, we note that while applying the procedure indicated, we do not separate other combinations of the form aa, since only those edges which are adjacent to the edge a, are separated, and they certainly are not equivalent to it.

(3) Assuming that the conditions of steps (1) and (2) are satisfied, we may show that if the letters a and a^{-1} in the word do not stand next to each other, then there are other letters b, b^{-1} such that the pairs a, a^{-1} and b, b^{-1} separate each other (Fig. 54).

Assume the opposite. If such a pair bb^{-1} does not exist, then for any letter c from the interval from a to a^{-1} its twin edge (which is glued with c), labelled by c or c^{-1}, also lies in the same interval. Indeed, if the twin edge for c is c then, due to (2), we can consider c and c to be adjacent. In the oother case, when the twin edge is c^{-1}, our statement is also true because of the assumption. Thus, no edge from the interval from a to a^{-1} is glued with an edge from outside of the interval. This means, that we can not glue together the vertices A and B of the edge a. This situation contradicts to the equivalence of all vertices of the development, since such a situation is possible only if the vertices A and B of the edge a are not equivalent (Fig. 55).

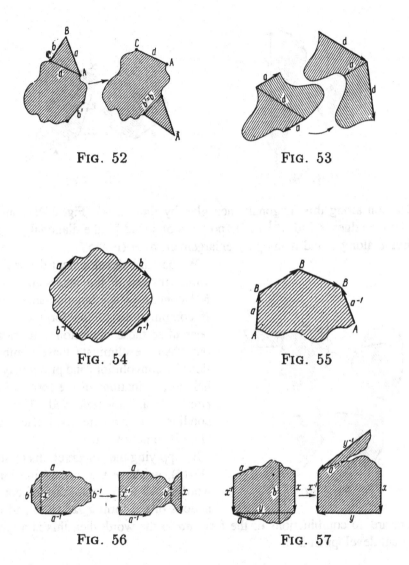

FIG. 52   FIG. 53

FIG. 54   FIG. 55

FIG. 56   FIG. 57

(4) Thus, we have two pairs in the word: a, a^{-1} and b, b^{-1}, that separate each other. We shall show that these four can be replaced by the combination of the form $xyx^{-1}y^{-1}$ while keeping the conditions of steps (1), (2). First, connect the initial points of the edges a and a^{-1} by the diagonal x and apply

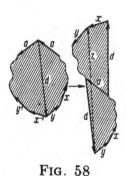

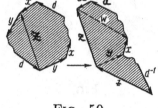

FIG. 58   FIG. 59

subdivision along this diagonal; then glue by the edge b (Fig. 56). Join the ends of the edges x and x^{-1} in the polygon obtained by the diagonal y, again subdivide along y and then apply enlargement by a (Fig. 57).

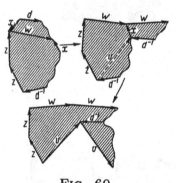

FIG. 60

We obtain a development the word of which instead of the letters a, b, a^{-1}, b^{-1} contains a combination $xyx^{-1}y^{-1}$. If combinations of the form cc^{-1} appear after these operations are carried out, then we eliminate these combinations by convolution, and previously existing combinations of the form dd and $cdc^{-1}d^{-1}$ are not separated. Thus, the conditions that are reached after steps (1), (2) remain valid.

By applying the constructions of steps (1)–(4), we have tranformed the initial word into a word consisting of combinations of the form $xyx^{-1}y^{-1}$ and aa.

If there are no combinations of the form aa in the word, then this is a type I canonical development.

(5) If combinations of the form $xyx^{-1}y^{-1}$ and aa occur simultaneously, then the word can be reduced to a type II canonical form in the following way. Connect the common vertex of the edges a and a with the common vertex of the edges y and x^{-1} by the diagonal d. Subdivide along the diagonal d

and enlarge by a (Fig. 58). The two pairs of divided edges x and x, y and y obtained turn into the combinations zz and ww by applying step (2) (Fig. 59, 60). These operations result in a separated pair d^{-1}, d^{-1} which is turned into a combination vv again using step (2) (Fig. 60). Thus, we obtain a word of the required canonical form.

Thus, the pair of combinations $xyx^{-1}y^{-1}$, aa is replaced in a word by the combination of three pairs of the form aa, and this does not disturb other $xyx^{-1}y^{-1}$ or aa type combinations. This process can be repeated untill all combinations of the form $xyx^{-1}y^{-1}$ disappear. □

Exercise 2°. Verify that two closed surfaces X, X' whose developments are equivalent to a canonical one of the same type and with the same number m, are homeomorphic.

4. The Euler characteristic and topological classification of surfaces. Let us turn to the geometrical interpretation of the theorem just proved.

In Ch. 1, § 3, it was shown that a combination of the form $xyx^{-1}y^{-1}$ in a word of the canonical development of a surface X correspond to a handle, and a combination of the form aa to a Möbius strip both of them being glued along its own boundary to the remaining part of the surface S. Thus, if the canonical development of a surface is of type I or II then this surface is glued together from a finite number of handles or a finite number of Möbius strips, respectively. This gluing can be easily represented as a result of gluing these handles or Möbius strips to the sphere S^2.

So we see that a surface with a type I canonical development is an orientable surface of type M_p, where p is the number of handles glued to the sphere (the *genus of the surface*). If, however, a canonical surface is of type II then it is a non–orientable N_q–type surface, $q \geq 1$, where q is the number of Möbius strips glued to the sphere (also the *genus of the surface*).

In the proof of the theorem it is shown that if p handles and $q \geq 1$ Möbius strips are glued to a sphere, then the surface obtained is non–orientable and is of the type N_{2p+q}.

The theorem on development classification leads to the conclusion that any closed surface is homeomorphic to some surface of the type M_p, N_q. In order to make this result more precise, consider the Euler characteristic of this surface. Let the decomposition of the surface X contain a_0 vertices, α_1 edges and α_2 images of polygons. The number $\chi(X) = \alpha_0 - \alpha_1 + \alpha_2$ is called the *Euler characteristic* of a surface. Obviously, this definition generalizes the previous one (see, Ch. 1, § 3), as the image of a polygon in this case is not necessarily topological (the sides of a polygon may be glued together).

If X is of the type M_p and P is its canonical development with word $a_1 b_1 a_1^{-1} b_1^{-1} \ldots a_p b_p a_p^{-1} b_p^{-1}$, then, evidently, $\alpha_0 = 1$, $\alpha_1 = 2p$, $\alpha_2 = 1$ and $\chi(X) = 2 - 2p$.

If X is of the type N_q and $a_1 a_1 a_2 a_2 \ldots a_q a_q$ is the word of its canonical development, then $\alpha_0 = 1$, $\alpha_1 = q$, $\alpha_2 = 1$ and $\chi(X) = 2 - q$.

If Q is an arbitrary development of the surface X then it can be transformed into a canonical development by elementary operations. It is easy to see that the elementary operations do not alter $\chi(X)$. Indeed, under subdivision the numbers α_1 and α_2 increase by 1, and α_0 does not change; under enlargement α_1 and α_2 decrease by 1 and α_0 is constant; under convolution α_0 and α_1 decrease by 1. Consequently, the alternating sum $\alpha_0 - \alpha_1 + \alpha_2$ is constant. Hence, we get an important conclusion: the canonical development does not depend on the choice of elementary transformations of the development Q. In fact, if Q could be reduced to two canonical developments P, P', say, of type I, i.e.

$$a_1 b_1 a_1^{-1} b_1^{-1} \ldots a_p b_p a_p^{-1} b_p^{-1}$$

and

$$\tilde{a}_1 \tilde{b}_1 \tilde{a}_1^{-1} \tilde{b}_1^{-1} \ldots \tilde{a}_{p_1} \tilde{b}_{p_1} \tilde{a}_{p_1}^{-1} \tilde{b}_{p_1}^{-1},$$

then the Euler characteristic calculated with respect to decomposition Q would be the same as that obtained with respect to decompositions P and P', and we would have the equality $2 - 2p = 2 - 2p'$, which implies $p = p'$, i.e. the coincidence of the words for P and P'. Similar reasonings are presented in the case of type II developments P, P'.

If, however, P is a type I development, and P' is of type II, then the equality $2 - 2p = 2 - q$ is fulfilled when $q = 2p$. Therefore, the reasoning presented above merely states that the development cannot have two canonical forms of type I and II with p and $q \neq 2p$. The general conclusion that the simultaneous reduction to both type I and II canonical forms is impossible follows from the property to preserve orientability (or non–orientability) of a development under elementary transformations (verify!).

Thus, we have proved the first part of the following central theorem on topological classification of surfaces.

Theorem 2. *Any closed surface is topologically equivalent to a surface of the type M_p or N_q. Surfaces of the types M_p, N_q, $q \geq 1$, are not topologically equivalent if p and q are not equal to zero; the surfaces M_p (or N_q) for different values p (or q) are not topologically equivalent either.*

The second part of the theorem was discussed in Ch. 1, § 3 (subsection 4) and above. This explanation could have been considered to be the proof had the topological invariance of the Euler characteristic $\chi(X)$ for an arbitrary closed surface X been proved (it is proved only for the case $X = S^2$) as well as the fact of M_p and N_q being not homeomorphic for $q = 2p$, $p > 0$. These facts will be established in Ch. 3, § 4, using the concept of the fundamental group of a space.

Exercise $3°$. Draw a diagram for gluing a surface whose canonical development has the word

$$a_1 b_1 a_1^{-1} b_1^{-1} a_2 b_2 a_2^{-1} b_2^{-1} a_3 b_3 a_3^{-1} b_3^{-1}.$$

Exercise $4°$. Draw a diagram for gluing a surface which is characterized by the word $a_1 a_1 a_2 a_2 a_3 a_3$. Indicate the type and genus of this surface.

Exercise $5°$. Verify that the following closed surfaces have the indicated type and genus:
  (1) the sphere $M_0 = N_0$;
  (2) the torus (sphere with one handle) M_1;
  (3) the double torus (sphere with two handles) M_2;
  (4) the projective plane N_1;
  (5) the Klein bottle N_2.
Draw diagrams of their decompositions.

Exercise $6°$. A topological space in which every point has a neighbourhood homeomorphic to an open interval of the number line, is called a *one–dimensional manifold* M^1. A decomposition of this manifold into *arcs* which are the topological images of the interval [0, 1] and whose ends are adjacent to each other (i.e. meets at vertices) is called a *triangulation* of M^1. We assume here that M^1 consists of a finite number of arcs. Prove that a triangulable manifold M^1 is homeomorphic to the circle S^1 or to several copies of it.

§ 5. Orbit spaces. Projective and lens spaces

1. The definition of an orbit space. We consider here important examples of quotient spaces which appear when groups act on topological spaces.

Let $H(X)$ be the set of all homeomorphisms of a topological space X into itself. A product of two homeomorphisms h_1 and h_2 : $(h_1 h_2)(x) = h_2(h_1(x))$ is defined; for each $h \in H(X)$ there is an inverse mapping $h^{-1} \in H(X)$, and, moreover, $hh^{-1} = h^{-1}h = 1_X$. Thus, $H(X)$ is a group with respect to multiplication (generally speaking, non–commutative) with the identity element 1_X.

Definition 1. We say that an abstract group G *acts* (*from the left*) *on a space* X if a homomorphism of the group G into the group $H(X)$ is given.

If G acts on X then, consequently, to each $g \in G$ there corresponds $h_g \in H(X)$: $g \mapsto h_g$; $g_1 g_2 \mapsto h_{g_1} h_{g_2}$; $g^{-1} \mapsto (h_g)^{-1}$; $1_G \mapsto 1_X$.
Evidently, the set $\{h_g\}_{g \in G}$ is a subgroup of the group $H(X)$.
Let $x \in X$ be an arbitrary point. The set $\bigcup_{g \in G} h_g(x)$ is called an *orbit* of this point and denoted by O_x.

Exercise 1°. Show that two orbits O_x and O_y either coincide or do not intersect.

The last statement allows us to introduce equivalence R in X: $x \sim y \Leftrightarrow O_x = O_y$, i.e. if and only if x, y belong to the same orbit.

Definition 2. The quotient space X/R is called the *orbit space of the group* G (*acting on* X) and denoted by X/G.

This method of constructing quotient spaces is important in modern topology. Let us consider some examples.

2. Projective spaces $\mathbf{R}P^n$, $\mathbf{C}P^n$. Consider the sphere $S^n \subset \mathbf{R}^{n+1}$. Let each point $x = (\xi_1, \ldots, \xi_{n+1}) \in S^n$ be associated with its diamterically opposite point $Ax = (-\xi_1, \ldots, -\xi_{n+1}) \in S^n$. The mapping $A : S^n \to S^n$ is a homeomorphism and is called the *central symmetry*. The relations $A = A^{-1}$, $A^2 = 1_{S^n}$ are obvious. Consequently, the set $\{A, 1_{S^n}\}$ is a group (with respect to multiplication) consisting of two elements; it is isomorphic to the group (with respect to addition) $\mathbf{Z}_2$ of residues mod 2. Thus, the action of $\mathbf{Z}_2$ on S^n is defined.

Definition 3. The space $S^n/\mathbf{Z}_2$ is *real projective space* of dimension n and is denoted by $\mathbf{R}P^n$.

Thus, $\mathbf{R}P^n$ is obtained from S^n by identifying opposite points x, $-x$.

Consider the set $G = \mathbf{R}\backslash 0$ (all real numbers except zero). This is a group with respect to multiplication; define its action on the space $X = \mathbf{R}^{n+1}\backslash\{0\}$ as

$$h_\lambda(x) = \lambda x, \quad \lambda \in G, \quad x \in \mathbf{R}^{n+1}\backslash\{0\}.$$

Obviously, O_x is the set of all the points of the line in $\mathbf{R}^{n+1}$ which passes through O and x, without the point 0. Therefore, $(\mathbf{R}^{n+1}\backslash\{0\})/G$ is the set of all lines in $\mathbf{R}^{n+1}$ which pass through the origin. The space $(\mathbf{R}^{n+1}\backslash\{0\})/G$ is homeomorphic to $\mathbf{R}P^n$. The homeomorphism is determined by the correspondence: the pair $(x, -x)$ corresponds to the line passing through the points x, $-x$.

Exercise 2°. Describe the topology of the space $(\mathbf{R}^{n+1}\backslash\{0\}/G$ and verify that the spaces $\mathbf{R}P^n$ and $(\mathbf{R}^{n+1}\backslash\{0\}/G$ are homeomorphic.

Exercise 3°. Glue the diametrically opposite points of the boundary of the disc $\bar{D}^n$ together. Show that the quotient space obtained is homeomorphic to $\mathbf{R}P^n$.

Consider now a complex space $\mathbf{C}^{n+1}$. Let $G = \mathbf{C}\backslash\{0\}$ be the group of complex numbers under multiplication. It acts on $\mathbf{C}^{n+1}\backslash\{0\}$ according to the rule $h_\lambda(x) = \lambda x$, $\lambda \in G$, $x \in \mathbf{C}^{n+1}\backslash\{0\}$. Consequently, $(\mathbf{C}^{n+1}\backslash\{0\}/G$ can be identified with a set of all complex lines in $\mathbf{C}^{n+1}$ passing through zero.

Definition 4. The space $(\mathbf{C}^{n+1}\backslash\{0\}/G$ is called *complex projective space* (of (complex) dimension n) and denoted by CP^n.

Let us construct another model of CP^n. Consider the unit sphere $S_C^n = \{x : |\xi_1|^2 + \ldots + |\xi_{n+1}|^2 = 1\}$ in C^{n+1}. The group $G = \{e^{i\alpha}, 0 \leq \alpha < 2\pi\}$ acts on it according the following rule

$$e^{i\alpha}x = (e^{i\alpha}\xi_1, e^{i\alpha}\xi_2, \ldots, e^{i\alpha}\xi_{n+1}).$$

This group G can be identified with the unit circle S^1 in the complex plane C. Then S^1 acts on the coordinate $\xi_i \in C$, and the orbit of the point ξ_i in C is the circle of radius $|\xi_i|$, if $|\xi_i| \neq 0$. Therefore, the orbit $O_x = \{e^{i\alpha}x\}$ ($0 \leq \alpha < 2\pi$) of each point $x \in S_C^n$ is a big circle on S_C^n. However, S_C^n can be identified with S^{2n+1} and O_x can be assumed to be a big circle on S_C^n; so an action of $G = S^1$ is defned on S^{2n+1}, and we have a homeomorphism $S_C^n/S^1 \rightarrow S^{2n+1}/S^1$.

Now, consider the homeomorphism $(C^{n+1}\setminus\{0\}/G \rightarrow S_C^n/S^1$. This homeomorphism is defined by associating to each complex line (a point in CP^n) that big circle S_C^n (a point in S_C^n/S^1) along which the complex line intersect with S_C^n. The superposition

$$(C^{n+1}\setminus\{0\})/G \rightarrow S_C^n/S^1 \rightarrow S^{2n+1}/S^1$$

determines a homeomorphism between CP^n and S^{2n+1}/S^1.

3. Lens spaces. At the end of Subsection 2, we considered the group S^1 of complex numbers of modulus 1 in the complex plane C.
Consider the finite subgroups of the group S^1; these are known to be finite cyclic groups isomorphic to the additive groups Z_k of residues   mod k.
   Let

$$\xi_j \mapsto e^{2\pi i \frac{k_{j-1}}{k}}\xi_j,$$

where k_{j-1} is an integer, $0 \leq k_{j-1} \leq k$. Then an action of Z_k on C^{n+1} and in S_C^n is defined as follows:

$$(\xi_1, \xi_2, \ldots, \xi_j, \ldots, \xi_{n+1}) \mapsto$$
$$\mapsto \left(e^{2\pi i \frac{1}{k}}\xi_1, e^{2\pi i \frac{k_1}{k}}\xi_2, \ldots, e^{2\pi i \frac{k_{j-1}}{k}}\xi_j, \ldots, e^{2\pi i \frac{k_n}{k}}\xi_{n+1}\right).$$

Definition 5. Let k_i and k be coprime for all i. Then the space S_C^n/Z_k is called a *generalized lens space* and is denoted by $L(k, k_1, \ldots n)$. When $n = 1$, the space $L(k, k_1)$ is called a *lens space*.

Exercise 4°. Show that under the condition of mutual primality of any k_i and k, each orbit of the action of the group Z_k described above consists of k points.

Exercise 5°. Show that the following formula determines an action of the group S^1 on a generalized lens space:

$$e^{i\alpha}(\xi_1, \ldots, \xi_{n+1}) = \left(e^{i\frac{\alpha}{k}}\xi_1, \ldots, e^{i\frac{\alpha}{k}}\xi_{n+1} \right).$$

Exercise 6°. Show that $L(k_1, k_2, \ldots, k_n)/S^1 = CP^n$.

§ 6. Operations on sets in a topological space

In this section we shall again investigate the properties of topological spaces and consider the closure operation, the operation of taking the interior part and boundary of a set and also the concept closely related to these operations, that is, the concept of limit and boundary points. All these ideas generalize well–known concepts of mathematical analysis.

1. Closure of a set. Let (X, τ) be a topological space.

Definition 1. The intersection of all closed sets containing A is called the *closure $\bar{A}$* of the set $A \subset X$.

The following statements are obvious:
  (1) The closure $\bar{A}$ is the smallest closed set containing A.
  (2) If A is closed then $\bar{A} = A$.

A closed set can be characterized by the concept of limit points described below.

Definition 2. A point $x \in X$ is called a *limit point* of a given set $A \subset X$, if in each neighbourhood $\Omega(x)$ of the point x there is at least one point $x' \in A$ different from x.

Exercise 1°. Verify that this definition only needs open neighbourhoods of the point x.

Examples. Consider the sets $A = \{n\}$, $B = \{1/n\}$, $n = 1, 2, \ldots$; $C = (0, 1)$, $D = [0, 1]$ in $\mathbf{R}^1$. The set A has no limit points, the set B has one limit point 0, and the limit points of the sets C and D fill the whole interval $[0, 1]$.

The concept of limit point in a topological space is, as can be easily seen, a generalization of the concept of limit point in mathematical analysis. We shall prove some useful statements which are related to the concept of limit points.

Theorem 1. *A set $A \subset X$ is closed iff it contains all its limit points.*

Proof. Let A be closed, x a limit point of A, and $x \notin A$. Then x belongs to the open set $\Omega(x) = X \backslash A$ which is a neighbourhood of the point x. But $\Omega(x) \cap A = \emptyset$, contradicting that x is a limit point.

Let A contain all its limits points. We shall demonstrate that it is closed, i.e. that its complement $U = X \backslash A$ is open. It is sufficient to show that for any point $x \in U$ there is a neighbourhood $\Omega(x)$ of x such that $\Omega(x) \subset U$. By assuming the opposite, we get that for some point $x_0 \in U$ and any neighbourhood $\Omega(x_0)$ of it there is a point $x' \in \Omega(x_0)$ such that $x' \notin U$. Then $x' \in X \backslash U = A$ so that x_0 is a limit point and hence $x_0 \in A$ contradicting the assumption that $x_0 \in U = X \backslash A$. $\qquad \qquad \square$

The set of all limit points of a set A is called the *derived set* of the set A and denoted by A'. Thus, a new operation has been defined which assigns to each set $A \subset X$ its derived set A'.

Theorem 2. *For any set $A \subset X$, the set $A \cup A'$ is closed.*

Proof. We shall show that the set $X \backslash (A \cup A')$ is open. Let x be an arbitrary point from $X \backslash (A \cup A')$. Then x is not a limit point of A; therefore, there is a neighbourhood $\Omega(x)$ of it such that $\Omega(x) \cap A = \emptyset$. Let $x' \in \Omega(x)$ be an arbitrary point. Then for any neighbourhood $V(x')$ of the point x' such that $V(x') \subset \Omega(x)$, we have $V(x') \cap A = \emptyset$. Therefore, x' is not a limit point for A and $\Omega(x) \cap A' = \emptyset$. Thus, $\Omega(x) \subset X \backslash (A \cup A')$. Because x is arbitrary the set $X \backslash (A \cup A')$ is open, and, hence, $A \cup A'$ is closed. $\square$

Exercise 2°. 1) Verify that $(A \cup B)' = A' \cup B'$; $(A \cap B)' \subset A' \cap B'$; $(A \backslash B)' \supset A' \backslash B'$.

2) Let $X = \{a, b\}$ be a space consisting of two elements with a trivial topology. Give an example of a set $A \subset X$ for which the inclusion $(A')' \subset A'$ is not valid.

Theorem 3. $\bar{A} = A \cup A'$ *for any set* A, $A \subset X$.

Proof. By theorem 2, the set $A \cup A'$ is closed. Therefore, by the definition of closure, $\bar{A} \subset A \cup A'$. On the other hand, any closed set containing A, obviously, contains all limit points of A, and also A'. Hence, $A \cup A' \subset \bar{A}$. Thus, $\bar{A} = A \cup A'$. $\square$

Exercise 3°. Let A be a set of rational points on the real line $\mathbf{R}^1$. Show that $\bar{A} = \mathbf{R}^1$.

If a topological space X has a countable subset A whose closure coincides with X, then this space is said to be *separable*. It is easy to verify that separability is a topological property.

Exercise 4°. Show that the space $\mathbf{R}^n$, the disc D^n and the sphere S^{n-1} are separable.

Exercise 5°. Verify the following properties of the closure operation: $\overline{A \cup B} = \bar{A} \cup \bar{B}$; $\bar{\bar{A}} = \bar{A}$; $\overline{A \cap B} = \bar{A} \cap \bar{B}$; $\bar{A} \backslash \bar{B} \subset \overline{A \backslash B}$; if $A \subset B$, then $\bar{A} \subseteq \bar{B}$.

Exercise 6°. Let Y be a subspace of a topological space X, and A a subset of X. Denote by $\bar{A}_Y$ the closure of the set A in the subspace Y, and by $\bar{A}$ the closure of A in X. Show that $\bar{A}_Y = \bar{A} \cap Y$.

Definition 3. A point $x \in A$ is said to be *isolated* if there exists a neighbourhood $\Omega(x)$ of the point x such that it does not contain any points of the set A different from x.

A point $x \in A$ is isolated iff $x \in A \backslash A'$.

Definition 4. A set A is called *discrete* if each point from it is isolated.

2. Interior of a set. Now consider two other important notions related to the concept of neighbourhood.

Definition 5. A point $x \in A$ is called an *interior point* of a set A if it has a neighbourhood $\Omega(x)$ such that $\Omega(x) \subset A$.

The set of all interior points of a set A is called the *interior* of A and denoted by Int A.

Example. Let $A = [0, 1]$ be an interval of the real line $\mathbf{R}^1$. Then Int $[0, 1] = (0, 1)$.

The operation Int is dual of the closure operation, which follows from its properties formulated in the theorem below.

Theorem 4. *For any set $A \subset X$ we have:* (1) *Int A an open set;* (2) *Int A the largest open set contained in A;* (3) *(A is open)* $\Leftrightarrow$ *(Int $A = A$);* (4) *($x \in$ Int A)* $\Leftrightarrow$ *($x \in A$ and x is not a limit point of $X \backslash A$);* (5) $\overline{X \backslash A} = X \backslash Int\ A$.

Proof. Properties (1)–(3) are almost obvious. Let us verify, for example, property (1). Let $x \in$ Int A. Then there is an open neighbourhood $U(x)$ of the point x such that $U(x) \subset A$. Therefore, Int A is a neighbourhood for each of its points and, thus, is an open set.

Let us check property (4). If $x \in$ Int A, then, evidently, $x \in A$ and $x \notin (X \backslash A)'$. Conversely, if $x \in A$ and $x \notin (X \backslash A)'$, then there is a neighbourhood $\Omega(x) \subset A$, and, hence, $x \in$ Int A.

The verification of property (5) is left to the reader. $\square$

The *exterior* of a set A is defined as the interior of its compliment: ext $A =$

int$(X\backslash A)$.

Exercise 7°. Show that $\bar{A} = X\backslash$ext A.

3. Boundary of a set. The next important concepts are those of boundary point and boundary of a set A. They are associated with the intuitive idea of a "separator" between a region of the Euclidean space and its exterior part.

Definition 6. A set $X\backslash(\text{Int } A \cup \text{ext } A)$ is called the *boundary* ∂A of a set A, and any point of the boundary is called a *boundary point* of a set A.

Thus, $x \in \partial A$ iff each neighbourhood of x contains a point both from A and $X\backslash A$.

Example. Let $X = \mathbf{R}^1$ and $A = (0, 1]$. Then Int $A = (0, 1)$, $X\backslash A = (-\infty, 0] \cup (1, +\infty)$, Int $(X\backslash A) = (-\infty, 0) \cup (1, +\infty)$. Therefore, $\partial A = \{0, 1\}$ is a set of two points 0 and 1.

Thus, we have the boundary operation ∂. Its relation with the closure and Int operations is clarified by the following theorem.

Theorem 5. *For any $A \subset X$ we have:* (1) $\partial A = \bar{A} \cap \overline{(X\backslash A)}$; (2) $\partial A = \bar{A}\backslash Int\, A$; (3) $\bar{A} = A \cup \partial A$; (4) *Int* $A = A\backslash\partial A$; (5) *(A is closed)* $\Leftrightarrow$ *($\partial A \subset A$)*; (6) *(A is open)* $\Leftrightarrow$ *((∂A)$\cap A = \emptyset$).*

Proof. We shall prove some of these statements leaving the rest as exercises.
(1) Let $x \in \partial A$; then any neighbourhood $U(x)$ of the point x contains points x_1, x_2 such that $x_1 \in A$, $x_2 \in X\backslash A$. Hence, $x \in \bar{A}$ and $x \in \overline{X\backslash A}$, i.e. $x \in \bar{A}\cap\overline{(X\backslash A)}$. Conversely, if $x \in \bar{A}\cup\overline{(X\backslash A)}$ then $x \in \bar{A}$, $x \in \overline{(X\backslash A)}$. Since $\overline{(X\backslash A)} = X\backslash\text{int } A$, $\bar{A} = X\backslash\text{ext } A$ (see item 5 of theorem 4 and exercise 7°), we get $x \notin \text{Int } A$, $x \notin \text{ext } A$, and, hence, $x \in \partial A$.
(2) According to definition

$$\partial A = X\backslash(\text{Int } A \cup \text{ext } A) = (X\backslash\text{ext } A)\backslash\text{Int } A = \bar{A}\backslash\text{Int } A.$$

(3) Since Int $A \subset \bar{A}$, it follows from (2) that $\bar{A} = \text{Int } A \cup \partial A \subset A \cup \partial A$; since $\partial A \subset \bar{A}$, we have $A \cup \partial A \subset A \cup \bar{A} = \bar{A}$.

(5) If A is closed then $\partial A \subset \bar{A} = A$. Conversely, if $\partial A \subset A$, then by (3), $\bar{A} = A \cup \partial A$ (see item 3), and, hence, $A = \bar{A}$, i.e. A is closed. $\square$

Exercise 8°. Let U be open in X, $A = \partial U$. Show that $\partial A = A$. Prove the inverse statement.

Exercise 9°. Let Y be a subspace of a topological space X and A a subset in Y. Denote by $\partial_Y A$ the boundary of the set A in Y, and by ∂A the boundary of A in X. Verify that $\partial_Y A = (\partial A) \cap Y$ is not always valid. Present examples.

§ 7. Operations on sets in a metric space. Sphere and ball. Completeness

1. Operations on sets in a metric space. Here we consider for metric spaces the concepts studied in the section above. Recall that a base for the topology in a metric space (X, ρ) consists of all possible balls $D_r(x_0)$, where $r > 0$ is the radius, x_0 is the centre of the ball. The metric ρ allows us to speak of convergent sequences in X (see Ch. 1, § 2). Let us express $\bar{A}$, A', Int A, ∂A in the following terms:

(a) the condition $x \in \text{Int}\, A$ is equivalent to the fact that for some $\epsilon > 0$ the ball $D_\epsilon(x)$ is wholly contained in A; this follows from the definition of the metric topology τ_ρ;

(b) the condition $x \in A'$ is equivalent to the existence of a sequence $\{a_n\}$ convergent to x, where $a_n \in A$, $a_n \neq x$.

Indeed, if $x \in A'$, then for any $r_1 > 0$ there is an element a_1 in A such that $a_1 \in D_{r_1}(x)$, $a_1 \neq x$. Let $0 < r_2 < \rho(x, a_1)$; then, again, an element $a_2 \in D_{r_2}(x)$, $a_2 \neq x$ can be found, etc. Thus, there are sequences $\{r_n\}$ and $\{a_n\} \subset A$ such that $\rho(a_n, x) < r_n$, $r_n \to 0$, $a_n \neq x$, i.e. $a_n \to x$.

Conversely, let there exist a sequence $a_n \to x$, where $a_n \neq x$, $a_n \in A$. Then for any neighbourhood $\Omega(x)$ of the point x we have a ball $D_\epsilon(x) \subset \Omega(x)$ and $N(\epsilon)$ such that $\rho(a_n, x) < \epsilon$ for $n \geq N(\epsilon)$. Hence $a_n \in \Omega(x)$, $n \geq N(\epsilon)$, and $a_n \neq x$, which completes the proof.

The definition of a limit point in terms of sequence convergent to this point presented above is always used in analysis as the definition of a limit point of a set;

c) the condition for a set A to be closed (A contains all its limit points) in a metric space is equivalent to the fact that from the existence of the sequence $\{a_n\} \subset A$ convergent to x there follows the condition $x \in A$. In fact, the condition for A to be closed is equivalent, for instance, to the condition $A' \subset A$ (see § 6), and this is equivalent to the previous statement;

d) the condition $x \in \partial A$ is equivalent to the fact that for any $r > 0$ we have $D_r(x) \cap A \neq \emptyset$ and $D_r(x) \cap (X\backslash A) \neq \emptyset$, i.e. any ball with a centre at the point x includes points of A and $X\backslash A$. This statement is obvious.

We shall present an equivalent definition which is often applied in analysis:

e) the condition $x \in \partial A$ is equivalent to the fact that there exist sequences $\{a'_n\} \subset X\backslash A$ and $\{a_n\} \subset A$ both convergent to x.

Indeed, let $x \in \partial A$. Then for any $r > 0$, the ball $D_r(x)$ includes points both of A (a point a_r) and $X\backslash A$ (a point a'_r). By assuming that $r = r_n$, $r_n \to 0$, we obtain sequences $a_{r_n} \in A$, $a'_{r_n} \in X\backslash A$ such that $a_{r_n} \to x$, $a'_{r_n} \to x$. Conversely, if $a_n \to x$, $\{a_n\} \subset A$ and $a'_n \to x$, $\{a'_n\} \subset X\backslash A$, then any ball $D_r(x)$ contains both the points a_n and a'_n for a sufficiently large $n = n(r)$; therefore, $x \in \partial A$.

2. Ball and sphere in $\mathbf{R}^n$. We shall investigate the sphere S^n, the open disc D^{n+1}, the closure $\overline{(D^{n+1})}$ of the open disc D^{n+1}, and the closed disc $\bar{D}^{n+1}$ in $\mathbf{R}^{n+1}$.

Theorem 1. *The following equalities are valid:* $\bar{D}^{n+1} = \overline{(D^{n+1})} = (D^{n+1})'$.

Proof. Consider the "ray" $\{tx_0\}$, $0 \leq t < +\infty$, which leaves the centre of a disc (the point 0) and passes through the point $x_0 \in \bar{D}^{n+1}$, $x_0 \neq 0$, then the points $x_n = \frac{n-1}{n} x_0$ of this ray tend to x_0 and are in D^{n+1} (verify this with the help of the metric on $\mathbf{R}^{n+1}$), while the points $y_n = \frac{1}{n} x_0$ also are in D^{n+1} and tend to 0. Therefore, $(D^{n+1})' \supset \bar{D}^{n+1}$. On the other hand, $\overline{(D^{n+1})} \subset \bar{D}^{n+1}$ (here $\overline{(D^{n+1})}$ is the topological closure of the disc D^{n+1}). Indeed, if $x_n \to y$, $x_n \in D^{n+1}$, i.e. if $y \in (D^{n+1})'$, then

$$\rho(y, 0) \leq \rho(y, x_n) + \rho(x_n, 0) < \rho(y, x_n) + 1,$$

hence, by assuming that $\rho(y, x_n) \to 0$ for $n \to \infty$, we have $\rho(y, 0) \leq 1$, i.e. $y \in \bar{D}^{n+1}$.

Combining the inclusions obtained with the obvious inclusion $(D^{n+1})' \subset \overline{(D^{n+1})}$, we obtain

$$\bar{D}^{n+1} \subset (D^{n+1})' \subset \overline{(D^{n+1})} \subset \bar{D}^{n+1},$$

from which the statemewnt of the theorem follows.   □

Theorem 2. *The sphere is the boundary of the disc:* $S^n = \partial(D^{n+1})$.

Proof. Let $x_0 \in S^n$ ($S^n \neq \emptyset$!). Then $x_n = \frac{n-1}{n}x_0 \in D^{n+1}$, and the sequence $\{x_n\}$ converges to x_0 when $n \to \infty$. Consequently, $S^n \subset \partial(D^{n+1})$. Conversely, let $x_0 \in \partial(D^{n+1})$. Then $x_0 \notin D^{n+1}$ since D^{n+1} consists of interior points, and there exists a sequence $\{x_n\} \in D^{n+1}$ convergent to x_0 (see Subsection 1, (e)). Therefore, $x_0 \in (D^{n+1})' = \bar{D}^{n+1}$, $x_0 \in S^n$.   □

Exercise 1°. Prove that $S^n = \partial(\bar{D}^{n+1})$.

Exercise 2°. Let $\phi : \mathbf{R}^n \to \mathbf{R}^1$ be a continuous function. Prove that the set $A = \{x \in \mathbf{R}^n : \phi(x) < \alpha\}$ is open, and the sets $B = \{x \in \mathbf{R}^n : \phi(x) \leq \alpha\}$, $C = \{x \in \mathbf{R}^n : \phi(x) = \alpha\}$ are closed for any $\alpha \in \mathbf{R}^1$ (these sets are called *Lebesgue sets* of the function ϕ).

Exercise 3°. Under the conditions of exercise 2° show that $\bar{A} \subset B$. Give an example for which $\bar{A} = B$ and $\partial A = C$ and also an example with $\bar{A} \neq B$ and $\partial A \neq C$.

3. Ball and sphere in an arbitrary metric space. Consider a metric space (X, ρ). Define the *closed ball* $\bar{D}_r(x_0)$ and the *sphere* $S_r(x_0)$ (with radius $r > 0$ and centre at the point x_0) by the equalities

$$\bar{D}_r(x_0) = \{x \in X : \quad \rho(x, x_0) \leq r\};$$
$$S_r(x_0) = \{x \in X : \quad \rho(x, x_0) = r\}.$$

Note that $\bar{D}_r(x_0)$ and $S_r(x_0)$ are closed sets in X. Indeed, if $\{x_n\} \in \bar{D}_r(x_0)$ and $x_n \to y$ then

$$\rho(x_0, y) \leq \rho(x_0, x_n) + \rho(x_n, y) \leq r + \rho(x_n, y),$$

and so $\rho(x_0, y) \leq r$, i.e. $y \in \bar{D}_r(x_0)$; $S_r(x_0)$ is closed as the complement in the closed set $\bar{D}_r(x_0)$ to the open set $D_r(x_0)$.

Are the theorems of Subsection 2 valid in the metric space? The following example refutes this idea.

Example 1. (Counterexample). Let X be a finite set. Take the metric $\rho(x, x) = 0$, $\rho(x, y) = 1$ when $x \neq y$. Then for $r < 1$

$$D_r(x_0) = \{x_0\}, \quad \bar{D}_r(x_0) = \{x_0\}, \quad S_r(x_0) = \emptyset$$

and

$$\overline{(D_r(x_0))} = \bar{D}_r(x_0) \neq (D_r(x_0))' = \emptyset.$$

However, $S_r(x_0) = \partial D_r(x_0) = \emptyset$. For $r = 1$, $D_1(x_0) = \{x_0\}$, $\bar{D}_1(x_0) = X$, $S_1(x_0) = X \setminus \{x_0\}$ and $\overline{(D_1(x_0))} \subset \bar{D}_1(x_0)$; while $\overline{(D_1(x_0))} \neq \bar{D}_1(x_0)$, $S_1(x_0) \neq \partial D_1(x_0) = \emptyset$. Finally, for $r > 1$

$$D_r(x_0) = \bar{D}_r(x_0) = X, \quad S_r(x_0) = \emptyset,$$

while $\overline{(D_r(x_0))} = \bar{D}_r(x_0) \neq (D_r(x_0))' = \emptyset$, $S_r(x_0) = \partial D_r(x_0) = \emptyset$.

The following theorem gives a sufficient and necessary condition for a sphere in a metric space to be the boundary of a ball.

Theorem 3. *In a metric space, the equality* $S_r(x_0) = \partial D_r(x_0)$ *holds iff* $\overline{(D_r(x_0))} = \bar{D}_r(x_0)$.

Proof. It follows from the equality $\overline{(D_r(x_0))} = \bar{D}_r(x_0)$ that

$$S_r(x_0) = \bar{D}_r(x_0) \setminus D_r(x_0) = \overline{(D_r(x_0))} \setminus D_r(x_0) = \partial D_r(x_0).$$

Conversely, if $S_r(x_0) = \partial D_r(x_0)$, then $\overline{(D_r(x_0))} = D_r(x_0) \cup \partial D_r(x_0) = D_r(x_0) \cup S_r(x_0) = \bar{D}_r(x_0)$. $\qquad \square$

Exercise $4°$. Let $X = C_{[0,1]}$ be the space of continuous functions on $[0, 1]$ with the standard metric (see Ch. 1, § 2). Give an interpretation of $D_r(x_0)$, $\bar{D}_r(x_0)$, $S_r(x_0)$ and show that $S_r(x_0) = \partial D_r(x_0)$.

4. Completeness of metric spaces. In analysis, the Cauchy criterion for the convergence of a numerical sequence (in the space $\mathbf{R}^1$) is used: the sequence $\{x_n\}$ converges to some point x_0 ($x_n \to x_0$) iff it is *fundamental*, i.e. for each $\epsilon > 0$ there is an integer $N(\epsilon)$ such that $|x_{n+m} - x_n| < \epsilon$ as soon as $n \geq N(\epsilon)$, $m \geq 1$.

If $x_n \xrightarrow{\rho} x_0$ in (X, ρ) then it can be easily shown that $\{x_n\}$ is a fundamental sequence, as in the case of $\mathbf{R}^1$, i.e. for each $\epsilon > 0$ there is an $N(\epsilon)$ such that

$$(1) \qquad \rho(x_{n+m}, x_n) < \epsilon, \quad n \geq N(\epsilon), \quad m \geq 1.$$

However, the opposite is not always true.

Definition 1. The metric space (X, ρ) in which any fundamental sequence has a limit is called a *complete metric space*.

Example 2. Let $X = \mathbf{Q} \subset \mathbf{R}^1$ be the set of rational numbers in $\mathbf{R}^1$. This metric space is not complete as there exist sequences of rational numbers covergent to an irrational number (i.e. there are fundamental sequences that have no limit in $\mathbf{Q}$).

Example 3. The space $X = \mathbf{R}^1$ is complete.

Example 4. The space $X = \mathbf{R}^n$ is complete. This follows from the fact that the fundamentality or convergence of sequences of ordered sets of numbers $\{(\xi_1^k, \ldots, \xi_n^k)\}$ is equivalent to the fundamentality or convergence of the n numerical sequences $\{\xi_1^k\}, \ldots, \{\xi_n^k\}$.

Exercise 5°. Prove that the space $X = \mathbf{C}^n$ is complete.

Example 5. The space $X = C_{[0,1]}$ is complete in the metric

$$\rho_1(x(t), y(t)) = \max_{0 \leq t \leq 1} |x(t) - y(t)|$$

and is not complete in the metric

$$(2) \qquad \rho_2(x(t), y(t)) = \left\{ \int_0^1 (x(t) - y(t))^2 dt \right\}^{1/2}.$$

The statements just formulated are proved in analysis.

These examples show that the property of completeness is not topological, i.e., generally speaking, it is not preserved under homeomorphisms of metric spaces. For instance, an interval $(a, b) \subset \mathbf{R}^1$ and $\mathbf{R}^1$ itself are homeomorphic spaces, but the space (a, b) is not complete, contrary to $\mathbf{R}^1$.

The following statement holds, but we leave the proof to the reader as an exercise.

Theorem 4. *Let (X, ρ) be a metric space and $X_1 \subset X$ a subspace. Then, if X_1 is complete it is closed in X; if X is complete and X_1 is closed in X, then X_1 is complete.*

§ 8. Properties of continuous mappings

1. Equivalent definitions of a continuous mapping. We shall express the property of continuity of the mapping $f : X \to Y$ of topological spaces X and Y in terms of other topological concepts, viz. of the neighbourhoods and closed sets.

Theorem 1. *Let $f : X \to Y$ be a mapping of topological spaces. The following statements are equivalent: (1) f is continuous; (2) for each $A \subset X$, $f(\bar{A}) \subset \overline{f(A)}$; (3) for each $B \subset Y$, we have $\overline{f^{-1}(B)} \subset f^{-1}(\bar{B})$.*

Proof. We shall prove a number of implications.

(1) $\Rightarrow$ (2): from the definition of continuity in terms of closed sets (§ 1, Exercise 6°), we conclude that the set $f^{-1}(\overline{f(A)})$ is closed in X; moreover, it contains A and, therefore, we have $\bar{A} \subset f^{-1}(\overline{f(A)})$, whence $f(\bar{A}) \subset \overline{f(A)}$;

(2) $\Rightarrow$ (1): from (2) it is evident that $\bar{A} \subset f^{-1}(\overline{f(A)})$ for any A. Choosing $A = f^{-1}(F)$, where F is an arbitrary closed set in Y, we obtain $\overline{f^{-1}(F)} \subset f^{-1}(\overline{f(f^{-1}(F))}) = f^{-1}(F)$. Consequently, $f^{-1}(F)$ is closed for any closed $F \subset Y$, i.e. f is continuous;

(1) $\Rightarrow$ (3): the continuity of f implies that $f^{-1}(\bar{B})$ is closed. From the inclusion $f^{-1}(B) \subset f^{-1}(\bar{B})$ immediately follows $\overline{f^{-1}(B)} \subset \overline{f^{-1}(\bar{B})} = f^{-1}(\bar{B})$, from which (3) is obtained;

(3) $\Rightarrow$ (1): for a closed B, the chain of inclusions $\overline{f^{-1}(B)} \subset f^{-1}(\bar{B}) = f^{-1}(B)$ follows from (3). Hence $f^{-1}(B)$ is closed, and, therefore, the mapping f is continuous. $\square$

Similarly to the definition of continuity of a mapping in a metric space we can define the continuity of a mapping of topological spaces as being continuous at every point by introducing the concept of continuity of a mapping at a point in a topological space.

Definition. A mapping $f : X \to Y$ of topological spaces is *continuous at a point* $x_0 \in X$ if for any neighbourhood $\Omega(f(x_0))$ of the point $f(x_0)$ there exists a neighbourhood $\Omega(x_0)$ of the point x_0 such that $f(\Omega(x_0)) \subset \Omega(f(x_0))$.

Exercise 1°. The following property of the mapping $f : X \to Y$ is equivalent to the continuity at a point: the complete preimage $f^{-1}(\Omega(f(x_0)))$ of any neighbourhood of the point $f(x_0)$ is a neighbourhood of the point x_0.

Theorem 2. *A mapping $f : X \to Y$ is continuous iff it is continuous at each point $x \in X$.*

Proof. Let $f : X \to Y$ be continuous, $x_0 \in X$ the arbitrary point, and $\Omega(f(x_0))$ an arbitrary neighbourhood of the point $f(x_0)$. Then there is an open set $V \subset Y$ such that $V \subset \Omega(f(x_0))$ and $f(x_0) \in V$. Set $U = f^{-1}(V)$. Then U is an open set, $x_0 \in U$. Further, $f(U) \subset \Omega(f(x_0))$, which proves the continuity of f at the point x_0.

Conversely, let f be continuous at every point $x \in X$. Let $V \subset Y$ be an arbitrary open set, and let $A = f^{-1}(V)$. Since V is a neighbourhood of each of its points, and f is continuous at each point, for any $x \in A$, there is a neighbourhood $\Omega(x)$ of the point x such that $f(\Omega(x)) \subset V$. Therefore, $\Omega(x) \subset A$ which proves that A is open. The continuity of f is proved.   □

Exercise 2°. Let $X = A \cup B$ be the union of two closed sets. Prove, that a mapping $f : X \to Y$ is continuous iff the mappings $f|_A$ and $f|_B$ are continuous. Give counterexamples for this statement when the condition that the sets A, B are closed, is not fulfilled.

2. Three problems on continuous mappings.

In topology and its applications the following types of problems have to be solved often:

(1) Given two topological spaces X, Y and a mapping $f : X \to Y$. Check if f is continuous.

(2) Given a topological space X, a set Y and a mapping $f : X \to Y$.

Introduce a topology on Y so that f is a continuous mapping.

(3) Given a topological space Y, a set X and a mapping $f : X \to Y$. Introduce a topology on X so as to make f a continuous mapping.

Problem (1) has been already considered for certain spaces and mappings. In order to solve it, additional information on X, Y and f is required.

Problem (2) can be solved without additional assumptions. Let $\{U\} = \tau$ be a topology on X. Introduce a topology on Y as follows: we say that a set $V \subset Y$ is open iff its preimage $f^{-1}(V) = U$ in X is open (also including the empty preimage). It is easy to check that the family of such sets $\{V\}$ forms a topology. In fact,

(1) $\emptyset \in \{V\}$ since $f^{-1}(\emptyset) = \emptyset \in \tau$, and $Y \in \{V\}$ since $f^{-1}(Y) = X = \tau$; further, let V_α be sets from $\{V\}$, then

(2) $\bigcup_\alpha V_\alpha \in \{V\}$ since $f^{-1}(\bigcup_\alpha V_\alpha) = \bigcup_\alpha f^{-1}(V_\alpha) = \bigcup_\alpha U_\alpha \in \tau$, where $U_\alpha = f^{-1}(V_\alpha) \in \tau$;

(3) $\bigcap_{i=1}^k V_{\alpha_i} \in \{V\}$ since $f^{-1}(\bigcap_{i=1}^k V_{\alpha_i}) = \bigcap_{i=1}^k f^{-1}(V_{\alpha_i}) = \bigcap_{i=1}^k U_{\alpha_i} \in \tau$.

The topology constructed on Y will be called the *topology induced by the mapping* f. This is the strongest topology on Y in which f is continuous[*].

Consider now the continuity of a mapping $g : X/R \to Y$, where R is a certain equivalence relation, and X/R is the quotient space.

Theorem 3. *Let X, Y be two topological spaces, $f : X \to Y$, $g : X/R \to Y$ certain mappings, and $\pi : X \to X/R$ the projection. Let the diagram*

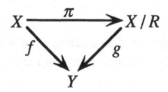

be commutative, i.e. $f(x) = (g\pi)(x)$, $x \in X$. *Then g is continuous iff f is*

---

[*] This method of introducing topology has occurred earlier when introducing quotient topologies (see § 3).

continuous.

Proof. Let f be continuous. Then if $V \subset Y$ is open, the $f^{-1}(V)$ is open in X. The set $\pi(f^{-1}(V)) = U$ is open in X/R since the set $\pi^{-1}(U) = f^{-1}(V)$ is open in V (a set W is open in X/R, iff its preimage $\pi^{-1}(W)$ is open in X). Since $f = g\pi$ we obtain

$$\pi(f^{-1}(V)) = \pi(\pi^{-1}g^{-1})(V) = g^{-1}(V).$$

Therefore, $g^{-1}(V)$ is open and, consequently, g is continuous. $\square$

Let us find out when the space Y with the topology described above is homeomorphic to the factor space of the space X with respect to the following equivalence relation (induced by the mapping $f : X \rightarrow Y$):

$$R_f : x_1 \sim x_2 \Leftrightarrow f(x_1) = f(x_2).$$

The class of equivalent points in X is the complete preimage $f^{-1}(y)$ of some value $y \in Y$. Let $\pi : X \rightarrow X/R_f$ be the projection, and $\widehat{f} : X/R_f \rightarrow Y$ the quotient mapping which takes a class of equivalent points $[x]$ into $f(x)$. We have the equality $\widehat{f}(\pi(x)) = f(x)$, $x \in X$, which signifies the commutativity of the diagram

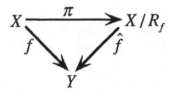

Theorem 4. *If the topology on Y is induced by a mapping $f : X \rightarrow Y$ and f is surjective, then $\widehat{f}$ is a homeomorphism of the spaces X/R_f and Y.*

Proof. Obviously, $\widehat{f}$ is bijective. Since the topology on Y is induced by the mapping f we get that f is continuous, and, therefore, according to theorem 3, $\widehat{f}$ is continuous. It remains to prove the continuity of $\widehat{f}^{-1}$ which is equivalent to the openness of $\widehat{f}$. We shall show that $\widehat{f}$ is open. Let U be an open set in

X/R_f, and $V = \widehat{f}(U)$ its image in Y. The set $\pi^{-1}(U)$ is open in X since π is continuous. Because

$$f^{-1}(V) = (\widehat{f}\pi)^{-1}(V) = \pi^{-1}(\widehat{f}^{-1}(V)) = \pi^{-1}(U),$$

$f^{-1}(V)$ is open. Since the sets W whose preimage $f^{-1}(W)$ is open in X, are open in Y, we get that V is an open set.   □

Exercise 3° Show that if a mapping f is not surjective, then the quotient space X/R_f is homeomorphic to the subspace $f(X) \subset Y$, where the topology on Y is induced by the mapping f.

When considering a continuous mapping $f : X \to Y$ of two topological spaces, the question may arise under which conditions the topology on Y is the one induced by the mapping f.

Theorem 5. *Let $f : X \to Y$ be a surjective mapping of topological spaces, and let f be continuous and open (or closed). Then the topology on Y is the quotient topology induced by f.*

Proof. Consider the case when f is open. Let $\{V\} = \sigma$ be the topology on Y induced by the mapping f; let $\tau = \{U\}$ be the original topology on Y. We shall show that they coincide. Indeed, let $V \in \sigma$, $V \neq \emptyset$. Then because of the surjectivity, $f^{-1}(V) \neq \emptyset$ and $f^{-1}(V)$ is open in X (by the construction of σ). Since f is open, we find that the set $f(f^{-1}(V)) = V$ is open in Y, i.e. $V \in \tau$. Conversely, let $U \in \tau$. Then from the continuity of f it follows that $f^{-1}(U)$ is open in X, and, therefore, $U \in \sigma$ by the definition of the topology σ.

The case when f is closed is dealt with similarly.   □

It remains to consider problem (3). Let $f : X \to Y$ be a mapping of a set X into a topological space Y. Let $\tau = \{V\}$ be a topology on Y. Set $\sigma = \{f^{-1}(V)\}_{V \in \tau}$. The system σ satisfies the topology axioms (verify!). Obviously, f is continuous as a mapping of topological spaces (X, σ), (Y, τ). It is clear that σ is the weakest of the topologies possessing this property.

It is useful to note the fact that if $X = A$ is a subset of a topological space Y, then for the inclusion $i : X \to Y$, the topology σ defined above by i coincides with the topology of a subspace $A \subset Y$ (obtained from the space Y).

§ 9. Product of topological spaces

1. Topology in the direct product of spaces. The operation of a direct product of topological spaces makes the construction of new topological spaces possible.

Recall that the *direct product* $X \times Y$ of the sets X, Y is the collection of ordered pairs (x, y), where $x \in X$, $y \in Y$. One may consider direct products of any number of factors. An element of such a product $\prod_{\alpha \in A} X_\alpha$ is a family $\{x_\alpha\}_{\alpha \in A}$, $x_\alpha \in X_\alpha$, or, in other words, the elements $\prod_{\alpha \in A} X_\alpha$ are functions $x \colon A \to \bigcup_{\alpha \in A} X_\alpha$ such that $x(\alpha) \in X_\alpha$. If $A = \{1, 2, \ldots, n\}$ is a finite set then the product $X_1, X_2, \ldots, X_n$ is often denoted by $X_1 \times X_2 \times \ldots \times X_n$, its elements being the ordered sets $(x_1, x_2, \ldots, x_n)$, where $x_i \in X_i$, $i = 1, 2, \ldots, n$

Let X, Y be topological spaces. We introduce a topology on the direct product $X \times Y$. A base of the topology is given by the system $\{U_\alpha \times V_\beta\}$, where $\{U_\alpha\}$, $\{V_\beta\}$ are bases for the topologies on, respectively, X and Y.

Exercise 1°. Verify that the covering $\{U_\alpha \times V_\beta\}$ of the set $X \times Y$ satisfies the criterion for being a base (see § 1).

The topology on $X \times Y$ determined by the base $\{U_\alpha \times V_\beta\}$ is called the *product topology*.

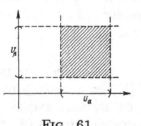

Example. The plane $\mathbf{R}^2$ is the direct product of two straight li nes: $\mathbf{R}^2 = \mathbf{R}^1 \times \mathbf{R}^1$. A base for the topology is the system of open rectangles of the form $U_\alpha \times V_\beta$, i.e. two–dimensional parallelepipeda (fig. 61), where U_α, V_β are intervals.

FIG. 61

Exercise 2°. Prove that the two–dimensional torus T^2 is homeomorphic to the product $S^1 \times S^1$.

Exercise 3°. Prove that the space $S^1 \times \mathbf{R}^1$ is homeomorphic to a cylinder.

Consider the projections

$$\rho_1 : X \times Y, \quad (x, y) \mapsto x;$$
$$\rho_2 : X \times Y, \quad (x, y) \mapsto y.$$

Theorem 1. *If X, Y are topological spaces and $X \times Y$ has the product topology, then the mappings ρ_1, ρ_2 are continuous. Moreover, this is the weakest of all the topologies on $X \times Y$ for which ρ_1, ρ_2 are continuous.*

Proof. We show that ρ_1 is continuous. Let U_α be a set from a base for the topology on X. It is sufficient to show that $\rho_1^{-1}(U_\alpha)$ is open. Since the space Y is representable in the form of a union $\bigcup_\beta V_\beta$ of all the sets of a base for the topology on Y,

$$\rho_1^{-1}(U_\alpha) = U_\alpha \times Y = U_\alpha \times \bigcup_\beta V_\beta = \bigcup_\beta (U_\alpha \times V_\beta),$$

and, consequently, $\rho_1^{-1}(U_\alpha)$ is open in $X \times Y$. The continuity of ρ_2 is checked in a similar way.

Now verify the second statement of the theorem. For ρ_1 to be continuous, it is necessary that the sets $\rho_1^{-1}(U_\alpha) = U_\alpha \times Y$ should be open. For the continuity of ρ_2, the sets $X \times V_\beta = \rho_2^{-1}(V_\beta)$ should be open. Then for both ρ_1 and ρ_2 to be continuous simultaneously, it is necassary that the sets $U_\alpha \times Y$, $X \times V_\beta$, and, thus, the sets $(U_\alpha \times Y) \cap (X \times V_\beta) = U_\alpha \times V_\beta$ should be open.

Thus, any topology on $X \times Y$ for which ρ_1 and ρ_2 are continuous, should contain the sets $U_\alpha \times U_\beta$ (and also the topology generated by them); consequently, it is stronger than the product topology on $X \times Y$. □

Consider the direct product $\prod_{\alpha \in A} X_\alpha$ with an arbitrary (possibly, infinite) number of factors. Let X_α, $\alpha \in A$, be topological spaces. We introduce the weakest of all those topologies on $\prod_{\alpha \in A} X_\alpha$ in which all projections $\rho_{\alpha'} : \prod_{\alpha \in A} X_\alpha \to X_{\alpha'}$ which associate to the function x the value $x(\alpha')$, are continuous. This topology on $\prod_{\alpha \in A} X_\alpha$ is called the *product topology* or the *Tikhonov topology* (it was introduced by A. N. Tikhonov), and the product with this topology is called also *Tikhonov product*.

We shall describe this topology. The simplest way is to characterize a pre-base of the Tihonov topology as follows: all possible sets in the product

$\prod_{\alpha \in A} X_\alpha$ which have the form $B_{\alpha_0} = \{x : x(\alpha_0) \subset U_{\alpha_0}\}$, where α_0 is an arbitrary element from A, and U_{α_0} is an arbitrary element of a base for the topology on the space X_{α_0}. It is easy to see that $B_{\alpha_0} = \rho_{\alpha_0}^{-1}(U_{\alpha_0})$. Thus, for fixed α_0, the sets $\{B_{\alpha_0}\}$ form the weakest of all topologies on $\prod_{\alpha \in A} X_\alpha$ in which the projection p_{α_0} is continuous. Consequently, declaring the system $\{B_\alpha\}_{\alpha \in A}$ to be a pre–base, we obtain the weakest of all the topologies on $\prod_{\alpha \in A} X_\alpha$ in which all the projections $p_{\alpha'}$ are continuous.

Hence, it follows that a base for the Tikhonov topology on $\prod_{\alpha \in A} X_\alpha$ is formed of the sets of the form

$$U = \rho_{\alpha_1}^{-1}(U_{\alpha_1}) \cap \rho_{\alpha_2}^{-1}(U_{\alpha_2} \cap \ldots \cap \rho_{\alpha_n}^{-1}(U_{\alpha_n}),$$

where $\alpha_1, \ldots, \alpha_n$ is an arbitrary finite collection of elements from A, and U_{α_i} is an arbitrary element of a base for the topology on X_{α_i}.

Exercise 4°. Show that

$$\rho_{\alpha_1}^{-1}(U_{\alpha_1}) \cap \rho_{\alpha_2}^{-1}(U_{\alpha_2}) \cap \ldots \cap \rho_{\alpha_n}^{-1}(U_{\alpha_n}) = \prod_{\alpha \in A} U_\alpha,$$

where $U_\alpha = X_\alpha$, if $\alpha \neq \alpha_1, \ldots, \alpha_n$. In other words, an open set of the base is a set of functions

$$\{x : x(\alpha_i) \in U_{\alpha_i}, \quad i = 1, 2,, \ldots, n\} =$$
$$= \{x : x(\alpha_1) \in U_{\alpha_1}\} \cap \ldots \cap \{x : x(\alpha_n) \in U_{\alpha_n}\}.$$

Note that products are considered, as a rule, together with the product topology.

Theorem 2. *For any $\alpha_0 \in A$ the projection $p_{\alpha_0} : \prod_{\alpha \in A} X_\alpha \to X_{\alpha_0}$ is a continuous and open mapping.*

Proof. The statement about the continuity of ρ_{α_0} needs no proof. The openesss of the image of the arbitrary open set from $\prod_{\alpha \in A} X_\alpha$ for the mapping p_{α_0} follows from the openness of the image of the set from the defining base for the topology on $\prod_{\alpha \in A} X_\alpha$.   □

Exercise 5°. Verify that $\mathbf{R}^n = \underbrace{\mathbf{R}^1 \times \ldots \times \mathbf{R}^1}_{n}$. Describe a base and pre–base for the Tikhonov topology on $\mathbf{R}^n$.

Exercise 6°. Verify that the n–dimensional cube I^n in $\mathbf{R}^n$ can be represented as $I^n = \underbrace{I \times \ldots \times I}_{n}$, where $I = [0, 1]$.

Exercise 7°. Consider the n–dimensional torus $T^n = \underbrace{S^1 \times \ldots \times S^1}_{n}$ and describe a pre–base and base of its topology.

2. Continuous mappings into product of spaces. We investigate the mappings $f : X \to \prod_{\alpha \in A} X_\alpha$ from a topological space X into a product. One can consider the components $f_\alpha : X \to X_\alpha$, $f_\alpha = p_\alpha f$ of the mapping f. There is a correspondence between every mapping f and a set $\{f_\alpha = p_\alpha f\}_{\alpha \in A}$ of mappings, i.e. its components. Conversely, any set of mappings $\{f_\alpha : X \to X_\alpha,\ \alpha \in A\}$ determines a mapping $f : X \to \prod_{\alpha \in A} X_\alpha$ in a unique way.

Thus, there exists a bijection between the set of mappings $f : X \to \prod_{\alpha \in A} X_\alpha$ and the set of the families of mappings $\{f_\alpha\}_{\alpha \in A}$.

Theorem 3. *A mapping f is continuous iff the mapping f_α is continuous for each $\alpha \in A$.*

Proof. Let all f_α be continuous. We shall show that f is continuous. It is sufficient to show that $f^{-1}(U)$ is open in X for any U from the base of the product topology on $\prod_{\alpha \in A} X_\alpha$. Let

$$U = p_{\alpha_1}^{-1}(U_{\alpha_1}) \cap p_{\alpha_2}^{-1}(U_{\alpha_2}) \cap \ldots \cap p_{\alpha_n}^{-1}(U_{\alpha_n}).$$

Then

$$f^{-1}(U) = \{x \in X : f_\alpha(x) \in X_\alpha,\quad \alpha \neq \alpha_1, \ldots, \alpha_n,$$
$$f_{\alpha_i}(x) \in U_{\alpha_i},\quad i = 1, 2, \ldots, n\} =$$
$$= \left[\bigcap_{\substack{\alpha \\ \alpha \neq \alpha_1, \ldots, \alpha_n}} f_{\alpha_1}^{-1}(X_\alpha) \right] \cap f_{\alpha_1}^{-1}(U_{\alpha_1}) \cap \ldots \cap f_{\alpha_n}^{-1}(U_{\alpha_n}) =$$
$$= X \cap V_1 \cap \ldots \cap V_n,$$

where $V_i = f_{\alpha_i}^{-1}(U_{\alpha_i})$ is an open set in X owing to the continuity of f_{α_i}. Therefore, $f^{-1}(U)$ is open in X. The proof of the opposite is left to the reader.   □

Consider now a mapping $f : \prod_{\alpha \in A} X_\alpha \to X$ which associates to each family $\{x(\alpha)\}_{\alpha \in A}$ a corresponding element from X.

Exercise 8°. Verify that if $A = \{1, 2, \ldots, n\}$ and $X_\alpha = X = \mathbf{R}^1$ for any $\alpha \in A$, then the mapping f is a numerical function in n arguments.

In the general case, the mapping f can be considered as a generalization of a numerical function in n arguments, assuming that it depends on the variables $x(\alpha) \in X_\alpha$. By fixing all the values $x(\alpha)$ with exeption of $x(\alpha_0)$, we obtain a function in one argument which runs over X_{α_0}. We make these ideas more precise.

Consider the subspace X'_{α_0} of the product $\prod_{\alpha \in A} X_\alpha$ consisting of all functions x which takes the value $x(\alpha) = y_\alpha$, $\alpha \neq \alpha_0$, where $y_\alpha \in X_\alpha$ is a fixed element.

Exercise 9°. Verify that X'_{α_0} is homeomorphic to X_{α_0}.

Let $\pi_{\alpha_0} : X_{\alpha_0} \to X'_{\alpha_0}$ be the natural homeomorphism (depending on the fixed y_α, $\alpha \neq \alpha_0$), and $f|_{X'_{\alpha_0}}$ the restriction of f to X'_{α_0}. The diagram

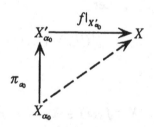

can be naturaly completed to a commutative one by taking the composition of the two given mappings (the dotted arrow), which we denote by $f_{\alpha_0}^{\{y_\alpha\}}$. This product characterizes the dependence of f on the argument $x(\alpha_0) \in X_{\alpha_0}$ for the given values y_α of the remaining values $x(\alpha)$.

Exercise 10°. Verify that if f is continuous, then the mapping $f_{\alpha_0}^{\{y_\alpha\}} : X_{\alpha_0} \to X$ is continuous for all $a_0 \in A$, $y_\alpha \in X_\alpha$, for $\alpha \neq \alpha_0$. The inverse statement is not true; give an example.

Consider another case of a mapping of products of topological spaces. Let $f_\alpha : X_\alpha \to Y_\alpha$, $\alpha \in A$ be a family of mappings of topological spaces. The mapping $\prod_{\alpha \in A} f_\alpha : \prod_{\alpha \in A} X_\alpha \to \prod_{\alpha \in A} Y_\alpha$ is determined naturally, if to each function $x \in \prod_{\alpha \in A} X_\alpha$ is associated the function $y \in \prod_{\alpha \in A} Y_\alpha$ according to the rule $y(\alpha) = f_\alpha(x(\alpha))$. This mapping is called the *product of the mappings* f_α. In the case A is finite, $A = \{1, 2, \ldots, n\}$, the product of the mappings $f_1, f_2, \ldots, f_n$ is often denoted by

$$f_1 \times f_2 \times \ldots f_n : X_1 \times X_2 \times \ldots \times X_n \to Y_1 \times Y_2 \times \ldots \times Y_n.$$

Exercise 11°. Prove that $\prod_{\alpha \in A} f_\alpha$ is continuous iff f_α is continuous for each $x \in A$.

Exercise 12°. The *graph of a mapping* $f : X \to Y$ is the subset $\Gamma_f \subset X \times Y$ of the form $\Gamma_f = \{(x, y) : x \in X, y = f(x)\}$. Verify that
  (1) Γ_f is the image of the mapping $\hat{f} : X \to X \times Y$, $\hat{f}(x) = (x, f(x))$;
  (2) (f is continuous) $\Leftrightarrow$ ($\hat{f}$ is continuous);
  (3) (f is continuous) $\Leftrightarrow$ (Γ_f is closed).

Exercise 13°. Let R be an equivalence relation on the topological space X. Consider the subspace R of the product $X \times X$ consisting of all pairs (x, y) of equivalent points $x, y \in X$. Show that (1) if X/R is Hausdorff, then the set R is closed; (2) if the projection $\pi : X \to X/R$ is open and the set R is closed, then X/R is a Hausdorff space.

Exercise 14°. Show that the product of Hausdorff spaces is also a Hausdorff space.

Exercise 15°. Show that a space X is Hausdorff iff the diagonal $\Delta = \{(x, x)\}$ is closed in $X \times X$.

§ 10. Connectedness of topological spaces

1. The concept of connectedness of a topological space. The concept of connectedness generalizes the intuitive idea of the "wholeness" or "non–separability" of a geometrical figure, and the concept of a disconnected space generalizes that of the negation of wholeness, i.e. separability. These concepts can be given a strict definition within the theory of topological spaces and are investigated in detail in this section.

Definition 1. The sets A and B are said to be *separated* from each other if $\bar{A} \cap B = A \cap \bar{B} = \emptyset$.

For instance, if $X = \mathbf{R}^1$, $A = (a, b)$, $B = (b, c)$ are intervals, $a < b < c$, then A and B are separated, but if $A = (a, b]$, $B = (b, c)$, then A and B are not separated $(A \cap \bar{B}) = \{b\}$).

Definition 2. A space X is said to be *disconnected* if it can be represented as the union of two nonempty separated sets.

A space not satisfying the condition mentioned is called *connected*. Thus, it is impossible to represent a connected space as the union of two nonempty separated sets.

We can talk about the connectedness (disconnectedness) of a subset A of a topological space X if we consider A as a topological space with the topology induced from X.

The simplest examples of connected spaces are: (1) a one–point space $X = \{*\}$; (2) an arbitrary set X with the trivial topology τ_0. The simplest example of a disconnected space is a two–point space X with the discrete topology τ_1 (verify!).

We shall present another definition of a disconnected space which is often used.

Definition 3. A topological space X is said to be *disconnected* if it can be represented as the union of two nonempty, non–intersecting and open sets.

Note that two mutually complementary open (closed) sets are simultaneously closed (respectively, open).

We shall prove the equivalence of definitions 2 and 3.

(1) Let X be disconnected in terms of definition 2. Then we have a decomposition $X = A \cup B$, where $\bar{A} \cap B = \emptyset$, $A \cap \bar{B} = \emptyset$, and A, B are nonempty. Therefore, $\bar{A} \subset X \backslash B$, $\bar{B} \subset X \backslash A$, i.e. $\bar{A} = A$, $\bar{B} = B$ which means that A

and B are closed. However, $A = X \backslash B$, $B = X \backslash A$, thus A, B are open and X is disconnected in terms of definition 3.

(2) Conversely, let X be disconnected in terms of definition 3. Then $X = A \cup B$; A, B are nonempty and open, $A \cap B = \emptyset$. Obviously, A and B are closed. Hence, $\bar{A} \cap B = \emptyset$ since $A = \bar{A}$; and $\bar{B} \cap A = \emptyset$ because $\bar{B} = B$.   □

The following theorem gives an important example of a connected space.

Theorem 1. *Any interval $[a, b]$ of the numerical line $\mathbf{R}^1$ is connected.*

Proof. Consider the topological space $X = [a, b]$ with its topology induced from $\mathbf{R}^1$. Assume that X is disconnected, i.e. $X = U \cup V$, $U \cap V = \emptyset$, where U, V are nonempty and open.

Let $a \in U$. Consider the semi–intervals $[a, x)$, where $x \in (a, b]$.

When x is near to a, then $[a, x) \subset U$ because U is open. The supremum of x such that $[a, x) \subset U$, we denote by a_* ($a_* \in X$); it is clear that $a_* \neq b$.

If $a_* \in U$ then due to the openness of U the points which are near to a_* (from the left and right) also lie in U, which contradicts the definition of a_*. Consequently, $a_* \notin U$. If $a_* \in V$ then because of the openness of V, the points near to a_* also lie in V. Therefore, $[a, a_* - \epsilon) \cap V \neq \emptyset$, for small $\epsilon > 0$, which contradicts the definition of a_*. Therefore, $a_* \notin V$. Thus, $a_* \notin U \cup V$, and this contradicts the assumption that $X = U \cup V$.   □

Now we can establish the connectedness of some more general spaces.

Theorem 2. *Any convex set $T \subset \mathbf{R}^n$ is connected.*

Proof. Let $T = U \cup V$, U, V be nonempty, non–intersecting open sets. Let $[a, b] = X$ be the interval joining a point $a \in U$ and $b \in V$. Then $U_X = X \cap U$ and $V_X = X \cap V$ are nonempty, non–intersecting open sets in X and $X = U_X \cup V_X$, a contradiciton with the connectedness of the interval X.   □

Corollary. *The space $\mathbf{R}^n$ and the discs $D_r^n(x_0)$, $\bar{D}_r^n(x_0)$ are connected.*

As an example of a disconnected space, consider the set of rational numbers $\mathbf{Q} = \{p/q\}$ with a topology induced from $\mathbf{R}^1$. Let $\alpha \in \mathbf{R}^1$ be an arbitrary irrational number. Then the sets

$$U_\alpha = \{x : x \in Q, \quad x < \alpha\}, \quad V_\alpha = \{x : x \in Q, \quad x > \alpha\}$$

are nonempty, open and do not intersect. Thus, $Q = U_\alpha \cup V_\alpha$, which means that Q is disconnected.

Exercise 1°. Prove that the set of all irrational numbers is disconnected.

Exercise 2°. (a) Show that the closure $\bar{A}$ of a subset A of a topological space is connected, if A is connected; (b) show that in a space with the discrete topology, any set, with an exeption of one–point sets, is disconnected.

2. **Properties of connected spaces.** To start with, note that connectedness (disconnectedness) is a topological property, i.e. it is preserved under homeomorphisms. Indeed, this follows from the fact that the separability of sets is preserved under homeomorphisms.

More generally, connectedness is preserved under continuous mappings.

Theorem 3. *Let* $f : X \to Y$ *be a continuous mapping of topological spaces. If* X *is connected then* $f(X)$ *is connected in* Y.

Proof. Assume the contrary: $f(X) = U_1 \cup V_1$, where $U_1 \cap V_1 = \emptyset$, U_1, V_1 being open in $f(X)$, $U_1 \neq \emptyset$, $V_1 \neq \emptyset$. The fact that U_1, V_1 are open in $f(X)$ implies that there exist sets U, V open in Y and such that $U \cap f(X) = U_1$, $V \cap f(X) = V_1$. Obviously, $X = f^{-1}(U_1) \cup f^{-1}(V_1)$, $f^{-1}(U_1) \cap f^{-1}(V_1) = \emptyset$ and $f^{-1}(U_1) \neq \emptyset$, $f^{-1}(V_1) \neq \emptyset$. Moreover, the sets $f^{-1}(U_1)$, $f^{-1}(V_1)$ are open because $f^{-1}(U_1) = f^{-1}(U)$, $f^{-1}(V_1) = f^{-1}(V)$ and f is continuous. Thus, X is disconnected; a contradiction to the assumption. $\square$

Exercise 3°. (a) Show that the graph Γ_f of a continuous mapping f from a connected space is connected.

(b) From (a) deduce the theorem that the numerical continuous function $f : [a, b] \to R^1$ which takes values of opposite signs at the ends of the interval $[a, b]$, has a zero in the interval (a, b), i.e. $\xi : f(\xi) = 0$.

Statement (b) of exercise 3 is the classical Bolzano–Cauchy theorem proved in analysis. This theorem is closely related to a more general classical intermediate–value theorem, viz. if a numerical function $f(x)$ is continuous on the interval $[a, b]$, $f(a) \neq f(b)$ and the number C is enclosed between the numbers $f(a)$ and $f(b)$, then there exists a point $c \in [a, b]$ such that $f(c) = C$. This theorem also follows from theorem 3. Indeed, the intermediate–value theorem is equivalent to the statement that the intersection of the graph Γ_f of the numerical function $f(x)$ with the line $y = C$ in the plane $\mathbf{R}^2$, is nonempty; this follows from the connectedness of the graph Γ_f and the choice of the number C.

We could prove the intermediate–value theorem without refering to the graph of the mapping f, by using as a basis the connectedness (in the plane $\mathbf{R}^1$) of the image $f([a, b])$ and the property of connected sets in $\mathbf{R}^1$ to contain all the intermediate points together with any two points (verify!).

Exercise 4°. Prove, that the circle S^1 is connected.

Hint. Consider the mapping $[0,1] \to S^1$ given by the formulas $x = \cos 2\pi t$, $y = \sin 2\pi t$.

The following theorem is obvious.

Theorem 4. *A space X is connected if any two of its points can be "joined" by a certain connected subset (i.e. lie in a connected subset).*

Proof. Assume the contrary. Write X as the union $X = U \cap V$ of nonempty, non–intersecting open sets U, V. Let $u_0 \in U$, $v_0 \in V$ be two points and $L \subset X$ a connected set containing u_0, v_0. Put $U_1 = U \cap L$, $V_1 = V \cap L$. The sets U_1, V_1 are nonempty and open in L, moreover, $L = U_1 \cup V_1$, $U_1 \cap V_1 = \emptyset$ and this contradicts the assumption. $\square$

Exercise 5°. Verify that: (a) $A \cup B$ is connected if A, $B \subset X$ are connected sets in X, and $A \cap B \neq \emptyset$; (b) $A \cup B \cup C$ is connected if A, B, $C \subset X$ are connected and $A \cap B \neq \emptyset$, $B \cap C \neq \emptyset$.

From exercise 5° follows, for instance, the connectedness of the sphere S^n, $n \geq 1$. Indeed, S^n consists of two closed semispheres $\bar{S}^n_+$, $\bar{S}^n_-$ which intersect along the equatorial sphere S^{n-1}, and each of these semispheres is connected as the continuous image of a disc (see § 2).

We shall establish a more general criterion of connectedness.

Theorem 5. *Let us have in X a family of connected sets $\{A_\alpha\}$ which are pairwise unseparated. Then the set $C = \bigcup_\alpha A_\alpha$ is connected in X.*

Proof. Assume the opposite. Let $C = D_1 \cup D_2$, $D_1 \cap D_2 = \emptyset$, D_1, D_2 nonempty and closed in C. Because of the connectedness of the sets A_α, each A_α is contained either in D_1 or D_2; and since D_1, D_2 are nonempty, there exist sets A_{α_1}, $A_{\alpha_2} \in \{A_\alpha\}$ such that $A_{\alpha_1} \subset D_1$, $A_{\alpha_2} \subset D_2$. Due to the closedness of the sets D_1, D_2 in C, the closures of the sets A_{α_1}, A_{α_2} in C are contained in D_1, D_2, respectively, and this is equivalent to the inclusion relations $\bar{A}_{\alpha_1} \cap C \subset D_1$, $\bar{A}_{\alpha_2} \cap C \subset D_2$ (here $\bar{A}_{\alpha_1}$, $\bar{A}_{\alpha_2}$ are the closures of the sets A_{α_1}, A_{α_2} in X). Therefore $(\bar{A}_{\alpha_1} \cap C) \cap A_{\alpha_2} = \emptyset$, $A_{\alpha_1} \cap (\bar{A}_{\alpha_2} \cap C) = \emptyset$. However, $(\bar{A}_{\alpha_1} \cap C) \cap A_{\alpha_2} = \bar{A}_{\alpha_1} \cap (C \cap A_{\alpha_2}) = \bar{A}_{\alpha_1} \cap A_{\alpha_2}$, $A_{\alpha_1} \cap (\bar{A}_{\alpha_2} \cap C) = (A_{\alpha_1} \cap C) \cap \bar{A}_{\alpha_2} = A_{\alpha_1} \cap \bar{A}_{\alpha_2}$. Consequently, $\bar{A}_{\alpha_1} \cap A_{\alpha_2} = \emptyset$, $A_{\alpha_1} \cap \bar{A}_{\alpha_2} = \emptyset$ which contradicts the non–separability of the sets A_{α_1}, A_{α_2}. $\square$

The condition of theorem 4 is satisfied, in particular, by a special class of spaces which are said to be pathwise connected (linearly connected). In order to describe them we introduce the concept of path in X.

Definition 4. The continuous mapping $S : [0,1] \to X$, $S(0) = a$, $S(1) = b$. is called a *path* connecting two points a, b of a topological space X

Exercise 6°. Verify that the image $S(I)$ of the interval $I = [0,1]$ is a connected set joining the points a and b.

Definition 5. A topological space X is said to be *pathwise connected* if any two points in it can be connected by a path.

A closed surface is an example of a pathwise connected space (see § 4).

From theorem 4 it follows that a pathwise connected space is necessarily connected. That the converse is not true can be seen from the following example. Consider the following subset in $\mathbf{R}^2$:

$$X = [(0,0),(1,0)] \cup \bigcup_{n=1}^{\infty} [(\frac{1}{n},0),(\frac{1}{n},1)] \cup (0,1),$$

where $[P,Q]$ denotes the interval connecting the points P and Q in $\mathbf{R}^2$; X is connected but not linearly connected (the point $(0, 1)$ cannot be connected by a path with any other point from X).

Exercise 7°. Verify that convex sets in $\mathbf{R}^n$ and the spheres S^n, $n \geq 1$ are linearly connected.

Exercise 8°. Prove that if a set $A \subset X$ is connected then any set B such that $A \subset B \subset \bar{A}$ is also connected. Give examples.

Finally, consider the product of connected spaces.

Theorem 6. *The product $X \times Y$ of connected spaces is connected.*

Proof. Assume the contrary. Let $X \times Y = U \cup V$, where U, V are nonempty, non–intersecting and open sets. Let $(x_0, y_0) \in U$. The set $x_0 \times Y$ is homeomorphic to Y and, consequently, connected; therefore, as it intersects U in the point (x_0, y_0), it lies wholly in U. The sets $X \times y$, $y \in Y$, are connected and intersect the set $x_0 \times Y$, and, consequently, U. Therefore they lie wholly in U. Thus, $\bigcup_{y \in Y}(X \times y) = X \times Y \subset U$. Therefore, $V = \emptyset$. We have arrived at a contradiction proving the theorem. $\qquad \square$

Exercise 9°. Prove theorem 6 for the product of n connected spaces ($n > 2$).

Exercise 10°. Prove the connectedness of the Tikhonov product $\prod_{\alpha \in A} X_\alpha = Y$ of connected spaces X_α.

Hint. Consider the set $\mathbf{R}$ of the points of the product that can be joined to a fixed point by connected sets, and verify that $\bar{R}=Y$.

3. Connected components. If a space is disconnected then it is natural to try to decompose it into connected pieces. We describe this decomposition. Let $x \in X$ be a point in a topological space X. Consider the largest connected set containing the point x : $L_x = \cup A_x$, where the A_x are all connected sets containing the point x. The set L_x is connected by theorem 5 and it is closed since the closure $\bar{L}_x$ of the connected set L_x is connected (exercise 2°), and, therefore, $\bar{L}_x \subset L_x$, i.e. $\bar{L}_x = L_x$.

Definition 6. The set L_x is called the *connected component* of the point x in a topological space X.

Let x, $y \in X$, $x \neq y$. Consider the sets L_x, L_y. Due to their connectedness and maximality, there are two possibilities: either (1) $L_x = L_y$, or (2) $L_x \cap L_y = \emptyset$. In the second case, L_x is separated from L_y, since $L_x \cap \bar{L}_y = \emptyset$, $L_y \cap \bar{L}_x = \emptyset$. In the evident equality $X = \cup L_x$, where the union is taken over all $x \in X$, we throw out repeated components.

Thus, the following theorem has been proved.

Theorem 7. *Any topological space can be represented as the union of its connected components which are closed and non–intersecting.*

Note. Generally speaking, connected components are not necessarily open (give examples!).

Exercise 11°. Verify that if a space has a finite number of connected components, then these components are open.

Exercise 12°. Verify that the number of connected components (being considered, in general case, as the cardinality of the correspondent set) of a space X is a topological characteristic of the space.

§ 11. Countability and separability axioms

Topological spaces encountered in various mathematical problems often possess additional properties. A number of properties are expressed by the system of the so–called countability and separability axioms.

1. Countability axioms. In § 1 the concept of a base for the topology has been introduced. While investigating topological spaces, it becomes clear, that spaces possessing a countable base, i.e. a base consisting of at most a countable number of sets, have a number of useful properties. In this connection, the following definition is introduced.

Definition 1. A topological space (X, τ) satisfies the *second countability* *axiom*, if its topology τ possesses a countable base.

Example 1. The space $\mathbf{R}^n$ satisfies the second countability axiom.

It is interesting to compare the spaces satisfying the second countability axiom and the separable spaces.

Theorem 1. *A topological space satisfiying the second countability axiom is separable.*

Proof. Let X be a topological space and $B = \{V_n\}$ be a countable base for its topology. In each of the sets $V_n \in B$, choose an element $a_n \in V_n$ and consider the set $A = \{a_1, \ldots, a_n, \ldots\}$. We shall show that $X = \bar{A}$. Since $\bar{A} = A \cup A'$, it is sufficient to show that any point from the set $X \backslash A$ is a limit point of the set A. Let x be an arbitrary point from $X \backslash A$, and $U(x)$ some neighbourhood of the point. Then there exists a set $V_k \in B$ such that $x \in V_k$, and $V_k \subset U(x)$; therefore, $a_k \in U(x)$; moreover $a_k \neq x$. Thus, $x \in A'$. $\quad\square$

The opposite statement, generally speaking, is not true, i.e. a separable space does not necessarily satisfy the second countability axiom as in the following example.

Example 2. Consider an uncountable set X, the topology of which consists of the complements to all possible finite subsets of the set X, the whole X and the empty set. (Verify that this system of subsets really forms a topology!). In this space, any infinite subset is dense since it intersects each open set. This implies the separability of X. $\quad\Diamond$

On the other hand, let X have a countable base. Then if $x \in X$ is a given point, the intersection of all open sets containing x is equal to $\{x\}$. Consequently, the countable intersection of the base elements containing x is also equal to $\{x\}$. But then the complement $X \backslash \{x\}$ is the union of at most a countable number of finite sets, and, thus, is at most countable. This is a contradiction to the assumption that X is uncountable.

It is important to note that for metric spaces the statement converse to the statement of theorem 1 is true, and the following theorem holds.

Theorem 2. *Any separable metric space* (X, ρ) *satisfies the second countability axiom.*

Proof. Let $A = \{a_1, a_2, \dots, a_n, \dots\}$ be an at most countable and everywhere dense set in X. As a base for the topology of the space X take the collection of open sets

$$B = \{V_{n,k} = D_{1/k}(a_n); \quad k = 1, 2, \dots\}.$$

It is easy to verify that it is indeed a base.

Indeed, due to the separability of X, for any $x \in X$ and a sufficiently small $\epsilon > 0$, there is an element $a_n \in A$ such that $a_n \in D_{\epsilon/3}(x)$; moreover, there is a number k such that $x \in V_{n,k} \subset D_\epsilon(x)$ (it is suficient to take k in such a way that $\epsilon/3 \le 1/k \le 2\epsilon/3$). Since any open set in X is representable as a union of balls, it can also be represented as a union of sets $V_{n,k}$ from B. □

In order to formulate the following statement, we need the concept of covering which was mentioned in § 1, and the concept of subcovering, i.e. a subset of covering which itself is a covering.

Theorem 3. (Lindelöf's theorem) *If a topological space* X *satisfies the second countability axiom, then every open covering* $\{U_\alpha\}$ *contains an at most countable subcovering.*

Proof. Let B be a countable base for the topology on X. Since any element of the covering $\{U_\alpha\}$ is a union of sets from B, a subfamily C can be singled out in B, which also covers X, and such that each element of C is contained in some element of the family $\{U_\alpha\}$. Then, by choosing for each element of the covering C a set $\{U_\alpha\}$ containing it, we obtain an at most countable subcovering of the covering $\{U_\alpha\}$. □

Besides the base for a topology as introduced in § 1, there exists the important concept of a base for the neighbourhood system of a point x in a topological space X.

Definition 2. A family $B(x) = \{V(x)\}$ of neighbourhoods of a point x is called a *base for the neighbourhood system* of the point x, if in each neighbourhood of the point x there is a neighbourhood from this family.

The family of all open neighbourhoods of a point, obviously, is a base for the neighbourhood system of this point.

Example 3. Let (X, ρ) be a metric space. Then

$$B(x) = \{V_k(x) = D_{1/k}(x)\}_{k=1}^{\infty}$$

is a base for the neighbourhood system of the point x. Indeed, any neighbourhood of a point x includes a spherical neigbourhood. For any spherical neighbourhood $D_\epsilon(x)$ we can choose a number k such that $1/k < \epsilon$; then $V_k(x) \subset D_\epsilon(x)$.

Definition 3. A topological space X satisfies the *first countability axiom* if the neighbourhood system of any of its points admits a countable base, i.e. a base consisting of at most a countable number of neighbourhoods.

Example 4. A metric space satisfies the first countability axiom.

Example 5. The space of continuous functions $C_{[0,1]}$ satisfies the first countability axiom.

Does the space $C_{[0,1]}$ satisfy the second countability axiom? The positive answer follows from the fact that $C_{[0,1]}$ is separable and from theorem 2 of this section.

The separability of the space $C_{[0,1]}$ follows from the Weierstrass theorem stating that any continuous function on the interval $[0, 1]$ can be uniformly approximated by polynomials to any degree of accuracy. Thus, a countable and everywhere dense set A in $C_{[0,1]}$ consists of the set of all polynomials $\{P_n(t)\}$ with rational coefficients.

Exercise 1°. Verify that a topological space satisfying the second countability axiom satisfies the first one as well.

The opposite is not true, which can be seen from the following example.

Example 6. Any uncountable space X with the discrete topology satisfies the first countability axiom. Indeed, any point $x \in X$ has a base for the neighbourhood system consisting of a single neighourhood $V = \{x\}$. But this space does not satisfy the second countability axiom, otherwise the coverings of the space formed by one–point sets $\{x\}$, according to theorem 3, would contain an at most countable subcovering, what is impossible.

Thus, the second countability axiom is a stronger condition on a topological space than the first countability axiom.

2. Separability properties of a space. Some important topological properties of spaces are characterized by the separability axioms. These axioms allow us to limit the class of the spaces investigated in order to consider deeper properties.

In order to formulate the axioms, the concept of an open neighbourhood of a set is needed.

Definition 4. An open set U containing A is called an *open neighbourhood* of the set A in a topological space X.

We shall present the main *separability axioms* T_0–T_4.

Axiom T_0 (Kolmogorov's axiom). At least one of any two different points in a topological space has a neighbourhood, which does not contain the other point.

Axiom T_1. Every point of any pair of different points in a topological space has a neighbourhood which does not contain the nother point.

Axiom T_2 (Hausdorff's axiom). Any two different points x, y in a topological space have neighbourhoods $U(x)$, $U(y)$ such that $U(x) \cap U(y) = \emptyset$.

Axiom T_3. For any point x of a topological space and any closed set F which does not contain x, there exist neighbourhoods $U(x)$ of the point x and $U(F)$ of the set F such that $U(x) \cap U(F) = \emptyset$.

Axiom T_4. Any two closed non–intersecting sets F_1, F_2 of a topological space have open neighbourhoods $U(F_1)$, $U(F_2)$ such that $U(F_1) \cap U(F_2) = \emptyset$.

We should stress that among axioms T_0–T_2 each subsequent axiom is a stronger condition than the previous one. The same is true for axioms T_2–T_4 provided axiom T_1 is satisfied, since T_1 does not follow from either T_3 or T_4.

A topological space X is called a T_i–*space* ($i = 0, 1, 2, 3, 4$), if it satisfies axiom T_i.

A topological space X is called a *regular space* if both axioms T_1 and T_3 are satisfied.

A topological space X is called a *normal space* if axioms T_1 and T_4 are fulfilled simultaneously.

T_0–spaces are also called *Kolmogorov* spaces, and T_2–spaces are called *Hausdorff* spaces (see also § 1).

An example of a topological space which is not a T_0–space can serve the space with the trivial topology (with more than one point). The other topological spaces considered above are T_0–spaces. It should be noted that topological spaces which are not T_0–spaces, are of no interest to investigate.

We shall present several examples.

Example 7. Let $\mathbf{R}^1$ be the real line with the topology whose base is formed by rays of the form $a < x < +\infty$. It is not difficult to show that the space $\mathbf{R}^1$ with this topology, satisfies the axiom T_0, but not the axiom T_1.

Example 8. Consider the interval $[0, 1]$ equipped with the topology whose open sets are the empty set and all the sets obtained from $[0, 1]$ by excluding at most countable number of points. The topological space obtained satisfies the axiom T_1, but not the axiom T_2.

Example 9. In the interval $[0, 1]$, the neighbourhoods of an arbitrary point, except zero, are ordinary neighbourhoods, and the neighbourhoods of zero are all the possible half–intervals $[0, \alpha)$ with the points $1/n$, $n = 1, 2, \ldots$ removed. It is easy to see that this topological space is Hausdorff but not regular, because the closed set $F = \{1/n\colon n = 1, 2, \ldots\}$ and the point zero, which is not in F, are unseparable in the sense of the axiom T_3.

We do not present examples of regular, but not normal spaces. Such examples are not trivial; this is related to the fact that the difference between regular and normal spaces is quite slight, which is demonstrated by the following theorem.

Theorem 4. (Tikhonov). *Any regular space satisfying the second countability* *axiom is normal.*

The proof of this theorem is not given here.

Topological spaces which do not satisfy the axiom T_1 are poorly organised from the point of view of analysis. A single point set in such a topology may not be closed, and a finite set may have limit points (present examples!). In a T_1–space such situations do not occur.

Exercise 2°. Show that a topological space is a T_1–space iff each of its single point sets is closed.

Theorem 5. *Each neighbourhood of a limit point of the set A in a T_1–space X contains an infinite number of points from A.*

Proof. Let x be a limit point of the set A, and $U(x)$ a neighbourhood of it. Assume that the set $U(x) \cap A$ is finite. Then the set $B = (U(x) \cap A) \backslash \{x\}$ is closed as the union of a finite number of closed single point sets, and, consequently, the set $U_1 = U(x) \backslash B$ is open. Thus, U_1 is a neighbourhood of the point x and $U_1 \cap A = \{x\}$, which contradicts the fact that x is the limit point of A. $\qquad\qquad\square$

Corollary. *A finite subset of a T_1–space has no limit points.*

We note that the class of normal spaces is rather extensive, and that it includes, for example, all metric spaces.

Theorem 6. *Any metric space is normal.*

Proof. Every metric space (X, ρ) satisfies, obviously, the axiom T_2, and, therefore, also the axiom T_1 (as the non–intersecting neighbourhoods of different points x, $y \in X$ one can take the balls $D_{r/2}(x)$, $D_{r/2}(y)$, where $r = \rho(x, y)$). We shall show that in a metric space the axiom T_4 is fulfilled. Let A, B be the arbitrary closed subsets in X. For any subset $C \subset X$ we denote by $\rho(x, C) = \inf_{y \in C} \{\rho(x, y)\}$ ($\rho(x, C)$ is said to be a distance between the point x and the set C). Let $U_1 = \{x \in X \colon \rho(x, A) < \rho(x, B)\}$, $U_2 = \{x \in X \colon \rho(x, A) > \rho(x, B)\}$. For a closed set C, $\rho(x, C) > 0$, if $x \notin C$ (prove!). Therefore, the sets U_1, U_2 contain A, B, respectively, and do not intersec-

t. The mappings $f(x) = \rho(x, A)$: $X \rightarrow \mathbf{R}^1$, $g(x) = \rho(x, B)$: $X \rightarrow \mathbf{R}^1$ are continuous (prove!), and, consequently, the mapping $f - g : X \rightarrow \mathbf{R}^1$ is continuous. Therefore, the sets U_1, U_2 are open (as the preimages of open sets $(-\infty, 0)$, $(0, +\infty)$ under the continuous mapping $f - g$). Thus, U_1, U_2 are the non–intersecting open neighbourhoods of the sets A, B. □

Exercise 3°. Show that a subspace of the T_i–space ($i = 0, 1, 2, 3$) is also the T_i–space.

Exercise 4°. Show that in a T_1–space, for any subset A, the inclusion $(A')' \subset A'$ holds.

Exercise 5°. Verify that a closed surface (see, Ch. 2, § 4) is a Hausdorff space.

3. Hausdorff spaces with the first countability axiom. For such spaces, the concept of a convergent sequence and of its limit is naturally defined, after which the definition of operations on sets and the concept of a continuous mapping copy the definitions for metric spaces.

Definition 5. A sequence $\{x_n\}$ of the points $x_n \in X$, $n = 1, 2, \ldots$ is said to *converge to a point* $x_0 \in X$ if for any neighbourhood $W_p(x_0)$ there exists a natural $N = N(x_0, p)$ such that for all $n > N$ we have $x_n \in W_p(x_0)$.

It is useful to note that the choice of N can depend both on p and x_0.

Exercise 6°. Prove that a sequence in a Hausdorff space can only converge to a unique point.

If the sequence $\{x_n\}$ converges to a point x_0, then x_0 is said to be the limit of the sequence $\{x_n\}$, and this is written $\lim x_n = x_0$, or $x_n \rightarrow x_0$.

Exercise 7°. Let $f : X \rightarrow Y$ be a mapping of Hausdorff spaces with the first countability axiom. Prove that the condition $\lim f(x_n) = f(x_0)$, for any sequence $\{x_n\}$, $x_n \rightarrow x_0$, is equivalent to the continuity of the mapping f at the point x_0.

Using the concept of a convergent sequence we can give the definitions of a limit point of a set $A \subset X$, of the derivative set A', of the boundary ∂A, and of the closure $\bar{A}$ in the same way as it was done for a metric space (see Subsection 1, § 7).

Exercise 8°. Verify that the definitions of the sets A', ∂A, $\bar{A}$ in a Hausdorff space with the first countability axiom considered above, are equivalent to the general definitions from § 6.

Exercise 9°. A mapping $f : X \to Y$ is called proper mapping if the preimage of any compact set from Y is compact in X. Show that if f is continuous, and X, Y are Hausdorff with the first countability axiom, then every proper mapping f is closed.

§ 12. Normal spaces and functional separability

1. An equivalent definition of normal spaces. The property of normal spaces formulated in the following lemma is quite often useful, and can be considered is an equivalent definition of a normal space.

Minor Uryson Lemma. *A T_1–space X is normal iff for any closed set $F \subset X$ and any of its open neighbourhoods U, there exists an open neighbourhood V of the set F such that $\bar{V} \subset U$.*

Proof. Let X be normal. Consider the two closed sets: F and $F_1 = X \backslash U$. Since the space is normal, there exist non–intersecting open neighbourhoods V and V_1 of the sets F and F_1. Then $V \subset X \backslash V_1$ and, consequently, $\bar{V} \subset X \backslash V_1$. However, $X \backslash V_1$ is closed, therefore $\overline{X \backslash V_1} = X \backslash V_1$. Thus, $\bar{V} \subset X \backslash V_1 \subset U$. Conversely, let the condition of the lemma be fulfilled, and let F_1, F_2 be non–intersecting closed sets. Consider the set $U_1 = X \backslash F_2$. Then $F_1 \subset U_1$, and, according to the condition, there exists an open neighbourhood V_1 of F_1 such that $\bar{V}_1 \subset U_1$. Setting $U_2 = X \backslash \bar{V}_1$, we obtain an open set U_2, $F_2 \subset U_2$, such that $V_1 \cap U_2 = \emptyset$. □

Corollary. *Two non–intersecting closed sets F_1, F_2 in a normal space X, possess open neighbourhoods U_1, U_2 such that $\bar{U}_1 \cap \bar{U}_2 = \emptyset$.*

Generally speaking, the normality of a subspace does not follow from the normality of the space. However, if any subspace in a normal space X is normal, then X is said to be a *hereditarily normal space*.

Exercise $1°$. Show that a metric space is hereditarily normal.

A condition for hereditary normality is given by the following theorem.

Theorem 1. (Uryson). *A space is hereditarily normal iff any two of separated sets in it possess non–intersecting open neighbourhoods.*

The proof of this theorem is not given here.

The image of a normal space under a continuous mapping is not necessarily normal. The simplest example is the identity mapping of the line $\mathbf{R}^1$ with the ordinary topology into the same line equipped by some non–Hausdorff, say, the trivial, topology. However, there exist sufficient conditions that the image of a normal space to be normal. For instance, the following statement is true.

Theorem 2. *Let X be a normal space, $f : X \to Y$ a continuous closed surjective mapping. Then the space Y is also normal.*

Proof. Let $A \subset Y$ be a closed subset. Set $A_1 = f^{-1}(A)$. Then the set A_1 is closed due to the continuity of f. Let U be an open neighbourhood of the set A in Y. Then the set $U_1 = f^{-1}(U)$ is open (because of the continuity of f) and contains A_1. Consequently, U_1 is an open neighbourhood of A_1, and according to the minor Uryson lemma, there exists an open neighbourhood V of the set A_1 such that $\bar{V} \subset U_1$.

We have the inclusions $A_1 \subset V \subset \bar{V} \subset U_1$. A closed surjective mapping is open, thus, $f(V)$ is open, and $f(\bar{V})$ is closed; moreover, we have the inclusions

$$A = f(A_1) \subset f(V) \subset f(\bar{V}) \subset f(U_1) = U,$$

from which the normality of Y can easily be perceived.   □

2. Functional separability. The Uryson theorems on extending numerical functions. The separability of a set was defined in terms of "neighbourhoods". Uryson introduced another concept of separability, so–called functional separability, which is highly convenient for studying normal spaces.

Definition. Two sets A, B in a topological space X are said to be *functionally separable* if there exists a continuous numerical function $\phi : X \to \mathbf{R}^1$ such that

$$\phi(x) = \begin{cases} 0, & \text{if } x \in A, \\ 1, & \text{if } x \in B, \end{cases}$$

and $0 \le \phi(x) \le 1$ at all points of X (Fig. 62).

The close relationship between the two concepts of separability can be clearly seen from the following simple fact.

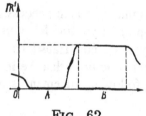

Lemma. *If two sets A and B are functionally separable in a topological space then they have non–intersecting open neighbourhoods.*

FIG. 62   The proof is left to the reader.

Thus, from the functional separability of any pair of closed non–intersecting sets of a T_1–space follows the normality of this space. It is interesting that the opposite statement is also valid!

Major Uryson lemma. *For any two closed, non–intersecting sets A, B of a normal space X, there exists a continuous function $\phi : X \to \mathbf{R}^1$ such that $\phi|_A \equiv 0$, $\phi|_B \equiv 1$ and $0 \le \phi(x) \le 1$ for any $x \in X$.*

Proof. Let A, B be arbitrary closed spaces in X, $A \cap B = \emptyset$. For each rational number of the form $r = k/2^n$, where $k = 0, 1, \ldots, 2^n$, we construct an open set $G(r)$ so that the following properties are fulfilled:
   (1) $A \subset G(0)$, $X \backslash B = G(1)$;
   (2) $\overline{G(r)} \subset G(r')$, if $r < r'$.
The existence of such a system of open sets will be proved with a help of induction on the index n. Let $n = 0$. Since X is normal, there exist non–intersecting open neighbourhoods $U(A)$, $U(B)$ of the sets A and B. Set $G(0) = U(A)$, $G(1) = X \backslash B$. Assume now that a system of sets $G(r)$ is

constructed for an index $n - 1$. We shall construct it for the index n. Since $2m/2^n = m/2^{n-1}$, it is sufficient to construct $G(r)$ for $r = k/2^n$ for an odd k.

Let $k = 2m + 1$. Then $(k + 1)/2^n = (m + 1)/2^{n-1}$, $(k - 1)/2^n = m/2^{n-1}$, and, consequently, according to the induction hypothesis, we already have the inclusion $\overline{G((k - 1)/2^n)} \subset G((k + 1)/2^n)$. Evidently, the sets $\overline{G((k - 1)/2^n)}$, $X \backslash G((k + 1)/2^n)$ are closed and do not intersect. Because of the normality of X, there exists an open neighbourhood V of the set $\overline{G((k - 1)/2^n)}$, which does not intersect a certain open neighbourhood of the set $X \backslash G((k + 1)/2^n)$. Set $G(k/2^n) = V$; it is clear that

$$\overline{G((k - 1)/2^n)} \subset G(k/2^n),$$
$$\overline{G(k/2^n)} \subset G((k + 1)/2^n).$$

The induction is complete.

Let us extend the family of sets $G(r)$, by putting

$$G(r) = \begin{cases} \emptyset, & \text{if } r < 0, \\ X, & \text{if } r > 1. \end{cases}$$

Now, define a function ϕ as follows: $\phi(x) = 0$, $x \in G(0)$ and $\phi(x) = \sup\{r : x \in X \backslash G(r)\}$. We shall prove the continuity of ϕ. To achieve this, we construct a neighbourhood $U_N(x_0)$ of the point x_0 for each $x_0 \in X$ and each $N > 0$, such that $|\phi(x_0) - \phi(x)| < 1/2^N$, $x \in U_N(x_0)$. Let r_0 (of the form $k/2^n$) be such that

(1) $$\phi(x_0) < r_0 < \phi(x_0) + 1/2^{N+1}.$$

Set $U_N(x_0) = G(r_0) \backslash \overline{G(r_0 - 1/2^N)}$. Then $x_0 \in U_N(x_0)$, since $r_0 > \phi(x_0)$ and $r_0 - 1/2^{N+1} < \phi(x_0)$. If $x \in U_N(x_0)$ then $x \in G(r_0)$, therefore $\phi(x) \leq r_0$. In addition,

$$x \in X \backslash \overline{G(r_0 - 1/2^N)} \subset X \backslash G(r_0 - 1/2^N),$$

therefore, $r_0 - 1/2^N \leq \phi(x)$. Thus,

(2) $$r_0 - 1/2^N \leq \phi(x) \leq r_0.$$

By comparing (1) and (2), we obtain

$$|\phi(x_0) - \phi(x)| < 1/2^N, \quad x \in U_N(x_0).$$

The latter means the continuity of ϕ.

According to the construction, it is clear that $\phi|_A \equiv 0$, $\phi|_B \equiv 1$ and $0 \leq \phi(x) \leq 1$. The function constructed is also called an *Uryson function*.   □

To apply this result, we shall consider the extension of a bounded function from a closed subset of a normal space to the whole space. Note first that the major Uryson lemma is equivalent to the statement about the existence of a continuous function $\phi_{a,b}(x)$, satisfying the conditions

$$\phi_{a,b}|_A \equiv a, \quad \phi_{a,b}|_B \equiv b, \quad a \leq \phi_{a,b}(x) \leq b, \quad x \in X,$$

where a, b ($a < b$) are arbitrary real numbers. Indeed, if $\phi(x)$ is an Uryson function then the function $\phi_{a,b}(x) = (b-a)\phi(x) + a$ does the job.

Theorem 3. (Tietze–Uryson). *For any bounded continuous function $\phi : A \to \mathbf{R}^1$ defined on a closed subset A of a normal space X, there exists a continuous function $\Phi : X \to \mathbf{R}^1$ such thatt $\Phi|_A \equiv \phi$ and $\sup_{(X)} |\Phi(x)| = \sup_{(A)} |\phi(x)|$.*

Proof. We shall construct the function Φ as the limit of a sequence of functions. Set $\phi_0 = \phi$ and

$$a_0 = \sup |\phi(x)|, \quad A_0 = \{x : \phi_0(x) \leq -a_0/3\},$$
$$B_0 = \{x : \phi_0(x) \geq a_0/3\}.$$

We can assume $a_0 > 0$. (Otherwise $\phi \equiv 0$ and we can take $\Phi \equiv 0$.) It is clear that the sets A_0, B_0 are closed and do not intersect. By the major Uryson lemma, there exists a continuous function $g_0 : X \to \mathbf{R}^1$ such that $|g_0(x)| \leq a_0/3$ and

$$g_0(x) = \begin{cases} -a_0/3, & \text{if } x \in A_0, \\ a_0/3, & \text{if } x \in B_0. \end{cases}$$

Now, define a function ϕ_1 on A by the equality $\phi_1 = \phi_0 - g_0$. Then the function ϕ_1 is continuous and $a_1 = \sup_{(A)} |\phi_1| \leq \frac{2}{3}a_0$. Similarly, by introducing

$$A_1 = \{x : \phi_1(x) \leq -a_1/3\}, \quad B_1 = \{x : \phi_1(x) \geq a_1/3\}$$

and taking an Uryson function g_1 such that $|g_1(x)| \leq a_1/3$ and

$$g_1(x) = \begin{cases} -a_1/3, & \text{if } x \in A_1, \\ a_1/3, & \text{if } x \in B_1, \end{cases}$$

we put $\phi_2 = \phi_1 - g_1$ and $a_2 = \sup_{(A)} |\phi_2| \leq \frac{2}{3}a_1$ on the set A.

Thus, we construct a sequence of functions $\phi_0 = \phi, \phi_1\phi_2, \ldots, \phi_n, \ldots$ that are continuous on A, and a sequence of functions $g_0, g_1, \ldots, g_n, \ldots$, that are continuous on X such that

$$\phi_{n+1} = \phi_n - g_n, \quad |g_n(x)| \leq a_n/3, \quad a_{n+1} \leq \frac{2}{3}a_n,$$

where $a_n = \sup_{(A)} |\phi_n(x)|$, $n = 0, 1, 2, \ldots$. Hence, we obtain that

$$|\phi_n(x)| \leq \left(\frac{2}{3}\right)^n a_0, \quad |g_n(x)| \leq \left(\frac{2}{3}\right)^n \frac{a_0}{3}.$$

Due to the latter inequality, the series $\sum_{n=0}^{\infty} g_n(x)$ converges absolutely and uniformly on X to a continuous function. Denoting this sum by $\Phi(x)$, we obtain the following estimation

$$|\Phi(x)| \leq \sum_{n=0}^{\infty} \left(\frac{2}{3}\right)^n \frac{a_0}{3} = a_0.$$

Let $x \in A$. Then the partial sum $S_n(x) = g_0(x) + \ldots + g_n(x)$ due to the construction of the function $\phi_{n+1}(x)$ is equal to $\phi_0(x) - \phi_n(x)$, and $\phi_n(x) \to 0$. Consequently, $\Phi(x) = \phi_0(x) = \phi(x)$ for every $x \in A$. $\qquad\square$

The Tietze–Uryson theorem can be generalized to the case of a mapping of the space X into an n–dimensional cube.

Corollary. *Any continuous mapping $\phi : A \to I^n$ of a closed subset A of a normal space X into an n–dimensional cube I^n can be extended to a continuous mapping $\Phi : X \to I^n$.*

Exercise. Prove the corollary.

Hint. Use a coordinate system in $\mathbf{R}^n$ and apply the Tietze–Uryson theorem to the components of the mapping ϕ.

§ 13. Compact, locally compact and paracompact spaces and their mappings

1. The concept of compact space. We turn now to the study of a quite important class of topological spaces that is characterized by a property of their open coverings. This property is an abstract (and convenient) analogue of the property of compactness of a numerical segment or an n–dimensional cube (ball) as known from analysis. Compact spaces and their mappings occur in many branches of mathematics.

To start with, we shall discuss some ideas related to coverings of topological spaces. Let $\sigma(A)$ be some system of subsets A of a set X. The union of all A from σ denote by $\tilde{\sigma}$ and call the *reach of the system* σ.

We shall extend the concept of a covering mentioned in § 1 after the definition of a base for a topology.

Definition 1. A system σ is called a *covering of a subspace Y* of a topological space X, if $\tilde{\sigma} \supseteq Y$.

In particular, σ is a covering of the space X if $\tilde{\sigma} = X$, and that agrees with the concept of covering used earlier in § 1.

Definition 2. It is said that a covering σ is a *refinement of a covering σ'* ($\sigma \succ \sigma'$), if each element of σ is contained in some element of the system σ'.

The refinement relation introduces a partial ordering on the set of all coverings of the space.

Coverings consisting of a finite or countable number of elements, are said to be *finite* or *countable*, respectively.

Definition 3. A covering σ of a space X is said to be *locally finite* if each point $x \in X$ has a neighbourhood which intersects with only a finite number of elements of σ.

Coverings consisting of open sets are of particular importance. These coverings are said to be *open coverings*.

Many important properties of spaces are related to the properties of open coverings. In connection with this, the following classes of spaces are singled out.

Definition 4. A topological space X is (A_1) *compact*, (A_2) *paracompact*, if for every open covering there is a refinement that is, respectively, (a_1) finite, (a_2) locally finite.

Exercise $1°$. Prove that an equivalent definition of (A_1) is obtained if we require that a covering of type (a_1) can be selected from any open covering of the space, and that a non–equivalent definition of (A_2) is obtained, if we require that a covering of type (a_2) can be selected from any open covering of the space.

Example 1. Let $X = [a, b] \subset \mathbf{R}^1$ be an interval and supply it with the topology induced from $\mathbf{R}^1$. The space X is compact, since, by the Heine–Borel theorem, a finite subcovering can be selected from any covering of X with intervals.

Example 2. Let $X = \mathbf{R}^1$. This is an example of a non–compact space. For instance, it is not possible to select a finite subcovering from the covering $\{(-n, n)\}_{n=1}^{\infty}$.

Analogous arguments show that the space $\mathbf{R}^n$ is non–compact as is any of its unbounded subsets. Hence, in particular, the requirement for a compact subset in $\mathbf{R}^n$ to be bounded is a necessary condition.

Example 3. The space $X = \mathbf{R}^1$ is paracompact. Indeed, let $\{U_\alpha\}$ be an open covering of $\mathbf{R}^1$. We have $\mathbf{R}^1 = \bigcup_{n=-\infty}^{+\infty} [n, n+1]$. Slightly extend each interval $[n, n+1]$ to the interval $(n - \epsilon, n+1 + \epsilon)$, and consider the covering $\{U_\alpha \cap (n - \epsilon, n+1+\epsilon)\}$ of the interval $[n, n+1]$. A finite covering $V_1^n, \ldots, V_{k_n}^n$ can be selected from it. The union of these subcoverings (for all n) produces a locally finite covering of $\mathbf{R}^1$ which is a refinement of $\{U_\alpha\}$.

If $Y \subset X$ is a subspace of a topological space X, then considering coverings of the space Y which are open in the topology inherited from X, we obtain, from definition 4, the concepts of a compact and paracompact subspace (one also often speaks of compact and paracompact sets Y in the space X). In a similar manner, we can consider coverings of a subspace Y which is open in X. In addition, it is useful to note that a closed set $Y \subset X$ inherits the properties

A_i, $i = 1, 2$ from the space X. Indeed, for any open covering $\sigma = \{V_\alpha\}$ of the space Y, where $V_\alpha = Y \cap U_\alpha$, and U_α is open in X, there exists an open covering $\sigma_* = \{U_\alpha, U_* = X \backslash Y\}$ of the space X. Now, we select a refinement $\widetilde{\sigma} \succ \sigma_*$ (of type a_i) of the space X. It is easy to obtain from the covering $\widetilde{\sigma}$ a covering $\widetilde{\sigma}_*$ of the subspace Y by intersecting the elements of $\widetilde{\sigma}$ with Y and discarding those contained in U_*. Obviously, $\widetilde{\sigma}_Y \succ \sigma$.

The following theorem is often applied in analysis.

Theorem 1. *Any infinite set $Z \subset X$ of a compact space X has a limit point in X.*

Proof. Assume the opposite, i.e. that $Z' = \emptyset$. Then $\bar{Z} = Z$, so Z is closed, and, consequently, compact. On the other hand, by assumption, each point $z \in Z$ is isolated in X. This implies that there exists an open neighbourhood $\Omega(z)$ in X with the condition $\Omega(z) \cap Z = \{z\}$. The open neighbourhoods $U(z) = \Omega(z) \cap Z$ in Z form an infinite covering of the space Z from which a finite subcovering cannot be selected. This contradicts the compactness of Z.
$\square$

The concept of compactness is narrowly related to the concept of closedness as the following statement demonstrates.

Theorem 2. *Let X be a compact subspace of a Hausdorff space Y. Then X is closed.*

Proof. Let $y \in Y \backslash X$. For any point $x \in X$, since Y is Hausdorff, there are open neighbourhoods $U_x(y)$, $U_y(x)$ of the points y, x such that $U_x(y) \cap U_y(x) = \emptyset$.

The system $\{U_y(x)\}_{x \in X}$ forms a covering of X. By the compactness of X, there is a finite subcovering $\{U_y(x_i)\}_{i=1}^k$. It is easy to see that the sets $U(X) = \bigcup_{i=1}^k U_y(x_i)$ and $\bigcap_{i=1}^k U_{x_i}(y) = U(y)$ are open and do not intersect. Thus, we have shown that in a Hausdorff space a compact set X and the point not in it, can be separated by non–intersecting neighbourhoods $U(X)$ and $U(y)$. Hence, the complement $Y \backslash X$ is open, and, therefore, X is closed.
$\square$

Now, we shall investigate the relation between the concepts of compactness and normality.

Theorem 3. *A compact Hausdorff space X is normal.*

Proof. First, we shall establish the axiom T_3 for X. Let $A \subset X$ be a closed subset, $x \in X \backslash A$. Because X is Hausdorff, for any $y \in A$, there exist neighbourhoods $U_x(y)$, $U_y(x)$ of the points x, y such that $U_x(y) \cap U_y(x) = \emptyset$. The system $\{U_x(y)\}_{y \in A}$ forms a covering of A; since A is compact, it is possible to select a finite subcovering $\sigma = \{U_x(y_i)\}_{i=1}^m$. Because $U_x(y_i)$ is included in the closed set $X \backslash U_{y_i}(x)$, $\bar{U}_x(y_i) \subset X \backslash U_{y_i}(x) \subset X \backslash \{x\}$, and, consequently, $\bigcup_{i=1}^m \bar{U}_x(y_i) \subset X \backslash \{x\}$. However, $\bigcup_{i=1}^m \bar{U}_x(y_i)$ is closed, therefore, $X \backslash \bigcup_{i=1}^m \bar{U}_x(y_i) = U_A(x)$ is an open neighbourhood of the point x. The union $\bigcup_{i=1}^m U_x(y_i) = V_x(A)$ is an open neighbourhood of the set A in X. It is evident that $V_x(A) \cap U_A(x) = \emptyset$, which shows that axiom T_3 holds for X. Since the axiom T_1 follows from the axiom T_2, X is regular.

Now we prove the normality of X. Let the sets A, B be closed in X, and $A \cap B = \emptyset$. Then for any point $x \in X$ there exists an open neighourhood $U(x)$ for which at least one of the conditions $\bar{U}(x) \cap A = \emptyset$, $\bar{U}(x) \cap B = \emptyset$ is true because X is regular. Consider the covering $\{U(x)\}_{x \in X}$ of the space X by such neighbourhoods. Select a finite subcovering $\{U_{\alpha_i}\}_{i=1}^m$ from it. For each U_{α_i}, at least one of the conditions $\bar{U}_{\alpha_i} \cap A = \emptyset$, $\bar{U}_{\alpha_i} \cap B = \emptyset$ is fulfilled. Let $P = \cup \bar{U}_{\alpha_i}$ be the union of those sets, for which $\bar{U}_{\alpha_i} \cap A = \emptyset$ and, similarly, $Q = \cup \bar{U}_{\alpha_i}$, where now α_i runs over those indices for which $\bar{U}_{\alpha_i} \cap B = \emptyset$. It is easy to see that the open sets $X \backslash P$, $X \backslash Q$ contain A, B, respectively, and do not intersect. Normality of the space X is proved. $\square$

Another definition of a compact space formulated only in terms of closed sets is often useful. First we give a preliminary definition.

Definition 5. A system $\{M_\alpha\}$ of subsets of a space X is said to be *centred* if any of its finite subsystems has a nonempty intersection.

Theorem 4. *A topological space X is compact iff each centred system of closed subsets of it has a nonempty intersection.*

Proof. Let $\sigma = \{M\}$ be an arbitrary centred system of closed subsets of the space, and let X be compact. We shall show that $\bigcap_{M \in \sigma} M \neq \emptyset$. Assume the opposite, i.e. $\bigcap_{M \in \sigma} M = \emptyset$. Then $\bigcup_{M \in \sigma}(X \backslash M) = X$, i.e. the system $\{X \backslash M\}_{M \in \sigma}$ is an open covering of X. By the compactness of X, there exists a finite subcovering $\{X \backslash M_k\}_{k=1}^n$. Therefore $\bigcup_{k=1}^n (X \backslash M_k) = X$, and, consequently, $\bigcap_{k=1}^n M_k = \emptyset$, which is a contradiction that the system σ is centred.

Assume that for any centred system $\sigma = \{M\}$ of closed subsets the intersection $\bigcap_{M \in \sigma} M$ is nonempty. Let $\{U_\alpha\}$ be an arbitrary open covering of X. Then the system $\{X \backslash U_\alpha\}$ has the empty intersection and, according to the assumption, is not centred.

Thus, for some $\alpha_1, \alpha_2, \ldots, \alpha_s$ the subsystem $\{X \backslash U_{\alpha_i}\}_{i=1}^s$ has an empty intersection and, hence, $\{U_{\alpha_i}\}_{i=1}^s$ is a finite subcovering of the covering $\{U_\alpha\}$. Therefore, the space X is compact.   □

Now, consider the property of paracompactness. It is interesting to examine the relation between the paracompactness and other properties of topological spaces. Consider the so–called locally compact spaces.

Definition 6. A space X is said to be *locally compact* if for each point $x \in X$ there exists a neighbourhood $U(x)$ whose closure is compact.

An example of a locally compact space is the space $\mathbf{R}^n$; another example is a two–dimensional manifold (see Ch. 2, § 4).

Theorem 5. *If a topological space X is Hausdorff and locally compact then it is regular.*

Proof. Let $a \in X$ be an arbitrary point, and $F \subset X$ a closed set not containing the point a. Then $X \backslash F$ is open, and $a \in X \backslash F$. Because the space X is locally compact, there is an open neighbourhood $V(a)$ such that $\bar{V}(a)$ is compact. Let $F_1 = \bar{V}(a) \cap F$; the F_1 is a closed set of the Hausdorff space X which lies in the compact set $\bar{V}(a)$, and, consequently, is compact in X. The point a and the compact set F_1 can be separated by non–intersecting neighbourhoods $W(a)$, $U(F_1)$ (see the proof of theorem 2). Now $W_1(a) = W(a) \cap V(a)$ is a new neighbourhood, for which we have $W_1(a) \cap U(F_1) = \emptyset$, $\bar{W}_1(a) \cap F_1 = \emptyset$. In particular, from the evident inclusion $\bar{W}_1(a) \subset \bar{V}$ we obtain $\bar{W}_1(a) \cap (F \backslash F_1) = \emptyset$, which together with the previous means $\bar{W}_1(a) \cap F = \emptyset$. A closed neighbourhood $\bar{W}_1(a)$ is compact, because $\bar{V}(a)$ is a compact space, and $\bar{W}_1(a) \subset \bar{V}(a)$. Consequently, for each point $x \in F$ there is an open neighbourhood $U(x)$ which does not intersect with $\bar{W}_1(a)$ (see the proof of theorem 2, in particular the remark about the "separation" of a compact set and a point in a Hausdorff space). Now take $U(F) = \bigcup_{x \in F}(U(x))$, and get $W_1(a) \cap U(F) = \emptyset$.   □

The paracompact space of example 3 is a spacial case of the spaces described by the following theorem.

Theorem 6. *Let X be a locally compact space, and $X = \bigcup_{n=1}^{\infty} C_n$, where each C_n is a compact set. Then X is paracompact.*

Proof. We represent X first as a countable union of nested open sets whose closures are compact. We shall construct these sets by induction. First, take $U_0 = \emptyset$. As U_1 take a neighbourhood of the set C_1 whose closure is compact. Now, once the set U_n is constructed, we select a neighbourhood of the set $\bar{U}_n \cup C_{n+1}$ whose closure is compact, and call it U_{n+1}. The existence of such neighbourhoods is guaranteed by the local compactness of the space X.

Now, let $\{V_\alpha\}_{\alpha \in M}$ be an arbitrary open covering of X. Denote the compact set $\bar{U}_n \setminus U_{n-1}$ by D_n. The open set $U_{n+1} \setminus \bar{U}_{n-2}$ is a neighbourhood of the set D_n. Then the system of sets

$$\{V_\alpha \cap (U_{n+1} \setminus \bar{U}_{n-2})\}_{\alpha \in M} = \{W_\alpha^n\}_{\alpha \in M}.$$

forms an open covering of the set D_n. Select finite subcovering $\{W_\alpha^n\}_{m=1}^{p_n}$. Having performed the described procedure for all n, we obtain a countable covering $\bigcup_{n=1}^{\infty}[\{W_m^n\}_{m=1}^{p_n}$ of the whole space X, which is a refinement of the covering $\{V_\alpha\}_{\alpha \in M}$.

We now show that this covering is locally finite. Let x be an arbitrary point from X, and $n_0 = \min\{n : x \in U_n\}$. Since $x \notin U_{n_0-1}$, there exists a neighbourhood $O(x)$ of x lying in U_{n_0} and such that $O(x) \cap \bar{U}_{n_0-2} = \emptyset$. Thus, $O(x)$ can only intersect with the sets W_m^k, where $1 \leq m \leq p_k$, $n_0 - 2 \leq k \leq n_0 + 1$. By the construction, the number of such sets is finite.   □

Corollary. *If a locally compact space X has a countable base, then it is paracompact.*

Indeed, if a space is locally compact, then it has a base $\{U^C\}$ of open sets such that $\bar{U}^C$ is compact. By choosing from some countable base those sets which compose the sets U^C, we obtain a countable base $\{V_i^C\}_{i=1}^{\infty}$ with the property that $\bar{V}_i^C$ is compact for any i. Then $X = \cup \bar{V}_i^C$, and the paracompactness of it follows from theorem 6. □

Remark. In some fields of mathematics (for example, in functional analysis), the setting of a convergent sequences is often used in studying the properties of compact sets. This setting can be introduced (see Subsection 3, § 11) for Hausdorff spaces which satisfy the first countability axiom. It is useful to formulate the general concept of compactness in terms of convergent sequences.

Exercise 2°. Prove that if a Hausdorff space X satisfies the second countability axiom, then the condition of X being compact is equivalent to the following condition: it is possible to select a convergent subsequence from each infinite sequence $\{x_n\}$ of elements from X.

2. Mappings of compact spaces. We shall study some important properties of continuous mappings of compact spaces.

Theorem 7. *Let X, Y be topological spaces, let X be compact, and $f : X \to Y$ a continuous mapping. Then the image $f(X)$ is a compact subspace in Y.*

Proof. Consider an arbitrary open covering $\{V_\alpha\}$ of the space $Z = f(X)$. Evidently, $f : X \to Z$ also is continuous; therefore, $\{f^{-1}(V_\alpha)\}$ is an open covering of X. Let us select a finite subcovering $\{f^{-1}(V_{\alpha_i})\}_{i=1}^{m}$ from it. Then $\{V_{\alpha_i}\}_{i=1}^{m}$ is a finite open covering of Z. □

Exercise 3°. Show that a closed surface (see Ch. 2, § 4) is a compact topological space.

Theorem 8. *Let $f : X \to Y$ be a continuous mapping, X compact, and Y Hausdorff. Then f is a closed mapping.*

Proof. Recall that any closed subset of a compact space is compact. Let $M \subset X$ be an arbitrary closed (and, therefore, compact) subset in X. By theorem 7, the set $f(M)$ is compact in Y and, consequently, is closed according to theorem 2. □

We shall derive an important test for a mapping to be a homeomorphism.

Theorem 9. *Let conditions of the previous theorem be fulfilled, and let the mapping f be bijective; then f is a homeomorphism.*

Proof. Consider the inverse mapping $f^{-1} : Y \to X$. We shall show that it is continuous. Let $A \subset X$ be an arbitrary closed subset. Since f is a closed mapping, $f(A) = (f^{-1})^{-1}(A)$ is closed in Y, which implies the continuity of the mapping f^{-1}. $\square$

Many examples of compact spaces are obtained when constructing quotient spaces.

Example 4. Let X be a quotient space of some compact space Y. Then X is compact. Indeed, X is a continuous image (with respect to the projection) of a compact space.

Consider a continuous numerical function $f : X \to \mathbf{R}^1$ on a compact space X. Then there is the following Weierstrass theorem which is important in analysis.

Theorem 10. *Any continuous function $f : X \to \mathbf{R}^1$ on a compact space X is bounded and attains its upper (lower) bound on X.*

Proof. By theorem 7, the set $f(X)$ is compact. According to theorem 2, any compact subspace in $\mathbf{R}^1$ is closed. As it has already been mentioned, any compact subspace in $\mathbf{R}^n$ is bounded. Consequently, $f(X)$ is bounded and closed. The boundedness of $f(X)$ implies the boundedness of the function f. Since a closed set contains all of its limit points, $\sup_{x \in X} f(x) \in f(X)$, and $\inf_{x \in X} f(x) \in f(X)$. This completes the proof of the theorem. $\square$

3. Products of compact spaces

Theorem 11. (A. N. Tikhonov). *The topological product $X = \prod_{\alpha \in M} X_\alpha$ of any system $\{X_\alpha\}_{\alpha \in M}$ of compact spaces is compact.*

Proof. We shall use the compactness criterion that any centred system of closed subsets has a nonempty intersection. Let $\{N^\lambda\} = \sigma_0$ be an arbitrary centred system of closed subsets in X. In the set of such systems, we consider the following partial ordering relation: $\sigma'' \succ \sigma'$, if each set from σ' is included in σ''. Let G be the set of all systems σ such that $\sigma \succ \sigma_0$. It is clear that any totally ordered subset from G possesses a maximal element (its union). Then, according to Zorn's lemma, there is a maximal centred system $\bar\sigma$ in G, i.e. a system such that for any system $\sigma \in G$, either $\bar\sigma \succ \sigma$, or $\bar\sigma$ and σ are incomparable.

Let $\bar\sigma = \{N^\lambda\}$. It is easy to show that any finite intersection of elements $\bar\sigma$ belongs to $\bar\sigma$, and also that any closed set M intersecting any N^γ, belongs to $\bar\sigma$. (Verify this property). Clearly, if it is shown that $\bar\sigma$ has a nonempty intersection, i.e. $\bigcap_{N^\gamma \in \bar\sigma} N^\gamma \neq \emptyset$, then the proof will be complete. Denote the projection on the factor X_α by $\pi_\alpha : X \to X_\alpha$. For a fixed α the system $\{\pi_\alpha(N^\gamma)\}_{\bar\sigma} = \{N_\alpha^\gamma\} = \bar\sigma_\alpha$ is a centred system (not necessarily consisting of closed sets) in X_α; therefore, the system $\{\bar N_\alpha^\gamma\}$ is also is centred. Because X_α is compact, there exists an element $x_\alpha \in X_\alpha$ such that for any of its neighbourhoods $U_\alpha = U(x_\alpha)$ the intersection $U_\alpha \cap N_\alpha^\gamma \neq \emptyset$ for any $N_\alpha^\gamma \in \bar\sigma_\alpha$.

Now consider the element $x = \{x_\alpha\} \in X$. Each of its neighbourhoods $U = U(x)$ contains the closure of a certain elementary neighbourhood of the form

$$p_{\alpha_1}^{-1}(U_{\alpha_2}) \cap p_{\alpha_2}^{-1}(U_{\alpha_2}) \cap \ldots \cap p_{\alpha_s}^{-1}(U_{\alpha_s}) = U_{\alpha_1,\ldots,\alpha_s},$$

which in turn is the intersection of a finite number of neighbourhoods of the form $p_{\alpha_i}^{-1}(U_{\alpha_i}) = V_{\alpha_i} \subset X$. It is clear that V_{α_i} intersects all the sets $N^\gamma \in \bar\sigma$, since $U_{\alpha_i} \cap N_{\alpha_i}^\gamma \neq \emptyset$ for all γ. Consequently, $\bar V_{\alpha_i} \in \bar\sigma$, and, therefore,

$$\bar U_{\alpha_1,\ldots,\alpha_s} = \bigcap_{i=1}^{s} \bar V_{\alpha_i} \in \bar\sigma.$$

Hence, we obtain that the neighourhood $U = U(x)$ intersects all $N^\gamma \in \bar\sigma$. Since U is arbitrary, we conclude that $x \in \bigcap_{N^\gamma \in \bar\sigma} N^\gamma$ (and, thus, $x \in \bigcap_{N^\gamma \in \bar\sigma} N^\gamma$). $\square$

We present some examples where, by using the Tikhonov theorem, the compactness of a space can be quickly determined.

Example 5. The cube $I^n = \underbrace{[0, 1] \times [0, 1] \times \ldots \times [0, 1]}_{n}$ is a compact space since it is the product of intervals.

Example 6. The boundedness and closedness of a set in $\mathbf{R}^n$ are equivalent to compactness. In fact, such a set in $\mathbf{R}^n$ can be included in a closed parallelepiped $[a_1, b_1] \times [a_2, b_2] \times \ldots \times [a_n, b_n]$, which is compact by example 5.

By this the sufficiency of the condition is proved. The necessity of the closedness condition has been proved in one of the previous theorems in this section, and that the boundedness condition is necessary was noted in example 2.

Exercise 4°. Prove that the sphere S^n is compact.

Exercise 5°. Verify that the n–dimensional torus $T^n = \underbrace{S^1 \times S^1 \times \ldots \times S^1}_{n}$ is compact.

Example 7. The projective space $\mathbf{R}P^n$ is compact because it is a quotient space of the sphere S^n.

Example 8. The lens space $S^n/\mathbf{Z}_p$ is compact for the same reason.

4. Compactness in a metric space. Compact metric spaces are often called *compacta*, and compact subspaces are called *compact sets* of a metric space.
   The property of compactness in a metric space can be expressed in terms of convergent sequences.

Definition 7. A set Y of a metric space (X, ρ) is said to be *sequentially compact* if any sequence consisting of its elements contains a subsequence covergent in X.

Theorem 12. *A set Y of a metric space X is compact iff it is closed and sequentially compact.*

Proof. Let Y be sequentially compact and closed. Then for any $\epsilon > 0$ there exists a finite set of points $A_\epsilon = \{x_k\}$ such that the balls $D_\epsilon(x_k)$ with centres in x_k and radius ϵ cover Y.[*] Indeed, if this were not true, then for some ϵ_0 there are points $x_1, x_2, \ldots, x_n, \ldots$ in Y such that $\rho(x_n, x_{n+p}) \geq \epsilon_0$ for all n, p. The presence of such a sequence contradicts the sequential compactness of Y. Thus, finite ϵ–nets exist for any $\epsilon > 0$.

Now, let $\{U\}$ be an arbitrary covering of Y. Assume that it is impossible to select a finite subcovering from it. Then, in any finite ϵ_1–net A_{ϵ_1} there is an element x_k such that the closed set $Y \cap \bar{D}_{\epsilon_1}(x_k) = Y_1$ cannot be covered by any finite subsystem from $\{U\}$. It is easy to see that the set Y_1 is closed and sequentially compact, and the diameter of it is not greater than $2\epsilon_1$. Applying similar arguments to Y_1, we construct a set $Y_2 \subset Y_1$ with the same properties, and with its diameter not greater that $2\epsilon_2 < 2\epsilon_1$

Thus, by taking into account that $\epsilon_n \to 0$, we construct a system $\{Y_n\}$ of closed sequentially compact sets $Y_{n+1} \subset Y_n$ with diameters convergent to zero.

Exercise 6°. Show that $\bigcap_{k=1}^\infty Y_k \neq \emptyset$.

From the latter fact it follows that there exists a point $x_0 \in \bigcap_{k=1}^\infty Y_k$. Since $\{U\}$ is a covering, $x_0 \in U_\alpha$ for one of its elements U_α. Because U_α is open, there exists an $\epsilon > 0$ such that $D_\epsilon(x_0) \subset U_\alpha$. By taking n large enough for the diameter Y_n to be less than ϵ, we obtain the inclusions $Y_n \subset D_\epsilon(x_0) \subset U_\alpha$, which is a contradiction with the assumption that Y_n cannot be covered by a finite number of elements from $\{U\}$.

Closedness and sequential compactness follow from the closedness of a compact set (see theorem 2) and the fact that for every infinite sequence there exists a limit point (see theorem 1). $\qquad\qquad\square$

We propose to the reader to prove the following useful statement.

Theorem 13. (On the *Lebesgue number*) *Let X be compact, and $\{U\}$ an arbitrary open covering of X. Then there exists a real number $\delta > 0$ such that any set in X of diameter less than δ, lies wholly in a certain element of the covering $\{U\}$.*

Exercise 7°. Let a metric space X be compact; let $f : X \to Y$ be a continuous mapping into a metric space Y. Prove that for any covering $U = \{U_\alpha\}$ of the

---

[*] The set A_ϵ is called a *finite ϵ–net* in X.

space Y, there exists a Lebesgue number $\delta = \delta(U)$ such that for any subset A in X of diameter less than δ, the image $f(A)$ is wholly contained in a certain element of the covering U.

In analysis, one of important questions is that concerning the compactness of sets in function spaces. There exists a number of special cirteria of compactness in concrete spaces. In particular, for the extensively used space $C_{[0,1]}$ in analysis, such a criterion is given by the Arzelà theorem [46].

§ 14. Compact extensions of topological spaces. Metrization

1. Compactifications. The property of compactness proves to be useful and convenient in many questions. For this reason, it is natural to find a construction which would allow us to construct, for some given non–compact space, a compact space containing the given one and to investigate the relationships between topologies, properties of functions on these spaces, etc.

Definition 1. A *compactification* of a topological space X is a compact space CX such that $X \subset CX$, and the topology of the space X is induced by the topology of the space CX.

As a classical example of compactification the extended complex plane $\tilde{C} = C \cup \{\infty\}$ homeomorphic to S^2 can serve (see Ch. 1, Subsection 2, § 3). In a similar way, we obtain compactifications CR^1, CR^n ($n > 1$) which are homeomorphic to S^1 and S^n, respectively.

For a more general concept of compactification of a topological space X, additional conditions such as CX being Hausdorff, denseness of X in CX, "maximality" or "minimality" of CX, etc. are imposed. Below, we shall consider the simplest method, the so–called "one–point compactification", which was first investigated by P. S. Alexandrov. This compactification is constructed by adjoining one additional element ξ (often denoted ∞) to X in such a way that $CX = X \cup \xi$.

Definition 2. The space $CX = X \cup \xi$ with the topology $\sigma = \{U\}$ consisting of all open sets $U = W$ of the space X and all sets of the form $U = V \cup \xi$, where V is an open set in X that is complementary to a closed compact subset of the set X, is called the *one–point compactification* of the space X.

Exercise 1°. Verify that the system of sets $\sigma = \{U\}$ satisfies the topology axioms.

The following proposition justifies the term "compactification" for $CX = X \cup \xi$.

Theorem 1. *The one–point compactification $CX = X \cup \xi$ is a compact space.*

Proof. Let $\{U_\alpha\}$ be an open covering of the space CX, where $U_\alpha = W_\alpha$ or $U_\alpha = V_\alpha \cup \xi$; let W_α, V_α are open sets in X, $V_\alpha = X \backslash K_\alpha$, K_α a compact and closed set in X. Let $U_{\alpha_0} = V_{\alpha_0} \cup \xi$ be a set from the covering; then $V_{\alpha_0} \cap K_{\alpha_0} = \emptyset$ and, consequently, $\{U_\alpha\}_{\alpha \neq \alpha_0}$ is an open covering of K_{α_0} in CX, and $\{V_\alpha\}_{\alpha \neq \alpha_0}$ is an open covering of K_{α_0} in X. Because K_{α_0} is compact, from the covering $\{V_\alpha\}_{\alpha \neq \alpha_0}$ can be selected a finite subcovering $\{V_{\alpha_1}, \dots, V_{\alpha_s}\}$, $a_i \neq a_0$, $i = 1, \dots, s$, of the set K_{α_0}. Then the system of sets $\{U_{\alpha_1}, \dots, U_{\alpha_s}, U_{\alpha_0}\}$ clearly forms an open covering of the space CX, which proves the compactness of the space CX. $\square$

Remark. If X is compact then the one–point compactification $CX = X \cup \xi$ has an isolated point ξ. Indeed, taking $K = X$, $V = X \backslash K = \emptyset$, we obtain an open neighbourhood of the point ξ: $U(\xi) = \xi$, consisting of one point. The inverse statement is also true: if the point ξ in CX is isolated then X is compact (prove!).

The examples of compactifications $\tilde{C}$, $C\mathbf{R}^1$, $C\mathbf{R}^n$ presented at the beginning of the subsection are also examples of one–point compactifications.

Theorem 2. *The one–point compactification CX is a Hausdorff space iff X is Hausdorff and locally compact.*

Proof. Let X be Hausdorff and locally compact. If $x, y \in CX$ are different points from the subspace X, then they are separable by open sets $W_1(x)$, $W_2(y)$ in X, and hence also in CX. If $x = \xi$, and $y \in X$, then by choosing a neighbourhood $W(y)$ with the compact closure $\bar{W}$ in X, we construct a set

$U(\xi) = (CX)\backslash \bar{W}(y)$, which is an open neighbourhood in CX of the point ξ, because $U(\xi) = (X\backslash \bar{W}(y)) \cup \xi$. It is evident that we have $U(\xi) \cap W(y) = \emptyset$. This concludes the proof that CX is Hausdorff.

Inversely, let CX be a Hausdorff space. Then in CX, for the points $x = \xi$, $y \in X$, there exist open neighbourhoods $U(\xi)$, $W(y)$ such that $U(\xi) \cap W(y) = \emptyset$. Moreover, $U(\xi) = V \cup \xi$, where $V = X\backslash K$ and K is a compact and closed set in X while $W(y)$ is an open set in X. From the condition $U(\xi) \cap W(y) = \emptyset$ it follows that $W(y) \subset K$, and, hence, $\bar{W}(y) \subset K$, where $\bar{W}(y)$ is the closure $W(y)$ in X. Since a closed subset of the compact space is compact, $\bar{W}(y)$ is compact in X, and this concludes the proof of local compactness of X. As a subspace of a Hausdorff space inherits the property of being Hausdorff, so the subspace X of the Hausdorff space CX is Hausdorff.   □

It is easy to verify that for a non–compact space X and its Hausdorff one–point compactification CX, the subspace X is dense in CX (i.e. $\bar{X} = CX$, where $\bar{X}$ is the closure of the set X in the space CX).

To conclude, it should be noted that for a non–compact X, the Hausdorff one–point compactification $CX = X \cup \xi$ is, under the inclusive ordering, the smallest one in the class of Hausdorff compactifications, and at the same time, there exists a maximal Hausdorff compactification (The Stone–Čech compactification) for X.

2. Metrizability of topological spaces. We now consider the question of how to introduce a metric on a topological space, which induces the same topology. Topological spaces that admit such a metric are said to be *metrizable*. In particular, we may speak of introducing another metric on a metric space so that it generates the original topology but is more convenient, e.g. such that the space is complete in it. Such metric spaces are said to be *topologically complete*.

One example of a topologically complete space is the interval $(a, b) \subset \mathbf{R}^1$, say, $X = (-1, 1)$. In addition to the standard metric $\rho(x, y) = |x - y|$ in which (X, ρ) is incomplete, we can introduce the topologically complete equivalent metric

$$\rho_*(x, y) = \left| \frac{x}{\sqrt{1 - x^2}} - \frac{y}{\sqrt{1 - y^2}} \right|$$

induced by the homeomorphism between the interval $(-1, 1)$ and the whole real line (the distance between the points of the interval in the metric ρ_*, is calculated as the distance in the usual metric between their images on the line). It is easy to verify that ρ_* is a metric, (X, ρ) and (X, ρ_*) are homeomorphic, and that (X, ρ_*) is a complete metric space. As an example of a topologically incomplete metric space, the set of rational numbers under the metric from $\mathbf{R}^1$ can be considered.

The *Tikhonov product* of a countable number of metric spaces (X_n, ρ_n) is metrizable. Indeed, if $x = (x_1, x_2, \ldots)$, $y = (y_1, y_2, \ldots)$ are elements from $\prod_{i=1}^{\infty} X_i$, then a metric can be defined by the formula

$$\bar{\rho}(x, y) = \sum_{n=1}^{\infty} \frac{1}{2^n} \cdot \frac{\rho_n(x_n, y_n)}{1 + \rho_n(x_n, y_n)}.$$

Exercise 2°. Verify that $\bar{\rho}$ is a metric and that the topology induced by it is equivalent to the Tikhonov topology. In particular, the Tikhonov cube I^{∞} (a countable product of intervals I) is a metrizable topological space.

Theorem 3. (Uryson) *A regular space with a countable base is metrizable.*

Theorem 4. (Stone) *A metrizable topological space is paracompact.*

3. Topology of spaces of subsets and many–valued mappings. In contemporary analysis, in the theory of optimal control, the theory of games and in mathematical economics, many–valued mappings (or, shortly, m–mappings) of topological spaces $f : X \to Y$, whose values $f(x)$ are non–empty subsets in Y, are intensively studied and applied. If we consider the family of all non–empty sets from Y then naturally one obtains a single–valued mapping $\tilde{f} : X \to P(Y)$ defined by $\tilde{f}(x) = f(x) \in P(Y)$; the inverse also holds. In addition, it is convenient to consider the family of all closed subsets $C(Y)$ or of all compact subsets $K(Y)$ as the domains of values of mappings $\tilde{f} : X \to C(Y)$ or $\tilde{f} : X \to K(Y)$. In order to apply the methods of topology to the study of such mappings $\tilde{f}$ (and, consequently, of f), it is necessary to introduce a topology on $C(Y)$, $K(Y)$. This can be achieved in different ways.

(1) The space $\kappa(Y)$ is $C(Y)$ with the topology defined by the base $\{C(U)\}$, where U is any open set in Y ($U \in \tau_Y$).

(2) The space $\lambda(Y)$ is $C(Y)$ with the topology defined by the prebase

$$\{C(Y) \backslash C(Y \backslash U)\},$$

where $U \in \tau_Y$.

(3) The space $\text{Exp}(Y)$ is $C(Y)$ with the topology prebase consisting of the union of the base $\{C(U)\}$ and the prebase $\{C(Y) \backslash C(Y \backslash U)\}$.

Exercise 3°. Verify that $C(U)$ consists of all closed subsets of Y lying completely in U, and that $C(U) \backslash C(Y \backslash U)$ consists of all closed subsets of Y intersecting with U.

The topologies κ, λ, Exp are called the *upper semifinite, lower semifinite* and *exponential* topology, repsectively.

(4) Let (Y, ρ) be a metric space, $C_0(Y)$ the family of all non–empty bounded closed subsets in Y.

Since $C_0(Y) \subset C(Y)$, each of the topologies κ, λ, and Exp can be induced on $C_0(Y)$. However, it is important to introduce a metric on $C_0(Y)$. Namely, we define the so–called Hausdorff metric $h(A, B)$, with $h(A, B) = \max\{\rho_*(A, B), \rho_*(B, A)\}$, which gives $C_0(Y)$ a metric topology. Here $\rho_*(A, B) = \sup \rho(a, B)$, $a \in A$, i.e. the deviation of the set A from the set B (here, $A, B \in C_0(Y)$).

Exercise 4°. Verify that $h(A, B)$ satisfies metric axioms.

Note that if Y is a compact metric space then the topology induced on $C_0(Y) = C(Y)$ by the Hausdorff metric is equivalent to the exponential topology on $C(Y)$, and this, by Subsection 2, means the metrizability of the space $\text{Exp}(Y)$. By the same token, we can consider m–mappings as mappings $f : X \to C(Y)$, $f : X \to C_0(Y)$ (we will not any more distinguish f from $\tilde{f}$) into the topological spaces $C(Y)$, $C_0(Y)$ with one or another topology.

Definition 3. An m–mapping $f : X \to Y$ is said to be (1) *upper semicontinuous*, (2) *lower semicontinuous*, (3) *continuous* if the mapping $f : X \to C(Y)$ is continuous in the topology (1) $\kappa(Y)$, (2) $\lambda(Y)$, (3) $\text{Exp}(Y)$, respectively.

An m–mapping $f : X \to Y$ with values in $C_0(Y)$, where (Y, ρ) is a metric space, is called continuous under the Hausdorff metric, if the corresponding mapping to the metric space $C_0(Y)$ is continuous.

The definitions of continuous m–mappings considered above can naturally be extended to mappings $f : X \to K(Y)$, because $K(Y) \subset C_0(Y) \subset C(Y)$

and $K(Y)$ inherits a corresponding topology.

Exercise 5°. Prove the equivalence of the continuity of m–mapping and the continuity with respect to the Hausdorff metric for a compact space Y.

Exercise 6°. Compare the definitions given above with the classic definitions in analysis in case $f : \mathbf{R}^1 \to \mathbf{R}^1$.

REVIEW OF THE RECOMMENDED LITERATURE

Systematic courses on general topolgy are in [2, 5, 45, 50, 78, 28].

Popular presentations of main topological concepts can be found in [14, 26, 82].

An outline of main concepts and constructions of general topology is given in [8].

Problem books on the material of chapter 2 are [7, 59, 63].

Initial information on metric spaces is in [46, 51].

The classification of closed two–dimensional surfaces can be found in [88, 54].

For further study of compact spaces we recommmend the primary source [4].

Further information on the topology of m–mappings and certain applications can be found, for instance, in the educational survey [15].

One of the main methods of topology centers around the study of the geometrical properties of topological spaces using algebraic properties. With that end in view, a number of methods which make it possible to associate to a topological space algebraic objects like groups, rings, etc., have been devised. The idea of such a correspondence (or functor) is the basis of algebraic topology; it associates to a collection of topological spaces a collection of certain algebraic objects, and continuous mappings of spaces with corresponding homomorphisms. This functorial approach enables the reduction of a topological problem to a corresponding algebraic one. Solvability of this "derived" algebraic problem in most cases allows us to determine the solvability of the original topological problem.

One of the first concepts which arouse in this way, is the idea of the fundamental group of a topological space; later the more general concept of homotopy groups has been introduced. The present chapter is devoted to the latter.

Figures associated with loops and spheroids are visible in this illustration. The study of the space properties, which are preserved under continuous deformations, results in the idea of loops and spheroids. In turn, the concepts of loops and spheroids lead to the idea of the fundamental group and homotopy group of spaces, which are the main tool in homotopic topology.

CHAPTER III

HOMOTOPY THEORY

One of the main methods in topology is to study the geometric properties of topological spaces algebraically. A number of approaches have been used in topology to associate to a topological space a number of algebraic objects, for instance, groups, rings, etc.. The basic idea of algebraic topology is the correspondence (or functor) associating a collection of certain algebraic objects to a collection of topological spaces, and continuous mappings of spaces to corresponding homomorphisms. This functorial approach allows us to transform topological problem into the corresponding algebraic one. The solvability of this "derived" algebraic problem in many cases implies the solvability of the initial topological problem.

One of the first concepts that has emerged in this way, is the concept of the fundamental group of a topological space; later the more general concept of homotopy groups arose. The present chapter is devoted to the latter concepts.

§ 1. Mapping space. Homotopy, retraction, and deformation

This section studies the set of all continuous mappings of one topological space into another. One can introduce various topologies on this set, and thereby turning it into various topological spaces. The connectedness of this space is of great importance, which naturaly leads to the concept of homotopic mappings. The consideration of special classes of mappings and their homotopies leads to the concepts of deformation of one space into another, of a retraction, etc.. All the concepts mentioned play important roles in homotopy theory.

1. The space of continuous mappings. Consider the set $C(X, Y)$ of all continuous mappings from a topological space X into a topological space Y. The properties of this set and many of those of the spaces X, Y are interrelated. A simple example: if X is a one–point space, then $C(X, Y) \simeq Y$, where the sign "$\simeq$" denotes a bijection.

One can introduce a topology on the set $C(X, Y)$, as on any set, using different methods. The question how to introduce a topology on $C(X, Y)$ in the most natural way arises. Here, the intuitive idea of the nearness of mappings helps considerably; the mappings f_1, f_2 are said to be near if the images $f_1(x)$ and $f_2(x)$ for any point $x \in X$ are near in Y. If Y is a metric space, then these notions are expressed in terms of the metric on Y. On this basis, different topologies are introduced on the set $C(X, Y)$, i.e., the topology of pointwise convergence, the topology of uniform convergence, etc.

If (Y, ρ) is a metric space, and X is compact, then the set $C(X, Y)$ is provided with a metric μ as follows:

$$\mu(f_1, f_2) = \sup_{x \in X} \rho(f_1(x), f_2(x)), \quad f_1, f_2 \in C(X, Y).$$

Definition 1. The topology τ_μ on $C(X, Y)$ determined by the metric μ is called the *topology of uniform convergence*.

Exercise 1°. Verify the metric axioms for μ.

Exercise 2°. Consider a convergent sequence $f_n^\mu \to f$ in $C(X, Y)$ and give an equivalent definition of the convergence in terms of the topology (instead of the metric) on Y. In the case of $X = [0, 1]$, compare this convergence with uniform convergence in $C_{[0,1]}$.

Definition 2. Consider in $C(X, Y)$ the sets

$$\{x_i, U_i\}_{i=1}^k = \{f \in C(X, Y) : f(x_i) \in U_i, \ i = 1, \dots, k\},$$

where $x_1, x_2, \dots, x_k \in X$, $U_1, \dots, U_k$ are open sets in Y. The topology τ_ρ generated by these sets as a prebase is called the *topology of pointwise convergence on $C(X, Y)$*.

Exercise 3°. Verify that the sets of the form $\{x_i, U_i\}_{i=1}^k$ and their finite intersections satisfy the criterion for being a base.

Exercise 4°. Consider a convergent sequence $\{f_n\}$ to f in the given topology and prove that its convergence is equivalent to the convergence of the sequences $f_n(x) \to f(x)$ for any point $x \in X$.

Exercise 5°. Consider the set $\{Y_x\}_{x \in X}$ of replicas of a space Y numbered by the elements $x \in X$, and the Tikhonov product $\prod_{x \in X} Y_x$. Show that the set $C(X, Y)$ can be identified with a subset of this product and that the topology of the product induces on $C(X, Y)$ the topology of pointwise convergence τ_ρ.

The following definition gives still another version of topology on the set $C(X, Y)$.

Definition 3. Consider all possible sets of mappings of the form

$$[K, U] = \{f \in C(X, Y) : f(K) \subset U\},$$

where K is a compact set in X, and U is an open set in Y. The topology τ_c generated by such sets $[K, U]$ as a prebase is called the *compact open* topology on $C(X, Y)$.

Exercise 6°. Verify that the system of sets $[K, U]$ and their finite intersections satisfy the criterion for being a base.

Exercise 7°. Show that $\tau_\rho \prec \tau_c$, and that for a metric space $\tau_c \prec \tau_\mu$.

Exercise 8°. Prove that if Y is a metric space, and X is compact then the compact open topology coincides with the topology of uniform convergence.

Exercise 9°. If X is a non–compact space, and Y is a metric space then sequences of mappings which are uniformly convergent on any compact subset of X are often considered. Show that this convergence is equivalent to the convergence in the compact open topology.

Exercise 10°. Prove that if (Y, ρ) is a complete metric space, then the space $C(X, Y)$ is a complete metric space in the metric μ.

Exercise 11°. Show that if X is locally compact, and Z is Hausdorff, then the spaces $C(X \times Z, Y)$ and $C(Z, C(X, Y))$ are homeomorphic in the compact open topology.

The space $C(X, Y)$ is often denoted by Y^X. Then the statement in exercise 11° can be written as $Y^{X \times Z} \simeq (Y^X)^Z$ (the *exponential law*).

As an example, we shall consider the space of ω–periodic continuous functions defined on the line of real numbers $\mathbf{R}^1$. By virtue of their periodicity, each such function f is completely determined by its values on the interval $[0, \omega]$, moreover, $f(0) = f(\omega)$. Therefore, in fact, we consider the set of functions on the interval $[0, \omega]$ with the ends "glued together" or, what is the same, functions on the circle S^1. This is the set $C(S^1, \mathbf{R}^1)$ on which any of the topologies τ_μ, τ_ρ, τ_c may be introduced.

In a similar way, we can consider the space of functions on the torus

$$T^n \; : \; C(T^n, \mathbf{R}^1) = C(\underbrace{S^1 \times \ldots \times S^1}_{n}, \mathbf{R}^1),$$

which can be interpreted as the space of periodic functions in n variables.

In homotopy theory, as a rule, $C(X, Y)$ is considered with a compact open topology.

2. Homotopy. In many problems, it is possible not to distinguish mappings, one of which can be "smoothly" changed, i.e. deformed into the other. A continuous deformation of one mapping into another can be naturally thought of as a path in the space $C(X, Y)$ which begins and ends at given points f_1 and f_2. Brouwer made the concept of continuous deformation more accurate with the help of the following concept of homotopy.

Definition 4. Two continuous mappings f_0, $f_1 \in C(X, Y)$ are said to be *homotopic* ($f_0 \sim f_1$) if there exists a continuous mapping $f : X \times [0, 1] \to Y$ such that $f(x, 0) = f_0(x)$, $f(x, 1) = f_1(x)$ for all $x \in X$.

The mapping f is often called a *homotopy* connecting the mappings f_0 and f_1.

Thus, if $f_0 \sim f_1$, then there is a family of mappings $f_t : X \to Y$ depending on a numerical parameter $t \in [0, 1]$ and connecting the mappings f_0 and f_1 so that the mapping $X \times [0, 1] \to Y$ induced by this family according to the rule $(x, t) \mapsto f_t(x)$ is continuous. The inverse is evident.

Exercise $12°$. Show that introduction of a homotopy $f : X \times [0, 1] \to Y$ is equivalent to introduction of a path s in $C(X, Y)$ (the topology is τ_c, X is locally compact).

Example 1. Let $X = Y = \mathbf{R}^n$, $f_0(x) = x$, $f_1(x) = 0$ for all $x \in \mathbf{R}^n$. Define $F : \mathbf{R}^n \times I \to \mathbf{R}^n$ according to the formula $F(x,t) = (1-t)x$, $t \in I$. It is easy to see that F is a homotopy between f_0 and f_1.

Example 2. Let X be an arbitrary space, Y a convex subset in $\mathbf{R}^n$, and f_0, $f_1 \in C(X,Y)$ arbitrary continuous mappings. Then the mapping $F : X \times I \to Y$ given by the formula $F(x,t) = tf_1(x) + (1-t)f_0(x)$, is a homotopy between f_0 and f_1. $\Diamond$

Note, that the concept of homotopy is related to the problem of extending a mapping. Indeed, let $f, g : X \to Y$ be continuous mappings. Define a mapping $\phi : X \times \{0\} \cup X \times \{1\} \to Y$ by the formulas $\phi(x,0) = f(x)$; $\phi(x,1) = g(x)$. It is easy to see that $f \sim g$ iff there exists an extension of ϕ to $X \times [0,1]$.

Theorem 1. *Homotopy is a relation of equivalence on the set $C(X,Y)$.*

Proof. Reflexivity ($f \sim f$) is seen with the help of the homotopy $F(x,t) \equiv f(x)$.

Symmetry: let $f_0 \sim f_1$ with homotopy $F(x,t)$. Then $\widetilde{F}(x,t) = F(x, 1-t)$ defines a homotopy from f_1 to f_0, i.e, $f_1 \sim f_0$.

Transitivity: let $f_0 \sim f_1$, $f_1 \sim f_2$ with homotopies $F_1(x,t)$, $F_2(x,t)$, respectively. Then the mapping

$$H(x,t) = \begin{cases} F_1(x, 2t), & 0 \le t \le 1/2; \\ F_2(x, 2t-1), & 1/2 \le t \le 1 \end{cases}$$

is continuous as its restriction to each of the closed sets $X \times [0, 1/2]$, $X \times [1/2, 1]$ is continuous. It can be easily seen that $H(x,t)$ is a homotopy between f_0 and f_1. $\square$

The equivalence classes of homotopic mappings are called *homotopy classes*. The quotient set $C(X,Y)/R$ is denoted by $\pi(X,Y)$. It is easy to see that $\pi(X,Y)$ is the set of components of pathwise connectednesss of the space $C(X,Y)$. The homotopy class of the mapping $f \in C(X,Y)$ is denoted by $[f]$.

Definition 5. A mapping $f \in C(X, Y)$ is called a *homotopy equivalence* if there exists a mapping $g \in C(X, Y)$ such that $gf \sim 1_X$, $fg \sim 1_Y$.

Definition 6. It is said that a space X is *homotopy equivalent* to a space Y or, that X and Y have *identical homotopy type*, if there exists a homotopy equivalence in $C(X, Y)$.

The concept of homotopy equivalence is a useful "coarsening" of the concept of homeomorphism of two spaces. Indeed, if $f : X \to Y$ is a homeomorphism then by putting $g = f^{-1} : Y \to X$, we get $gf = 1_X$, $fg = 1_Y$, i.e., we obtain the homotopy equivalence of X and Y. In view of this, the mapping g in the definition of a homotopy equivalence is said to be *homotopy inverse to* f.

The simplest example of a (non–empty) topological space is a singleton (one–point space). The question arises which spaces possess the homotopy type of a point.

Definition 7. A space X is said to be *contractible* if the identity mapping $f_X : X \to X$ is homotopic to a constant mapping (i.e., a mapping of X into a point $x_0 \in X$). A homotopy between them is called a *contraction* of the space X (into the point x_0).

Exercise 13°. Prove that any two mappings of a space X into a contractible space Y are homotopic to one another.

Theorem 2. *A space is contractible iff it has the same homotopy type as a point.*

Proof. Let X be contractible, and $\Phi : X \times I \to Y$ be a contraction of X into a point $x_0 \in X$. Denote by Q the one–point space consisting of the point x_0. Let $\phi : X \to Q$ be the mapping to the point x_0; let $j : Q \to X$ be the imbedding. Then $\phi j = 1_Q$, and Φ is a homotopy connecting 1_X with $j\phi$. Thus, ϕ is a homotopy equivalence between X and Q. The proof of the inverse statement is left to the reader.   □

Exercise 14°. Prove that any convex subset in $\mathbf{R}^n$ (in particular, $\mathbf{R}^n$ itself) is contractible.

Exercise 15°. Prove that the space $X \times Y$ is contractible if X and Y are contractible spaces.

3. Extending mappings. Consider now the problem of extending a mapping. It can be formulated as follows: is it possible to extend a given mapping $f : A \to Y$ defined on a subspace A of a space X to the whole space X, i.e., is there a mapping $\phi : X \to Y$ such that its restriction $\phi|_A : A \to Y$ coincides with the mapping f? Such a mapping ϕ is called an *extension of the mapping f*.

The solution of this problem has been found only for some special cases. A complete theory of extension does not yet exist. One of the examples of a partial solution of this problem is the Tietze–Uryson theorem for normal spaces presented in Ch. 2, § 12.

The following theorem connects the problem of extending mappings to the concept of homotopy.

Theorem 3. *Let* $\phi : S^n \to Y$ *be a continuous mapping of the unit sphere. Then the following two conditions are equivalent:*

   (i) the mapping ϕ *is homotopic to a constant mapping;*
   (ii) the mapping ϕ *can be extended to the whole ball* $\bar{D}^{n+1} \subset \mathbf{R}^{n+1}$.

Proof. $(i) \Rightarrow (ii)$. Let $f \sim c$, where c is the constant mapping of S^n into a point $p \in Y$. Let $F : S^n \times I \to Y$ be a homotopy between f and c. Define an extension f' of the mapping f to the ball $\bar{D}^{n+1}$ in the following way:

$$f'(x) = \begin{cases} p, & 0 \le \|x\| \le 1/2; \\ F\left(\frac{x}{\|x\|}, 2 - 2\|x\|\right), & 1/2 \le \|x\| \le 1. \end{cases}$$

It is easy to see that $f'|_{S^n} = f$; f' is continuous since its restrictions to each of the closed sets

$$\{x \in \bar{D}^{n+1} : 0 \le \|x\| \le 1/2\}, \quad \{x \in \bar{D}^{n+1} : 1/2 \le \|x\| \le 1\}$$

are continuous.

$(ii) \Rightarrow (i)$. Let f', an extension of f to the whole ball $\bar{D}^{n+1}$, be given. Let $y_0 \in S^n$. Define the mapping $\Phi : S^n \times I \to Y$ as

$$\Phi(x, t) = f'[(1 - t)x + ty_0].$$

It is clear that $\Phi(x,0) = f'(x) = f(x)$, $\Phi(x,1) = f'(y_0) = p \in Y$, and therefore $\Phi(x,t)$ is the required homotopy.   □

Exercise 16°. Show that any mapping f of a space X into a contractible space Y is homotopic to a constant mapping (compare with exercise 13°).

Exercise 17°. Using the result of the previous exercise, derive from theorem 3 that any mapping of the sphere S^n into a contractible space can be extended to the whole ball $\bar{D}^{n+1}$.

4. Retraction. A special case of the extension problem is that of a retraction formulated as follows.

Definition 8. Let A be a subspace in X, $1_A : A \to A$ the identity mapping. If there exists a mapping $r : X \to A$ such that $r|_A = 1_A$ then it is called a *retraction* of X onto A, and the space A a *retract* of X.

Exercise 18°. Verify that any point of a topological space X is a retract of X.

Exercise 19°. Verify that any linear subspace in $\mathbf{R}^n$ is a retract of $\mathbf{R}^n$.

Exercise 20°. If $Z = X \times Y$ is the Tikhonov product of two spaces and $p \in X$, $q \in Y$ are fixed points, then $A = X \times q$, $B = p \times Y$ are retracts of the space $X \times Y$, and the mappings $r_X : (x,y) \mapsto (x,q)$, $r_Y : (x,y) \mapsto (p,y)$ are the corresponding retractions.

Exercise 21°. Show that the zero–dimensional sphere $S^0 = \{-1, 1\}$ is not a retract of the one–dimensional disc $\bar{D}^1 = [-1, 1]$.

Hint. Use the properties of connected spaces.

Definition 9. If there exists a mapping $r : X \to A$ such that $r|_A \sim 1_A$, then A is called a *weak retract* of X, and r a *weak retraction* of X on A.

It is easy to see that a retract is always a weak retract. The inverse, generally speaking, is not true, which is demonstrated by the following exercise.

Exercise 22°. Consider the square $I^2 = [0,1] \times [0,1]$ and its subset A which is a "comb" consisting of the vertical intervals whose bases are at the points $(1/n, 0)$, $n = 1, 2, \ldots$, the point $(0,0)$, and the base of the square (Fig. 63). Show that (1) the set A is not a retract of the square I^2, (2) A is a weak retract of I^2, (3) if in the "comb" A only a finite number of teeth are left, then the set A' obtained is a retract of I^2.

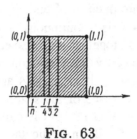

FIG. 63

Definition 10. A *deformation* of a space X into the subspace A, $A \subset X$, is a homotopy $D : X \times I \to X$ such that $D(x, 0) = x$, $D(x, 1) \in A$ for all $x \in X$.

Definition 11. If there exists a deformation $D : X \times I \to X$ of X into A such that $D(x, t) = x$ for $x \in A$, $t \in I$, then A is called a *strong deformation retract* of X, and D a *strong deformation retraction*.

Example 3. A point is a strong deformation retract of any convex subset of $\mathbf{R}^n$ containing it.

Other examples of strong deformation retracts are presented in the following exercises.

Exercise 23°. Let a space X be contractible to a point $x_0 \in Y$. Show that $x_0 \times X$ is a strong deformation retract of the product $X \times Y$. In particular, consider a two–dimensional cylinder and show that its base is a strong deformation retract.

Exercise 24°. Verify that the vertex of a cone in three-dimensional space is a strong deformation retract of the cone.

Exercise 25°. Show that a strong deformation retract A of a space X is homotopy equivalent to X.

Hint. The imbedding $i : A \to X$ and the retraction $D(x, 1)$ of the space X onto A are homotopy inverses of each other.

5. Mapping cylinder. Consider first some operations on topological spaces.

The *topological (disconnected* or *disjoint) sum* $X \sqcup Y$ *of two spaces* X, Y is defined as the union of non–intersecting copies of X and Y.

The topology on $X \sqcup Y$ is defined as follows: V is open in $X \sqcup Y$ iff $V \cap X$ and $V \cap Y$ are open in X and Y, respectively.

If $f : A \to Y$ is a continuous mapping, where $A \subset X$, then it is possible to glue X and Y together with respect to the mapping f. For this purpose, we introduce the equivalence relation $R : x \sim y$ on $X \sqcup Y$, if $x \in A$, $y \in Y$, and $f(x) = y$; $x_1 \sim x_2$, if $x_1, x_2 \in A$ and $f(x_1) = f(x_2)$.

The quotient space of the space $X \sqcup Y$ with respect to the equivalence R is denoted by $X \cup_f Y$ and is called the *gluing of the spaces* X and Y *by the mapping* f. If, in particular, A is a point $x_0 \in X$, and the mapping $f : A \to Y$ takes x_0 into $y_0 = f(x_0)$, then the gluing $X \cup_f Y$ is called the *wedge of the spaces* X, Y, and is denoted by $X_{x_0} \vee_{y_0} Y$. It is easy to see that it is the quotient space of the disjoint sum $X \sqcup Y$ with respect to the equivalence relation gluing the points $x_0 \in X$ and $y_0 \in Y$ together.

Exercise 26°. Show that the homotopy type of the wedge $X_{x_0} \vee_{y_0} Y$ coincides with the homotopy type of the space Y if X is contractible to the point $x_0 \in X$.

Exercise 27°. Prove that the interval $I = [0, 1]$ and the wedge $I_0 \vee_{p_0} S^1$, where $p_0 \in S^1$, $0 \in I$, have different homotopy types.

Definition 12. Let $f : X \to Y$ be a continuous mapping. Then we may assume that the mapping $\phi : X \times \{1\} \to Y$, $\phi(x, 1) = f(x)$, where $X \times \{1\}$ is obviously a subspace of $X \times I$, is given. The gluing $(X \times I) \cup_\phi Y$ of the spaces $X \times I$ and Y with respect to the mapping ϕ, is called the *cylinder* Z_f *of the mapping* $f : X \to Y$.

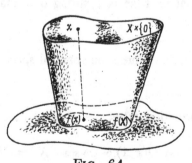

The mapping cylinder can be represented as it is shown in Fig. 64.

The concept of a mapping cylinder is important because we can consider X, Y as subspaces of Z_f.

FIG. 64

Thus, the mapping f is replaced, in a sense, by the imbedding of X into Z_f. It should also be noted that Y is a strong deformation retract of Z_f, and the imbedding of Y into Z_f is a homotopy equivalence (verify!).

Definition 13. The cylinder of a constant mapping $c : X \to p$ is called the *cone over the space* X and is denoted by CX.

Theorem 4. *A mapping* $f : X \to Y$ *is homotopic to a constant one iff there exists an extension* $\tilde{f} : CX \to Y$ *of the mapping* f.

Proof. If f is homotopic to the constant mapping $c_0 : X \to (*)$, then

$$F : X \times I \to Y, \quad F(x,0) = x, \quad F(x,1) = (*).$$

Thus, F is constant on the upper base of the cylinder $X \times I$, and consequently, it induces a mapping $\bar{F}$ of the quotient space $(X \times I)/R$, where R denotes the gluing of the upper base into a point. However the space $(X \times I)/R$ is homeomorphic to CX (verify!).

The proof of the inverse statement we leave to the reader.

Exercise 28°. Let a mapping $f : A \to Y$ be continuous, $A \subset X$ closed in X; let X, Y be normal spaces. Prove that $X \cup_f Y$ is normal.

Exercise 29°. Prove that $f : A \to Y$ can be extended to the whole X ($A \subset X$) iff Y is a retract of $X \cup_f Y$.

§ 2. Category, functor, and algebraization of topological problems

Taking description in terms of categories implies an approach to a mathematical object in which this object, for instance, a group or a space, is not considered as isolated, but as a member of a collection of similar objects. Intuitively, a category can be seen as a collection of sets (possibly with additional structure) and mappings compatible with this structure. Correspondences between elements of different categories obeying special rules are called *functors*.

1. Category

Definition 1. A *category* $\mathcal{A}$ is given by (1) a certain collection of objects, (2) a set $\mathrm{Mor}_{\mathcal{A}}(X, Y)$ of morphisms[*] from X to Y for each ordered pair of objects, (3) a mapping associating to any ordered triple of objects X, Y, Z and any pair of morphisms $f \in \mathrm{Mor}_{\mathcal{A}}(X, Y)$, $g \in \mathrm{Mor}_{\mathcal{A}}(Y, Z)$ their composition $gf \in \mathrm{Mor}_{\mathcal{A}}(X, Z)$. Thus, we get the following commutative diagram of morphisms in the given category (if we denote morphisms by arrows):

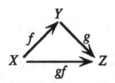

In addition, two axioms are fulfilled:

A. Associativity. If

$$f \in \mathrm{Mor}_{\mathcal{A}}(X, Y), \quad g \in \mathrm{Mor}_{\mathcal{A}}(Y, Z), \quad h \in \mathrm{Mor}_{\mathcal{A}}(Z, W),$$

then $h(gf) = (hg)f$ *in* $Mor_{\mathcal{A}}(X, W)$.

B. The existence of identity elements. For any object Y there exists in $Mor_{\mathcal{A}}(Y, Y)$ *a morphism* 1_Y *such that for any*
$f \in \mathrm{Mor}_{\mathcal{A}}(X, Y), \quad g \in \mathrm{Mor}_{\mathcal{A}}(Y, Z),$
$1_Y f = f$ *and* $g 1_Y = g$.

Note that from the presented axioms there follows the uniqueness of the element 1_Y. This element is called the *identity morphism* of the object Y. If for any two morphisms $f \in \mathrm{Mor}_{\mathcal{A}}(X, Y)$, $g \in \mathrm{Mor}_{\mathcal{A}}(Y, X)$, the equality $gf = 1_X$ holds, then the morphism g is said to be *left inverse* to f, and f *right inverse* to g. A morphism which is both left and right inverse to f is said to be *two–sided inverse* to f.

---

[*] Here it is understood that $\mathrm{Mor}_{\mathcal{A}}(X, Y) \cap \mathrm{Mor}_{\mathcal{A}}(X', Y') = \emptyset$ if $X \neq X'$ or $Y \neq Y'$.

Definition 2. A morphism $f \in \text{Mor}_A(X, Y)$ is called an *equivalence* ($f : X \approx Y$) if there exists for f a two–sided inverse morphism $f^{-1} \in \text{Mor}_A(Y, X)$.

Exercise 1°. Prove that if a morphism $f \in \text{Mor}_A(X, Y)$ has a left inverse and a right inverse, then they coincide.

From the exercise follows that if $f : X \approx Y$, then $f^{-1} : Y \approx X$.

We present several important examples of categories.

Examples.
1. The collection of sets and their mappings.
2. The collection of metric spaces and their continuous mappings.
3. The collection of topological spaces and their continuous mappings.
4. The collection of linear spaces and their linear mappings.
5. The collection of groups and their homomorphisms.
6. The collection of pairs of topological spaces and their continuous mappings.

By a *pair of topological spaces* (X, A) we mean a space X and a subspace A. A mapping of pairs $f : (X, A) \to (Y, B)$ is a mapping $f : X \to Y$ such that $f(A) \subset B$.   ◊

Exercise 2°. Show that in categories of examples 2, 3, and 6, the homeomorphisms are precisely the equivalences.

Exercise 3°. Verify that in the category of example 1 the equivalences are the bijective mappings of sets.

Exercise 4°. Show that in examples 4, 5 the equivalences are the isomorphisms of linear spaces and groups, respectively.

2. Functors. We will be interested in natural 'mappings' of one category into another, i.e., in mappings which preserve identity elements and compositions of morphisms. We will formulate this concept more presisely.

Definition 3. Let A, and B be the categories. A *covariant functor* T from A into B is a mapping which associates to each object X from A an object

$T(X)$ from B, and to each morphism $f : X_1 \to X_2$ in A a morphism $T(f)$: $T(X_1) \to T(X_2)$ in B, such that moreover, the following relations hold:

(1) $T(1_X) = 1_{T(X)}$,

(2) $T(gf) = T(g)T(f)$.

Properties (1) and (2) of a functor can be visually presented as follows: any commutative diagram of the category A is mapped by a functor into the corresponding commutative diagram of the category B:

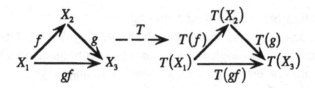

Example 7. An example of a covariant functor is the correspondence associating to a topological space the set of all points that make it up, and to a continuous mapping of spaces the underlying mapping of sets. This is a functor from the category of example 3 into the category of example 1, which is said to be *forgetful*, since it "forgets" the structure of a topological space.

Analogously, there is the covariant functor from the category of metric spaces into the category of topological spaces associating to a metric space the same space considered as a topological space with the topology induced by the metric; this functor "forgets" the metric.

Definition 4. A *contravariant functor* T from a category A into a category B is a mapping associating to each object X from A an object $T(X)$ from B, and to each morphism $f : X_1 \to X_2$ a morphism $T(f) : T(X_2) \to T(X_1)$ from $\mathrm{Mor}_A(T(X_2), T(X_1))$ such that the following relations hold:

(1) $T(1_X) = 1_{T(X)}$,

(2) $T(gf) = T(f)T(g)$.

In other words, a contravariant functor transforms a commutative diagram of a category A into a commutative diagram of a category B with reversed arrows:

$$X_1 \xrightarrow[f]{} X_2 \xrightarrow{g} X_3 \quad X_1 \xrightarrow{gf} X_3 \quad \xrightarrow{\;T\;} \quad T(X_1) \xrightarrow[T(f)]{} T(X_2) \xrightarrow{T(g)} T(X_3) \quad T(X_1) \xrightarrow{T(gf)} T(X_3)$$

The functors of homology groups and of homotopy groups can serve as important examples of functors studied in algebraic topology. These are functors from the category of topological spaces into the category of groups. We will consider in detail the functors of homotopy groups in the following section, whereas the functor of homology group will be investigated in Ch 5.

Let us consider an example of how a functor to the category of groups is applied to the investigation of certain topological problems. In the previous section, we formulated the mapping extension problem. Now, we formulate it as follows: let $A \subset X$ be a subspace of topological space X, $i : A \to X$ be the natural mapping associating any point $a \in A$ with itself, but in the space X (i is the imbedding mapping); let $\phi : A \to Y$ be a mapping of the space A into the space Y. A mapping $\bar{\phi} : X \to Y$ extends the mapping ϕ iff the diagram

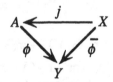

is commutative.

With a help of a functor T (covariant, for example) we derive an algebraic problem: is there a homomorphism $T(\bar{\phi})$ such that the diagram

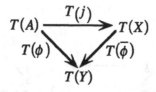

is commutative?

It is clear that the solvability of the original problem implies the solvability of our algebraic problem. Thus, the existence of the homomorphism $T(\bar{\phi})$ is a necessary condition that an extension $\bar{\phi}$ of the mapping ϕ exists. For instance, if the homomorphism $T(i)$ is zero, and $T(\phi)$ is nonzero, then the homomorphism $T(\bar{\phi})$ does not exist (otherwise, the commutativity of the diagram would be violated). In this case, an extension $\bar{\phi}$ of the mapping ϕ does not exist.

§ 3. Homotopy group functors

In this section, we turn to the study of questions concerning mappings of spaces. In some cases, the set $\pi(X, Y)$ turns out to be a group, sometimes Abelian, and it may be helpful in the construction of various algebraic functors on the category of topological spaces and their continuous mappings. The construction and applications of these functors are the basis of homotopic topology.

1. The homotopy group of a space. First note that for each topological space Y and continuous mapping $f : X_1 \rightarrow X_2$ of topological spaces X_1, X_2 there exists the natural mapping

$$\pi^Y(f) : \pi(X_2, Y) \rightarrow \pi(X_1, Y).$$

More exactly, if $[\phi] \in \pi(X_2, Y)$, then in $\pi(X_1, Y)$, there is the element $[\phi f]$ uniquely corresponding to $[\phi]$. Similarly, to any topological space X and continuous mapping $g : Y_1 \rightarrow Y_2$ there are associated the mapping

$$\pi_X(g) : \pi(X, Y_1) \rightarrow \pi(X, Y_2).$$

Exercise 1°. Describe the construction of $\pi_X(g)$ and prove the correctness of the definitions of $\pi^Y(f)$ and π_X.

Exercise 2°. Using the remarks made above, show that for a fixed Y, the correspondence $X \mapsto \pi(X, Y)$ is a contravariant functor into the category of sets, and the correspondence $Y \mapsto \pi(X, Y)$ (for a fixed X) is a covariant functor.

The correspondence $(X, Y) \mapsto \pi(X, Y)$ is called a *bifunctor* from the category of topological spaces into the category of sets, which is covariant with respect to the second argument and contravariant with respect to the first one.

In a similar way, one can consider the bifunctor π on the category of pairs of topological spaces determined by the correspondence $(X, A; Y, B) \rightarrow \pi(X, A; Y, B)$. Note that the homotopy $F(x, t)$ between mappings f and $g : (X, A) \rightarrow (Y, B)$ of pairs of spaces is understood to be a mapping of pairs $F : (X \times I, A \times I) \rightarrow (Y, B)$ such that $F(x, 0) = f(x)$, $F(x, 1) = g(x)$.

Exercise 3°. Describe the construction of the mapping

$$\pi_{(X,A)}(f) : \pi(X, A; Y_1, B_1) \rightarrow \pi(X, A; Y_2, B_2),$$

which is naturally induced by a continuous mapping of pairs $f : (Y_1, B_1) \rightarrow (Y_2, B_2)$, and verify that the correspondence $(Y, B) \mapsto \pi(X, A; Y, B)$ is a covariant functor.

Definition 1. The pair (X, x_0) is called a *space with a distinguished point* $x_0 \in X$.

Now, we fix the pair $(I^n, \partial I^n)$, where I^n is the n–dimensional cube, $n \geq 1$, and ∂I^n is its boundary; we consider the correspondence between the pair (X, x_0) and the set $\pi(I^n, \partial I^n; X, x_0)$.

Recall that the elements of $\pi(I^n, \partial I^n; X, x_0)$ are homotopy classes of mappings of pairs $\phi : (I^n, \partial I^n) \rightarrow (X, x_0)$. They are sometimes called *spheroids*. Each of these mappings takes I^n into X, and ∂I^n into the point x_0; moreover, this property is supposed to continue to hold when the mapping ϕ is changed in the process of homotopy. The sets $\pi(I^n, \partial I^n; X, x_0)$ and $\pi(S^n, p_0; X, x_0)$ coincide (correspond bijectively to each other). Here p_0 is a distinguished point of the sphere S^n. Indeed, we have noted above that the quotient space $I^n / \partial I^n$ is homeomorphic to the sphere S^n; moreover, the interior Int I^n of the cube I^n corresponds bijectively to the set $S^n \backslash p_0$ under this homeomorphism

θ, and the boundary ∂I^n is transformed into the point p_0 of the sphere S^n. In such cases, it is said that a *relative homeomorphism* is given:

$$\theta : (I^n, \partial I^n) \to (S^n, p_0).$$

Then to any mapping $f : (S^n, p_0) \to (X, x_0)$ there corresponds the mapping $f\theta : (I^n, \partial I^n) \to (X, x_0)$, and vice versa, to a mapping $g : (I^n, \partial I^n) \to (X, x_0)$ there corresponds the mapping $\bar{g} : (S^n, p_0) \to (X, x_0)$ which coincides with $g\theta^{-1}$ on $S^n \backslash p_0$, and transforms the point p_0 into x_0.

Exercise 4°. Show that this correspondence between mappings provides a bijection between $\pi(S^n, p_0; X, x_0)$ and $\pi(I^n, \partial I^n; X, x_0)$.

Thus, we have given another interpretation of the set $\pi(I^n, \partial I^n; X, x_0)$ which enables us to consider also the case of $n = 0$.

Exercise 5°. Show that the set $\pi(S^0, p_0; X, x_0)$ is the set of components of path connectedness of the space X.

Thus, we have defined a covariant functor $(X, x_0) \mapsto \pi(I^n, \partial I^n; X, x_0)$ from the category of spaces with a distinguished point into the category of sets.

The structure of the set $\pi(I^n, \partial I^n; X, x_0)$ is of great interest in homotopy topology. First, we will consider the case $n > 1$.

Theorem 1. *The set* $\pi(I^n, \partial I^n; X, x_0)$, $n > 1$, *is an abelian group.*

Proof. Let $[\phi]$, $[\psi] \in \pi(I^n, \partial I^n; X, x_0)$. We define the sum $[\phi] + [\psi]$ as $[\phi] + [\psi] = [\phi + \psi]$, where the mapping $\phi + \psi$ is defined as follows: let $t = (t_1, t_2, \ldots, t_n) \in I^n$, $t_i \in I = [0, 1]$, $i = 1, \ldots, n$; then

$$(\phi + \psi)(t) = \begin{cases} \phi(2t_1, t_2, \ldots, t_n), & \text{if } 0 \le t_1 \le 1/2, \\ \psi(2t_1 - 1, t_2, \ldots, t_n), & \text{if } 1/2 \le t_1 \le 1. \end{cases}$$

The definition can be visually illustrated by the following picture, where the square represents the side (t_1, t_2) of the cube I^n (Fig. 65).

We define the zero element as the class of the constant mapping $\theta : (I^n, \partial I^n) \to (X, x_0)$, for which $\theta(I^n) = x_0$. We shall show that $[\phi] + [\theta] = [\phi]$ for any

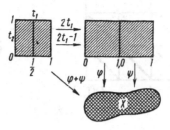

FIG. 65

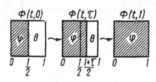

FIG. 66

$[\phi]$, i.e., $\phi + \theta$ is homotopic to ϕ. Indeed, the required homotopy is given by the mapping

$$\Phi : (I^n \times I, \partial I^n \times I) \to (X, x_0),$$

where

$$\Phi(t, \tau) = \begin{cases} \phi(\frac{2t_1}{\tau+1}, t_2, \ldots, t_n), & \text{if } 0 \leq t_1 \leq \frac{\tau+1}{2}, \\ x_0, & \text{if } \frac{\tau+1}{2} \leq t_1 \leq 1, \quad \tau \in I. \end{cases}$$

The idea of the homotopy $\Phi(t, \tau)$ is represented in Fig. 66.

Exercise 6°. Verify that the equality $[\theta] + [\phi] = [\phi]$ also holds.

Exercise 7°. Explain the reason why the construction which gives the proof of the equality $[\phi] + [\theta] = [\phi]$ does not establish the equality $[\phi] + [\psi] = [\phi]$ if $[\psi] \neq [\theta]$.

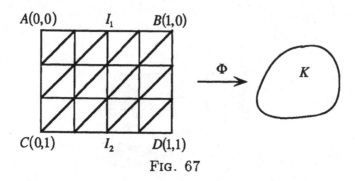

FIG. 67

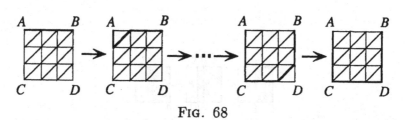

FIG. 68

Each $[\phi]$ has the inverse element $(-[\phi])$ in $\pi_n(X, x_0)$ which is the class $[\phi\eta]$, where $\eta : I^n \to I^n$ is defined by the formula $\eta(t) = (1 - t_1, t_2, \ldots, t_n)$; thus, $(\phi\eta)(t) = \phi(1 - t_1, t_2, \ldots, t_n)$.

We will verify that $[\phi] + [\phi\eta] = [\theta]$. In fact, a homotopy between the mappings $\phi + \phi\eta$ and θ is given by the mapping

$$\Phi(t, \tau) = \begin{cases} x_0, & 0 \le t_1 \le \tau/2, \\ \phi(2t_1 - \tau, t_2, \ldots, t_n), & \tau/2 \le t_1 \le 1/2, \\ \phi(-2t_1 + 2 - \tau, t_2, \ldots, t_n) & 1/2 \le t_1 \le 1 - \tau/2, \\ x_0, & 1 - \tau/2 \le t_1 \le 1. \end{cases}$$

In Fig 67, this homotopy is represented in the form of a diagram.

Exercise 8°. Verify that the mapping $\Phi(t, \tau)$ indicated is the homotopy of mappings of pairs of spaces $(I^n, \partial I^n)$, (X, x_0).

It remains to verify the associativity and commutativity of addition in $\pi_n(X, x_0)$, $n > 1$. We first prove the associativity. Let $[\phi]$, $[\psi]$, $[\mu] \in$

$\pi_n(X, x_0)$. We show that $([\phi] + [\psi]) + [\mu] = [\phi] + ([\psi] + [\mu])$. It is easy to verify that the required homotopy is given by the mapping

$$\Phi(t, \tau) = \begin{cases} \phi(\frac{4t_1}{\tau+1}, t_2, \dots, t_n), & 0 \le t_1 \le \frac{\tau+1}{4}, \\ \psi(4t_1 - \tau - 1, t_2, \dots, t_n), & \frac{\tau+1}{4} \le t_1 \le \frac{\tau+2}{4}, \\ \mu(\frac{4t_1-2-\tau}{2-\tau}, t_2, \dots, t_n), & \frac{\tau+2}{4} \le t_1 \le 1. \end{cases}$$

In terms of pictures, this homotopy is quite simple (Fig. 68).

We now show that $[\phi] + [\psi] = [\psi] + [\phi]$. Recall that

$$(\phi + \psi)(t) = \begin{cases} \phi(2t_1, t_2, \dots, t_n), & \text{if } 0 \le t_1 \le 1/2, \\ \psi(2t_1 - 1, t_2, \dots, t_n), & \text{if } 1/2 \le t_1 \le 1, \end{cases}$$

$$(\psi + \phi)(t) = \begin{cases} \psi(2t_1, t_2, \dots, t_n), & \text{if } 0 \le t_1 \le 1/2, \\ \phi(2t_1 - 1, t_2, \dots, t_n), & \text{if } 1/2 \le t_1 \le 1. \end{cases}$$

We shall satisfy ourselves that the mappings $\phi + \psi$ and $\psi + \phi$ are homotopic to the same mapping. (Hence it follows that they are homotopic to one another.) Consider the homotopy $\Phi_1(t, \tau)$:

$$\Phi_1(t, \tau) =$$
$$= \begin{cases} \left. \begin{array}{ll} x_0, & 0 \le t_2 \le \tau/2 \\ \phi(2t_1, \frac{2t_2-\tau}{2-\tau}, t_3, \dots, t_n), & \tau/2 \le t_2 \le 1 \end{array} \right\} & 0 \le t_1 \le 1/2, \\ \left. \begin{array}{ll} \psi(2t_1 - 1, \frac{2t_2}{2-\tau}, t_3, \dots, t_n), & 0 \le t_2 \le 1 - \tau/2 \\ x_0, & 1 - \tau/2 \le t_2 \le 1 \end{array} \right\} & 1/2 \le t_1 \le 1. \end{cases}$$

It is easy to see that $\Phi_1(t, 0) = \phi + \psi$, and

$$\Phi_1(t, 1) =$$
$$= \begin{cases} \left. \begin{array}{ll} x_0, & 0 \le t_2 \le \tau/2 \\ \phi(2t_1, 2t_2 - 1, t_3, \dots, t_n), & 1/2 \le t_2 \le 1 \end{array} \right\} & 0 \le t_1 \le 1/2, \\ \left. \begin{array}{ll} \psi(2t_1 - 1, 2t_2, t_3, \dots, t_n), & 0 \le t_2 \le 1/2 \\ x_0, & 1/2 \le t_2 \le 1 \end{array} \right\} & 1/2 \le t_1 \le 1. \end{cases}$$

Consider another homotopy Φ_2:

$$\Phi_2(t, s) = \begin{cases} \left. \begin{array}{ll} \phi(\frac{2t_1}{1+s}, 2t_2 - 1, t_3, \dots, t_n), & 0 \le t_1 \le \frac{1+s}{2} \\ x_0, & \frac{1+s}{2} \le t_1 \le 1 \end{array} \right\} & 1/2 \le t_2 \le 1, \\ \left. \begin{array}{ll} x_0, & 0 \le t_1 \le \frac{1-s}{2} \\ \psi(\frac{2t_1-1+s}{1+s}, 2t_2, t_3, \dots, t_n), & \frac{1-s}{2} \le t_1 \le 1 \end{array} \right\} & 0 \le t_2 \le 1/2. \end{cases}$$

It is not difficult to verify that $\Phi_2(t,0) = \Phi_1(t,1)$, and

$$\Phi_2(t,1) = \begin{cases} \phi(t_1, 2t_2 - 1, t_3, \ldots, t_n), & 0 \le t_1 \le 1, \quad 1/2 \le t_2 \le 1, \\ \psi(t_1, 2t_2, t_3, \ldots, t_n), & 0 \le t_1 \le 1, \quad 0 \le t_2 \le 1/2. \end{cases}$$

The homotopies Φ_1, Φ_2 are illustrated in Fig. 69 in terms of pictures.

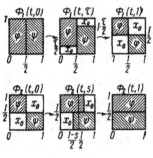

FIG. 69

Thus, we have

(1) $$\phi + \psi \sim \Phi_1(t,1) = \Phi_2(t,0) \sim \Phi_2(t,1).$$

We shall verify a similar construction for the sum $\psi + \phi$. For this purpose, we write down the homotopies:

$$\Psi_1(t,\tau) =$$

$$= \begin{cases} \psi(2t_1, \frac{2t_2}{2-\tau}, t_3, \ldots, t_n), & 0 \le t_2 \le 1 - \tau/2 \\ x_0, & 1 - \tau/2 \le t_2 \le 1 \end{cases} \; 0 \le t_1 \le 1/2, \\ \begin{cases} x_0, & 0 \le t_2 \le \tau/2 \\ \phi(2t_1 - 1, \frac{2t_2 - \tau}{2-\tau}, t_3, \ldots, t_n), & \tau/2 \le t_2 \le 1 \end{cases} \; 1/2 \le t_1 \le 1.$$

It is easy to see that $\Psi_1(t,0) = \psi + \phi$, and

$$\Psi_1(t,1) =$$

$$= \begin{cases} \psi(2t_1, 2t_2, t_3, \ldots, t_n), & 0 \le t_2 \le 1/2 \\ x_0, & 1/2 \le t_2 \le 1 \end{cases} \; 0 \le t_1 \le 1/2, \\ \begin{cases} x_0, & 0 \le t_2 \le 1/2 \\ \phi(2t_1 - 1, 2t_2 - 1, t_3, \ldots, t_n), & 1/2 \le t_2 \le 1 \end{cases} \; 1/2 \le t_1 \le 1.$$

Let us construct another homotopy

$$\Psi_2(t, s) =$$
$$= \begin{cases} \phi(\frac{2t_1 - 1 + s}{1 + s}, 2t_2 - 1, t_3, \dots, t_n), & \frac{1-s}{2} \le t_1 \le 1 \\ x_0, & 0 \le t_1 \le \frac{1-s}{2} \end{cases} \Bigg\} \quad 1/2 \le t_2 \le 1,$$
$$\begin{cases} x_0, & \frac{1+s}{2} \le t_1 \le 1 \\ \psi(\frac{2t_1}{1+s}, 2t_2, t_3, \dots, t_n), & 0 \le t_1 \le \frac{1+s}{2} \end{cases} \Bigg\} \quad 0 \le t_2 \le 1/2.$$

Thus, $\Psi_2(t, 0) = \Psi_1(t, 1)$, and $\Psi_2(t, 1) = \Phi_2(t, 1)$.

In Fig. 70, the homotopies $\Psi_1(t, \tau)$, $\Psi_2(t, s)$ are illustrated by pictures. We find that

$$\psi + \phi \sim \Psi_1(t, 1) = \Psi_2(t, 0) \sim \Psi_2(t, 1) = \Phi_2(t, 1).$$

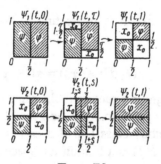

FIG. 70

From the last chain of homotopies and the chain (1) we get that $\phi + \psi \sim \Phi_2(t, 1)$, $\psi + \phi \sim \Phi_2(t, 1)$; therefore $\phi + \psi \sim \psi + \phi$. □

Remark. The attentive reader will have noticed that in the proof of the commutativity of the algebraic operation in $\pi_n(X, x_0)$ the condition $n > 1$ was used essentially.

The group $\pi(I^n, \partial I^n; X, x_0)$, $n > 1$, is called the *n–dimensional homotopy group* of the space X with distinguished point x_0 and is denoted by $\pi_n(X, x_0)$.

Theorem 2. *Any continuous mapping $f : (X, x_0) \to (Y, y_0)$ induces a homomorphism of groups $\pi_{(I^n, \partial I^n)}(f) : \pi_n(X, x_0) \to \pi_n(Y, y_0)$.*

The proof is left to the reader.

Hint. Show that the mapping of homotopy classes indicated in exercise $3°$ is a homomorphism. Use the definition of the sum $(\phi + \psi)(t)$ (see Fig. 65).

The homomorphism $\pi_{(I, \partial I^n)}(f)$ is denoted by f_n and is called the n–dimensional homotopy group *homomorphism induced by the continuous mapping* f.

Thus, the functor π_n, $n > 1$, acts from the category of spaces with distinguished point and their continuous mappings into the category of Abelian groups and their homomorphisms. Consequently, if

$$f : (X, x_0) \rightarrow (Y, y_0), \quad g : (Y, y_0) \rightarrow (Z, z_0)$$

are continuous mappings then $(gf)_n = g_n f_n$, where f_n, g_n, $(gf)_n$ are the corresponding homomorphisms of n–dimensional homotopy groups.

2. The fundamental group. It is interesting to consider separately the set $\pi_1(X, x_0)$ defined by

$$\pi_1(X, x_0) = \pi(I, \partial I; X, x_0) = \pi(S^1, p_0; X, x_0),$$

which is endowed with a group structure in the same manner as π_n, $n > 1$, and is applied in solving many problems. By the general definition, every element of $\pi_1(X, x_0)$ is the homotopy class $[\phi]$ of a certain mapping $\phi : (I, \partial I) \rightarrow (X, x_0)$, where the image $\phi(I)$ is a loop in the space X starting and ending at the point x_0 (Fig. 71).

The orientation of the loop is defined by the parameter $t \in I$. The product $\phi \cdot \psi$ of two loops ϕ and ψ is defined as the loop in X such that the image $(\phi \cdot \psi)(t)$ runs over the loop ϕ as the parameter t changes from

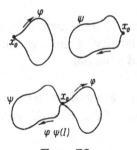

FIG. 71   FIG. 72

0 to 1/2, and the image($\phi \cdot \psi$)(t) runs over the loop ψ for the values of the parameter from 1/2 to 1 (Fig. 72); more exactly,

$$(\phi \cdot \psi)(t) = \begin{cases} \phi(2t), & 0 \le t \le 1/2, \\ \psi(2t - 1), & 1/2 \le t \le 1. \end{cases}$$

As one can see, the product of loops is defined in much the same way as the sum of elements of $\pi_n(I^n, \partial I^n; X, x_0)$. The product $[\phi] \cdot [\psi] = [\phi \cdots \psi]$ is defined on the set $\pi_1(X, x_0)$, and, generally speaking, it is not commutative (but it is associative).

Proposition. *The set $\pi_1(X, x_0)$ is a group with respect to the described product operation.*

Proof. Note that in the proof of theorem 1, the condition $n > 1$ has been used only to prove the commutativity of the group π_n, where the second coordinate in I^n plays a role in the necessary homotopies. Therefore all the previous steps of the proof of theorem 1 can be transferred for the case of $\pi_1(X, x_0)$ (and are considerably simplified). The unit and inverse elements in $\pi_1(X, x_0)$ are defined exactly in the same way as θ and $(-[\phi])$ in $\pi_n(X, x_0)$ for $n > 1$: $e = [\phi_0]$, where $\phi_0(I) = x_0$ is the constant loop; for each $[\phi] \in \pi_1(X, x_0)$, $[\phi]^{-1} = [\phi^{-1}]$, where $\phi^{-1}(t) = \phi(1 - t)$ is the same loop run through in the reverse direction. Thus, the required statement follows directly from the proof of theorem 1. □

Definition 2. The group $\pi_1(X, x_0)$ is called the *fundamental group* of a space X with distinguished point x_0.

Exercise 9°. Prove that the fundamental group of the disc $D_r^n(x_0)$ with distinguished point x_0 is trivial.

We shall elucidate the difference between the groups $\pi_1(X, x_0)$ and $\pi_1(X, x_1)$ of the same space with different distinguished points $x_0 \in X$, $x_1 \in X$. Several concepts will be necessary for this purpose.

The product $\omega_1 \cdot \omega_2$ of the paths[*] ω_1 and ω_2 such that $\omega_2(0)\omega_1(1)$ we define

---

[*]Remember that a path in a space X is a continuous mapping of an interval: $\omega : I \to X$.

in the same way as the product of loops:

$$(\omega_1 \cdot \omega_2)(t) = \begin{cases} \omega_1(2t), & 0 \le t \le 1/2, \\ \omega_2(2t - 1), & 1/2 \le t \le 1. \end{cases}$$

Evidently, $\omega_1 \cdot \omega_2$ is a path in the space X. A constant path in X is a path $C_{x_0} : I \to X$ such that $C_{x_0}(t) \equiv x_0$ for $t \in [0,1]$. The reverse path of a path ω is the path $\omega^{-1} : I \to X$ such that $\omega^{-1}(t) = \omega(1 - t)$. Since $(\omega \cdot \omega^{-1})(0) = (\omega \cdot \omega^{-1})(1)$, the path $(\omega \cdot \omega^{-1})(t)$ is a loop at the point $\omega(0)$.

Exercise 10°. Draw the path $(\omega \cdot \omega^{-1})(t)$. Show that $[\omega \cdot \omega^{-1}] = e$ in $\pi_1(X, x_0)$, $x_0 = \omega(0)$.

Note that the product of paths is associative just like the product of loops: $(\omega_1 \cdot \omega_2) \cdot \omega_3 = \omega_1 \cdot (\omega_2 \cdot \omega_3)$.

Theorem 3. *Any path* $\omega : I \to X$ *connecting the points* x_0 *and* x_1, *i.e.,* $\omega(0) = x_0$, $\omega(1) = x_1$, *induces an isomorphism of groups*

$$S_1^\omega : \pi_1(X, x_0) \to \pi_1(X, x_1),$$

which depends only on the homotopy class of the path ω.

Proof. Let $[\phi] \in \pi_1(X, x_0)$. Consider the mapping $\psi : (I, \partial I) \to (X, x_1)$ given by the formula $\psi = \omega^{-1} \cdot \phi \cdot \omega$. The path $\psi(t)$ can be represented visually as the loop in Fig. 73. Consider the class $[\psi] \in \pi_1(X, x_1)$.

By associating to each element $[\phi] \in \pi_1(X, x_0)$ the element $[\psi] \in \pi_1(X, x_1)$, we obtain a mapping $S_1^\omega : \pi_1(X, x_0) \to \pi_1(X, x_1)$. Indeed, if ϕ_τ is a homotopy of the loop ϕ in the class $[\phi] \in \pi_1(X, x_0)$, $0 \le \tau \le 1$, then $\psi_\tau =$

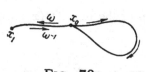

FIG. 73

$\omega^{-1} \cdot \phi_\tau \cdot \omega$ is a homotopy of the loop ψ in the class $[\psi] \in \pi_1(X, x_0)$. This shows that S_1^ω is well defined. It turns out that S_1^ω is a homomorphism of groups. We shall prove this. Let $[\phi_1], [\phi_2] \in \pi_1(X, x_0)$, and let $S_1^\omega[\phi_i] \in \pi_1(X, x_1)$, $i = 1, 2$. The product of loops $\phi_1 \cdot \phi_2$ defines the product $[\phi_1] \cdot [\phi_2]$ of the classes $[\phi_1]$ and $[\phi_2]$ (by definition). Applying associativity of the product of paths and exercise 10° as well, we obtain $S_1^\omega([\phi_1] \cdot [\phi_2]) = S_1^\omega[\phi_1 \cdot \phi_2] = [\omega^{-1} \cdot \phi_1 \cdot \phi_2 \cdot \omega] = [\omega^{-1} \cdot \phi_1 \cdot \omega \cdot \omega^{-1} \cdot \phi_2 \cdot \omega] = [\omega^{-1} \cdot \phi_1 \cdot \omega] \cdot [\omega^{-1} \cdot \phi_2 \cdot \omega] = (S_1^\omega[\phi_1]) \cdot (S_1^\omega[\phi_2])$; for the identity class $e_{x_0} = [\phi_0]$, where ϕ_0 is a constant loop, we have $S_1^\omega e_{x_0} = [\omega^{-1} \cdot \phi_0 \cdot \omega] = [\omega^{-1} \cdot \omega] = e_{x_1}$, i.e., the identity class in $\pi_1(X, x_1)$. Thus, S_1^ω is a homomorphism.

Similarly, we associate to element $[\psi] \in \pi_1(X, x_1)$ the an element $[\phi] \in \pi_1(X, x_0)$, where $\phi = \omega \cdot \psi \cdot \omega^{-1}$. We thus obtain a mapping

$$S_1^{\omega^{-1}} : \pi_1(X, x_1) \to \pi_1(X, x_0).$$

Exercise 11°. Show that $S_1^{\omega^{-1}}$ is a homomorphism of groups and that the homomorphisms $S_1^{\omega^{-1}}$ and S_1^ω are mutually inverse, i.e.,

$$(S_1^\omega)^{-1} = (S_1^{\omega^{-1}}).$$

Thus, S_1^ω is an isomorphism. It is clear from its definition that it remains unaltered under a homotopy of the path ω (with fixed ends). □

Exercise 12°. Prove that if $f : X \to Y$ is a continuous mapping, then for any path ω which connects the points x_0 and x_1 the diagram

$$
\begin{array}{ccc}
\pi_1(X, x_0) & \xrightarrow{f_1} & \pi_1(Y, f(x_0)) \\
\downarrow{\scriptstyle S_1^\omega} & & \downarrow{\scriptstyle S_1^{\bar\omega}} \\
\pi_1(X, x_1) & \xrightarrow{f_1} & \pi_1(Y, f(x_1))
\end{array}
$$

is commutative. Here, $\bar\omega = f\omega$ is a path connecting the points $f(x_0)$ and $f(x_1)$.

From theorem 3 we have that if a space X is path–connected then, at different points $x_0 \in X$, the groups $\pi_1(X, x_0)$ are isomorphic to each other and can be considered as one abstract group $\pi_1(X)$. This group $\pi_1(X)$ is called the fundamental group of the path–connected space X.

We shall present one more fact that follows from theorem 3.

Corollary. *Any element* $[\alpha] \in \pi_1(X, x_0)$ *defines an automorphism* $S_1^{[\alpha]}$ *of the group* $\pi_1(X, x_0)$ *under which* $[\beta] \to [\alpha]^{-1}[\beta][\alpha]$.

Proof. In virtue of theorem 3, there is an isomorphism $S_1^{[\alpha]} : \pi_1(X, x_0) \to \pi_1(X, x_0)$, since α is a loop at the point x_0. In addition, the isomorphism S_1^α depends only on the homotopy class of the path α.   □

The following definition singles out an important class of spaces.

Definition 3. A path–connected space X is said to be *simply connected* (*1–connected*) if any two paths $\omega_1 : I \to X$ and $\omega_2 : I \to X$ such that $\omega_1(0) = \omega_2(0) = x_0$, $\omega_1(1) = \omega_2(1) = x_1$, belong to the same homotopy class in $\pi(I, \partial I; X, x_0 \cup x_1)$, i.e. they are homotopic in the class of paths starting at x_0 and ending at x_1.

The following theorem characterizes simply connected spaces in terms of their fundamental grous.

Theorem 4. *A path–connected space X is simply connected iff $\pi_1(X) = 0$.*

Proof. Let the path–connected space X be simply connected. Consider an arbitrary class $[\phi] \in \pi_1(X, x_0)$ and the identity class $e = [\phi_0]$. Consider the two paths $\omega_1 = \phi : (I, \partial I) \to (X, x_0)$, $\omega_2 = \phi_0 : (I, \partial I) \to (X, x_0)$, $\phi_0(I) = x_0$, for which the starting and ending points coincide: $\phi(0) = \phi(1) = \phi_0(0) = \phi_0(1) = x_0$. The condition of being simply connected says that these path–loops ω_1, ω_2 are homotopic. Consequently, the loop ϕ is homotopic to the loop ϕ_0, and thus $[\phi] = e$. By the arbitrariness of $[\phi]$, we conclude $\pi_1(X, x_0) = 0$, and also $\pi_1(X) = 0$.

Inversily, let $\pi_1(X, x_*) = 0$ at the point $x_* \in X$; it can be assumed to be arbitrary because X is simply connected. Consider two paths ω_1, ω_2 in X with common starting point and ending point: $\omega_1(0) = \omega_2(0) = x_0$, $\omega_1(1) = \omega_2(1) = x_1$. We will show that they are homotopic as mappings $\omega_1 : (I, \partial I) \to (X, x_0 \cup x_1)$, $\omega_2 : (I, \partial I) \to (X, x_0 \cup x_1)$. Let us construct the loop $\phi = \omega_1 \cdot \omega_2^{-1}$ at the point x_0. Since the path ω_2^{-1} is given by the equality $\omega_2^{-1}(s) = \omega_2(1 - s)$, $0 \leq s \leq 1$, by the definition of the product of two paths, we obtain

$$\phi(t) = (\omega_1 \cdot \omega_2)^{-1}(t) = \begin{cases} \omega_1(2t), & 0 \leq t \leq 1/2, \\ \omega_2(2 - 2t), & 1/2 \leq t \leq 1. \end{cases}$$

Take $x_* = x_0$, we have $\pi_1(X, x_0) = 0$ by the condition. Consequently, the loop ϕ is homotopic to the constant loop ϕ_0 at the point x_0; it is useful to present a homotopy in the form

$$\Phi(t,\tau) = \begin{cases} \Omega_1(2t,\tau), & 0 \le t \le 1/2, \\ \Omega_2(2-2t,\tau), & 1/2 \le t \le 1, \end{cases}$$

where $\Omega_1(2t,\tau)$, $\Omega_2(2-2t,\tau)$ are homotopies of paths ω_1, ω_2 into a constant path at the point x_0, and τ is the parameter of homotopy, $0 \le \tau \le 1$, and $\Phi(t,1) = \phi(t)$, $\Phi(t,0) = \phi_0$. Let us fix the points $\omega_1(s)$, $\omega_2(s)$, where $0 \le s \le 1$. These points on the loop ϕ correspond to the values of the parameter: $t = s/2 \le 1/2$ and $t = 1 - s/2 \ge 1/2$. We define a path $\psi(t)$ of moving the point $\omega_1(s)$ to the point $\omega_2(s)$ as follows:

$$\psi_s(\tau) = \begin{cases} \Phi(s/2, 1-2\tau), & 0 \le \tau \le 1/2, \\ \Phi(1-s/2, 2\tau-1), & 1/2 \le \tau \le 1. \end{cases}$$

Geometrically this means that the point $\omega_1(s)$ moves along the trajectory given by the homotopy $\Phi(t,\tau)$ to the point x_0, and further, in a similar way, to the point $\omega_2(s)$. Since s can be chosen arbitrarily (from the interval $0 \le s \le 1$), the previous formula, in fact, defines a homotopy of the path ω_1 into ω_2 (the function $\psi_s(\tau)$ also depends on s : $\psi_s(\tau) = \psi(s,\tau)$). Evidently, this dependence $\psi(s,\tau)$ is continuous in (s,τ), and thus, ω_1, ω_2 belong to the same homotopy class in $\pi_1(I, \partial I; X, x_0 \cup x_1)$. $\qquad \square$

Exercise 13°. Verify that the Euclidean space $\mathbf{R}^n$ is simply connected, and that S^1 and the torus $S^1 \times S^1$ are not simply connected.

Exercise 14°. Construct an example of a connected space with non–isomorphic groups $\pi_1(X, x_0)$ at different points x_0.

Hint. Use the example of a connected but not path–connected space from Ch. 2, § 10.

Now, we shall investigate the dependence of the higher homotopy groups on variation of the base point. The homotopy group $\pi_n(X, x_0)$ turns out to vary similarly to the fundamental group $\pi_1(X, x_0)$ when its distinguished point changes.

Theorem 5. *Any path* $\omega : I \to X$ *connecting points* x_0 *and* x_1 *determines the isomorphism*

$$S_n^\omega : \pi_n(X, x_1) \to \pi_n(X, x_0),$$

which depends only on the homotopy class $[\omega] \in \pi(I, \partial I; X, x_0 \cup x_1)$. *In addition, for any mapping* $f: X \to Y$, *the diagram*

$$\begin{array}{ccc} \pi_n(X, x_1) & \xrightarrow{\;f_n\;} & \pi_n(Y, y_1) \\[2mm] s_n^\omega \downarrow & & s_n^{\tilde\omega} \downarrow \\[2mm] \pi_n(X, x_0) & \xrightarrow{\;f_n\;} & \pi_n(Y, y_0) \end{array}$$

in which $S_n^{\tilde\omega}$ *is the isomorphism defined by the path* $\tilde\omega = f\omega$ *between the points* $y_0 = f(x_0)$, $y_1 = f(x_1)$, *is commutative.*

We give only the idea of the proof of this theorem. Let $[\phi] \in \pi_n(X, x_1)$. As in the case of the fundamental group, the element $[\phi]$ gives rise to an element $[\psi] \in \pi_n(X, x_0)$. Visually this procedure can be seen as pulling a "whisker" out of the spheroid ϕ at the point x_1 to the point $\omega(t)$ and stretching it along the path ω up to the point x_0 (Fig. 74).

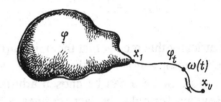

FIG. 74

Thus, we obtain a mapping $S_n^\omega : \pi_n(X, x_1) \to \pi_n(X, x_0)$ which is an isomorphism with the required properties. Here we omit the details.

As a consequence of theorem 5, we obtain that any element $[\alpha] \in \pi_1(X, x_0)$ determines an automorphism of the group $\pi_n(X, x_0)$.

Thus, the group $\pi_1(X, x_0)$ acts on the group $\pi_n(X, x_0)$ as a group of automorphisms (more exactly, as a subgroup of the group of all automorphisms).

Now, it is natural to define the following generalization of simply connected spaces.

Definition 4. If the isomorphism $S_n^\omega : \pi_n(X, x_1) \to \pi_n(X, x_0)$ for a space X and for any points $x_0, x_1 \in X$ which are in the same path connectedness component, does not depend on the choice of the path ω connecting x_0 with x_1, then the space X is said to be n–*simple* (or *homotopy simple in dimension* n).

We suggest to prove the following statement.

Theorem 6. *A space* X *is* n–*simple iff for any point* $x_0 \in X$ *the group* $\pi_1(X, x_0)$ *acts trivially on* $\pi_n(X, x_0)$, *i.e. does not alter the elements of* $\pi_n(X, x_0)$.

From theorem 4, immediately follows that a simply connected space is n–simple for all $n \geq 1$.

§ 4. Computing the fundamental and homotopy groups of some spaces

In this section, we compute the fundamental group of the circle and of an arbitrary closed surface of the type M_p or N_q. For this we use a combinatorial technique which is based on the results of Ch. 2, § 4 and presented at the beginning of this section (see Subsections 1, 2). Along the way, the topological invariance of the Euler characteristic of a closed surface is established (see Subsection 5). Further, the problem of computing higher homotopy groups is discussed, and a supplement to the problem concerning fixed points of a continuous mapping is given (the Brouwer theorem, the fundamental theorem of algebra).

1. Line paths on a surface and their combinatorial homotopies

Consider a closed surface X given, as in Ch 2, § 4, by its decomposition. This means that a certain development Π is given, and the surface X is homeomorphic to the quotient space Π/R, where R is an equivalence defined by the gluing homeomorphisms of the development.

Denote the product of the quotient mappings $\alpha : \Pi \to \Pi/R$ and the homeomorphism $\beta : \Pi/R \to X$ by κ. Then κ is a mapping $\Pi \to X$, and it determines a decomposition of X into the images of polygons, edges and vertices of the development (we call the κ–images of edges the *edges*, and the κ–images of vertices the *vertices of the decomposition*). An edge of a decomposition is the κ–image of two edges, a and a^{-1}, or a and a; we agree to denote it by a; the κ–image of a vertex A we denote by the same letter A; the points of an edge which are different from the vertices we call the *interior points of an edge*.

We shall require the following elementary operations on decompositions: (a) addition of a new vertex: an interior point of an edge is a new vertex of the decomposition; (b) addition of a new edge: one of the polygons of the development is decomposed into two by a diagonal; the κ–image of this diagonal in X is a new edge of the decomposition.

Consider an edge a in the development Π, and let $\gamma : I \to a$ be an affine mapping (linear path) under which the points 0 and 1 are mapped into the vertices of the edge. Then the mapping $\bar{\gamma} = \kappa\gamma : I \to X$ determines a path on the surface X which we call an *elementary path*. Obviously, the image of an elementary path either coincides with one of the vertices of the edge a of the decomposition of the surface or completely covers this edge.

In the first case, an elementary path is constant and is set to be zero ($\widetilde{\gamma} = 0$). In the second case, the beginning of the linear path γ either coincides with the beginning of the oriented edge a, or with its end. In view of this, we denote an elementary path by a or a^{-1} ($\widetilde{\gamma} = a$ or $\widetilde{\gamma} = a^{-1}$, respectively). According to the same rule, we denote $\widetilde{\gamma}$, if $\gamma : I \to a^{-1}$, assuming $(a^{-1})^{-1} = a$.

Thus, to each oriented edge $a(a^{-1})$ of the development, there corresponds an elementary path $a(a^{-1})$ in the decomposition.

Definition 1. A *line path* in a decomposition Π of a surface X is a finite product of elementary paths. A closed line path is called a *line loop*.

By definition 1, a line path λ can be written in the form of the product of elementary paths $\lambda = \lambda_1 \lambda_2 \ldots \lambda_s$, where $\lambda_i = a_i^{\pm 1}$ or $\lambda_i = 0$. By omitting zeroes, we associate with the path λ a word $\omega(\lambda) = a_{i_1}^{\pm 1} \ldots a_{i_s}^{\pm 1}$ which indicates the order and the direction of the orientation of the edges of the decomposition of the surface X along the path λ.

Consider the boundary Γ_i of a polygon Q_i of a development Π. By associating each edge of the boundary with an elementary path as described above, we associate to the whole boundary a line path λ_i in X which is determined by the word $\omega(\lambda_i) = \omega(Q_i)$. The word $\omega(Q_i)$, in turn, describes the scheme of gluing the polygon Q_i (see Ch. 2, § 4, subsection 2).

For instance, the line path λ corresponding to the oriented boundary of the rectangle Q which represents the development of a torus (see Fig. 50), is determined by the word $\omega(\lambda) = aba^{-1}b^{-1}$.

Definition 2. A *combinatorial deformation* of type I (resp. type II) of a line loop λ is a deletion or insertion of a combination of the form aa^{-1} to the word $\omega(\lambda)$ (resp. the deletion or insertion of a word $\omega(Q_i)$ corresponding to the oriented boundary of the polygon Q_i of the development Π) which determines the line loop in X.

Definition 3. The line loops γ and γ' in Π are said to be *combinatorially homotopic* in Π if one loop can be obtained from the other with a help of a finite number of combinatorial deformations of the type I or II.

Note that any line path in the decomposition Π of the surface X can be considered as a line path in some decomposition Π_1 which is obtained from Π by applying a finite number of operations of the type (a) or (b) defined above.

Lemma 1. *Let a decomposition Π_1 be obtained from a decomposition Π by applying a finite number of operations of the form (a) or (b). Then for any line loop λ in Π_1 there exists a line loop λ' in Π, which is combinatorially homotopic to the loop λ in Π.*

Proof. Obviously, it is sufficient to consider the case when Π_1 is derived from Π by applying one of the operations (a) or (b). Let Π_1 be obtained from Π by decomposing an edge a into two new edges b and c (the operation of adding a new vertex is applied). If the loop λ in Π_1 contains one of the combinations bb^{-1}, cc^{-1}, $b^{-1}b$, $c^{-1}c$, then it can be omitted obtaining by this a loop which is homotopic to λ. By omitting all such combinations, we obtain a loop either not containing $b^{\pm 1}$, $c^{\pm 1}$ or containing them in the form bc $(= a)$ or $c^{-1}b^{-1}$ $(= a^{-1})$; in either case, this is the sought for line path λ' in Π.

FIG. 75

Now, let Π_1 be obtained from Π by adding a new edge d which devides a certain polygon of Π into parts E and F. Let the boundary paths of E and F be ud^{-1} and dv, respectively (Fig. 75). If the line loop λ includes the edge $d^{\pm 1}$, then we replace it by the path $v^{\mp 1}$ (or $u^{\pm 1}$). The loop λ' obtained is combinatorially homotopic to λ and is a line loop in Π. □

Lemma 2. *Let Π_1 be obtained from Π by one of the operations of the form (a) or (b). Then any line loop λ in Π which is combinatorially homotopic to zero in Π_1, will also be combinatorially homotopic to zero in Π.*

Proof. According to the condition of the lemma, there exists a sequence of line loops $\lambda = v_0, v_1, \ldots, v_r = 0$ in Π_1, where v_{i+1} is obtained from v_i by one of the combinatorial deformations. In addition, $v_1, \ldots, v_r$ are not, generally speaking, loops in Π. For each loop v_i, $i = 1, \ldots, r$, we construct a line loop ω_i homotopic to it in Π so that in the sequence of the loops $\lambda, \omega_1, \ldots, \omega_r = 0$, each loop ω_{i+1} is obtained from ω_i with a help of one or several combinatorial deformations.

Assume that Π_1 is obtained from Π by decomposing an edge a into edges b and c (the operation of type (a)). Associate then to each loop v_i a loop ω_i by assigning to an edge which is different from $b^{\pm 1}$, $c^{\pm 1}$, the same edge, to the edge $b^{\pm 1}$ the edge $a^{\pm 1}$, and nothing to the edge $c^{\pm 1}$. It is easy to verify that then the transfer from ω_i to ω_{i+1}, $i = 1, \ldots, r$, becomes a combinatorial deformation of type I or II.

If, however, Π_1 is obtained from Π by an operation of type (b), then we associate any edge different from the decomposing edge d with itself, and replace $d(d^{-1})$ by the path $u(u^{-1})$. If, now, we add or delete the combination dd^{-1} in v_i, in order to obtain v_{i+1}, then it is required to add or delete the combination uu^{-1} in ω_i, respectively. The deformations of the type II in Π_1 correspond to the deformations of the type I or II in Π.   □

2. Combinatorial approximations of paths and homotopies.
Here, we shall show that any continuous path in a triagulation K is homotopic to a line path, and also investigate the relationship between combinatorial and continuous homotopies.

Everywhere below we consider homotopies of paths and loops with fixed ends.

Lemma 3. *Let a triangulation K of a surface X be given. Let $\lambda : I \to K$ be a continuous path in K, moreover, let $\lambda(0)$, $\lambda(1)$ be vertices of the triangulation. Then there exists a line path in K, which is homotopic to it.*

Proof. Let us divide the interval $I = [0, 1]$ by a finite number of points $\{t_k\}_{k=0}^n$ ($t_0 = 0$, $t_n = 1$) into sufficiently small intervals so that for each interval (t_{k-1}, t_{k+1}), $k = 1, \ldots, n - 1$, there is a vertex $A_k \in K$ such that the image $\lambda(t_{k-1}, t_{k+1})$ of this interval lies wholly in the star $S(A_k)$ which is the union of open triangles and edges of the triangulation K adherent to that vertex A_k and the vertex A_k itself. Since $S(A_k)$ is an open set in X, and λ is a continuous mapping, this can always be achieved (see Ch. 2, § 13, Exercise 7°).

Now, we associate to each point $t_k \in I$ the vertex $A_k \in K$. It should be noted that for any $k = 1, \ldots, n - 1$

$$\lambda((t_k, t_{k+1})) \subset S(A_k) \cap S(A_{k+1}),$$

where $S(A_k) \cap S(A_{k+1})$, evidently, contains a triangle adherent to both vertices A_k and A_{k+1}. Consequently, if $A_k \neq A_{k+1}$ then they are connected in K by

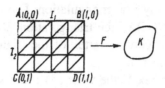

FIG. 76

an edge which we denote by l_k. Let $\lambda'_k : [t_k, t_{k+1}] \to l_k$ be the elementary path which is the extension of the indicated correspondence of the vertices A_k, A_{k+1} and the points t_k, t_{k+1}. In the case of $A_k = A_{k+1}$, we take λ'_k to be zero. The product of elementary paths λ'_k determines a line path $\lambda' : I \to K$ which is called a *line approximation of the path* λ.

The paths λ and λ' are homotopic to one another. Indeed, due to the construction of the path λ', for any point $t \in I$, the images $\lambda(t)$ and $\lambda'(t)$ lie in the same closed topological triangle from K; therefore, they can be connected by an "interval" which is the homeomorphic image of an interval in a triangle of the development; consequently, there is given a natural linear deformation of the point $\lambda(t)$ into the point $\lambda'(t)$ which defines the required homotopy. To this end, note that any point $\lambda(t)$ does not leave that closed triangle, edge or vertex, in which it was at the initial moment of the process of the homotopy.   □

It is necessary to distinguish whether the line loop is homotopic to a constant one in a topological or combinatorial sense. A loop homotopic to a constant one is said to be *contractible* or *combinatorially contractible*, respectively.

Lemma 4. *A contractible line loop* λ *in a triangulation* K *is combinatorially contractible in* K.

Proof. Let a line loop λ be given by a mapping of an interval $\psi : I_1 \to K$. Let $F : I_1 \times I_2 \to K$ be the contraction of the loop to a vertex $x_0 \in K$, i.e.,

$$F|_{I_1 \times \{0\}} = \psi, \quad F|_{I_1 \times \{1\}} = c_0 : I_1 \to x_0 \in K.$$

It is clear that $F|_{\{0\} \times I_2} : I_2 \to x_0$ and $F|_{\{1\} \times I_2} : I_2 \to x_0$.

Since F is a contraction keeping the ends of the loop fixed, the edges AC, CD, and BD (Fig. 76) are mapped into one point x_0. Mark those points on AB whose images are vertices of K, and draw vertical lines through them. Then, by drawing additional vertical and horizontal lines and diagonals (Fig. 76), we obtain a sufficiently fine triangulation Σ of the square $ABCD$ such that the image of a star $S(V)$ of the triangulation Σ under the mapping F to lies in the star $S(W)$ of a certain vertex of the triangulation K (this follows from Ch. 2, § 13, Exercise 7°).

Associate now to the vertex V the vertex W; in a similar way, we deal with all the vertices of the triangulation Σ. Afterwards, extend this mapping to the edges of the triangulation Σ exactly in the same manner as it was done in the proof of the lemma on a line approximation of a path. The mapping obtained $F_1 : \Sigma_1 \to K$, where Σ_1 is the union of the edges of the triangulation Σ, transforms the decomposed side AB into a line loop $\bar{\lambda}$ in K.

Now, we show that $\bar{\lambda}$ is combinatorially deformable into λ. In fact, under the line approximation of a path, no point of the path leaves the triangle, edge or vertex, where it was situated. Therefore, the loop $\bar{\lambda}$ consists of the same elementary paths as λ (if no attention is paid to the zero paths which can be omitted). However, generally speaking, some edges can be run through several times in different directions. Thus, we can go from $\bar{\lambda}$ to λ by combinatorial deformations of the type I.

Note now that the decomposed side AB in the triangulation Σ can be transformed into the decomposed broken line $ACDB$ with a help of combinatorial deformations of the type I and II, by a successive "squeezing out" one triangle at a time (Fig. 77). However, due to the construction of the mapping F_1, each of these combinatorial deformations applied to AB determines a combinatorial deformation of the type I or II of the loop $\bar{\lambda}$ in K (verify this!).

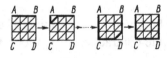

FIG. 77

Thus, we have shown that with a help of combinatorial deformations of the type I and II, the line loop λ can be tranformed into the loop $\bar{\lambda}$, and afterwards, into the F_1–image of the path $ACDB$. But this image is the point x_0. Therefore, λ is combinatorially homotopic to a constant. $\qquad\square$

We recommend the reader to prove two other uncomplicated statements which will be used later.

Exercise 1°. Prove that a line path λ in a decomposition Π determined by the word $\omega(\lambda) = aa^{-1}$, is homotopic to a constant path.

Exercise 2°. Prove that a line path in a decomposition Π which is equal to the image of the boundary of some polygon of the development Π is homotopic to a constant path in X.

From exercises $1°$, $2°$ it follows that any combinatorial homotopy determines a usual continuous homotopy between line paths.

Note. In the following subsection, we shall need a rather special case of the combinatorial technique developed above, namely, the decomposition of the circle S^1.

Fix a finite number of points A', B', C',... on S^1 and define a homeomorphism ϕ of the boundary of a convex polygon ABC... in S^1 so that $\phi(A) = A'$, $\phi(B) = B'$, $\phi(C) = C'$,.... We say that the homeomorphism ϕ determines a decomposition of S^1 with the edges $A'B' = \phi(\overline{AB})$, $B'C' = \phi(\overline{BC})$, $C'A' = \phi(\overline{CA})$... and vertices A', B', C',.... The line paths and combinatorial deformations of the type I are defined naturally. It is easy to see that Lemmas $1-4$ hold for such decompositions with the only change that the operations on decompositions of the type (b) and combinatorial deformations of type II vanish.

3. The fundamental group of a circle. Now we can calculate the group $\pi_1(S^1)$.

Theorem 1. *The group $\pi_1(S^1)$ is Abelian and isomorphic to the group $\mathbf{Z}$.*

To prove this theorem, we need the following auxiliary statement which will be strengthened later (see § 4, theorem 4 of this chapter).

Lemma 5. *The fundamental groups of homeomorphic spaces are isomorphic.*

Proof. Let X, Y be topological spaces with distinguished points x_0 and y_0, respectively, and let $\phi : (X, x_0) \to (Y, y_0)$ be a homeomorphism. Then the homomorphisms of the fundamental groups

$$\pi(\phi) : \pi_1(X, x_0) \to \pi_1(Y, y_0),$$
$$\pi(\phi^{-1}) : \pi_1(Y, y_0) \to \pi_1(X, x_0),$$

are defined; moreover, due to the functorial property, we have:

$$\pi(\phi^{-1})\pi(\phi) = \pi(\phi^{-1}\phi) = 1_{\pi_1(X, x_0)},$$
$$\pi(\phi)\pi(\phi^{-1}) = \pi(\phi\phi^{-1}) = 1_{\pi_1(Y, y_0)};$$

therefore, $\pi(\phi) = [\pi(\phi^{-1})]^{-1}$.   □

Proof of theorem 1. By the last lemma, it is sufficient to calculate the fundamental group of a plane triangle. Let $\triangle$ be the triangle with vertices A, B, C, oriented edges a, b, c and distinguished vertex A (Fig. 78).

We first compute the group $\pi_1(\triangle, A)$. Let λ be an arbitrary loop in $\triangle$ with origin at the point A. Then by lemma 3, there exists a line loop λ' in the homotopy class of the loop λ. (It is clear that the triangle $\triangle$ is a decomposition.) Associate to each edge a, b, c loops $\tilde{a}$, $\tilde{b}$, $\tilde{c}$ according to the following rule: $\tilde{a} = cab$, $\tilde{b} = b^{-1}b$, $\tilde{c} = cc^{-1}$. We show that the classes of the loops $\tilde{a}$, $\tilde{b}$, $\tilde{c}$, which are not a priori different, are generators of the group $\pi_1(\triangle, A)$. Indeed, any line loop λ' consists of elementary paths corresponding to the

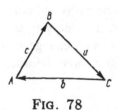

FIG. 78

edges, i.e., $\lambda' = \phi(a, b, c)$. By replacing each edge with the corresponding loop in this expression, we obtain a new loop $\lambda'' = \phi(\tilde{a}, \tilde{b}, \tilde{c})$. It is easy to see that the loops λ' and λ'' are combinatorially homotopic. Indeed, the replacement of an edge with a loop makes us first to "reach" the origin of this edge from the fixed vertex A and then "run through" this edge, "return" to A along the shortest path (Fig 78). Therefore, under each successive replacement of an edge with a loop, after returning to A from the end P of the previous edge, we have to "start" for the origin of the next edge, i.e., for the same point P. By this replacement, a path of the form $\triangle\triangle^{-1}$ is also introduced between each two adjacent edges of the loop, i.e a path which is combinatorially homotopic to zero. Thus, in the homotopy class of the loop λ', there always exists a line loop λ represented by a finite product of loops $\tilde{a}$, $\tilde{b}$, $\tilde{c}$ and of the loops inverse to it.

Now, note that the loops $\tilde{b}$, $\tilde{c}$ are homotopic to constant ones. Therefore the loop $\tilde{a}$ (or, more exactly, the homotopy class determined by it in $\pi_1(\triangle, A)$) is a unique generator in the group $\pi_1(\triangle, A)$. The element $\tilde{a}$ is non–trivial. In fact, if the loop $\tilde{a}$ were contractible then, according to lemma 4, it would also be combinatorially contractible, i.e. would be reduced to zero by a finite number of combinatorial deformations of the type I, which is obviously impossible. Consequently, the loop $\tilde{a}$ is not combinatorially contractible, and, thus, it determines a non–trivial element $[\tilde{a}] \in \pi_1(\triangle, A)$. Similarly, any element $[\tilde{a}^l] \in \pi_1(\triangle, A)$, where $l > 1$, is non–trivial.

Thus, $\pi_1(\triangle, A)$ is the free cyclic group generated by the element $[\tilde{a}]$, i.e. it is an Abelian group isomorphic to **Z**. $\qquad\square$

Exercise 3°. Generalizing the construction of the proof of theorem 1, prove that the fundamental group of the wedge of m circles is a free group with m generators.

A useful tool for computing fundamental groups of more complicated spaces is the following theorem.

Theorem 2. (Seifert–Van Kampen). *Let X be a topological space which is the union $X=X_1\cup X_2$ of open subsets X_1 and X_2 such that the spaces X_1, X_2 and $X_0=X_1\cap X_2$ are path–wise connected and non–empty, and let $p\in X_0$. Consider the commutative diagram generated by the imbedding mappings:*

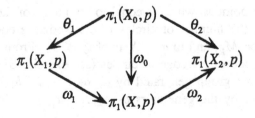

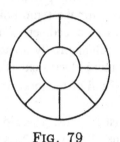

Then the group $\pi_1(X,p)$ is the quotient group of the free product $\pi_1(X_1,p)*\pi_1(X_2,p)$ by the normal divisor which is generated by the set $\{\Theta_1*\Theta_2\alpha^{-1}:\alpha\in\pi_1(X_0,p)\}$. In other words, the group $\pi_1(X,p)$ is generated by the images of the elements from $\pi_1(X_i,p)$, $i=1,2$, and all the relations between generators are derived relations obtained as the ω_i-images of relations in each of the groups $\pi_i(X_i,p)$, $i=1,2$, and the relations $\omega_1\Theta_1\alpha=\omega_2\Theta_2\alpha$, where $\alpha\in\pi_1(X_0,p)$.

Fig. 79

Exercise 4°. Using the Seifert–Van Kampen theorem, obtain the statement of exercise 3°.

Exercise 5°. Compute the fundamental group of the space consisting of two circles connected by intervals (Fig. 79).

4. The fundamental group of a surface.

Now we turn to the computation of the fundamental group of a surface. On the basis of Lemma 5, we may assume that a closed surface is given as the decomposition which is determined by a canonical development.

Theorem 3. *Let X be a closed surface of the type I or II determined by a word ω of the form $a_1b_1a_1^{-1}b_1^{-1}\ldots a_pb_pa_p^{-1}b_p^{-1}$ or $a_1a_1a_2a_2\ldots a_qa_q$, respectively; let $x_0 \in X$ be a fixed point on a surface (a triangulation vertex). Then $\pi_1(X, x_0)$ is a group with generators $a_1, a_2, \ldots, a_p, b_1, b_2, \ldots, b_p$, or $a_1, a_2, \ldots, a_q$, respectively, and with one relation $\omega = e$, where e is the identity element.*

Proof. Let X_1 be a closed surface, $\mathcal{P}$ its canonical development determined by a polygon Q and the word $\omega(Q)$. Let $X_1 = \kappa(Q_1)$, where Q_1 is the union of all edges of the polygon Q. Due to the equivalence of all vertices of Q in the development $\mathcal{P}$, their images under the mapping κ coincide in X. Therefore, the image of each edge is homeomorphic to a circle, and X_1 is the wedge of circles glued at the point x_0 which is equal to the image of the vertices of the polygon Q. Thus, the number of circles in the wedge is equal to $2p$ if the surface X is of type M_p, and to q, if X is of type N_q. From the result about the fundamental group of a wedge of circles (see exercise 3°) we obtain that $\pi_1(X_1, x_0)$ is the free group generated by $a_1, a_2, \ldots, a_p, b_1, b_2, \ldots, b_p$, if X has the type M_p, or by the generators $a_1, a_2, \ldots, a_q$, if X has the type N_q.

We denote this group by G.

Consider now the imbedding mapping $i : X_1 \to X$ and the homeomorphism of the fundamental groups induced by it:

$$i_* : \pi_1(X_1, x_0) \to \pi_1(X, x_0).$$

We shall calculate the group $\pi_1(X, x_0)$ in the following manner. First, we prove that i_* is an epimorphism. Then using the homomorphism theorem we obtain

$$\pi_1(X, x_0) \simeq \pi_1(X_1, x_0)/\text{Ker } i_* = G/\text{Ker } i_*.$$

By computing kernel Ker i_*, we shall complete the proof.

Proof that i_* is an epimorphism. Let $\alpha \in \pi_1(X, x_0)$ and K a triangulation of the surface X. Then, by lemma 3 about line approximations, there is a line loop λ (in the decomposition K) in the homotopy class α. One may assume K to be obtained from the canonical decomposition $\mathcal{P}$ of the surface X with the help of a finite number of operations of the form (a) or (b).

Consequently, on the basis of lemmas 1, 2, and exercises 1°, 2°, in the same class of $\alpha \in \pi_1(X, x_0)$, there is a line loop λ' (i.e. a composition of edges of X_1) in the decomposition $\mathcal{P}$. By the same token, a certain class $\beta_\alpha \in \pi_1(X_1, x_0)$ for which, evidently, $i_*(\beta_\alpha) = \alpha$, is defined. This proves that i_* is epimorphic.

Now we compute the kernel Ker i_* of the epimorphism i_*. Let $\gamma \in \text{Ker } i_*$, and λ a line loop of the decomposition $\mathcal{P}$ from the class γ. Then λ is contractible to a point in X. By lemma 4, there exists in $\mathcal{P}$ a combinatorial contraction of λ to the vertex x_0. In other words, the word $\omega(\lambda)$ which determines the loop λ is reduced to the zero word with the help of a finite number of combinatorial deformations of the type I or II. It is then clear that the word $\omega(\lambda)$ can consist only of combinations of the form:

(1) aa^{-1};

(2) more complex one of the form $\omega_4 h_2 \omega_1 h \omega^l h^{-1} \omega_2 h_1 \omega^m h_1^{-1} \omega_3 h_2^{-1} \omega_5$, where $\omega_4 \cdot \omega_5 = \omega$, $\omega_1 \cdot \omega_2 \cdot \omega_3 = \omega$; h_1, h_2, h are words of the given decomposition; l, m are integral exponents (positive or negative);

(3) combinations analogous to (2), but with the word ω differently decomposed into components. This follows from the fact that with an exeption of ω there are no other bounding words.

It is easy to see that combinatorial deformations of the type I (additions or deletions of a combination of the type aa^{-1}) do not take the loop λ out of its homotopy class, since the loop aa^{-1} is homotopic to zero in X_1. Owing to this, one may assume that there are no combinations of the form (1) in the word $\omega(\lambda)$. The combinations of the form (2) are simplified by combinatorial deformations of the type I as follows:

$$\omega_4 h_2 \omega_1 h \omega^l h^{-1} \omega_2 h_1 \omega^m h_1^{-1} \omega_3 h_2^{-1} \omega_5 \rightarrow$$

By inserting combinations $\omega_1^{-1}\omega_1$, $\omega_3\omega_3^{-1}$, we obtain

$$\rightarrow \omega_4 h_2 \omega_1 h \omega^l h^{-1} \omega_1^{-1} \omega_1 \omega_2 \omega_3 \omega_3^{-1} h_1 \omega^m h_1^{-1} \omega_3 h_2^{-1} \omega_5 \rightarrow$$

Use condition $\omega_1\omega_2\omega_3 = \omega$ and write $\omega_1 h = g$, $\omega_3^{-1} h = \alpha$, to obtain

$$\rightarrow \omega_4 h_2 g \omega^l g^{-1} \omega \alpha \omega^m \alpha^{-1} h_2^{-1} \omega_5 \rightarrow$$

Now insert combinations $h_2^{-1}\omega_4^{-1}\omega_4 h_2$, $h_2^{-1}\omega_5\omega_5^{-1} h_2$ and get

$$\rightarrow \omega_4 h_2 g \omega^l g^{-1} h_2^{-1} \omega_4^{-1} \omega_4 h_2 \omega h_2^{-1} \omega_5 \omega_5^{-1} h_2 \alpha \omega^m \alpha^{-1} h_2^{-1} \omega_5 \rightarrow$$

Writing $\omega_4 h_2 g = f$, $\omega_5^{-1} h_2 \alpha = \beta$, to obtain

$$\rightarrow f \omega^l f^{-1} \omega_4 h_2 \omega h_2^{-1} \omega_5 \beta \omega^m \beta^{-1} \rightarrow$$

And, finally, inserting the combination $\omega_5\omega_5^{-1}$ and writing $\omega_5 h_2 = \gamma$ we obtain

$$\rightarrow f \omega^l f^{-1} \omega \gamma \omega \gamma^{-1} \beta \omega^m \beta^{-1}.$$

Thus, we have shown that with the help of combinatorial deformations of the type I, any line loop representing an element from the kernel Ker $_*$ is reduced to a loop whose word consists only of combinations of the form $\alpha \omega^m \alpha^{-1}$, where m is an integral exponent, and α is an arbitrary word which is empty or consists of symbols of edges of the development.

Inversily, it is evident that if a line loop λ has a word $\omega(\lambda)$ consisting only of combinations of the form $\alpha \omega^m \alpha^{-1}$, then it determines an element from Ker i_*.

Exercise 6°. Prove that the set of words of the form described is a normal divisor N of the group G and is generated by the element $\omega = \omega(Q)$.

Thus, Ker $i_* = N$, and, consequently, $\pi_1(X, x_0) = G/N$. The latter equality is equivalent to the fact that the unique relation $\omega = e$ is imposed on the generators of the group G.   □

We shall indicate several corollaries of theorem 3.

Corollary 1. *The fundamental group* $\pi_1(\mathbf{R}P^2, p)$ *of the projective plane* $\mathbf{R}P^2$ *is a cyclic group of order two.*

Proof. The surface $X = \mathbf{R}P^2$ has a canonical development with word $a_1 a_1$, therefore, $\pi_1(X, x_0)$ is a cyclic group with one generator a_1 and the relation $a_1^2 = e$.   □

Using more powerful tools we shall show in Ch 4, § 9, that $\pi_1(\mathbf{R}P^n) \simeq \mathbf{Z}_2$, $n \geq 2$, in particular, $\pi_1(\mathbf{R}P^3) \simeq \mathbf{Z}_2$.

Corollary 2. *The fundamental group of the torus* $\pi_1(T^2, x_0)$ *is the free Abelian group on two generators.*

Proof. The torus T^2 has a canonical development with word $aba^{-1}b^{-1}$, therefore, we obtain that the group $\pi_1(T^2, x_0)$ is generated by a, b. The relation $aba^{-1}b^{-1} = e$ gives its commutativity $ab = ba$.   □

Geometrically, to the generator a_1 of the fundamental group of the projective plane, there corresponds its absolute (see the models for $\mathbf{R}P^2$ in Ch. 2, § 4). To the generators a_1, b_1 of the fundamental group of the torus T^2, there correspond its parallel and meridian, i.e., two principal non–contractible loops on the torus.

Exercise 7°. Explain the geometrical meaning of the generators of the fundamental groups for the surfaces M_p, N_q.

The fundamental group of the complement of a knot plays an important role in the knot classification problem.

Exercise 8°. Prove that the trivial knot is not equivalent to either the "trefoil" or "figure eight" knot.

Hint. Show that the fundamental groups of the complements to these knots in $\mathbf{R}^3$ are not isomorphic.

5. The topological invariance of the Euler characteristic of a surface.

Let X, X' be two homeomorphic closed surfaces with decompositions Π, Π'; let $\chi(\Pi)$, $\chi(\Pi')$ be their Euler characteristics computed by the decompositions Π, Π', respectively. We shall prove that $\chi(\Pi) = \chi(\Pi')$.

The development $\Pi(\Pi')$ is reduced to a canonical development by equivalent transformations of the type I or II (determined by the number of handles $p(p')$ or Möbius strips $q(q')$ that are glued to the sphere). The numbers $2p(2p')$, $q(q')$ are the numbers of generators (connected by the defining relation $\omega = e$) of the fundamental group of the surface. Due to the homeomorphism of X and X', the groups $\pi_1(X)$ and $\pi_1(X')$ are isomorphic. Consequently, if the canonical type of the developments Π, Π' corresponds to the type I, then $2p = 2p'$, whence $\chi(\Pi) = \chi(\Pi')$ in virtue of the equalities $\chi(\Pi) = 2 - 2p$, $\chi(\Pi') = 2 - 2p'$. The case of canonical developments of the type II is considered similarly. Thus, two different surfaces of the type M_p of different genus (as well as surfaces of type N_q) are not homeomorphic.

Two surfaces of the type M_p and N_q, $q \geq 1$, are not homeomorphic either. This follows from the fact that the fundamental groups of an orientable surface M_p of genus p and a non–orientable surface N_q, $q \geq 1$, of genus q are not isomorphic. Indeed, $\pi_1(M_p)$ is a group with generators $a_1, \ldots, a_p, b_1, \ldots, b_p$ and defining relation $a_1 b_1 a_1^{-1} b_1^{-1} \ldots a_p b_p a_p^{-1} b_p^{-1} = e$, while the group $\pi_1(N_q)$ is a group with generators $a_1, \ldots, a_q$ and defining relation $a_1 a_1 a_2 a_2 \ldots a_q a_q = e$. It is clear that these groups are not isomorphic if $2p \neq q$. If we assume that $2p = q$, then $\pi_1(M_p)$ is not isomorphic to $\pi_1(N_q)$ because of the following reason. The quotient group $\pi_1(N_q)/[\pi_1(N_q), \pi_1(N_q)]$ of the group $\pi_1(N_q)$ with respect to its commutant $[\pi_1(N_q), \pi_1(N_q)]$ has a coset of order two which contains the element $a_1 a_2 \ldots a_q$. There are no elements of order two in the quotient group $\pi_1(M_p)/[\pi_1(M_p), \pi_1(M_p)]$, since the element $a_1 b_1 a_1^{-1} b_1^{-1} \ldots a_p b_p a_p^{-1} b_p^{-1}$ of the free group with the generators $a_1, \ldots, a_p$, $b_1, \ldots, b_p$ is contained in the commutant of this group.

Note that the classification theorem 2 (Ch 2, § 4) is now proved completely. The genus of a surface and its property of orientability completely determine its topological type.

6. On calculating higher homotopy groups. Computing homotopy groups of spaces is an important but a difficult problem. Methods for these computations have been developed, but even their application to concrete cases encounter considerable difficulties. Nevertheless, certain homotopy groups are computed for sufficiently "good" spaces and play an important role in many problems.

The following theorem enables us to reduce the computation of the homotopy groups of a space X to the computation of the corresponding groups of a space Y which is homotopy equivalent to X.

Theorem 4. *If $f : X \to Y$ is a homotopy equivalence then for any point $x \in X$ the homeomorphism*

$$f_n : \pi_n(X, x) \to \pi_n(Y, f(x)),$$

which is induced by f, is an isomorphism.

Proof. Let a mapping g be a homotopy inverse to f, and ϕ a representative of some class $[\phi] \in \pi_n(X, x_0)$. Then $(gf)\phi$ is a representative of its image $(gf)_n[\phi]$. The spheroid ϕ is "attached" to the point x_0, and the spheroid $(gf)\phi$ to the point $(gf)(x_0) = z_0$; moreover, the first spheroid is homotopic to the latter in virtue of $gf \sim 1_X$. Suppose that the point x_0 shifts to the point z_0 under this homotopy, drawing a path $\omega(t)$ (Fig. 80).

Let $\omega(t)$ induce an isomorphic mapping $\pi_n(X, z_0) \underset{n}{\overset{S^\omega}{\longrightarrow}} \pi_n(X, x_0)$ (see § 3, theorem 5). The homotopy of the spheroids ϕ and $(gf)\phi$ generates a homotopy of the spheroids $gf\phi$ and α from $S_n^{\omega^{-1}}[\phi]$. Consequently, $g_n f_n[\phi] = [gf\phi] = [\alpha] = S_n^{\omega^{-1}}[\phi]$ which means the commutativity of the diagram

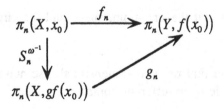

Analogously, we can show (try it yourself) that the diagram

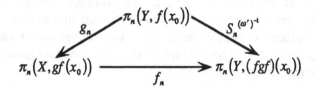

is commutative, where $\omega' : I \to Y$ is a path between the points $f(x_0)$ and $(fgf)(x_0)$, equal to $f\omega$. From the commutativity of these diagrams and the fact that $S_n^{\omega^{-1}}$, $S_n^{(\omega')^{-1}}$ are isomorphisms it follows that f_n, g_n are isomorphisms. $\quad\square$

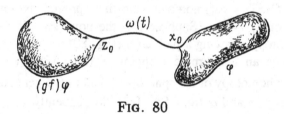

FIG. 80

Exercise 9°. Prove that the one–point space and the circle S^1 have different homotopy types.

Exercise 10°. Prove that the two–dimensional disc and the two–dimensional cylinder over a circle have different homotopy types.

The computation of the groups $\pi_k(S^n)$ has stimulated the development of many branches of contemporary topology, though this problem has not been completely solved yet. Here are two very distinct cases: $k \leq n$ and $k > n$. The first case is elementary enough, though it requires the development of special methods. We present the following results without proof[*]:

$$\pi_1(S^n) = \pi_2(S^n) = \ldots = \pi_{n-1}(S^n) = 0 \quad (n > 1),$$
$$\pi_n(S^n) \simeq \mathbf{Z} \quad (n \geq 1).$$

Hence, it follows, in particular, that the sphere S^n is noncontractible to any of its points.

The second case has not been completely settled, and the difficulties increase with the growth of n and $k - n$. We shall present several of the more simple results:

$$\pi_3(S^2) \simeq \mathbf{Z}, \quad \pi_4(S^3) \simeq \mathbf{Z}_2, \ldots, \pi_{n+1}(S^n) \simeq \mathbf{Z}_2 \ (n \geq 3).$$

This disproves the intuitive idea that $\pi_k(S^n) = 0$ for $k > n$.

Thus, the groups $\pi_n(S^n)$ for $n = 1, 2, \ldots$, are free Abelian groups with one generator γ_n, and γ_n being the homotopy class of the identity mapping $1_{S^n} : S^n \to S^n$. One can imagine the multiples $l \cdot \gamma_n$ as the homotopy classes of mappings $\phi : S^n \to S^n$ which "pile up" the sphere S^n onto itself l times. In addition, if $l > 0$, then the orientation of the sphere under the mapping ϕ is said to be preserved, and if $l < 0$, then the orientation is said to be changed (compare with the homotopy classes from $\pi_1(S^1)$).

Exercise 11°. Let S^n be a sphere in the space $\mathbf{R}^{n+1}$ with its centre at zero. Show that the mapping of S^n into itself given by the correspondence

$$(x_1, x_2, \ldots, x_{n+1}) \to (-x_1, x_2, \ldots, x_{n+1}),$$

defines a homotopy class which is equal to $(-\gamma_n)$.

It is quite easy to prove that $\pi_n(X, x_0) = 0$, $n \geq 1$, if the space X is contractible to a point (theorem 4 should be applied). In particular, for the disc D^n and the space $\mathbf{R}^n$ we obtain:

$$\pi_k(\bar{D}^n) = 0, \quad \pi_k(D^n) = 0, \quad \pi_k(\mathbf{R}^n) = 0, \quad k = 1, 2, \ldots$$

---

[*]The group $\pi_1(S^1)$ is computed in theorem 1.

In Ch 4, § 9, will be shown that for $k \geq 2$

$$\pi_k(\mathbf{R}P^n) \simeq \pi_k(S^n) \simeq = \begin{cases} 0, & k < n, \\ \mathbf{Z}, & k = n. \end{cases}$$

For certain applications (see Ch. 1, § 6), it is necessary to know the groups π_1 and π_2 for the group of orthogonal matrices with determinant 1 (denoted as $SO(3)$) which is considered to be a subspace in the topological space $\mathbf{R}^9$ of all 3×3 matrices. The answer is immediate, if one notes that $SO(3)$ is homeomorphic to $\mathbf{R}P^3$. Indeed, each orthogonal matrix from $SO(3)$ represents a rotation of the standard basic frame in $\mathbf{R}^3$. From the canonical form of this matrix, we may conclude that the rotation is characterized by a certain axis l which passes through 0, and the rotation of the whole space over an angle α, $|\alpha| \leq \pi$; besides, rotations over π and $(-\pi)$ are equivalent. If one takes the sphere S^2 of identity radius, then the axis l gives a pair of diametrically opposite points $(x, -x)$ on S^2, i.e., the points of intersection of l with S^2, and the angle α, gives a point on the diameter $[-x, +x]$ with the coordinate $x(\alpha) = \alpha/\pi$. Thus, the set of all matrices from $SO(3)$ bijectively corresponds to the points of the unit disc $\bar{D}^3$ with diametrically opposite points on the boundary identified, i.e., to the points of $\mathbf{R}P^3$. It is not difficult to verify that the indicated correspondence is a homeomorphism.

Thus, $\pi_1(SO(3)) \simeq \pi_1(\mathbf{R}P^3) = \mathbf{Z}_2$, $\pi_2(SO(3)) \simeq \pi_2(\mathbf{R}P^3) = 0$, and also $\pi_3(SO(3)) \simeq \pi_3(\mathbf{R}P^3) \simeq \mathbf{Z}$.

7. Some applications. We shall first prove an important property of the sphere S^n.

Theorem 5. *The sphere S^n (the boundary of the disc $\bar{D}^{n+1}$) is not a retract of $\bar{D}^{n+1}$.*

Proof. In § 2, a necessary condition for the existence of an extension of a mapping was indicated in terms of a functor to the category of groups. We shall apply this remark using the functor π_n. We know already that $\pi_n(S^n) = \mathbf{Z}$, $\pi_n(\bar{D}^{n+1}) = 0$. Further, if the sphere S^n were a retract of $\bar{D}^{n+1}$, then there would be a commutative diagram (1)

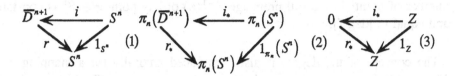

where i is the imbedding of a sphere into the ball, and r is a sought for retraction. Since π_n is a covariant functor, it would transform diagram (1) into the commutative diagram (2) which has the form (3). The latter fact contradicts its commutativity. Consequently, the assumption that a retraction r exists is not true. $\square$

The following theorem which is interesting and important in applications, is proved with the help of theorem 5.

Brouwer fixed–point theorem. *Any continuous mapping $g : \bar{D}^{n+1} \to \bar{D}^{n+1}$ of an $(n+1)$–dimensional closed ball (disc) into itself has at least one fixed point, i.e., there exists a point $x_* \in \bar{D}^{n+1}$ such that $g(x_*) = x_*$.*

Proof. Indeed, if such a point does not exist, i.e., for any point $x \in \bar{D}^{n+1}$, $g(x) \neq x$, then the interval joining the points $g(x)$ and x can be extended beyond the point x up to the intersection with the sphere S^n at some point $r(x)$. Then the mapping $r : \bar{D}^{n+1} \to S^n$, $x \mapsto r(x)$ is an extension to $\bar{D}^{n+1}$ of the identity mapping of the sphere S^n. But we have already proved that such an extension does not exist. The contradiction proves the theorem. $\square$

8. The degree of a mapping. The group $\pi_n(S^n) = \mathbf{Z}$ is closely related to the concept of the degree of a continuous mapping $f : S^n \to S^n$, which is often used in analysis. Let γ_n be a generator of the group $\pi_n(S^n)$. Then $f_*(\gamma_n) = \alpha\gamma_n$, where α is an integer, and f_* the homomorphism of the group $\pi_n(S^n)$ induced by the mapping f. The number α is called the *degree of the mapping* f and is denoted by $\deg f$ (the sign of $\deg f$ is not dependent on the choice of generator).

Exercise 12°. Consider the mapping of the unit circle $S^1 = \{z : |z| = 1\}$ of the complex plane given by the formula $f(z) = z^n$. Show that $\deg f = n$.

Exercise 13°. Show that if $f : S^1 \to S^1$ is a local homeomorphism then the number of points in the full preimage $f^{-1}(x)$ of any point $x \in S^1$ is constant and equal to $|\deg f|$.

The concept of the degree is also introduced naturally for a mapping $f : S^n_1 \to S^n_2$ from one copy of the sphere into another. (To obtain this, it is necessary to fix the basis classes γ^1_n in $\pi_n(S^n_1)$ and γ^2_n in $\pi_n(S^n_2)$; then $f_*(\gamma^1_n) = \deg f \cdot \gamma^2_n$.) Since γ_n is the homotopy class $[1_{S^n}]$ of the identity mapping, for a mapping $f : S^n \to S^n$ we have

$$f_*(\gamma_n) = f_*[1_{S^n}] = [f 1_{S^n}] = [f],$$

therefore, $\deg f \cdot \gamma_n$ is the homotopy class of the mapping f, and so the degree, $\deg f$, is the "number" of the homotopy class $[f]$.

If $f = 1$ is the identity mapping, then $\deg f = 1$; if $f \sim 0$ (is homotopic to the constant mapping), then $\deg f = 0$; if $f : S^n \to S^n$, $g : S^n \to S^n$ are two mappings, then they are homotopic iff their degrees are equal: $\deg f = \deg g$. We also present the useful formula $\deg(fg) = (\deg f) \cdot (\deg g)$, which follows from the relation $[fg] = f_*[g]$.

The concept of the degree is applied in the investigations concerning extending continuous mappings $f : S^n \to \mathbf{R}^{n+1} \backslash \{0\}$ to the ball $\bar{D}^{n+1}$ which is bounded by the sphere S^n. Since the space $\mathbf{R}^{n+1} \backslash \{0\}$ is homotopy equivalent to S^n, their homotopy groups are isomorphic and, consequently, one may consider the degree of a given mapping, usually called the characteristic (or rotation) of the vector field f; denote it by $\chi_{S^n}(f)$.

Lemma 6. *The condition* $\chi_{S^n}(f) = 0$ *is necessary and sufficient for the existence of the extension* $\tilde{f} : \bar{D}^{n+1} \to \bar{R}^{n+1} \backslash \{0\}$ *of the mapping* f.

The proof is obvious from the remark that an extension $\tilde{f}$ determines a homotopy $f \sim 0$ by the formula

$$f(x, t) = \tilde{f}(tx), \quad x \in S^n, \quad t \in [0, 1]$$

(if S^n is the sphere of radius 1 and centre at 0), and vice versa.

Exercise 14°. Construct the extension $\widetilde{f}$ when $f \sim 0$.

The lemma implies the following evident corollary.

Corollary. *If* $\chi_{S^n}(f) \neq 0$ *then any extension* $\widetilde{f} : \bar{D}^{n+1} \to \mathbf{R}^{n+1}$ *has a zero, i.e., there exists a point* $x_0 \in D^{n+1}$ *such that* $\widetilde{f}(x_0) = 0$.

This corollary is often used to prove the existence of a solution of the equation $\widetilde{f}(x) = 0$, where $\widetilde{f} : \bar{D}^{n+1} \to \mathbf{R}^{n+1}$ is a given mapping.

Example 1. It is easy to verify that under the conditions of the Brouwer fixed point theorem, the mapping $\widetilde{f} = -g(x) + x$ either has zero on S^n, or $\chi_{S^n}(\widetilde{f}) = 1$ ($\widetilde{f} : S^n \to \mathbf{R}^{n+1} \backslash \{0\}$ admits a homotopy to the identity mapping $\widetilde{f}(x, t) = -tg(x) + x$, $x \in S^n$, $0 \leq t \leq 1$). Consequently, $\widetilde{f}$ has zero in $\bar{D}^{n+1}$.

Example 2. The fundamental theorem of algebra: for a complex polynomial $\widetilde{f}(z) = z^m + a_1 z^{m-1} + \ldots + a_{m-1} z + a_m$ there exists a root in the complex plane.

Denote the circle in the z–plane: $\{z : |z| = \rho\}$, by S_ρ^1.

Lemma 7. *For a sufficiently large* ρ, *we have*

$$\widetilde{f} : S_\rho^1 \to \mathbf{R}^2 \backslash \{0\},$$

and, in addition, $\chi_{S_\rho^1}(\widetilde{f}) = m$.

Proof. Consider the homotopy

$$\widetilde{f}(z, t) = z^m + t(a_1 z^{m-1} + \ldots + a_{m-1} z + a_m), \quad t \in [0, 1].$$

We have an estimation

$$|\widetilde{f}(z, t) \geq$$

$$\geq |z^m| \left[1 - t \left(a_1 \frac{1}{|z|} + \ldots + a_{m-1} \frac{1}{|z|^{m-1}} + \right. \right.$$

$$\left. \left. + a_m \frac{1}{|z|^m} \right) \right], \quad |z| \neq 0.$$

It is evident that there can be found $\rho > 0$ large enough that $|\tilde{f}(z,t)| > 0$ for $|z| = \rho$, $t \in [0,1]$. Consequently, $\tilde{f} : S^1_\rho \to \mathbf{R}^2\backslash\{0\}$ is homotopic to the mapping $g : S^1_\rho \to \mathbf{R}^2\backslash\{0\}$, $g(z) = z^m$. According to exercise 12°, $\chi_{S^1_\rho}(g) = m$; therefore $\chi_{S^1_\rho}(f) = m$ as well. To conclude the proof, the corollary from lemma 6 is used. $\square$

9. Some results on homotopy groups of concrete spaces. First we shall add more information about homotopy groups of spheres (subsection 6). The relations $\pi_{n+2}(S^n) \simeq \mathbf{Z}_2$ for $n \geq 4$, $\pi_{n+3}(S^n) \simeq \mathbf{Z}_{24}$ for $n \geq 5$, $\pi_{n+4}(S^n) = 0$ for $n \geq 6$, $\pi_{n+5}(S^n) = 0$ for $n \geq 7$, together with the Freudenthal theorem about the isomorphism of all groups $\pi_{n+k}(S^n)$ for $n \geq k+2$, and with the relations from subsection 6, enable us to calculate a large number of groups $\pi_k(S^n)$ also including the isomorphic groups $\pi_{n+k}(S^n) = \pi^s_k$, $n \geq k+2$, which are called the *stable homotopy groups of the spheres*.

In applications, the groups from the series of the classical matrix groups are of interest, i.e., the groups of real $n \times n$ matrices: $O(n)$ the group of orthogonal matrices, $SO(n)$, consisting of a special orthogonal matrices (with determinant 1), $Sp(n)$ the group of symplectic matrices, and their complex analogues $U(n)$, $SU(n)$ the unitary and special unitary matrices, respectively. All the groups indicated are imbedded in the linear space of $n \times n$ matrices (real or complex, respectively) and inherit an Euclidean topology from this space.

We present the following table:

$\pi_k(O(n)) \simeq \pi_k(SO(n)) \oplus \pi_k(SO(n))$ for $n \geq 2$, $k \geq 1$; $\pi_k(O(1)) = 0$;

$\pi_1(SO(n)) \simeq \mathbf{Z}_2$, $n \geq 3$; $\pi_1(SU(n)) \simeq \pi_1(Sp(n)) = 0$, $n \geq 1$; $\pi_1(SO(2)) \simeq \mathbf{Z}$; $\pi_1(SO(3)) \simeq \mathbf{Z}_2$.

$\pi_2 = 0$ for all the groups considered.

$$\begin{cases} \pi_3(SO(n)) \simeq \mathbf{Z}, & n = 3, \quad n \geq 5; \\ \pi_3(SO(4)) \simeq \mathbf{Z} \oplus \mathbf{Z}, & \pi_3(SU(n)) \simeq \mathbf{Z} \text{ for } n \geq 2; \\ \pi_3(Sp(n)) \simeq \mathbf{Z}, & n \geq 1. \end{cases}$$

$$\begin{cases} \pi_4(SO(3)) \simeq \pi_4(SU(2)) \cong \mathbf{Z}_2, & \pi_4(SO(4)) \simeq \mathbf{Z}_2 \oplus \mathbf{Z}_2, \\ \pi_4(SO(5)) \simeq \mathbf{Z}_2, & \pi_4(SO(n)) = 0, \quad n \geq 6; \\ \pi_4(SU(n)) = 0, & n \geq 3; \quad \pi_4(Sp(n)) \simeq \mathbf{Z}_2, \quad n \geq 1. \end{cases}$$

$$\begin{cases} \pi_k(SO(3)) \simeq \pi_k(SU(2)) \simeq \pi_k(Sp(1)) \simeq \pi_k(S^3), & k > 1; \\ \pi_k(SO(4)) \simeq \pi_k(S^3) \oplus \pi_k(S^3), & k > 1. \end{cases}$$

$\pi_1(U(1)) \simeq \mathbf{Z}$, $\pi_k(U(1)) = 0$, $k > 1$.

$\pi_1(U(n)) \simeq \mathbf{Z}$, $\pi_k(U(n) \simeq \pi_k(SU(n))$, $k > 1$.

The groups $\pi_k(SO(n))$, $\pi_k(U(n))$ do not depend on n for $1 \leq k \leq n - 2$ and $1 \leq k \leq 2n - 1$, respectively; they are called *stable homotopy groups* and denoted by $\pi_k(SO)$, $\pi_k(U)$. The Bott periodicity $\pi_{k+8}(SO) \simeq \pi_k(SO)$, $\pi_{k+2}(U) \simeq \pi_k(U)$ is valid for them and together with the relations $\pi_1(SO) \simeq \mathbf{Z}_2$, $\pi_2(SO) = 0$, $\pi_3(SO) \simeq \mathbf{Z}$, $\pi_4(SO) = 0$, $\pi_5(SO) = 0$, $\pi_6(SO) = 0$, $\pi_7(SO) \simeq \mathbf{Z}$, $\pi_8(SO) \simeq \mathbf{Z}_2$, $\pi_1(U) \simeq \mathbf{Z}$, $\pi_2(U) = 0$, this gives all the stable homotopy groups for the groups $SO(n)$, $U(n)$.

REVIEW OF THE RECOMMENDED LITERATURE

A systematic summary of the homotopy theory is given in the books [71, 74, 84, 79, 33, 90, 42].

An informal introduction to the homotopy theory and its applications can be found in [25].

From an elementary study of the idea of the fundamental group we recommned the books [47, 54].

When studying the concepts of a category and a functor, it is useful to look at the monograph [52].

The theory of the degree of a mapping and the characteristic of a vector field can be found in the review of the literature of Ch 4, 5.

Applications of the theory of degree are presented in [48].

Problem books in homotopy theory are [59, 63].

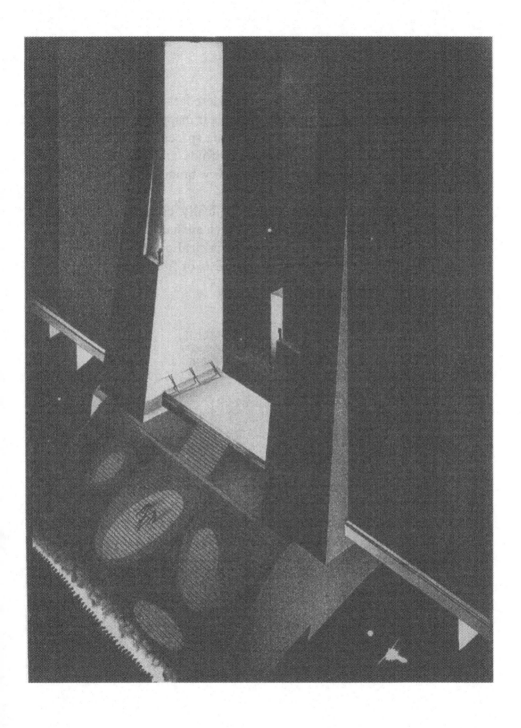

In the previous chapters, we considered the more general properties of topological spaces and their mappings. However, in topology and its applications there are spaces with additional structure, for instance, smooth manifolds and fibre spaces, which play an important role in many branches of modern mathematics.

In the present chapter, we systematically consider smooth manifolds and the tangent bundles which are naturally related to them. The elements of the theory of critical points of smooth functions on manifolds, and the elements of fibre space theory are also presented.

Topological spaces with a more complicated, specific "architecture", such as manifolds, fibre bundles, and coverings, possess a number of more subtle inherent properties. The complicated constructions presented in this drawing create a fluid impression because of the intermixture of a global view (from far away) and local views (details).

CHAPTER IV

MANIFOLDS AND FIBERINGS

In the previous chapters, we considered general properties of topological spaces and their mappings. However, in topology and its applications there appear spaces with additional structures, e.g., smooth manifolds and fibre bundles which are of great importance in many branches of modern mathematics.

In this chapter, we study smooth manifolds and the tangent vector bundles which are naturally related to them. The elements of the theory of critical points of smooth functions on manifolds and the elements of the theory of fibre bundles are presented.

§ 1. General concepts of differential calculus in n–dimensional space

1. Smooth mappings. Recall that $\mathbf{R}^n$ is the space of ordered sets $x = (x_1, \dots, x_n)$ consisting of real numbers (see Ch. 2, § 2) called *points* or *vectors*. We assume that $\mathbf{R}^n$ is *standardly imbedded* in $\mathbf{R}^{n+k}$, i.e., a point $(x_1, \dots, x_n)$ from $\mathbf{R}^n$ is identified with the point $(x_1, \dots, x_n, 0, \dots, 0)$ from $\mathbf{R}^{n+k}$. The numbers $x_1, \dots, x_n$ from the set $(x_1, \dots, x_n)$ are called the *standard coordinates* of the point $x = (x_1, \dots, x_n)$ in $\mathbf{R}^n$.

Let $U \subset \mathbf{R}^n$ be an open set. Any mapping $f : U \to \mathbf{R}^m$ can be represented (see Ch. 2, § 2) as an ordered set of m functions:

$$f(x_1, \dots, x_n) =$$
$$= (f_1(x_1, \dots, x_n), f_2(x_1, \dots, x_n), \dots, f_m(x_1, \dots, x_n)).$$

Definition 1. The mapping $f : U \to \mathbf{R}^m$ is said to be *smooth* (or *differentiable*) *of class* C^r, $r \geq 1$, on U, if each function f_k, $k = 1, \dots, m$ has all

219

continuous partial derivatives $\frac{\partial^s f_k}{\partial^{s_1} x_1 \ldots \partial^{s_n} x_n}$, $s_1 + \ldots + s_n = s$, on U up to the order $s = r$.

The smooth mappings f of class C^r are also called C^r-*mappings* and this is written as $f \in C^r$.

If all the functions f_k have continuous partial derivatives of any order then the mapping f is said to be *infinitely smooth* ($f \in C^\infty$). The continuous mappings are called C^0-*mappings*. Obviously, the following inclusions

$$C^0 \supset C^1 \supset \ldots \supset C^r \supset \ldots \supset C^\infty.$$

are valid.

In case all the functions f_k are analytic (a function is said to be *analytic* if its Taylor series converges to it in a neighbourhood of each point), the mapping f is said to be *analytic* ($f \in C^\omega$). The inclusion $C^\infty \supset C^\omega$ is valid.

Definition 2. The matrix

$$\begin{pmatrix} \frac{\partial f_1}{\partial x_1} & \cdots & \frac{\partial f_1}{\partial x_n} \\ \cdots & \cdots & \cdots \\ \frac{\partial f_m}{\partial x_1} & \cdots & \frac{\partial f_m}{\partial x_n} \end{pmatrix}$$

of first derivatives of the mapping f calculated at the point x_0 is called the *Jacobian matrix of the mapping* f at x_0 and is denoted by $\left(\frac{\partial f}{\partial x} \right) \Big|_{x_0}$.

The Jacobian matrix determines a linear mapping $\mathbf{R}^n \to \mathbf{R}^m$:

$$y_1 = \frac{\partial f_1}{\partial x_1} \Big|_{x_0} x_1 + \frac{\partial f_1}{\partial x_2} \Big|_{x_0} x_2 + \ldots + \frac{\partial f_1}{\partial x_n} \Big|_{x_0} x_n,$$

$$y_m = \frac{\partial f_m}{\partial x_1} \Big|_{x_0} x_1 + \frac{\partial f_m}{\partial x_2} \Big|_{x_0} x_2 + \ldots + \frac{\partial f_m}{\partial x_n} \Big|_{x_0} x_n,$$

which is called the *derivative of the mapping* f *at the point* x_0 and is denoted by $D_{x_0} f$. The derivative is a "*linearization*" of the mapping f, i.e., the affine mapping $f(x_0) + (D_{x_0} f)(x - x_0)$ coincides with $f(x)$ up to infinitesimals of higher order than $\|x - x_0\|$ (which is sometimes also expressed as "up to order 1 in $\|x - x_0\|$"). More exactly, $D_{x_0} f$ is the unique linear mapping of $\mathbf{R}^n$ into $\mathbf{R}^m$ for which

$$\frac{f(x) - f(x_0) - (D_{x_0} f)(x - x_0)}{\|x - x_0\|} \to 0 \quad \text{for } x \to x_0.$$

As a corollary of the theorem on derivatives of composite functions the following statement (the "chain rule") can be considered: *under a composition of mappings f and g, their Jacobian matrices are multiplied, i.e.,*

$$\left(\frac{\partial(fg)}{\partial x}\right)\Big|_{x_0} = \left(\frac{\partial f}{\partial y}\right)\Big|_{g(x_0)}\left(\frac{\partial g}{\partial x}\right)\Big|_{x_0}.$$

The proof is recommended as an exercise.

2. The rank of a mapping. Let $U \subset \mathbf{R}^n$ be an open set, $f : U \to \mathbf{R}^m$ a mapping of class C^r, $r \geq 1$. The *rank of the mapping* f at a point x_0 is the rank of its Jacobian matrix calculated at the point x_0; it is denoted by $\mathrm{rank}_{x_0} f$. It is equal to the dimension of the image of $\mathbf{R}^n$, in $\mathbf{R}^m$ under the linear mapping $D_{x_0} f$. Since the rank of a matrix cannot exceed the number of rows or columns, $\mathrm{rank}_{x_0} f \leq \min(n, m)$. The points at which $\mathrm{rank}_{x_0} f = \min(n, m)$, are said to be *regular* (sometimes *noncritical, nonsingular*). The points at which $\mathrm{rank}_{x_0} f < \min(n, m)$, are said to be *irregular* (also *critical, singular*).

The set of regular points of a mapping f is open in $\mathbf{R}^n$ (due to the continuity of the partial derivatives. Indeed, if a subdeterminant of the Jacobian is different from zero at x_0 than it is different from zero in a neighbourhood of x_0. The set of regular points may be empty. (Give some examples.)

3. The implicit function theorem. We present the *implicit function theorem* as it is proved in analysis. The points of the space $\mathbf{R}^{n+m} = \mathbf{R}^n \times \mathbf{R}^m$ we represent in the form (x, y), where $x = (x_1, \ldots, x_n) \in \mathbf{R}^n$, $y = (y_1, \ldots, y_m) \in \mathbf{R}^m$. Let $U \subset \mathbf{R}^n$, $V \subset \mathbf{R}^m$ be open sets and $(x_0, y_0) \in U \times V \subset \mathbf{R}^{n+m}$.

Theorem 1. *If $f : U \times V \to \mathbf{R}^m$ is a C^1-mapping, $f(x_0, y_0) = 0$, and* $\det\left(\frac{\partial f}{\partial y}\right)\Big|_{(x_0, y_0)} \neq 0$, *then there exists an open neighbourhood $W(x_0) \subset U$ of the point x_0 and a mapping $g : W(x_0) \to V$ such that $g(x_0) = y_0$ and $f(x, g(x)) = 0$ for any $x \in W(x_0)$; moreover the mapping g is unique. In addition, $g \in C^1$ and*

(1)
$$\left(\frac{\partial g}{\partial x}\right) = -B^{-1}A,$$

where the matrices B and A are obtained from the matrices $\left(\frac{\partial f}{\partial y}(x, y)\right)$ and $\left(\frac{\partial f}{\partial x}(x, y)\right)$, respectively, by replacing the argument y with $g(x)$.

Remark. If $f \in C^r$, $r \geq 1$, then $g \in C^r$. This statement follows from equality (1).

A corollary of the implicit function theorem is the *inverse mapping theorem*.

Theorem 2. *Let* $U \subset \mathbf{R}^n$ *be an open set,* $f : U \to \mathbf{R}^n$ *a mapping of class* C^r, $r \geq 1$; *let* $x_0 \in U$ *be a regular point of the mapping* f. *Then there exist open neighbourhoods* $V(x_0)$, $W(f(x_0))$ *of the points* x_0 *and* $f(x_0)$ *such that* f *is a homeomorphism* $V(x_0) \xrightarrow{f} W(f(x_0))$ *and* $f^{-1} \in C^r$.

Proof. Consider the mapping $F : \mathbf{R}^n \times U \to \mathbf{R}^n$ defined by the rule $F(y, x) = y - f(x)$ ($y \in \mathbf{R}^n$, $x \in U$). Let $y_0 = f(x_0)$. Obviously, $F \in C^r$ and $F(y_0, x_0) = 0$. Since $\operatorname{rank}_{x_0} f = n$, it follows that $\det \left(\frac{\partial F}{\partial x} \right) \big|_{(y_0, x_0)} \neq 0$. According to the implicit function theorem, there exists an open neighbourhood $W(y_0) \subset \mathbf{R}^n$ of the point y_0 and a unique mapping $g : W(y_0) \to U$ such that $g(y_0) = x_0$, and for any $y \in W(y_0)$

$$(2) \qquad\qquad F(y, g(y)) = y - f(g(y)) = 0.$$

Put $V(x_0) = g(W(y_0))$. Since $V(x_0) = f^{-1}(W(y_0))$, we find that $V(x_0)$ is open because f is continuous. Thus, $g : W(y_0) \to V(x_0)$ is a mapping of open sets, and by (2), $g = f^{-1}$. Besides, because of the remark following theorem 1, we have $f^{-1} \in C^r$. $\qquad\square$

Definition 3. A mapping $f : U \to V$ of an open set $U \subset \mathbf{R}^n$ into an open set $V \subset \mathbf{R}^n$ is called a C^r–*diffeomorphism*, $r \geq 1$, if 1) f is a homeomorphism of U to V, 2) $f \in C^r$, and 3) $f^{-1} \in C^r$.

Exercise 1°. Construct a C^∞–diffeomorphism $f : D^n_\rho(x) \to \mathbf{R}^n$.

Now, we can formulate the inverse mapping theorem in the following way. *If* $f : U \to \mathbf{R}^n$ *is a* C^r–*mapping,* $r \geq 1$, *of an open set* $U \subset \mathbf{R}^n$ *into* $\mathbf{R}^n$, *and* x_0 *is a regular point of* f, *then there exist open neighbourhoods* $V(x_0)$, $W(f(x_0))$ *of the points* x_0, $f(x_0)$ *such that the mapping* $f : V(x_0) \to W(f(x_0))$ *is a* C^r–*diffeomorphism.*

Exercise 2°. Prove that a diffeomorphism has no nonregular points.

Hint. Use the remark about the Jacobian matrix of the composition of mappings f, f^{-1}.

Exercise $3°$. Let $f : \mathbf{R}^n \to \mathbf{R}^1$ be a mapping of class C^r, $r \geq 1$. Show that the nonregular points of f are characterized by the fact that at these points the first partial derivatives of f with respect to all variables are equal to zero.

4. "Curvilinear" coordinate system. Let U, $V \subset \mathbf{R}^n$ be open sets, and $f : U \to V$ a homeomorphism. The position of each point $y \in V$ can be specified by means of the standard coordinates $y_1, \ldots, y_n$ of the point y, but it can be also done by the standard coordinates of the point $x = f^{-1}(y) \in U$.

Definition 4. The standard coordinates of the point $f^{-1}(y) \in U$ are called *"curvilinear" coordinates of the point* $y \in V$.

In other words, instead of the coordinate planes y_i, $i = 1, \ldots, n$, in V, we consider the images of coordinate planes $x_i = a_i$, $i = 1, \ldots, n$, in U, under the homeomorphism f, their intersection determining the position of the point y. The term "curvilinear" coordinates merely reflects the fact that the new coordinate "planes" in V are , generally speaking, "curvilinear" (Fig. 81).

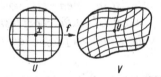

FIG. 81

It should be noted that in analysis, curvilinear coordinates are not usually introduced by a homeomorphism, but by a C^r–diffeomorphism, where the order of smoothness r depends on the problem considered.

If a function g on V is given as a function of the standard coordinates of the point y, then it can be considered as a function of the standard coordinates of the point x, i.e., as a function of the curvilinear coordinates of the point y. In analysis, this operation is called *substitution of variables*. In other words, we change the coordinates in the source space of the function g. This is equivalent to the fact that instead of the function g the function gf is considered. Of course, one may also consider mappings and perform similar substitutions of variables. Such substitutions are also carried out in the image space of a mapping: if $g : W \to V$ is a mapping of a set $W \subset \mathbf{R}^m$, then instead of the standard coordinates of the point $g(z)$, $z \in W$, we consider curvilinear coordinates of the point $g(z)$, which are defined by a homeomorphism f. Such a substitution is equivalent to considering the mapping $f^{-1}g$ instead of the mapping g.

Note that the rank of a smooth mapping does not change under a smooth substitution of variables.

5. Rectification theorem.

5. Rectification theorem. The *standard imbedding* of $\mathbf{R}^n$ into $\mathbf{R}^{n+k}$ is the mapping $\mathbf{R}^n \to \mathbf{R}^{n+k}$ given by $(x_1, \dots, x_n) \mapsto (x_1, \dots, x_n, 0, \dots, 0)$.

The *standard projection* of $\mathbf{R}^{n+k}$ on $\mathbf{R}^n$ is the mapping $\mathbf{R}^{n+k} \to \mathbf{R}^n$ given by $(x_1, \dots, x_n, x_{n+1}, \dots, x_{n+k}) \mapsto (x_1, \dots, x_n)$.

Theorem 3. (On the rectification of a mapping in a neighbourhood of a regular point). *Let $U \subset \mathbf{R}^n$ be an open set, $f : U \to \mathbf{R}^m$ a C^r-mapping, $r \geq 1$, x_0 a regular point f. Then*

(A) If $n \leq m$, then there exist an open neighbourhood $W(f(x_0))$ of the point $f(x_0)$, an open set $W_1 \subset \mathbf{R}^m$, and a C^r-diffeomorphism $F : W(f(x_0)) \to W_1$ such that Ff for some open neighbourhood $V(x_0) \subset \mathbf{R}^n$, is the standard imbedding of $\mathbf{R}^n$ into $\mathbf{R}^m$ (Fig. 82).

(B) If $n \geq m$ then there exist an open neighbourhood $V(x_0)$ of the point x_0, an open set $W \subset \mathbf{R}^n$, and a C^r-diffeomorphism $F : V(x_0) \to W$ such that on the set W fF^{-1} is the standard projection of $\mathbf{R}^n$ onto $\mathbf{R}^m$ (Fig. 83).

We shall explain the essence of this theorem from the point of view of coordinate substitutions. In case (A), the diffeomorphism F^{-1} determines curvilinear coordinates $\{\xi_1, \dots, \xi_m\}$ in the space $\mathbf{R}^m$ in which the mapping f is of the form

$$\xi_1 = x_1, \dots, \xi_n = x_n, \xi_{n+1} = 0, \dots, \xi_m = 0.$$

In case (B), the diffeomorphism F^{-1} determines curvilinear coordinates $\{\xi_1, \dots, \xi_n\}$ in the space $\mathbf{R}^n$, in which the mapping f has the form $y_1 = \xi_1, \dots, y_m = \xi_m$.

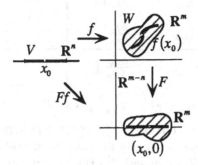

FIG. 82

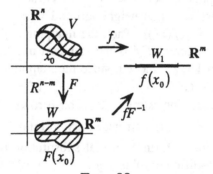

FIG. 83

The idea of the proof is to extend the given mappings to mappings of spaces of the same dimension, and then, by applying the inverse function theorem, to obtain the necessary substitutions of coordinates.

Proof. (A) We shall represent the points of the space $\mathbf{R}^m = \mathbf{R}^n \times \mathbf{R}^{m-n}$ in the form (x, y), where $x = (x_1, \dots , x_n) \in \mathbf{R}^n$, $y = (y_1, \dots , y_{m-n}) \in \mathbf{R}^{m-n}$. According to the condition, $\operatorname{rank}_{x_0} f = n$, i.e., $\operatorname{rank} \left(\frac{\partial f}{\partial x} \right) \Big|_{x_0} = n$. Assume, at first, that the determinant consisting of first n rows of the Jacobian matrix $\left(\frac{\partial f}{\partial x} \right) \Big|_{x_0}$ differs from zero. Consider the mapping $F_1 : U \times \mathbf{R}^{m-n} \to \mathbf{R}^m$ which is given by the rule $F_1(x, y) = f(x) + (0, y)$. The Jacobian matrix of the mapping F^1 at the point $(x_0, 0)$ has the form

$$\begin{pmatrix} \frac{\partial f_1}{\partial x_1}\big|_{x_0}, & \cdots & \frac{\partial f_1}{\partial x_n}\big|_{x_0} & & 0 & \\ \cdots\cdots\cdots\cdots\cdots\cdots & & & & & \\ \frac{\partial f_n}{\partial x_1}\big|_{x_0}, & \cdots & \frac{\partial f_n}{\partial x_n}\big|_{x_0} & & & \\ \hline \frac{\partial f_{n+1}}{\partial x_1}\big|_{x_0}, & \cdots & \frac{\partial f_{n+1}}{\partial x_n}\big|_{x_0} & 1 & & 0 \\ \cdots\cdots\cdots\cdots\cdots\cdots & & & & \ddots & \\ \frac{\partial f_m}{\partial x_1}\big|_{x_0}, & \cdots & \frac{\partial f_m}{\partial x_n}\big|_{x_0} & 0 & & 1 \end{pmatrix}.$$

By assumption, the determinant in the upper left–hand corner differs from zero, therefore $\operatorname{rank}_{(x_0,0)} F_1 = m$. By the inverse function theorem, there exist open neighbourhoods $W(x_0, 0)$ and $W_1(f(x_0))$ of the points $(x_0, 0)$ and $f(x_0)$, respectively, such that the mapping $F_1|_{W(x_0,0)} : W(x_0, 0) \to W_1(f(x_0))$ is a C^r–diffeomorphism. Thus, the mapping $F_1^{-1} : W_1(f(x_0)) \to W(x_0, 0)$ is also a C^r–diffeomorphism. Set $F = F_1^{-1}$. In virtue of the continuity of the mapping f, there exists an open neighbourhood $V(x_0)$ of the point $x_0 \in \mathbf{R}^n$ such that $f(V(x_0)) \subset W_1(f(x_0))$. Then the mapping $Ff : V(x_0) \to W(x_0, 0)$ is well defined. The mapping Ff is the standard imbedding of $\mathbf{R}^n$ into $\mathbf{R}^m$ on the neighbourhood $V(x_0)$. Indeed, since the mapping F_1^{-1} is bijective and $F_1(x, 0) = f(x)$, we get $(Ff)(x) = F(f(x)) = F_1^{-1}(f(x)) = (x, 0)$.

In case the determinant consisting of the first n rows of the Jacobian matrix $\left(\frac{\partial f}{\partial x}\right)\big|_{x_0}$ is equal to zero, it is first necessary to renumber the coordinates in $\mathbf{R}^m$ (in other words, to perform a special substitution of coordinates in $\mathbf{R}^m$ by means of the C^∞–diffeomorphism $g : \mathbf{R}^m \to \mathbf{R}^m$ in such a a way that the determinant consisting of first n rows of the Jacobian matrix $\left(\frac{\partial (g^{-1} f)}{\partial x}\right)\big|_{x_0}$ is nonzero). We then construct a C^r–diffeomorphism F for the mapping $g^{-1} f$ using the method described above; then Fg^{-1} will be the sought for C^r–diffeomorphism for the mapping f.

(B) We represent the elements of the space $\mathbf{R}^n = \mathbf{R}^m \times \mathbf{R}^{n-m}$ in the form (x, y), where $x = (x_1, \ldots, x_m) \in \mathbf{R}^m$, $y = (y_1, \ldots, y_{n-m}) \in \mathbf{R}^{n-m}$. Let $x_0 = (x^0, y^0)$. According to the condition, we have $\operatorname{rank}_{(x^0, y^0)} f = m$, i.e. $\operatorname{rank} \left(\frac{\partial f}{\partial (x,y)}\right)\big|_{(x^0, y^0)} = m$. Assume first that the determinant consisting of the first m columns of the Jacobian matrix $\left(\frac{\partial f}{\partial (x,y)}\right)\big|_{(x^0, y^0)}$ differs from zero. Consider the mapping $F : U \to \mathbf{R}^m \times \mathbf{R}^{n-m}$ given by the rule $F(x, y) = (f(x, y), y)$. The Jacobian matrix of the mapping F at the point (x^0, y^0) is of the form

$$\begin{pmatrix} \frac{\partial f_1}{\partial x_1}\big|_{(x^0,y^0)}, & \cdots & \frac{\partial f_1}{\partial x_m}\big|_{(x^0,y^0)} & \frac{\partial f_1}{\partial y_1}\big|_{(x^0,y^0)}, & \cdots & \frac{\partial f_1}{\partial y_{n-m}}\big|_{(x^0,y^0)} \\ \cdots\cdots\cdots & & \cdots\cdots\cdots & \cdots\cdots\cdots & & \cdots\cdots\cdots \\ \frac{\partial f_m}{\partial x_1}\big|_{(x^0,y^0)}, & \cdots & \frac{\partial f_m}{\partial x_m}\big|_{(x^0,y^0)} & \frac{\partial f_m}{\partial y_1}\big|_{(x^0,y^0)}, & \cdots & \frac{\partial f_m}{\partial y_{n-m}}\big|_{(x^0,y^0)} \\ & & & 1 & & 0 \\ & 0 & & & \ddots & \\ & & & 0 & & 1 \end{pmatrix}.$$

By the assumption, the determinant in the upper left–hand corner differs from zero, therefore, $\mathrm{rank}_{(x^0,y^0)} F = n$. According to the inverse function theorem, there exist open neighbourhoods $V(x^0, y^0)$ and $W(F(x^0, y^0))$ of the points (x^0, y^0) and $F(x^0, y^0)$, respectively, such that the mapping

$$F|_{V(x^0,y^0)} : V(x^0, y^0) \to W(f(x^0, y^0), y^0)$$

is a C^r–diffeomorphism. The mapping fF^{-1} on the neighbourhood $W(f(x^0, y^0), y^0)$ is the standard projection of $\mathbf{R}^n$ on $\mathbf{R}^m$. Indeed, let $z \in W(f(x^0, y^0), y^0)$. Since F^{-1} is bijective, there is a unique point (ξ, η) in the neighbourhood $V(x^0, y^0)$, for which $z = (f(\xi, \eta), \eta)$; then

$$[fF^{-1}](f(\xi, \eta), \eta) = f[F^{-1}(f(\xi, \eta), \eta)] = f(\xi, \eta).$$

In the case when the determinant consisting of first m columns of the Jacobian matrix $\left(\frac{\partial f}{\partial(x,y)}\right)\big|_{(x^0,y^0)}$ is equal to zero, it is necessary first to renumber the coordinates in $\mathbf{R}^n$ (i.e., to make a substitution of coordinates by using a C^∞–difeomorphism $g : \mathbf{R}^n \to \mathbf{R}^n$) so that the determinant consisting of first m columns of the Jacobian matrix $\left(\frac{\partial f}{\partial(x,y)}\right)\big|_{(x^0,y^0)}$ differs from zero. Then construct a C^r–diffeomorphism F for the mapping fg in the manner described above; and then Fg^{-1} is the sought for C^r–diffeomorphism for the mapping f. $\qquad\qquad\qquad\qquad\qquad\qquad\qquad\qquad\qquad\qquad\qquad\square$

6. Lemma on the representation of smooth functions. We shall present one more result which is necessary for further investigations.

Lemma 1. *Let f be a C^{r+1}–function $(r \geq 0)$ given on a convex neighbourhood $V(x^0)$ of a point x^0 in $\mathbf{R}^n$. Then there exist C^r–functions $g_i : V(x^0) \to \mathbf{R}^1$,*

$i = 1, \ldots, n$, such that $f(x) = f(x^0) + \sum_{i=1}^{n} g_i(x)(x_i - x_i^0)$, and, moreover,

$g_i(x^0) = \frac{\partial f}{\partial x_i}(x^0)$

Proof. Set

$$g_i(x) = \int_0^1 \frac{\partial f}{\partial x_i}(x^0 + t(x - x^0))dt.$$

By applying elementary transformations from analysis, we obtain

$$f(x) - f(x^0) = \int_0^1 \frac{df(x^0 + t(x - x^0))}{dt}dt =$$

$$= \int_0^1 \left(\sum_{i=1}^{n} (x_i - x_i^0)\frac{\partial f}{\partial x_i}(x^0 + t(x - x^0)) \right) dt =$$

$$= \sum_{i=1}^{n}(x_i - x_i^0)\int_0^1 \frac{\partial f}{\partial x_i}(x^0 + t(x - x_0))dt =$$

$$= \sum_{i=1}^{n}(x_i - x_i^0)g_i(x).$$

$\square$

Exercise 4^0. Let f be a function of class C^{r+2}, $r \geq 0$, given on a convex neighbrouhood $V(x^0)$ of a point x^0 in $\mathbf{R}^n$. Show that

$$f(x) = f(x^0) + \sum_{i=1}^{n}(x_i - x_i^0)\frac{\partial f}{\partial x_i}(x^0) +$$

$$+ \sum_{i,j=1}^{n}(x_i - x_i^0)(x_j - x_j^0)A_{ij}(x),$$

where the $A_{ij}(x)$ are functions of class C^r on $V(x^0)$, and, moreover, $A_{ij}(x^0) = \frac{\partial^2 f}{\partial x_i \partial x_j}(x^0)$.

§ 2. Smooth submanifolds in Euclidean space

1. The concept of a smooth submanifold in $\mathbf{R}^N$. In analysis and analytical geometry, smooth surfaces in three–dimensional Euclidean space are considered; they are given by an equation $z = f(x, y)$, where f is a smooth function of two variables defined on a domain D of the plane (x, y). More complex surfaces (for instance, closed ones) which are (locally) defined on separate regions by one of the following equations: $z = p(x, y)$, $y = q(x, z)$, $x = r(y, z)$ are also considered. The simplest example of such a surface is the sphere S^2. Other objects studied in analysis and analytical geometry are smooth curves locally defined by one of the systems of equations:

$$\begin{cases} x = \phi(z), \\ y = \psi(z); \end{cases} \quad \begin{cases} x = \phi(y), \\ z = \psi(y); \end{cases} \quad \begin{cases} y = \phi(x), \\ z = \psi(x). \end{cases}$$

All these objects are special cases of the general concept of a smooth submanifold in Euclidean space.

Consider a subset M in $\mathbf{R}^N$ as a topological space with the topology induced from $\mathbf{R}^N$. Let x be a point of M, and $U(x)$ an open neighbourhood (in M).

Definition 1. If there is a homeomorphism $\phi : \mathbf{R}^n \to U(x)$, $n \leq N$, satisfying the conditions:

(1) $\phi \in C^r$, $r \geq 1$, as a mapping from $\mathbf{R}^n$ to $\mathbf{R}^N$,

(2) $\mathrm{rank}_y\, \phi = n$ for any point $y \in \mathbf{R}^n$,

then the pair $(U(x), \phi)$ is called a *chart at the point x in M of class C^r*, or a C^r–*chart* on M.

Remark. From the definition it follows that a C^r–chart $(U(x), \phi)$ at a point x is a C^r–chart at any point $y \in U(x)$. This helps to explain the fact that the pair $(U(x), \phi)$ is also called a C^r–chart on M.

Thus, the idea of a chart has the meaning of a local representation of the set M (representation of the neighbourhood $U(x)$) in the form

(1)

$$\begin{aligned} x_1 &= \phi_1(y_1, \dots, y_n), \\ x_2 &= \phi_2(y_1, \dots, y_n), \\ &\dots\dots\dots\dots \\ x_N &= \phi_N(y_1, \dots, y_n), \end{aligned}$$

where ϕ_i, $i = 1, \ldots, N$, are the functions of class C^r determining the homeomorphism ϕ.

The neighbourhood $U(x)$ is often called a *coordinate neighbourhood* in view of the fact that the homeomorphism (1) defines curvilinear coordinates $y_1, \ldots, y_n$ on the set $U(x)$, which, in general, are not connected to the standard coordinates of the ambient (containing) space $\mathbf{R}^N$.

Note that in literature, often charts are considered that come with homeomorphisms ϕ defined not on the whole space $\mathbf{R}^n$ but on an open connected sets U. In this case, the pair (U, ϕ) is called a C^r–chart on M as above (for more details concerning such charts, see § 3).

Definition 2. A mapping $f : A \to \mathbf{R}^n$ of a subset $A \subset \mathbf{R}^N$ to the space $\mathbf{R}^n$ is called a C^r–*mapping*, $r \geq 1$, on A ($f \in C^r(A)$) if for every point $x \in A$, there exist an open neighbourhood $\widetilde{U}(x)$ in the space $\mathbf{R}^N$ and a C^r–mapping $\widetilde{f} : \widetilde{U}(x) \to \mathbf{R}^n$ such that $\widetilde{f}\big|_{A \cap \widetilde{U}(x)} = f$.

Lemma 1. *The homeomorphism ϕ^{-1} from definition 1 is of class C^r.*

Proof. Let $x \in U(x_0) \subset \mathbf{R}^N$, then $\phi^{-1}(x) \in \mathbf{R}^n$. Since $\mathrm{rank}_{\phi^{-1}(x)}\phi = n$, according to the rectification theorem for a mapping (see § 1), there exist an open neighbourhood $W(x) \subset \mathbf{R}^N$ of the point x, an open set $W_1 \subset \mathbf{R}^N$ and a C^r–diffeomorphism $F : W(x) \to W_1$ such that $F\phi$ on a certain neighbourhood $V(\phi^{-1}(x)) \subset \mathbf{R}^n$ of the point $\phi^{-1}(x)$ is the standard imbedding of $\mathbf{R}^n$ in $\mathbf{R}^N$. Let $g : \mathbf{R}^N \to \mathbf{R}^n$ be the standard projection; evidently, $g \in C^r$. Consider the mapping $gF : W(x) \to \mathbf{R}^n$, then we obtain $gF \in C^r$ and $gF\big|_{W(x) \cap U(x_0)} = \phi^{-1}$. $\qquad\square$

Definition 2 enables us to generalize the concept of a diffeomorphism of open sets of the space $\mathbf{R}^n$ (see § 1, subsection 3) in the following way: a homeomorphism $f : A \to B$ of subsets $A \subset \mathbf{R}^n$, $B \subset \mathbf{R}^m$ is called a C^r–*diffeomorphism* if $f \in C^r(A)$, $f^{-1} \in C^r(B)$.

From lemma 1 it follows that conditions (1), (2) from definition 1 of a C^r–chart $(U(x), \phi)$ are equivalent to the fact that ϕ is a C^r–diffeomorphism. Now we shall present the basic definition.

Definition 3. A set $M \subset \mathbf{R}^N$ is called an *n–dimensional submanifold in $\mathbf{R}^n$ of class C^r* or a C^r–*submanifold*, if each of its points admits a C^r–chart.

We shall denote such a submanifold by M^n and write $M^n \in C^r$, indicating that it belongs to the class C^r. In other words, a set M in $\mathbf{R}^N$ is an n–dimensional submanifold if for each of its points a coordinate system can be constructed; each coordinate system is determined locally (and called a *local system of coordinates*), but the whole set of coordinate systems "encompasses" the whole submanifold.

The definition of a chart can be extended to the case when $r = 0$, omitting condition (2) from definition 1. Naturally, in this case there is nothing that can be said about the differentiability of the homeomorphism (1). Then M^n is called a *topological manifold* and this is written $M^n \in C^0$.

Note that for any point x of a submanifold M^n of class C^r, there is, generally speaking, an infinite number of charts. An *atlas for a submanifold* M^n is a collection of charts $\{(U_\alpha, \phi_\alpha)\}$ of class C^r, whose open sets $\{U_\alpha\}$ cover M^n. An atlas $\{(U_\alpha, \phi_\alpha)\}$ of the manifold M^n determines a set of coordinate systems which "served" for the whole submanifold. In order to define a submanifold, it is sufficient to define an atlas.

Exercise $1°$. Show that if there are given two charts (U, ϕ), (V, ψ) of a C^r–submanifold M^n such that $U \cap V \neq \emptyset$, then the mapping $\psi^{-1}\phi : \phi^{-1}(U \cap V) \to \psi^{-1}(U \cap V)$ of open sets in $\mathbf{R}^n$ is a C^r–diffeomorphism.

Exercise $2°$. Show that if a sufficient number of copies of $\mathbf{R}^n$ are "glued together" with the help of the homeomorphisms $\psi^{-1}\phi$ that are determined by the charts of a certain atlas, then a topological space homeomorphic to M^n is obtained. (Compare this with the idea of a development of a two–dimensional surface, see Ch 2, § 4.)

Thus, the choice of an atlas determines a "gluing" of a submanifold M^n from n–dimensional spaces by means of a system of charts.

2. Examples of submanifolds.

Example 1. The pair $(\mathbf{R}^1, 1_{\mathbf{R}^1})$, where $1_{\mathbf{R}^1} : \mathbf{R}^1 \to \mathbf{R}^1$ is the identity mapping, determines one C^∞–chart for all $x \in \mathbf{R}^1$ and forms an atlas for a one–dimensional submanifold in $\mathbf{R}^1$ of class C^∞.

Exercise $3°$. Show that the pair $(\mathbf{R}^1, \phi)$, where $\phi : \mathbf{R}^1 \to \mathbf{R}^1$, is defined by the formula $\phi(x) = x^3$, is not a C^r–chart ($r \geq 1$) at the point $x = 0$, but it does constitute an atlas for a one–dimensional submanifold in $\mathbf{R}^1$ of class C^0.

Example 2. Similarly to example 1, the pair $(\mathbf{R}^n, 1_{\mathbf{R}^n})$, where $1_{\mathbf{R}^n} : \mathbf{R}^n \to \mathbf{R}^n$ is the identity mapping, determines a C^∞–chart for all $x \in \mathbf{R}^n$ and forms an atlas for an n–dimensional submanifold in $\mathbf{R}^n$ of class C^∞.

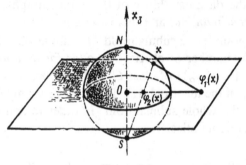

FIG. 84

Example 3. We define an atlas on the sphere $S^2 \subset \mathbf{R}^3$, consisting of two charts. We shall use the stereographic projection (Fig. 84). The sets $U_1 = S^2 \backslash \{N\}$ and $U_2 = S^2 \backslash \{S\}$ form an open covering of the sphere. The stereographic projections from the North and South poles have the form

$$\phi_1 = \left(\frac{x_1}{1 - x_3}, \frac{x_2}{1 - x_3} \right),$$

$$\phi_2 = \left(\frac{x_1}{1 + x_3}, \frac{x_2}{1 + x_3} \right),$$

respectively, and are homeomorphisms from U_1, U_2 onto $\mathbf{R}^2$.

Exercise 4°. Verify that the mappings ϕ_1^{-1}, ϕ_2^{-1} are of class C^∞ and that in each point $y \in \mathbf{R}^2$, $\operatorname{rank}_y \phi_1^{-1} = \operatorname{rank}_y \phi_2^{-1} = 2$.

Thus, the sphere S^2 with the atlas consisting of the two charts (U_1, ϕ_1^{-1}) and (U_2, ϕ_2^{-1}), is a two–dimensional submanifold in $\mathbf{R}^3$ of class C^∞.

Example 4. By using stereographic projections (see example 3) one may define on the sphere S^n an atlas consisting of two charts (U_1, ϕ_1^{-1}), (U_2, ϕ_2^{-1}), where

$$\phi_1(x) = \left(\frac{x_1}{1 - x_{n+1}}, \dots, \frac{x_n}{1 - x_{n+1}} \right),$$

$$\phi_2(x) = \left(\frac{x_1}{1 + x_{n+1}}, \dots, \frac{x_n}{1 + x_{n+1}} \right).$$

The sphere S^n with this atlas is an n–dimensional submanifold in $\mathbf{R}^{n+1}$ of class C^∞.

Example 5. The graph of a mapping. Let a mapping $f : \mathbf{R}^n \to \mathbf{R}^m$, $f \in C^r$, be given. Consider the graph $\Gamma(f) = \{x, f(x)\} \subset \mathbf{R}^n \times \mathbf{R}^m$ of the mapping (see Ch. 2, § 9, Exercise 12°). We define an atlas on $\Gamma(f)$ consisting of one chart $(\mathbf{R}^n, \phi)$, where $\phi : \mathbf{R}^n \to \mathbf{R}^{n+m}$ is determined by the formula $\phi(x) = (x, f(x))$. Thus, $\Gamma(f)$ is an n–dimensional submanifold in $\mathbf{R}^{n+m}$ of class C^r.

Exercise 5°. Show that the set of points $\left(x, \sin \frac{1}{x}\right)$ of the plane $\mathbf{R}^2$, $x \in \mathbf{R}^1$, $x \neq 0$, is a one–dimensional submanifold in $\mathbf{R}^2$ of class C^∞.

Exercise 6°. Show that any set in $\mathbf{R}^n$ consisting of isolated points is a zero-dimensional submanifold in $\mathbf{R}^n$.

Example 6. The solution set of a system of equations. Consider a system of equations

$$f_1(x_1, \dots, x_n) = 0,$$

(2) $\qquad\qquad \dots\dots\dots\dots\dots$

$$f_m(x_1, \dots, x_n) = 0,$$

where $f_1, \dots, f_m : \mathbf{R}^n \to \mathbf{R}^1$ are functions of class C^r, $r \geq 1$. Let $n \geq m$. The system of functions $f_1, \dots, f_m$ defines a C^r–mapping $f : \mathbf{R}^n \to \mathbf{R}^m$. The solution set of the system we denote by M. It is clear that $M = f^{-1}(0)$.

Theorem 1. *Let the set M be nonempty. If for each point $x \in M$ the rank of the Jacobian matrix $\left(\frac{\partial f}{\partial x}\right)\Big|_x$ is equal to m, then M is an $(n-m)$–dimensional submanifold in $\mathbf{R}^n$ of class C^r.*

Proof. Let x_0 be an arbitrary point of M. According to the condition of the theorem, x_0 is a regular point of the mapping f. By the rectification theorem for a mapping (see § 1), there exist an open neighbourhood $V(x_0) \subset \mathbf{R}^n$ of the point x_0, an open set $W \subset \mathbf{R}^m$, and a C^r–diffeomorphism $F : V(x_0) \to W$ such that on the set W, fF^{-1} is the standard projection of $\mathbf{R}^n$ onto $\mathbf{R}^m$. Without loss of generality, one may assume that W is a certain open disc $D_\rho^n(y_0)$, $y_0 = F(x_0)$; then

$$(fF^{-1})^{-1}(0) \cap D_\rho^n(y_0) = \mathbf{R}^{n-m} \cap D_\rho^n(y_0) = D_\rho^{n-m}(y_0).$$

(Here $\mathbf{R}^{n-m} = \{x \in \mathbf{R}^n : x_1 = \ldots = x_m = 0\}$ is a subspace in $\mathbf{R}^n$.) It is clear that the open disc $D_\rho^{n-m}(y_0)$ in $\mathbf{R}^{n-m}$ is the image of the set $M \cap V(x_0)$ under the diffeomorphism F. Thus, the open neighbourhood $V(x_0) \cap M$ of the point x_0 in M is C^r–diffeomorphic to the open disc $D_\rho^{n-m}(y_0)$, and, therefore, to the space $\mathbf{R}^{n-m}$. $\square$

As an example, we shall consider again the sphere $S^n \subset \mathbf{R}^{n+1}$ as given by the equation $x_1^2 + \ldots + x_{n+1}^2 - 1 = 0$. Here the rank $\left(\frac{\partial f}{\partial x}\right)$ at any point of the sphere is equal to one, so the conditions of theorem 1 are fulfilled (for any $r \geq 1$). Thus, we have proved once more that the sphere S^n is an n–dimensional submanifold in $\mathbf{R}^{n+1}$ of class C^∞.

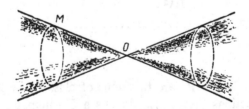

FIG. 85

Consider a case when the conditions of theorem 1 are not fulfilled. Let the set $M \subset \mathbf{R}^3$ be given by the equation $x_1^2 - x_2^2 - x_3^2 = 0$ (Fig. 85). A structure of a two–dimensional C^∞–submanifold may be defined on the set $M\backslash 0$ as before. All the minors of the Jacobian matrix at the point 0 are zero, and its rank is not maximal. The set M is a simple example of an algebraic manifold, and the point 0 is a singular point of this manifold.

§ 3. Smooth manifolds

1. The concept of a smooth manifold. This concept is one of the central concepts of smooth topology and modern analysis. The method of introducing coordinates on a set may be generalized without assuming that it lies in a space $\mathbf{R}^N$. The development of this idea leads us to the notion of a smooth manifold.

Let M be a topological space, $U \subset M$ an open set, and $\phi : \mathbf{R}^n \to U$ a homeomorphism. Then the standard coordinates $\{\xi_1(x), \ldots, \xi_n(x)\}$ of the point $\phi^{-1}(x)$ in $\mathbf{R}^n$ are taken to be the coordinates of the point $x \in U$. Thus, the homeomorphism ϕ determines coordinates on a part U of the space M; the pair (U, ϕ) is called a *chart* on M. For any point $x \in U$ the chart (U, ϕ) is also called a *chart at the point x*.

Let $(U, \phi)(\phi : \mathbf{R}^n \to U)$, $(V, \psi)(\psi : \mathbf{R}^n \to V)$ be two charts on M, and $U \cap V \neq \emptyset$. Then to each point $x \in U \cap V$, there correspond two coordinate systems: $\{\xi_1(x), \ldots, \xi_n(x)\}$ and $\{\eta_1(x), \ldots, \eta_n(x)\}$, víz., the coordinates of the points $\phi^{-1}(x) \in \phi^{-1}(U \cap V)$ and $\psi^{-1} \in \psi^{-1}(U \cap V)$, which, in general, are different. Both coordinate systems are "equally good" in the sense that there exists a transition homeomorphism

$$\psi^{-1}\phi : \phi^{-1}(U \cap V) \to \psi^{-1}(U \cap V),$$

connecting the two coordinate systems and enabling us to express the first coordinates continuously in terms of the second set:

$$\xi_1 = \chi_1(\eta_1, \ldots, \eta_n),$$

(1)
$$\cdots\cdots\cdots\cdots\cdots$$

$$\xi_n = \chi_n(\eta_1, \ldots, \eta_n)$$

and, inversely, the second coordinates can be expressed continuously through the first:

$$\eta_1 = \kappa_1(\xi_1, \dots, \xi_n),$$

(2)

$$\dots\dots\dots\dots$$

$$\eta_n = \kappa_n(\xi_1, \dots, \xi_n).$$

In formulae (1) and (2), the coordinate functions of the mappings $\phi^{-1}\psi$, $\psi^{-1}\phi$ are denoted by $\chi_1, \dots, \chi_n$; $\kappa_1, \dots, \kappa_n$, respectively.

When solving problems in analysis, sometimes it is necessary that the dependences (1) and (2) should be differentiable r times, $r = 0, 1, \dots, \infty$. This means that the homeomorphism $\psi^{-1}\phi$ is a C^r–diffeomorphism. (For convenience, we call a homeomorphism a C^0–diffeomorphism.)

Definition 1. The charts (U, ϕ) and (V, ϕ) on M are called C^r–compatible if one of the following conditions is fulfilled (1) $U \cap V = \emptyset$; (2) $U \cap V \neq \emptyset$ and the homeomorphism $\psi^{-1}\phi : \phi^{-1}(U \cap V) \to \psi^{-1}(U \cap V)$ is a C^r–diffeomorphism.

Definition 2. A set of charts $\{(U_\alpha, \phi_\alpha)\}$ on M is called a C^r–*atlas* or an *atlas of class* C^r if any two of its charts are C^r–compatible and $\bigcup_\alpha U_\alpha = M$.

Remark. All homeomorphisms ϕ_α in the definition of a C^r–atlas are defined on the same space $\mathbf{R}^n$.

Thus, by giving an atlas on M, we thereby introduce coordinates in the neighbourhood of each point $x \in M$; these are called local coordinates.

Let a C^r–atlas $\{(U_\alpha, \phi_\alpha)\}$ be given on M, and, consequently, let local coordinates be introduced in the neighbourhood of each point. In many problems of analysis, in the course of an argument it is convenient to introduce new local coordinates (by using a certain chart (V, ψ)), which should be "equally good" to the original coordinates given by the charts of the C^r–atlas $\{(U_\alpha, \phi_\alpha)\}$. Thus, the chart (V, ψ) should be C^r–compatible with each chart (U_α, ϕ_α) of the given C^r–atlas. If this condition is fulfilled then the set of charts $\{(U_\alpha, \phi_\alpha)\} \cup (V, \psi)$ is a C^r–atlas. Since a C^r–atlas $\{(U_\alpha, \phi_\alpha)\} \cup (V, \psi)$ is obtained from the C^r–atlas $\{(U_\alpha, \phi_\alpha)\}$ by adding an "equally good" chart, it is natural to assume these atlases to be equivalent.

Definition 3. Two C^r–atlases $\{(U_\alpha, \phi_\alpha)\}$, $\{(V_\beta, \psi_\beta)\}$ are said to be *equivalent* if any two charts (U_α, ϕ_α) and (V_β, ψ_β) are C^r-compatible. In other words, two C^r-atlases are equivalent if their union is a C^r-atlas.

Exercise 1°. Show that the relation introduced on the set of C^r-atlases is an equivalence relation.

From exercise 1° follows that the set of C^r-atlases on M is decomposed into nonintersecting classes of equivalent atlases.

Definition 4. An equivalence class of C^r-atlases on M is called a C^r-*structure* on M.

Each equivalence class of C^r-atlases on M is defined by any of its representatives, i.e., a given C^r-structure can be recovered from any of its C^r-atlases. This remark means that a C^r-structure on M is determined by specifying on it one of the C^r-atlases from the given C^r-structure.

The union of all C^r-atlases from a given C^r-structure is also a C^r-atlas called *maximal*. Defining a C^r-structure is equivalent to specifying the maximal atlas. Sometimes the maximal atlas is called a C^r-structure.

C^0-structures are called *topological structures*; C^r-structures ($r = 1, \ldots, \infty$) are called *smooth* (or *differential*) structures.

Definition 5. A topological space M with a given C^r-structure on it is called a C^r-*manifold* (or a *manifold of class* C^r), and the dimension of the space $\mathbf{R}^n$ on which the homeomorphisms of the charts act, is called the *dimension* of the C^r-manifold.

Similarly to C^r-structures, C^0-manifolds are said to be topological, C^r-manifolds ($r = 1, \ldots, \infty$) smooth. Sometimes (for brevity's sake) C^r-manifolds are simply called manifolds, and C^r-atlases just atlases.

If in condition 2 of definition 1, the homeomorphisms $\psi^{-1}\phi$, $\phi^{-1}\psi$ are analytic mappings ($\psi^{-1}\phi$, $\phi^{-1}\psi \in C^\omega$), then the charts (U, ϕ) and (V, ψ) in M are said to be C^ω-compatible. The C^ω-*atlases*, C^ω-*structures*, and C^ω-*manifolds* are defined in a natural way. The C^ω-structures and C^ω-manifolds are called *analytic structures* and *analytic manifolds*, respectively. In order to indicate the dimension of a manifold, we write M^n, and also $\dim M = n$.

Remark. The dimension of a C^0-manifold is its invariant, i.e., it does not depend on the choice of an atlas. Indeed, if M allows the atlases

$$\{(U_\alpha,\phi_\alpha)\}(\phi_\alpha :\mathbf{R}^n \to U_\alpha), \quad \{(V,\psi_\beta)\}(\psi_\beta :\mathbf{R}^m \to V_\beta)$$

and $n \neq m$, then there can be found sets U_α, V_β such that $U_\alpha \cap V_\beta \neq \emptyset$ and a mapping

$$\psi_\beta^{-1}\phi_\alpha :\phi_\alpha^{-1}(U_\alpha \cap U_\beta) \to \psi_\beta^{-1}(U_\alpha \cap U_\beta)$$

that is a homeomorphism. This contradicts the Brouwer theorem which states that nonempty open sets $U \subset \mathbf{R}^n$, $V \subset \mathbf{R}^m$ can be homeomorphic only in the case when $n = m$. (This theorem will be proved independently on the subject of this chapter in Ch. 5, § 6.) For C^r-manifolds, $r \geq 1$, the correctness of the definition of dimension is evident.

Note that a C^0-structure on any space M is unique (this follows from the definition); but if $r \neq 0$, then M may have several different C^r-structures. Indeed, the atlas consisting of one chart (U, ϕ), where $U = \mathbf{R}^1$, and $\phi : \mathbf{R}^1 \to \mathbf{R}^1$ is the identity mapping, determines a structure of a C^∞-manifold on $\mathbf{R}^1$. The atlas consisting of one chart $(\mathbf{R}^1, \phi)$, where $\phi(x) = x^3$, also determines the structure of a C^∞-manifold on $\mathbf{R}^1$. It is easy to verify that the atlases considered are not equivalent, and, therefore, the C^∞-structures defined by them differ.

Moreover, it can be proved that if there exists on M at least one C^r-strucutre $(r \geq 1)$, then there exists an infinite number of C^r-structures on M.

Exercise 2°. Show that the atlases

$$\{(\mathbf{R}^1, \phi_0)\}, \dots , \{(\mathbf{R}^1, \phi_k)\}, \dots$$
$$\text{where } \phi_k(x) = x^{2k+1}, \quad k = 0, 1, \dots ,$$

specify different C^∞-structures on $\mathbf{R}^1$.

Exercise 3°. Show that any C^r-submanifold in $\mathbf{R}^N$ is a C^r-manifold (see § 2, exercise 1°).

We shall indicate one formal generalization of the concept of a chart (U, ϕ) where the homeomorphism ϕ is defined on some open connected set of $\mathbf{R}^r$ which, in general, does not coincide with the whole space. In this case, one can define all the concepts introduced above by repeating the same formulations.

However, this will not lead to generalization of the concept of a C^r–manifold. In fact, in this C^r–structure, one may select a C^r–atlas $\{U_\alpha, \phi_\alpha)\}$ such that all homeomorphisms ϕ_α are defined on open discs D_α in $\mathbf{R}^n$. Since there exists a C^r–diffeomorphism $f_\alpha : \mathbf{R}^n \to D_\alpha$, the C^r–atlas $\{(U_\alpha, \phi_\alpha f_\alpha)\}$ is contained in our C^r–structure and consists of standard charts.

In some cases, it is simpler to define an atlas consisting of generalized charts. We shall use this if necessary without reserve.

Example 1. Any open set V of a manifold M^n of class C^r is itself a manifold of class C^r with its structure defined by the atlas $\{(U_\alpha \cap V,\ \phi_\alpha|_{\phi_\alpha^{-1}(U_\alpha \cap V)})\}$, where $\{(U_\alpha, \phi_\alpha)\}$ is a certain atlas from the C^r–structure given on M^n.

Example 2. Define a C^∞–atlas on $S^2 \subset \mathbf{R}^3$ consisting of six charts. Set

$$U_k^+ = \{(x_1, x_2, x_3) \in S^2 : x_k > 0\},$$
$$U_k^- = \{x \in S^2 : x_k < 0\}, \quad k = 1, 2, 3.$$

Define homeomorphisms $\phi_k^+ : D^2 \to U_k^+, \phi_k^- : D^2 \to U_k^-$:

$$\phi_1^+, \phi_1^- : (x_2, x_3) \mapsto (\pm\sqrt{1 - x_2^2 - x_3^2}, x_2, x_3),$$

$$\phi_2^+, \phi_2^- : (x_1, x_3) \mapsto (x_1, \pm\sqrt{1 - x_1^2 - x_3^2}, x_3),$$

$$\phi_3^+, \phi_3^- : (x_1, x_2) \mapsto (x_1, x_2, \pm\sqrt{1 - x_1^2 - x_2^2}),$$

where the sign of the right–hand side is chosen in accordance with the sign "+" or "−" on the left–hand side.

In a similar way, one may define on the sphere S^n a C^∞–atlas consisting of $2(n + 1)$ charts.   ◊

To make a great number of constructions possible in the study of topological spaces, it is necessary that the topology is Hausdorff and has a countable base. These properties do not, in general, follow from the definition of a manifold. This can be illustrated by the following example.

Example 3. A non–Hausdorff manifold M^1 of class C^∞. Consider the interval $(0, 3)$ and break it into three sets $(0, 1]$, $(2, 3)$, $(1, 2]$. On their formal (dis-

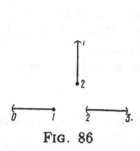

FIG. 86

joint) union (Fig. 86) we introduce a topology in the following way: the neighbourhoods of points in the set $(0, 1) \cup (1, 2) \cup (2, 3)$ are the same as in the topology induced by the real straight line. The sets $(1 - \epsilon, 1] \cup (2, 2 + \epsilon), (2 - \epsilon, 2] \cup (2, 2 + \epsilon)$ are neighbourhoods of the points $x_1 = 1$, $x_2 = 2$, respectively. Then the points x_1, x_2 are not separable.

We recommend as an exercise to show that a structure of a one–dimensional C^∞–manifold can be defined on the obtained space in a natural way and that this manifold has a countable base.

Example 4. A manifold M^1 of class C^∞ not possessing a countable base. Consider the set $M = \mathbf{R}^1 \times \mathbf{R}^1$. We define the topology on M as the topology of Cartesian product were the first factor $\mathbf{R}^1$ has usual topology, and the second factor $\mathbf{R}^1$ has a discrete one. It is not difficult to show that this is a Hausdorff one–dimensional manifold of class C^∞ whose topology does not have a countable base.

Combining two last examples, it is easy to construct a non–Hausdorff manifold without a countable base (by taking their Cartesian product).

Note that the absence of a countable base for the topology of the manifold in example 4 lead us to the "pathology", viz., the plane is a manifold of dimension 1, and not 2.

Usually, a manifold M^n is assumed to be Hausdorff and to satisfy the second countability axiom. We also accept this without further notice. Then it is easy to show that a manifold M^n is a locally compact and even a paracompact space.

Indeed, the local compactness follows from the following simple exercise.

Exercise 4°. Show that if (U, ϕ) is a chart on M^n, $x \in U$ and $D^n(\phi^{-1}(x))$, $\bar{D}^n(\phi^{-1}(x))$ are the open and closed discs in $\mathbf{R}^n$ with centre at the point $\phi^{-1}(x)$ and radius 1, then $\phi(D^n(\phi^{-1}(x)))$ is an open neighbourhood of the point x in M^n of which the closure (in M^n) is compact and equal to $\phi(\bar{D}^n(x)))$.

The paracompactness of the manifold M^n follows from its local compactness and the existence of a countable of a base (due to the corollary of theorem 6, Ch 2, § 13).

Note that from the existence of a countable base for a manifold, there immediately follows that any C^r-manifold M^n, $r \geq 0$, has a countable atlas $\{U_\alpha, \phi_\alpha\}$, $\phi_\alpha : \mathbf{R}^n \to U_\alpha$, i.e., an atlas consisting of at most a countable set of charts.

2. Projective spaces. The definition and various topologically equivalent interpretations of the projective spaces $\mathbf{R}P^{n-1}$, $\mathbf{C}P^{n-1}$, $n \geq 2$, were given in Ch. 2, § 5, subsection 2 (see also Ch. 1, § 3, subsection 1). There are natural structures of C^∞-manifolds that can be defined on the spaces $\mathbf{R}P^{n-1}$ and $\mathbf{C}P^{n-1}$. We shall illustrate the idea of introducing local coordinates on $\mathbf{R}P^{n-1}$. Consider $\mathbf{R}P^{n-1}$ as the set $L = \{l\}$ of all straight lines in the space $\mathbf{R}^n$ passing through the origin. Each straight line intersects one or several hyperplanes of the form $x_j = 1$. We fix one of these hyperplanes $x_i = 1$ and take the union U_i of all straight lines from L which intersect the hyperplane $x_i = 1$. Then the position of the straight line $l \in U_i$ is determined by the Cartesian coordinates $(\xi_1, \ldots, \xi_{i-1}, 1, \xi_i, \ldots, \xi_{n-1})$ of its point p of intersection with the hyperplane $x_i = 1$. It is natural to take the coordinates $(\xi_1, \ldots, \xi_{i-1}, \xi_i, \ldots, \xi_{n-1})$ as the local coordinates of the straight line l (see Fig. 87). Thus, we have the homeomorphisms

$$\psi_i(l) = (\xi_1, \ldots, \xi_{i-1}, \xi_i, \ldots, \xi_{n-1}) : U_i \to \mathbf{R}^{n-1},$$
$$i = 1, \ldots, n.$$

The local coordinates $\xi_1, \ldots, \xi_{n-1}$ are also called the projective coordinates of the straight line l. It is not difficult to express the local coordinates of the straight line l by means of the coordinates of an arbitrary point $x = (x_1, \ldots, x_n)$ of the straight line l : $\xi_1 = x_1/x_i, \ldots, \xi_{i-1} = x_{i-1}/x_i$, $\xi_i = x_{i+1}/x_i, \ldots, \xi_{n-1} = x_n/x_i$.

The atlas of n charts (U_i, ϕ_i), $i = 1, \ldots, n$, where $\phi_i = \psi_i^{-1}$, determines the structure of a C^∞-manifold of dimension $n-1$ on $\mathbf{R}P^{n-1}$. We shall show that the charts of the atlas constructed are C^∞-compatible. Indeed, let $l \in U_i \cap U_j$, and $\eta_1 = x_1/x_j, \ldots, \eta_{j-1} = x_{j-1}/x_j$, $\eta_j = x_{j+1}/x_j, \ldots, \eta_{n-1} = x_n/x_j$ be the local coordinates of the straight line l in the chart (U_j, ϕ_j). Note that $\eta_i \neq 0$, because $l \in U_i$. To be definite, let $i < j$. Then the following relations are evident:

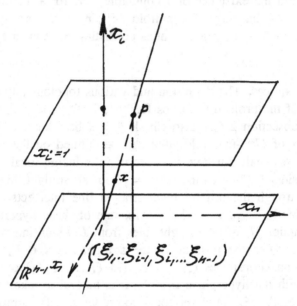

FIG. 87

$$\eta_1/\eta_i = \xi_1, \ldots, \eta_{i-1}/\eta_i = \xi_{i-1},$$
$$\eta_{i+1}/\eta_i = \xi_i, \ldots, \eta_{j-1}/\eta_i = \xi_{j-2},$$
$$1/\eta_i = \xi_{j-1}, \eta_j/\eta_i = \xi_j, \ldots, \eta_{n-1}/\eta_i = \xi_{n-1}.$$

which shows obviously, that the local coordinates $\xi_1, \ldots, \xi_{n-1}$ depend infinitely smoothly on the coordinates $\eta_1, \ldots, \eta_{n-1}$.

Exercise 5°. Verify, that for the projective space $\mathbf{R}P^{n-1}$ considered as the collection of pairs of diametrically opposite points of the sphere S^{n-1}, local coordinates may also be given in the way described above.

Exercise 6°. Show that the complex projective space $\mathbf{C}P^{n-1}$ has a C^∞–atlas converting it into a C^∞–manifold of real dimension $2n - 2$.

Hint. Considering CP^{n-1} as the set of complex straight lines in C^n, define the atlas by formulae analogous to those in the case of RP^{n-1}.

More generally, one can consider certain manifolds M^n of class C^r on which the group Z_k acts (see Ch. 2, § 5). We assume that under this action the orbit of each point consists of k different elements.

Exercise 7°. Let $h : Z_k \to H(M^n)$ be a homomorphism of the group Z_k (where k is prime) to the group $H(M^n)$ of homeomorphisms of M^n, which defines the given action of Z_k on M^n, and let g be generator of the group Z_k. Show that the condition $h_g(x) \neq x$, for any $x \in M^n$, is equivalent to the assumption that under this action the orbit of each point consists of k different elements. In this case, it is said that the group Z_k *acts without fixed points*.

Further, we suppose that the charts of the form $(h_g U_\alpha, h_g \phi_\alpha)$ are C^r-compatible with the charts (U_β, ϕ_β) of an C^r-atlas on M^n. Consider the quotient space M^n/Z_k. It is also a C^r-manifold of dimension n. The atlas is given as follows: let O_x be the orbit of the point x, $U(O_x)$ the neighbourhood of the orbit in M^n/Z_k consisting of all the orbits O_y which pass through the points y of a sufficiently small neighbourhood $V(x)$ of a point x in M^n ($V(x)$ should not contain any pair of points y, $h_g(y)$, and it has to lie wholly in some chart of the manifold M^n). Then the local coordinates in $V(x)$ of the point $y \in V(x)$ can serve as the local coordinates of the orbit $O_y \subset U(O_x)$. This can be verified to be an C^r-atlas. The condition of compatibility of charts $(h_g U_\alpha, h_g \phi_\alpha)$, (U_β, ϕ_β) is not difficult to check (see theorem 2, § 5, below).

Exercise 8°. Verify that the generalized lens space $L(k, k_1, \dots, k_n)$ is a C^∞-manifold of dimension $2n + 1$.

3. Induced structures. Let M^n be a C^r-manifold, and $f : M^n \to N$ a homeomorphism of topological spaces M^n and N. A structure of a C^r-manifold, called the *structure induced by* f, may be introduced on the topological space N in a natural way. Indeed, if $\{(U_\alpha, \phi_\alpha)\}$ is a C^r-atlas for the manifold M^n, then $\{(f(U_\alpha), f\phi_\alpha)\}$ is a C^r-atlas on N.

Exercise 9°. Verify that $\{(f(U_\alpha), f\phi_\alpha)\}$ is, in fact, an atlas determining the structure of a C^r-manifold of dimension n on N.

The described method of specifying a structure is quite useful for defining the structure of a C^r–manifold on a topological space N: we can give a structure of a C^r–manifold on a "simpler" space M homeomorphic to N, and afterwards induce on N the structure of a C^r–manifold. Thus, for instance different models $\mathbf{R}P^{n-1}$ are supplied with a C^r–structure.

Example 5. It is easy to see that any one–dimensional compact C^0–manifold is triangulable. Hence any connected one–dimensional compact C^0–manifold is homeomorphic to the circle S^1 (see, Ch. 2, § 4, exercise 6°), therefore, there is a natural C^∞–structure induced on it.

Example 6. A two–dimensional oriented closed surface, as shown in Ch. 2, § 4, is homeomorphic to a surface of type M_p (i.e., a sphere with p handles), which can be realized in $\mathbf{R}^3$ as a C^∞–submanifold (this is assumed to be intuitively obvious). Thus, the orientible closed surfaces get a structure of a C^∞–manifold.

Exercise 10°. Specify the structure of a C^∞–manifold on the boundary of the cube $I^n = \{x = (x_1, \ldots, x_n) : |x_i| \leq 1, i = 1, \ldots, n\}$ by inducing it from the sphere S^{n-1}.

Exercise 11°. Show that the mapping

$$(x_1, x_2, x_3) \mapsto (x_1^2, x_2^2, x_3^2, x_1 x_2, x_1 x_3, x_2 x_3)$$

is a homeomorphism of the projective plane $\mathbf{R}P^2$ onto a subset in $\mathbf{R}^6$. When inducing the structure of a smooth manifold on $\mathbf{R}P^2$, by this homeomorphism we realize $\mathbf{R}P^2$ as a submanifold in $\mathbf{R}^6$.

Exercise 12°. Construct a realization $\mathbf{R}P^3$ in $\mathbf{R}^{10}$.

4. Manifolds of matrices. We endow the set $M(m, n)$ of all $m \times n$–matrices with elements from $\mathbf{R}$ with the topology induced by the natural mapping $i :$ $\mathbf{R}^{mn} \to M(m, n)$:

$$(x_1, \ldots, x_{mn}) \mapsto \begin{pmatrix} x_1 & \cdots & x_n \\ x_{n+1} & \cdots & x_{2n} \\ \cdots\cdots\cdots\cdots\cdots \\ x_{(m-1)n+1} & \cdots & x_{mn} \end{pmatrix}.$$

Then the homeomorphism i induces on $M(m,n)$ the structure of a C^∞–manifold of dimension mn.

Denote by $M(m,n;k)$ the subspace in $M(m,n)$ of matrices of the fixed rank k. We are going to define on $M(m,n;k)$ the structure of a C^∞–manifold of dimension $k(m+n-k)$. Note beforehand that if $Y \in M(m,n)$ and rank $Y \geq k$, then by interchanging rows and interchanging columns, the matrix Y can be transformed into a matrix of the following form

$$\left(\frac{A_Y | B_Y}{C_Y | D_Y}\right),$$

where A_Y is a non–singular square matrix of order k. In other words, there exist non–singular square matrices $P_Y \in M(m,m)$, $Q_Y \in M(n,n)$ such that

$$P_Y Y Q_Y = \left(\frac{A_Y | B_Y}{C_Y | D_Y}\right).$$

We show that rank $Y = k$ iff $D_Y = C_Y A_Y^{-1} B_Y$. Indeed, from the equality

$$\left(\frac{I_k}{-C_Y A_Y^{-1}} \left|\frac{0}{I_{m-k}}\right.\right) \left(\frac{A_Y | B_Y}{C_Y | D_Y}\right) = \left(\frac{A_Y}{0} \left|\frac{B_Y}{-C_Y A_Y^{-1} B_Y + D_Y}\right.\right)$$

follows that

$$\text{rank}\, Y = \text{rank} \left(\frac{A_Y}{0} \left|\frac{B_Y}{-C_Y A_Y^{-1} B_Y + D_Y}\right.\right).$$

From the latter equality it can be seen that rank $Y = k$ iff $D_Y = C_Y A_Y^{-1} B_Y$.

Now let $X_0 \in M(m,n;k)$. Let X be an arbitrary matrix from $M(m,n;k)$. Denote

$$P_{X_0} X Q_{X_0} = \left(\frac{A_{X,X_0} | B_{X,X_0}}{C_{X,X_0} | D_{X,X_0}}\right),$$

where A_{X,X_0} is a square matrix of order k. Consider an open neighbourhood

$$V(X_0) = \{X \in M(m,n) : \det A_{X,X_0} \neq 0\}$$

of the matrix X_0 in $M(m,n)$. Then $U(X_0) = V(X_0) \cap M(m,n;k)$ is an open neighbourhood of the matrix X_0 in $M(m,n;k)$ and the mapping

$$\phi_{X_0} : U(X_0) \to \mathbf{R}^{mn-(m-k)(n-k)},$$

given by

$$X \mapsto \left(\frac{A_{X,X_0} | B_{X,X_0}}{C_{X,X_0} | D_{X,X_0}} \right) \mapsto$$

$$\mapsto \left(\frac{A_{X,X_0} | B_{X,X_0}}{C_{X,X_0} | 0} \right) \xrightarrow{i} \mathbf{R}^{mn-(m-k)(n-k)},$$

is a homeomorphism (i is the natural mapping). Consequently, $(U(X_0), \phi_{X_0}^{-1})$ is a chart. By defining in such a way a chart for every matrix $X_0 \in M(m, n; k)$ we obtain a C^∞-atlas on $M(m, n; k)$.

Exercise $13°$. Show that the charts of the atlas constructed are C^∞–compatible.

Note that $M(k, n)$ can be interpreted as the set of ordered sets of k vectors in $\mathbf{R}^n$, and $M(k, n; k)$ as a set of ordered collections of k linearly independent vectors in $\mathbf{R}^n$; $M(n, n)$ is denoted by $L(n, \mathbf{R})$, and $M(n, n; n)$ by $GL(n, \mathbf{R})$ (the group of invertible $n \times n$ matrices called the general linear group).

5. Grassmann manifolds. A natural generalization of the projective space $\mathbf{R}P^{n-1}$ is the Grassmann manifold $G_k(\mathbf{R}^n)$ consisting of all k–dimensional subspaces, $k \geq 1$, of the space $\mathbf{R}^n$ (when $k = 1$, this is a projective space). We equip the set $G_k(\mathbf{R}^n)$ with the topology induced by the natural mapping $M(k, n; k) \to G_k(\mathbf{R}^n)$, associating to each matrix

$$\begin{pmatrix} x_{11} & \cdots & x_{1n} \\ \cdots\cdots\cdots\cdots \\ x_{k1} & \cdots & x_{kn} \end{pmatrix}$$

the subspace in $\mathbf{R}^n$ spanned by the vectors

$$x_i = (x_{i1}, \ldots, x_{in}), \quad i = 1, \ldots, k.$$

Note that $G_k(\mathbf{R}^n)$ is homeomorphic to the orbit space of the space $M(k, n; k)$ by the action (from left) of the group $GL(k, \mathbf{R})$; the element $C \in GL(k, \mathbf{R})$ acts on the element $Y \in M(k, n; k)$ according to the rule $Y \mapsto CY$ (the product of matrices). In other words, the space $G_k(\mathbf{R}^n)$ is homeomorphic to the quotient space $M(k,n;k)/R$ where R is the following equivalence relation: $X \overset{R}{\sim} Y$, iff there exists a square non–singular matrix C of order k such that $X = CY$. In order to define the structure of a C^∞–manifold on $G_k(\mathbf{R}^n)$, we equip $M(k,n;k)/R$ with a C^∞–structure and induce the C^∞–structure on $G_k(\mathbf{R}^n)$ by the homeomorphism $f : M(k,n;k)/R \to G_k(\mathbf{R}^n)$. The manifold obtained is called a *Grassmann manifold*.

Local coordinates on $M(k,n;k)/R$ can be defined similarly to that for the projective space $\mathbf{R}P^{m-1}$ by considering $M(k,n;k)$ instead of $\mathbf{R}^m\backslash 0$, instead of the straight lines $l = \{tx\}$, $t \in \mathbf{R}^1\backslash 0$, $x \in \mathbf{R}^m\backslash 0$, take the subspace $L = \{TX\}$, $T \in GL(k,\mathbf{R})$, $X \in M(k,n;k)$, and instead of the hyperplane $x_i = 1$ take the set $H_{i_1,\dots,i_k}$ of matrices from $M(k,n;k)$ for which the submatrix composed of columns $i_1,\dots,i_k$ is a unit matrix. The subspace $\{TX\}$ intersects the set $H_{i_1,\dots,i_k}$ iff the submatrix $X_{i_1,\dots,i_k}$ consisting of columns $i_1,\dots,i_k$ of the matrix X is non–singular (i.e., $\det X_{i_1,\dots,i_k} \neq 0$); in case $X_{i_1,\dots,i_k}$ is non–singular, the "intersection point", as can be easily seen, is the matrix $Y = X_{i_1,\dots,i_k}^{-1} X$. In order to determine charts, fix the set $H_{i_1,\dots,i_k}$ (i.e., fix the numbers of k columns in the matrix $X \in M(k,n;k)$) and consider the set $U_{i_1,\dots,i_k}$ of all subspaces $\{TX\}$ whose intersections with $H_{i_1,\dots,i_k}$ is non–empty. In other words, $U_{i_1,\dots,i_k}$ is the set of subspaces $\{TX\}$ for which the generator X of the subspace $\{TX\}$ has a non–singular submatrix $X_{i_1,\dots,i_k}$. It is natural to take the elements of the matrix $Y_{j_1,\dots,j_{n-k}}$ formed by the columns $j_1,\dots,j_{n-k}$ different from $i_1,\dots,i_k$ of the matrix Y as local coordinates. More precisely, the pair $(U_{i_1,\dots,i_k}, \phi_{i_1,\dots,i_k})$, where $\phi_{i_1,\dots,i_k} : M(k,n;k)/R \to \mathbf{R}^{k(n-k)}$ is a homeomorphism determined by the correspondence

$$X \mapsto Y_{j_1,\dots,j_{n-k}} = \begin{pmatrix} y_{11} & \cdots & y_{1(n-k)} \\ \cdots\cdots\cdots\cdots\cdots\cdots \\ y_{k1} & \cdots & y_{k(n-k)} \end{pmatrix} \overset{i}{\mapsto}$$

$$\overset{i}{\mapsto} (y_{11},\dots,y_{1(n-k)}, y_{21},\dots,y_{k(n-k)}),$$

is a chart in $M(k,n;k)/R$.

Exercise 14°. Verify that the mappings $\phi_{i_1,\dots,i_k}$ are homeomorphisms.

The atlas $\{(U_{i_1,\dots,i_k}, \phi_{i_1,\dots,i_k}^{-1})\}$ consisting of C_n^k (the binomial coefficient) charts determines the structure of a C^∞–manifold of dimension $k(n - k)$ on $M(k, n; k)/R$.

Exercise 15°. Show the C^∞–compatibility of the charts of the atlas constructed.

Exercise 16°. Show that the manifold $G_k(\mathbf{R}^n)$ is homeomorphic to the manifold $G_{n-k}(\mathbf{R}^n)$.

6. Stiefel manifolds. In the manifold $M(n, k)$, consider the subset $V_k(\mathbf{R}^n)$ of matrices with the elements satisfying the system of $k(k+1)/2$ linear equations

$$\sum_{s=1}^{n} x_{si}x_{sj} = \delta_{ij}, \quad 1 \le i \le j \le k$$

(here δ_{ij} is the Kronecker symbol). Provide $V_k(\mathbf{R}^n)$ with the topology induced by $M(n, k)$. Consider the natural mapping $i : V_k(\mathbf{R}^n) \to \mathbf{R}^{nk}$:

$$\begin{pmatrix} a_{11} & \cdots & a_{1k} \\ \cdots\cdots\cdots\cdots \\ a_{n1} & \cdots & a_{nk} \end{pmatrix} \longmapsto$$

$$\longmapsto (a_{11},\dots,a_{1k},a_{21},\dots,a_{2k},\dots,a_{n1},\dots,a_{nk}).$$

Provide the set $i(V_k(\mathbf{R}^n))$ with the topology induced by $\mathbf{R}^{nk}$. Then i homeomorphically maps $V_k(\mathbf{R}^n)$ to its own image $i(V_k(\mathbf{R}^n))$. Show that it is possible to define the structure of C^∞–manifold on $i(V_k(\mathbf{R}^n))$. The rank of the Jacobian matrix of the mapping $f : \mathbf{R}^{nk} \to \mathbf{R}^{k(k+1)/2}$ with as components the functions $\sum_{s=1}^{n} x_{si}x_{sj} - \delta_{ij}, 1 \le i \le j \le k$, is, evidently, equal to $k(k + 1)/2$ for any point $x \in i(V_k(\mathbf{R}^n))$. Therefore, according to theorem 1, § 2, $i(V_k(\mathbf{R}^n))$ is a submanifold in $\mathbf{R}^{nk}$ of class C^∞ of dimension $kn - k(k + 1)/2$, and, consequently it is a C^∞–manifold. The homeomorphism i^{-1} induces on $V_k(\mathbf{R}^n)$ the structure of a C^∞–manifold of dimension $kn - k(k+1)/2$; it is called a *Stiefel manifold*. The Stiefel manifold $V_k(\mathbf{R}^n)$ can be interpreted geometrically as the set of orthonormal k–frames of the space $\mathbf{R}^n$, since the coordinates of the vectors identically determine the matrix from $V_k(\mathbf{R}^n)$.

The elements of the Stiefel manifold $V_n(\mathbf{R}^n)$ are the orthogonal matrices, and this manifold is denoted by $O(n, \mathbf{R})$. The subset in $O(n, \mathbf{R})$ consisting of matrices with the determinant $+1$, is open in $O(n, \mathbf{R})$, therefore, it is a C^∞–manifold of dimension $(n^2 - 1)/2$; it is denoted by $SO(n, \mathbf{R})$.

7. Product of manifolds. If M^n, N^m are two C^r–manifolds, then one can naturally define the structure of a C^r–manifold of dimension $m + n$ on the topological product $M \times N$. We leave it as an exercise to the reader.

Examples of products of manifolds are the cylinder $\mathbf{R}^1 \times S^1$ and the k–dimensional torus $T^k = S^1 \times \ldots \times S^1$ (of k factors). According to what was said above, they are C^∞–manifolds of dimensions 2 and k, respectively.

8. Lie groups. Consider the special class of manifolds which at the same time are groups; they are called Lie groups after the Norwegian mathematician Sophus Lie.

A group G provided with the structure of a smooth manifold in such a way that the mapping $G \times G \rightarrow G$ given by the rule $(g, h) \mapsto g \cdot h^{-1}$ is smooth, is called a *Lie group*.

Obviously, the mappings $G \rightarrow G : g \mapsto g^{-1}$ and $G \times G \rightarrow G : (f, g) \mapsto f \cdot g$ for a Lie group are also smooth. Indeed, the first mapping is the composition of the smooth mappings $g \mapsto (e, g)$ and $(e, g) \mapsto e \cdot g^{-1} = g^{-1}$, and the second one is the composition of the mappings $(g, h) \mapsto (g \cdot h^{-1})$, $(g, h^{-1}) \mapsto g(h^{-1})^{-1} = g \cdot h$. Further, if G_e is the connected component of the group G containing identity e, then $g \cdot G_e \subset G_e$ for any $g \in G_e$ (due to the connectedness of the image $g \cdot G_e$ and $(g \cdot G_e) \cap G_e \neq \emptyset$); analogously, we get $g^{-1} \cdot G_e \subset G_e$. Thus, the connected component G containing identity is also a Lie group.

Some simple examples of Lie groups are: the space $\mathbf{R}^n$ with respect to addition of vectors; $\mathbf{C} \backslash 0$, i.e., the complex numbers $\neq 0$ with respect to multiplication; the identity circle S^1 considered as a subset in $\mathbf{C} \backslash 0$ with respect to multiplication; the manifolds $GL(n, \mathbf{R})$, $O(n, \mathbf{R})$, $SO(n, \mathbf{R})$ with respect to multiplication of matrices.

If G_1 and G_2 are two Lie groups, then $G_1 \times G_2$ is the Lie group with respect to the multiplication $(g_1, h_1) \cdot (g_2, h_2) = (g_1 \cdot g_2, h_1 \cdot h_2)$ (with the smooth structure of the product of the manifolds G_1, G_2).

Hence, it immediately follows that the n–dimensional torus $T^n = S^1 \times \ldots \times S^1$ is a Lie group with a component–wise multiplication.

The group of affine transformations of the space $\mathbf{R}^n$: $x \mapsto Ax + v$, where $v, x \in \mathbf{R}^n$, A is a non–singular $n \times n$–matrix, is also a Lie group; its underlying manifold $L^m = GL(n, \mathbf{R}) \times \mathbf{R}^n$, $m = n^2 + n$, consists of pairs (A, v) with the smooth structure of a product, and the group operation is given by the formula: $(A_1, v_1) \cdot (A_2, v_2) = (A_1 A_2, A_2 v_1 + v_2)$; the identity element is $e = (I, 0)$, where I is the identity matrix, 0 the zero vector.

Exercise 17°. Prove that L^m is a Lie group.

9. Riemann surfaces. Consider an example which is important in the function theory of a complex variable. Let M^2 be a two–dimensional smooth manifold. Consider $\mathbf{R}^2$ as the complex z–plane. Let $\{(U_\alpha, \phi_\alpha)\}$ be an atlas on M^2 such that the transition diffeomorphisms

$$\phi_\beta^{-1}\phi_\alpha(U_\alpha \cap U_\beta) \to \phi_\beta^{-1}(U_\alpha \cap U_\beta)$$

are complex analytic functions of z in the domains $\phi_\alpha^{-1}(U_\alpha \cup U_\beta)$. A manifold M^2 with such an atlas is called an (abstract) *Riemann surface*. Its complex analytic structure is determined by the equivalence of atlases under which the diffeomorphisms of transition are complex analytic functions.

In particular, the complex z–plane C is a Riemann surface, i.e., its complex analytic structure is given by the atlas consisting of a unique chart $(\mathbf{C}, 1_\mathbf{C})$, where $1_\mathbf{C}$ is the identity mapping.

The sphere $S^2 = \{x \in \mathbf{R}^3 : x_1^2 + x_2^2 + x_3^2 = 1\}$ is also a Riemann surface. We shall define on S^2 an analytic structure using the covering $U_1 = S^2 \setminus \{N\}$, $U_2 = S^2 \setminus \{S\}$; the local coordinates of the point $P(x_1, x_2, x_3)$ in U_1, U_2, respectively, have the form

$$z_1 = \frac{x_1 + ix_2}{1 - x_3}, \quad z_2 = \frac{x_1 - ix_2}{1 + x_3}.$$

The coordinate z_1 appears under stereographical projection from the N–pole of the sphere S^2 (see, Fig. 84) onto the equatorial plane, and z_2 is obtained under the projection from the South pole S. If $P \in U_1 \cap U_2$ then $z_1 \neq 0$, $z_2 \neq 0$ and, evidently, $z_1 z_2 = 1$; hence the diffeomorphism of transition, $z_1 = 1/z_2$, is an analytic function. The extended z–plane (z–sphere) $\tilde{C}$ is endowed with the analytic structure by means of this homeomorphism with S^2.

The double–sheeted Riemann surface of the function $w = \sqrt{z}$ (see Ch. 1, § 4) is a complex analytic manifold, and the analytic structure on it is introduced by means of the homeomorphism with the z–sphere.

Exercise 18°. Describe the corresponding atlas of the double–sheeted Riemmann surface of the function $w = \sqrt{z}$.

In the theory of functions of a complex variable, it is proved that any analytic function on the z–plane has an abstract Riemann surface and that any compact abstract Riemann surface can be realized as the Riemann surface of a certain algebraic function.

10. Configuration space.
The various examples considered of smooth manifolds emerge naturally in various problems of mathematics. The concept of a manifold is also naturally used in applied sciences (mechanics, physics) to

describe the set of positions (i.e., the *configuration space*) of a system. We present a very simple example.

Consider a pendulum with a hinge swinging in the vertical plane. The point of attachment we denote by O, the hinge by O_1, and the end of the pendulum by O_2. Each position of the given system is determined by the directions of the rod OO_1 and of the rod O_1O_2, i.e., by the pair of angles ϕ, ψ (Fig. 88) running independently over the intervals $0 \leq \phi < 2\pi$, $0 \leq \psi < 2\pi$. The configuration space of the given system is, thus, the Cartesian product of two circles $S^1 \times S^1$, i.e., the two–dimensional torus T^2.

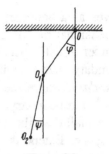

FIG. 88

Exercise 19°. Describe the configuration space of a flat pendulum on two hinges.

More complicated configuration spaces appear in the study of more complex mechanical systems consisting of a greater number of point masses and under more complex patterns of movement. These conditions usually are given in the form of equations which the coordinates of all the point masses should satisfy (these equations are called geometrical relations). The geometrical relations (under the corresponding conditions) determine a smooth manifold in the space $\mathbf{R}^{3n}$, where n is a number of point masses (see § 2, example 6). An ordered set of coordinates of point masses in $\mathbf{R}^{3n}$ determines a position of a mechanical system in configuration space.

11. Manifolds with boundary. The concept of manifold introduced above does not cover, however, a number of familiar and important geometric objects, for instance, the n–dimensional closed disc, surfaces with boundary, etc. Indeed, it is impossible to indicate the neighbourhood homeomorphic to the space $\mathbf{R}^n$ (or an open part of $\mathbf{R}^n$) of points on the boundary of the disc $\bar{D}^n$. This gap is filled by the concept of manifold with boundary.

Consider in $\mathbf{R}^n$ the subspace $\mathbf{R}^{n-1}$. The latter subdivides the space $\mathbf{R}^n$ into two half–spaces:

$$\mathbf{R}^n_+ = \{x \in \mathbf{R}^n : x_n \geq 0\} \quad \text{and} \quad \mathbf{R}^n_- = \{x \in \mathbf{R}^n : x_n \leq 0\},$$

for which the subspace

$$\mathbf{R}^{n-1} = \{x \in \mathbf{R}^n : x_n = 0\}$$

serves as a boundary. The half–space $\mathbf{R}_+^n$ is the simplest example of an n–dimensional manifold with the boundary $\mathbf{R}^{n-1}$. If now we "glue" a certain number of half–spaces $\mathbf{R}_+^n$ taking care that a boundary is "glued" with another boundary, then we will obtain an object called an n–*dimensional manifold with boundary*, where the boundary is the result of "gluing" copies of subspaces $\mathbf{R}^{n-1}$; the boundary itself is an $(n-1)$–dimensional manifold.

We shall describe a manifold with boundary precisely. Let M be a topological space. Extend the concept of a chart for M by assuming that the charts of homeomorphisms can act not only on the space $\mathbf{R}^n$ but also on the half–space $\mathbf{R}_+^n$, i.e., any pair (U, ϕ), where U is an open set in M, and ϕ is a homeomorphism $\phi : \mathbf{R}^n \to U$ or $\phi : \mathbf{R}_+^n \to U$ is called a chart. Two such charts (U, ϕ), (V, ϕ) for M are said to be C^r–compatible, if either $U \cap V = \emptyset$ or the homeomorphism $\psi^{-1}\phi : \phi^{-1}(U \cap V) \to \psi^{-1}(U \cap V)$ is a C^r–diffeomorphism; in the case when $\psi^{-1}\phi$ acts between sets which are open in $\mathbf{R}_+^n$ but not open in $\mathbf{R}^n$, the smoothness is understood in the sense of definition 2, § 2. Proceeding from such a concept of chart for M one may introduce the concept of C^r–atlas, equivalent C^r–atlases, C^r–structures on M and maximal atlas by repeating literally the definitions of the analogous concepts from subsection 1. A topological space M with a given C^r–structure on it is called a C^r–manifold with boundary. We also introduce the concept of dimension for a manifold with boundary similarly to that in subsection 1 (to indicate the dimension, we also write M^n).

A point x of a manifold with boundary M^n is called a *boundary point* if there exists a chart (U, ϕ), $\phi : \mathbf{R}_+^n \to U$, $x \in U$, in M^n such that $\phi^{-1}(x) \in \mathbf{R}^{n-1}$.

The definition of a boundary point does not depend on the choice of charts. In fact, if for a certain chart a point x is not a boundary point then the homeomorphism $\phi^{-1}\psi : \psi^{-1}(U \cap V) \to \phi^{-1}(U \cap V)$ would transform an internal point of a half–space into a boundary point. From the inverse mapping theorem follows that the latter statement is impossible for a manifold with boundary of class C^r, $r \geq 1$, (show it!), and the difficult classical Brouwer theorem on the invariance of domains stating that a subspace in $\mathbf{R}^n$ which is homeomorphic to an open subset of this space is also open in $\mathbf{R}^n$, proves that the statement is impossible for the C^0–manifold with boundary as well.

Exercise 20°. Show that the definition of a boundary point for a C^0–manifold is independent of the chart used.

We will not present the proof of the Brouwer theorem referring the interested reader to the literature.

The set of boundary points of a manifold with the boundary M^n is called the *boundary* and denoted by ∂M^n.

Thus, the concept of a C^r–manifold without boundary introduced in subsection 1 is a special case of a C^r–manifold with boundary when the boundary is the empty set.

Note that if the boundary ∂M^n of a C^r–manifold with boundary M^n is not empty, then it is an $(n-1)$–dimensional C^r–manifold (without boundary), and $M^n \backslash \partial M^n$ is an n–dimensional C^r–manifold (without boundary). Indeed, if $\{(U_\alpha, \phi_\alpha)\}$ is a C^r–atlas on M^n then, evidently, $\{(U_\alpha \cap (M^n \backslash \partial M^n),$ $\phi|_{\mathbf{R}^n_+ \backslash \mathbf{R}^{n-1}})\}$ is a C^r–atlas on $M^n \backslash \partial M^n$, and the set of charts $\{(U_\alpha \cap \partial M^n,$ $\phi_\alpha|_{\mathbf{R}^{n-1}})\}$ for which $U_\alpha \cap \partial M^n \neq \emptyset$, is a C^r–atlas on ∂M^n.

We may consider charts for manifolds with boundary similarly as in the case of manifolds (without boundary), that act not from the whole space $\mathbf{R}^n$ or the half–space $\mathbf{R}^n_+$, but from their open connected sets, thus making it easier to define a chart. Manifolds with boundary are usually assumed to be Hausdorff and to satisfy the second countability axiom.

Example 7. A structure of a C^∞–manifold with boundary can be defined on the half–space $\mathbf{R}^n_+$ by the atlas consisting of a single chart $(\mathbf{R}^n_+, 1_{\mathbf{R}^n_+})$.

Exercise 21°. Prove that the sets in $\mathbf{R}^n$ defined by the inequalities $x_1^2 + \ldots + x_n^2 \leq 1$, $x_1^2 + \ldots + x_n^2 \geq 1$ are n–dimensional C^∞–manifolds with boundary.

Exercise 22°. Show that the product $M^n \times N^m$ of a C^r–manifold M^n and a C^r–manifold with boundary N^m is an $(n+m)$–dimensional C^r–manifold with boundary; moreover, $\partial(M^n \times N^m) = M^n \times \partial N^m$.

Exercise 23°. Show that if a manifold with boundary is compact then the boundary of this manifold is also compact.

Exercise 24°. Show that if $f : M^n \to N^m$ is a homeomorphism of manifolds with boundary then $f(\partial M^n) = \partial N^n$.

Manifolds with boundary may be used to construct manifolds without boundary. We present the following construction.

Let M^n be a C^0–manifold with boundary. The double DM^n of the manifold M^n is a topological space which is obtained by joining two copies of the manifold M^n $(M^n \times 0) \cup (M^n \times 1)$ by identifying the points $(x, 0)$ and $(x, 1)$ for any $x \in \partial M^n$.

Exercise 25°. Prove that DM^n is a C^0–manifold (without boundary) of dimension n.

12. The existence of smooth structures. We will make several remarks regarding the possibility of introducing smooth structures. Whitney has proved that if there exists a C^r–structure ($r \geq 1$) on a space M then there also exists a C^∞–structure (and even a C^ω–structure) on it; morover, a C^∞–atlas can be chosen from the maximal atlas for the given C^r–structure. The exception is the case when $r = 0$. It is known that on any C^0–manifold of dimension $n < 4$ one may introduce a C^1–structure (and, consequently, a C^∞–structure), but for any $n \geq 4$ there exist C^0–manifolds which do not permit the introduction of a C^1–structure.

§ 4. Smooth functions in manifolds and smooth partition of unity

This and subsequent sections are devoted to the construction of the fundaments of analysis on smooth manifolds.

1. The concept of a smooth function on a manifold. A function defined on a manifold M^n can be considered locally as a function of the local coordinates of a point $x \in M^n$, i.e., as a function of the standard coordinates of the point $\phi_\alpha^{-1}(x)$ in $\mathbf{R}^n$ that are given by a certain chart (U_α, ϕ_α), $x \in U_\alpha$. Thus, we enter the range of the concepts of analysis, and, in particular, can define and investigate the concept of smooth function.

Definition 1. Let M^n be a manifold of class C^r, $r \geq 1$. A mapping $f : M^n \to \mathbf{R}^1$ is called a C^r–*function* (a *function of class* C^r) *in a neighbourhood of a point* $x \in M^n$, if there is a chart (U_α, ϕ_α), $(x \in U_\alpha)$ for M^n such that the mapping $f\phi_\alpha : \mathbf{R}^n \to \mathbf{R}^1$ is a C^r–mapping to $\mathbf{R}^n$.

Exercise 1°. Show that the definition of the C^r–function in a neighbourhood of a point is not dependent on the choice of a chart.

Definition 2. A function $f : M^n \to \mathbf{R}^1$ is called a C^r–*function on a certain set* $A \subset M^n$, if it is a C^r–function in a neighbourhood of each point $x \in A$.

One often has to consider a function determined not on the whole manifold M^n, but only on a subset of it. Definitions 1 and 2 are extended naturally to the case of functions $f : U \to \mathbf{R}^1$ defined on an open subset $U \subset M^n$ while choosing the charts (U_α, ϕ_α) in such a way that $U_\alpha \subset U$. However, these definitions should be extended to the case of functions $f : A \to \mathbf{R}^1$ which are defined on an arbitrary subset $A \subset M^n$.

Definition 3. A function $f : A \to \mathbf{R}^1$ ($A \subset M^n$) is called a C^r–*function on* A, if for any point $y \in A$ there exist an open neighbourhood $U(y) \subset M^n$ of the point y and a C^r–function $\phi_y : U(y) \to \mathbf{R}^1$ such that $\phi_y|_{U(y) \cap A} = f|_{U(y) \cap A}$.

It is easy to verify that each of the local coordinates $\xi_i(x)$, $i = 1, \ldots, n$, of a C^r–manifold is a C^r–function in its domain of definition.

In the special case of two–dimensional manifolds, i.e., an (abstract) Riemann surface, the complex–valued functions form an important class.

Let M^2 be an (abstract) Riemann surface and $f : M^2 \to \mathbf{C}$ a function on it with values in the field $\mathbf{C}$ of complex numbers. The function f is said to be *regular analytic* or *holomorphic at a point* $P_0 \in M^2$, if, when expressed in local coordinates $z = \Phi(P)$, $0 = \Phi(P_0)$ in a neighbourhood of the point P_0, it is a regular analytic function of z in a certain circle $|z| < r$, i.e.

$$f(\Phi^{-1}(z)) = \sum_{n=0}^{\infty} a_n z^n,$$

where the power series on the right–hand side converges in the circle $|z| < r$.

A function f is said to be *analytic in a certain open set* $U \subset M^2$ if it is a regular analytic function at every point $P_0 \in U$.

The functions on the z–sphere are usually specified in the local coordinates of the open set U_1 (see § 3, subsection 9), i.e. as functions $w = w(z)$ on the z–plane. In order to investigate a function in a neighbourhood of ∞, it is necessary to have its expression in terms of the local coordinates on the set U_2. The latter is achieved by replacing z by $1/z$; we then obtain the function $w = w(1/z) = w_1(z)$ which we investigate in the neighbourhood of zero.

Exercise 2°. Verify that the function $w=1/z$ is defined in the neighbourhood of the point $z=\infty$ of the z–sphere and is holomorphic at this point. The same holds for the function $w=\sum_{k=0}^{n} a_k z^{-k}$.

2. Partition of unity.

A main tool in the theory of manifolds when turning local statements into the global ones, is a partition of unity.

Let M^n be a C^r–manifold, and $\{U_\alpha\}$ an open covering of it.

Defintion 4. If $\phi : M^n \to \mathbf{R}^1$ is a function then the closure of the set $\{x : \phi(x) \neq o\}$ is called its *carrier* or support; it is denoted $\operatorname{supp} \phi$.

Definition 5. A family of C^r–functions $\{\phi_\beta : M^n \to [0,1]\}$ is called a *partition of unity of class* C^r *subordinate to a covering* $\{U_\alpha\}$, if (1) each of the sets $\operatorname{supp} \phi_\beta$ is compact and is contained in one of the U_α, (2) the family $\{\operatorname{supp} \phi_\beta\}$ forms a locally finite covering of M^n, (3) $\sum_\beta \phi_\beta(x) = 1$ for any point $x \in M^n$.

The summation in condition (3) makes sense since at each point x only a finite number of functions ϕ_β differs from zero in view of condition (2).

Theorem 1. *For any open covering of a* C^r*–manifold* M^n, $r = 1,\dots,\infty$, *there exists a* C^r*–partition of unity subordinate to it.*

To prove this fundamental theorem, several lemmas are needed.

Lemma 1. *For any* $s \in \mathbf{R}^1$ *there exists a* C^∞*–function* $h_s : \mathbf{R}^1 \to [0,1]$ *such that* $\operatorname{supp} h_s \subset [s,\infty)$.

Proof. It is easy to verify that the function

$$h_s(x) = \begin{cases} e^{-1/(x-s)}, & \text{if } x > s, \\ 0, & \text{if } x \le s, \end{cases}$$

does the job (Fig. 89). $\qquad\qquad\qquad\qquad\qquad\qquad\qquad\qquad\qquad\qquad\quad \square$

Exercise 3°. Draw the graphs of the functions $h_{-s}(x)$, $h_{-s}(-x)$, $h_{s/2}(x) + h_{s/2}(-x)$.

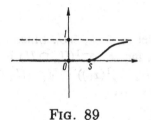

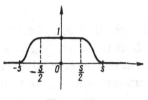

FIG. 89   FIG. 90

Lemma 2. *For any $s > 0$ there exists a C^∞-function $g_s : \mathbf{R}^n \to [0, 1]$ such that*

$$g_s(x) = \begin{cases} 1, & \text{if } x \in \bar{D}_{s/2}(0), \\ 0, & \text{if } x \in \mathbf{R}^n \backslash D_s(0). \end{cases}$$

Proof. Consider the function $\tilde{g}_s : \mathbf{R}^1 \to [0, 1]$ (Fig. 90) given by the equality

$$\tilde{g}_s(x) = \frac{h_{-s}(x)h_{-s}(-x)}{h_{-s}(x)h_{-s}(-x) + h_{s/2}(x) + h_{s/2}(-x)}.$$

The function $g_s(x) = \tilde{g}_s(\|x\|) : \mathbf{R}^n \to [0, 1]$ is clearly as desired. □

Note that if $x_0 \in \mathbf{R}^n$ is an arbitrary point, then

$$g_s(x - x_0) = \begin{cases} 1, & \text{if } x \in \bar{D}_{s/2}(x_0), \\ 0, & \text{if } x \in \mathbf{R}^n \backslash D_s(x_0). \end{cases}$$

Lemma 3. *Let M^n be a C^r-manifold ($r = 1, \ldots, \infty$), and $U \subset M^n$ an open set. Then for any point $x_0 \in U$ there exist its open neighbourhoods $V_1(x_0)$, $V_2(x_0)$ and a C^r-function $f : M^n \to [0, 1]$ such that*

$$(1) \quad \bar{V}_1(x_0) \subset V_2(x_0) \subset U,$$

$$(2) \quad f(x) = \begin{cases} 1, & \text{if } x \in \bar{V}_1(x_0), \\ 0, & \text{if } x \in M^n \backslash V_2(x_0). \end{cases}$$

Proof. Let $\{(U_\alpha, \phi_\alpha)\}$ be a C^r-atlas M^n and let $x_0 \in U_\alpha \cap U$. Since the set $U_\alpha \cap U$ is open, and ϕ_α is a homeomorphism, the set $\phi_\alpha^{-1}(U_\alpha \cap U)$ is open in $\mathbf{R}^n$, and, consequently, there exists a disc $D_s(\phi_\alpha^{-1}(x_0)) \subset \phi_\alpha^{-1}(U_\alpha \cap U)$. Furthermore, since ϕ_α is a homeomorphism, we have

$$\overline{\phi_\alpha(D_{s/2}(\phi_\alpha^{-1}(x_0)))} = \phi_\alpha(\bar{D}_{s/2}(\phi_\alpha^{-1}(x_0))) \subset$$
$$\subset \phi_\alpha(D_s(\phi_\alpha^{-1}(x_0))) \subset U,$$

therefore, the sets $V_1(x_0) = \phi_\alpha(D_{s/2}(\phi_\alpha^{-1}(x_0)))$, $V_2(x_0) = \phi_\alpha(D_s(\phi_\alpha^{-1}(x_0)))$ satisfy condition 1. According to lemma 2, there exists a C^r-function $g_{s,x_0}(x) = g_s(x - x_0)$ such that

$$g_{s,x_0}(x) = \begin{cases} 1, & \text{if } x \in \bar{D}_{s/2}(\phi_\alpha^{-1}(x_0)), \\ 0, & \text{if } x \in \mathbf{R}^n \backslash D_s(\phi_\alpha^{-1}(x_0)). \end{cases}$$

Then the function

$$f(x) = \begin{cases} g_{s,x_0}\phi_\alpha^{-1}(x), & \text{if } x \in V_2(x_0), \\ 0, & \text{if } x \in M^n \backslash V_2(x_0). \end{cases}$$

obviously satisfies condition 2. $\qquad\qquad \square$

Lemma 4. *Let M^n be a manifold of class C^r, $r = 1, \ldots, \infty$, $K \subset U \subset M^n$, where the set K is compact, and the set U is open. Then there exists a C^r-function $f : M^n \to [0, 1]$ such that*

$$f(x) = \begin{cases} 1, & \text{if } x \in K, \\ 0, & \text{if } x \in M^n \backslash U. \end{cases}$$

Proof. According to lemma 3, for every point $y \in K$ there exist open neighbourhoods $V_1(y)$, $V_2(y)$ such that $\bar{V}_1(y) \subset V_2(y) \subset U$, and there exists a C^r-function $f_y(x) : M^n \to [0, 1]$ such that

$$f_y(x) = \begin{cases} 1, & \text{if } x \in \bar{V}_1(y), \\ 0, & \text{if } x \in M^n \backslash V_2(y). \end{cases}$$

Due to the compactness of K, in the open covering $\{V_1(y)\}_{y \in K}$ of the set K, there is the finite subcovering $V_1(y_1), \ldots, V_1(y_p)$. Put $g(x) = \prod_{i=1}^{p}(1 - f_{y_i}(x))$, then

$$g(x) = \begin{cases} 0, & \text{if } x \in K, \\ 1, & \text{if } x \in M^n \backslash U, \end{cases}$$

$$f(x) = 1 - g(x) = \begin{cases} 1, & \text{if } x \in K, \\ 0, & \text{if } x \in M^n \backslash U. \end{cases}$$

Clearly, $f \in C^r$.   □

Lemma 5. *For any open covering $\{U_\alpha\}$ of a C^r–manifold M^n, $r = 1, \ldots, \infty$, one can construct its refinement which is an open locally finite covering $\{U'_\beta\}$ such that each set $\bar{U}'_\beta$ is compact and is contained in one of the sets U_α.*

Proof. Let $\{(V_\gamma, \phi_\gamma)\}$ be a C^r–atlas on the manifold M^n. The sets $\{V_\gamma \cap U_\alpha\}$ form an open covering that refines $\{U_\alpha\}$. Since ϕ_γ is a homeomorphism, the set $\phi_\gamma^{-1}(V_\gamma \cap U_\alpha)$ is open in $\mathbf{R}^n$, and, consequently, for any point $x \in V_\gamma \cap U_\alpha$, the point $\phi_\gamma^{-1}(x)$ is contained in the set $\phi_\gamma^{-1}(V_\gamma \cap U_\alpha)$ together with a certain disc $D_{s(x)}(\phi_\gamma^{-1}(x))$. Furthermore, since ϕ_γ is a homeomorphism, we have

$$(1) \qquad \begin{aligned} \overline{\phi_\gamma(D_{s/2}(\phi_\gamma^{-1}(x_0)))} &= \phi_\gamma(\bar{D}_{s(x)/2}(\phi_\gamma^{-1}(x_0))) \subset \\ &\subset \phi_\gamma(D_{s(x)}(\phi_\gamma^{-1}(x_0))) \subset V_\gamma \cap U_\alpha; \end{aligned}$$

moreover, since the set $\bar{D}_{s(x)/2}(\phi_\gamma^{-1}(x))$ is compact, and M^n is Hausdorff, $\phi_\gamma(\bar{D}_{s(x)/2}(\phi_\gamma^{-1}(x)))$ is compact (see Ch 2, § 13). In virtue of paracompactness of the manifold M^n (see § 3), one can find an open locally finite refinement covering $\{U'_\beta\}$ of the open covering $\{\phi_\gamma(D_{s(x)/2}(\phi_\gamma^{-1}(x)))\}$. Then each U'_β is contained in a certain $\phi_\gamma(D_{s(x)/2}(\phi_\gamma^{-1}(x)))$. Since

$$\bar{U}'_\beta \subset \overline{\phi_\gamma(D_{s(x)/2}(\phi_\gamma^{-1}(x)))},$$

we derive from (1) that $\bar{U}'_\beta \subset \phi_\gamma(D_{s(x)}(\phi_\gamma^{-1}(x))) \subset V_\gamma \cap V_\alpha$, therefore, $\bar{U}'_\beta \subset U_\alpha$. The compactness of $\bar{U}'_\beta$ follows from the fact that $\bar{U}'_\beta$ is a closed subset of the compact space $\phi_\gamma(\bar{D}_{s(x)/2}(\phi_\gamma^{-1}(x)))$ (see Ch 2, § 13). $\square$

Proof of theorem 1. Applying lemma 5 twice, we refine the given covering $\{U_\alpha\}$ to a locally finite covering $\{U'_\beta\}$, and $\{U'_\beta\}$ is refined to an open locally finite covering $\{U''_\gamma\}$ so that each $\bar{U}''_\gamma$ is compact and contained in a certain U_α. For each U''_γ fix one of the sets of the system $\{U'_\beta\}$ containing $\bar{U}''_\gamma$, and relabel it as U'_γ. Now, by applying lemma 4 to the set $K = \bar{U}''_\gamma$, we find a corresponding function $f_\gamma : M^n \to [0,1]$ such that

$$f_\gamma(x) = \begin{cases} 1, & \text{if } x \in \bar{U}''_\gamma, \\ 0, & \text{if } x \in M^n \backslash U'_\gamma. \end{cases}$$

Consider the function

$$\phi_\gamma(x) = \frac{f_\gamma(x)}{\sum_\nu f_\nu(x)}, \quad x \in M^n$$

(the summation in the denominator is over the set of indices of the covering $\{U''_\gamma\}$ and makes sense, since the covering $\{U'_\gamma\}$ is locally finite and, consequently, at each point $x \in M^n$ there is only a finite number of functions f_γ that are different from zero at x). Since supp $\phi_\gamma = $ supp f_γ and the set supp f_γ is compact (as a closed subset of the compact space $\bar{U}'_\gamma$), the set supp ϕ_γ is also compact. It is not difficult to see that $\phi_\gamma \in C^r$, and thus we obtain the required partition of unity. $\square$

Definition 6. A function $f : A \to \mathbf{R}^1$ ($A \subset M^n$) is called a C^r–*function on the set A*, if it is the restriction to A of a C^r–function given on an open subset U of the manifold M^n which contains A.

Exercise 4°. By applying the partition of unity theorem, show that definition 3 of a smooth function given on a set $A \subset M^n$ is equivalent to definition 6.

3. The algebra of C^r–functions on a manifold. Now, consider the set $\mathcal{O}(M^n)$ of all C^r–functions on a C^r–manifold M^n. The functions from $\mathcal{O}(M^n)$ can be added and multiplied by real numbers in a natural way, i.e., if $f, g \in \mathcal{O}(M^n)$, $\alpha \in \mathbf{R}^1$, then for each point $x \in M^n$, we put $(f+g)(x) =$

$f(x) + g(x)$ and $(\alpha f)(x) = \alpha f(x)$. Thus, $\mathcal{O}(M^n)$ becomes a vector space. Moreover, the usual multiplication $(f \cdot g)x = f(x) \cdot g(x)$, $x \in M^n$, makes $\mathcal{O}(M^n)$ into an algebra over the field $\mathbf{R}$.

Let x be a point from the manifold M^n. Consider the following equivalence relation on the algebra $\mathcal{O}(M^n)$: $f_1 \sim f_2$ if for a point x there exists a neighbourhood $U(x)$ such that $f_1|_{U(x)} = f_2|_{U(x)}$. An equivalence class is called a C^r–germ (of functions) at the point x, and the collection of all C^r–germs at the point x is denoted by $\mathcal{O}(x)$. Evidently, $\mathcal{O}(x)$ is also an algebra. We give also a second definition of $\mathcal{O}(x)$.

Consider the quotient algebra $\mathcal{O}(M^n)/\mathcal{O}_0(x)$, where $\mathcal{O}_0(x)$ is the ideal of all those functions from the ring $\mathcal{O}(M^n)$ which are zero in a certain (dependent on the function) neighbourhood of the point x, then it is natural to identify its elements with the germs of the functions at the point x. It is easy to verify that $\mathcal{O}(M^n)/\mathcal{O}_0(x) = \mathcal{O}(x)$. The set of germs $\mathcal{O}(x)$ can also be defined as the set of C^r–functions defined on some neighbourhood of the point x and which are factorized with respect to the same equivalence relation as in the definition of $\mathcal{O}(x)$. On the face of it, we obtain a new object because we now consider functions that are not necessarily defined on the whole manifold M^n. From the following exercise it follows that this makes no difference.

Exercise 5°. Let f be a function of class C^r given in a certain open neighbourhood $U(x)$ of a point x of a manifold M^n of class C^r. Show that there exist a closed neighbourhood $\bar{V}(x)$ of the point x, with $\bar{V}(x) \subset U(x)$ and a C^r–function $\tilde{f}$ defined in the whole manifold M^n such that $\tilde{f}|_{\bar{V}(x)} = f|_{\bar{C}(x)}$. *Hint.* Use lemma 4.

Thus, on a smooth manifold we have the algebras of smooth functions $\mathcal{O}(M^n)$ and the algebras of germs of smooth functions $\mathcal{O}(x)$. An interesting question is whether it is possible, conversely, to recover the structure of the manifold with the help of the algebras $\mathcal{O}(M^n)$ and $\mathcal{O}(x)$, arises. We show below that this can be done.

First of all, we fix the main properties of algebras of functions on a smooth manifold in axiomatic form. Consider a topological space M and some real functions $f, f_1, \dots, f_k$ defined on M. We say that f C^r–*smoothly depends on the functions* $f_1, \dots, f_k$ $(r \geq 1)$, if there exists a C^r–function $U(t_1, \dots, t_k)$ of real variables $t_1, \dots, t_k$ defined on $\mathbf{R}^k$ such that

(2) $$f(x) = U(f_1(x), \dots, f_k(x)), \quad x \in M.$$

If equality (2) is valid only for the points of a certain set $V \subset M$, then we say that the function f

smoothly depends on the functions $f_1, \ldots, f_k$ *on the set* V. A nonempty set $\mathcal{O}(M)$ of real functions on M is called C^r-*smoothness on a topological space* M if it satisfies the following conditions:

(1) any C^r-function which smoothly depends on functions from $\mathcal{O}(M)$, belongs to $\mathcal{O}(M)$;

(2) any function on M, which coincides with a certain function from $\mathcal{O}(M)$ in a certain neighbourhood of each point $x \in M$, belongs to $\mathcal{O}(M)$.

Exercise $6°$. Verify that for a C^r-manifold M^n, the algebra of C^r-functions $\mathcal{O}(M^n)$ satisfies the conditions for C^r-smoothness.

From condition (1) it follows that the set $\mathcal{O}(M)$ is an algebra with natural operations of summation and multiplication of functions and multiplication of functions by a real number. The concepts of the C^r-germ f_x of a function $f \in \mathcal{O}(M)$ at a point x, of the ideal $\mathcal{O}_0(x)$ and of the set of C^r-germs $\mathcal{O}(x) = \mathcal{O}(M)/\mathcal{O}_0(x)$ are defined in the natural way.

Now we move to the construction of a C^r-structure on M. Let M be a topological space with a C^r-smooth set of functions $\mathcal{O}(M)$. Assume that the following conditions are fulfilled: (1) for any point $x \in M$, there are germs $\widehat{f}_x^1, \ldots, \widehat{f}_x^n \in \mathcal{O}(x)$, a neighbourhood $V(x)$, representatives $f^i : V(x) \to \mathbf{R}^1$, $i=1, \ldots, n$, of the germs $\widehat{f}_x^i$ such that the mapping $\psi_V : y \mapsto \{f^1(y), \ldots, f^n(y)\}$, $y \in V(x)$, is a homeomorphism $V(x)$ onto the space $\mathbf{R}^n$; (2) for any point $x \in M$, the germs $\widehat{f}_y^1, \ldots \widehat{f}_y^n$ of the functions $f^1, \ldots, f^n$ belong to $\mathcal{O}(y)$; (3) for any germ $\widehat{g}_y \in \mathcal{O}(y)$, any representative g C^r-smoothly depends on $f^1, \ldots, f^n$ in a neighbourhood of the point y. Thus, by specifying the coordinate system in $V(x)$ by means of the homeomorphism $\phi_V = \psi_V^{-1} : \mathbf{R}^n \to V(x)$, we obtain a system of charts $\{(V(x), \phi_V)\}$ which, as it can be easily verified by properties (2), (3), form a C^r-atlas on M.

Exercise $7°$. Show that the system of charts $\{(V(x), \phi_V)\}$ forms an atlas.

Thus, there is a differential structure of a C^r-manifold defined on M, induced by the algebras $\mathcal{O}(M)$, $\mathcal{O}(x)$.

Exercise $8°$. Show that if M^n is a C^r-manifold and $\{\mathcal{O}(x)\}_{x \in M^n}$ are the corresponding algebras of the germs of the C^r-functions on M^n, then the differential structure defined by them coincides with the structure of the manifold M^n.

Note. Conditions (1) and (3) leads to the fact that the considered smooth set of functions on M consists of continuous functions. One can also consider C^r-smooth set of functions on an abstract set M and induce the weakest topology on it so that all the functions from the given smooth set of functions will be continuous.

§ 5. Mappings of manifolds

1. The concept of a smooth mapping. We will define and study smooth mappings of smooth manifolds; these represent a natural generalization of differentiable functions in analysis. Let M^n, N^m be two C^r–manifolds, $r \geq 1$. Regarding M^n, N^m as topological spaces, we can speak about continuous mappings $f : M^n \to N^m$. The structures of class C^r given on M^n, N^m, permit to introduce a narrower class of mappings. It is natural to specify a mapping $f : M^n \to N^m$ in terms of local coordinates. Namely, if $x \in M^n$ is an arbitrary point, (U, ϕ), (V, ψ) are charts on the manifolds M^n, N^m, respectively, such that $x \in U$, $f(x) \in V$, and $W(x)$ is an open neighbourhood of the point x such that $W(x) \subset U$, $f(W(x)) \subset V$, then the mapping

$$\psi^{-1} f \phi : \phi^{-1}(W(x)) \to \psi^{-1}(V)$$

is called the *coordinate representation of the mapping* f in a neighbourhood of the point x. This representation allows us to invoke the concept of a smooth mapping of $\mathbf{R}^n$ to $\mathbf{R}^m$ studied in analysis (see § 1).

Definition 1. A mapping $f : M^n \to N^m$ is called a C^r–*mapping (a mapping of class C^r) in a neighbourhood of a point $x \in M^n$*, if a certain coordinate representation of the mapping f in a neighbourhood of the point x is a C^r–mapping.

Exercise $1°$. Show that the definition of a C^r–mapping in a neighbourhood of a point does not depend on the choice of a coordinate representation.

Defining a smooth mapping in a neighbourhood of a point, it is natural to consider mappings given not on the whole M^n, but only in an open neighbourhood of a point.

For the case of submanifolds in $\mathbf{R}^N$, definition 1 can be expressed in other terms. Let M^n, N^m be submanifolds in $\mathbf{R}^{N_1}$ and $\mathbf{R}^{N_2}$, respectively.

Definition 2. A mapping $f : M^n \to N^m$ is called a C^r–*mapping (a mapping of class C^r) in a neighbourhood of a point $x \in M^n$*, if there exists an open set $U \subset \mathbf{R}^{N_1}$, $x \in U$ and a C^r–mapping $\widetilde{f} : U \to \mathbf{R}^{N_2}$ that coincides with f on $U \cap M^n$.

Exercise $2°$. Show that for the case of submanifolds in $\mathbf{R}^N$ definitions 1 and

2 are equivalent.

Hint. Use the property that mappings of charts are diffeomorphisms (see § 2, lemma 1).

Now, using these local definitions, we turn to the global ones.

Definition 3. A mapping $f : M^n \to N^m$ is called a C^r-*mapping (a mapping of class C^r)* if it is a C^r-mapping in a neighbourhood of each point $x \in M^n$.

It is evident that the concept of a C^r-mapping is a generalization of the concept of a C^r-function.

Similarly, the concept of a complex analytic function on a Riemann surface is extended to the concept of a complex analytic mapping between Riemann surfaces (under the assumption that the coordinate representation is analytic).

Exercise $3°$. Verify that the mapping $w=\sqrt{z}$ of the (corresponding) double-sheeted Riemann surface onto the z-sphere is analytic.

Exercise $4°$. Verify that the mappings $w=1/z$ and $w=\sum_{k=0}^{n} a_k/z^k$ considered as mappings of the z-sphere onto itself are analytic.

Remark 1. The concept of smooth mapping can be extended to mappings of manifolds with boundary. If $f : M^n \to N^m$ is a continuous mapping of a C^r-manifold with boundary, $r \geq 1$, then as in the case of mappings of manifolds (without boundary), we may speak of a coordinate representation of the mapping f in a neighbourhood of the point $x \in M^n$. A mapping f is called a C^r-mapping in a neighbourhood of a point $x \in M^n$, if a certain coordinate representation $\psi^{-1} f \phi : \phi^{-1}(W(x)) \to \psi^{-1}(V)$ of the mapping f in a neighbourhood of the point x is a C^r-mapping in the sense of definition 2, § 2, i.e., it is a C^r-mapping (in the usual sense), if $\phi^{-1}(W(x))$ is open in $\mathbf{R}^n$, and it is extendible as a C^r-mapping onto some open neighbourhood of the point x in $\mathbf{R}^n$, if $\phi^{-1}(W(x))$ is open in $\mathbf{R}_+^n$, but not open in $\mathbf{R}^n$. The mapping f is called a C^r-mapping if it is a C^r-mapping in the neighbourhood of every point $x \in M^n$.

Definition 4. A mapping $f : M^n \to N^n$ of manifolds of class C^r is called a

C^r–*diffeomorphism*, if (1) f is bijective, (2) f, f^{-1} are C^r–mappings.

Exercise 5°. Why is it impossible to have a diffeomorphism between manifolds of different dimensions?

Two C^r–manifolds M^n, N^m are said to be C^r–*diffeomorphic* if there exists a C^r–diffeomorphism $f : M^n \to N^n$.

Exercise 6°. Verify that diffeomorphism of manifolds is an equivalence relation.

Theorem 1. *If M^n is a manifold of class C^r and N^n a manifold of class C^r with the structure induced by the homeomorphism $f : M^n \to N^n$, then M^n and N^n are C^r–diffeomorphic.*

Proof. It is easy to see that the required diffeomorphism is f. $\qquad\square$

Thus, by inducing the structure of a C^r–manifold on a topological space N by means of a homeomorphisn $f : M^n \to N$, we transform f into a C^r–diffeomorphism.

Theorem 2. *Let $f : M^n \to N^n$ be a C^r–diffeomorphism of C^r–manifolds M^n, N^n. Then the mapping $f : M^n \to N^n$, as a homeomorphism of topological spaces induces on N^n a C^r–structure coinciding with the initial one.*

Proof. Let $\{(U_\alpha, \phi_\alpha)\}$ and $\{(V_\beta, \psi_\beta)\}$ be C^r–atlases on M^n and N^n, respectively. We show that any chart of the atlas $\{(f(U_\alpha), f\phi_\alpha)\}$ is C^r–compatible with any chart of the atlas $\{(V_\beta, \psi_\beta)\}$, i.e., that the mapping

(1) $\qquad \psi_\beta^{-1}(f\phi_\alpha) : (f\phi_\alpha)^{-1}(f(U_\alpha) \cap V_\beta) \to \psi_\beta^{-1}(f(U_\alpha) \cap V_\beta)$

is a C^r–diffeomorphism for any α and β.

Indeed, since f is a C^r–diffeomorphism, its representation in local coordinates of the open sets

$$f^{-1}(f(U_\alpha) \cap V_\beta) \subset U_\alpha, \quad f(U_\alpha) \cap V_\beta \subset V_\beta,$$

$$\psi_\beta^{-1} f \phi_\alpha : \phi_\alpha^{-1}[f^{-1}(f(U_\alpha) \cap V_\beta)] \to \psi_\beta^{-1}(f(U_\alpha) \cap V_\beta)$$

is a C^r–diffeomorphism. But

$$\phi_\alpha^{-1}[f^{-1}(f(U_\alpha) \cap V_\beta)] = (f\phi_\alpha)^{-1}(f(U_\alpha) \cap V_\beta);$$

therefore, the mapping (1) is a C^r–diffeomorphism, and that proves the statement.   □

From the point of view of general topology, we do not distinguish homeomorphic spaces. It is natural to take the point of view that homeomorphic manifolds M^n and N^n should not be distinguished; here the manifold N^n is endowed with the structure of a smooth manifold induced by the homeomorphism $f : M^n \to N^n$. But then by theorem 1, M^n and N^n are diffeomorphic. Inversely, if the manifolds M^n and N^n are diffeomorphic ($f : M^n \to N^n$), then they are homeomorphic, and according to theorem 2, f as the homeomorphism, induces on N^n the structure of a smooth manifold which coincides with the initial one. Thus, the point of view elucidated above is equivalent to the one that the diffeomorphic manifolds should not be distinguished.

Exercise 7°. Show that the collection of C^r–manifolds of all dimensions $n \geq 1$ forms a category with as morphisms the C^r–mappings of manifolds. Show that in this category the C^r–diffeomorphisms of manifolds are the isomorphisms.

Exercise 8°. Show that all the C^∞–manifolds determined by the various C^∞–structures on $\mathbf{R}^1$ which are indicated in exercise 2°, § 3, are C^∞–diffeomorphic.

The last exercise reflects the general situation. As it was already noted in § 3, if there exists at least one C^r–structure ($r \geq 1$) on the manifold M^n, then there exist infinitely many C^∞–structures on M^n, however the C^∞–manifolds defined by them on M^n mostly are diffeomorphic. For instance, it is known that all C^∞–structures on $\mathbf{R}^n$, $n \neq 4$, define C^∞–diffeomorphic manifolds; all C^∞–structures on S^n, $n = 1, 2, 3, 4, 5, 6, 12$ define C^∞–diffeomorphic manifolds. It is also known that if the dimension is lower than 4, then home-omorphic differentiable manifolds are diffeomorphic, i.e. for the manifolds of dimension lower than 4, the differentiable and topological classifications coincide.

The question naturally arises whether there exist homeomorphic but not d-iffeomorphic manifolds. This problem was solved by G. Milnor in 1956. He showed that there are exactly 28 smooth manifolds (the Milnor spheres) homeomorphic to S^7, but not diffeomorphic to each other[*]

The discovery that there exist homeomorphic but not diffeomorphic manifolds was the beginning of differential topology as an independent field of mathe-matics.[**]

Important examples of diffeomorphisms appear in the form of actions of groups of transformations on manifolds. We considered already the affine group of transformations of the space $\mathbf{R}^n$.

The Lie groups $GL(n, \mathbf{R})$, $O(n, \mathbf{R})$, $SO(n, \mathbf{R})$ also are groups of transforma-tions of the space $\mathbf{R}^n$ (as it is said, they *act* on $\mathbf{R}^n$). The concept of an action of an abstract group on a topological space is discussed in Ch 2, § 5. We will generalize it to the case of Lie groups.

Definition 5. Let M be a smooth manifold, and G a Lie group. An action (from the right) of G on M is a smooth mapping $f : M \times G \to M$ which associates to the pair $(x, g) \in M \times G$ the point $y = f(x, g) \in M$ (denoted by $x \cdot g$) and which satisfies the following conditions: (1) for $g_1, g_2 \in G$, $x \in M$, we have $x(g_1 \cdot g_2) = (xg_1)g_2$ for any g_1, g_2, x; (2) for the identity element $e \in G$ we have $x \cdot e = x$ for any x.

If for a fixed g we denote the smooth mapping $y = f(\cdot, g)$ by h_g, then from axioms (1), (2) of action of G on M we may conclude that $h_{g_1 \cdot g_2} = h_{g_2} \cdot h_{g_1}$, $h_g \cdot h_{g^{-1}} = h_e = h_{g^{-1}} \cdot h_g$. Since $h_e = 1_M$, we conclude that h_g is invertible,

---

[*] The Milnor spheres can be given as submanifolds in $\mathbf{R}^{10} = \mathbf{C}^5 = \{(z_1, \dots, z_5)\}$ by the system of two equations

$$z_1^{6k-1} + z_2^3 + z_3^2 + z_4^2 + z_5^2 = 0, \quad |z_1|^2 + \dots + |z_5|^2 = 1,$$

$$k = 1, 2, \dots, 28.$$

[**] In 1987, C. Taubes showed that there exists an uncountable set of smooth manifolds homeo-morphic to $\mathbf{R}^4$, but not diffeomorphic to each other.

and $(h_g)^{-1} = h_g^{-1}$ is a smooth mapping; consequently, h_g is a diffeomorphism of M, and the action of the group G means that there is given a homomorphism $G \to \mathrm{Diff}\,(M)$ into the group of diffeomorphisms of the manifold M.

Having the action from the right, we can define an *action from the left of G on M* by the equality $g \cdot x = f(x, g^{-1})$; then $(g_1 \cdot g_2)x = g_2(g_1 \cdot x)$, $h_{g_1 \cdot g_2} = h_{g_1} \cdot h_{g_2}$.

Exercise 9°. Show that the action of the groups $GL(n, \mathbf{R})$, $O(n, \mathbf{R})$, $SO(n, \mathbf{R})$ on the space $\mathbf{R}^n$ given by the formula $y = Ax$, where A is a matrix from the corresponding group, and x, y are coordinate vector–columns, is an action from the left; the same is valid for the group of affine representations $y = Ax + v$.

When considering algebraic homomorphisms of Lie groups G_1 into G_2, it is useful to take into account the structure of smooth manifolds. The concepts that are formulated below are the development of the corresponding concepts of smooth manifolds.

If $\phi : G_1 \to G_2$ is an algebraic homomorphism of the groups G_1 and G_2, which are Lie groups, and if ϕ is a smooth mapping of the manifolds G_1 and G_2, then it is called a *homomorphism of Lie groups*. A homomorphism which is a diffeomorphism is called an *isomorphism of Lie groups*.

Isomorphisms of a Lie group into itself are called *automorphisms* of this group. If $G_2 = \mathrm{Aut}\,(V)$ (the set of non–singular linear transformations of a vector space V, i.e., of its automorphisms), then a homomorphism $\phi : G \to G_2$ is called a *representation of the Lie group G*.

Each element g from the Lie group G generates automorphisms G: *the left shift $L_g : h \mapsto g \cdot h$; the right shift $R_g : h \mapsto h \cdot g$; an inner automorphism $L_g \cdot R_g^{-1} : h \mapsto g \cdot h \cdot g^{-1}$* denoted by α_g.

Exercise 10°. Verify the relations: $L_{g \cdot h} = L_g \cdot L_h$, $L_{g^{-1}} = L_g^{-1}$, $R_{g \cdot h} = R_h \cdot R_g$, $R_{g^{-1}} = R_g^{-1}$, $L_g \cdot R_h = R_h \cdot L_g$, $\alpha_{gh} = \alpha_g \alpha_h$, $\alpha_{g^{-1}} = \alpha_g^{-1}$.

2. Classification of one–dimensional manifolds. Any manifold (of arbitrary dimension) consists of no more than countable quantity of connected components (see § 3). Therefore it is sufficient to classify only connected manifolds.

The zero–dimensional and one–dimensional manifolds are the easiest to classify. Obviously, a connected zero–dimensional manifold is a point. We consider one–dimensional manifolds.

Lemma 1. *If a connected topological manifold M^1 has an atlas consisting of* *two charts* (U_i, ϕ_i), $\phi_i : \mathbf{R}^1 \to U_i$, $i = 1, 2$, *then* M^1 *is homeomorphic to* $\mathbf{R}^1$ *or* S^1.

Proof. If $U_1 \subset U_2$ or $U_2 \subset U_1$ then M^1, evidently, is homeomorphic to $\mathbf{R}^1$. Now, consider the case when none of the sets U_1, U_2 are contained in each other. By the connectedness of M^1, it follows that $U_1 \cap U_2 \neq \emptyset$. Since the sets $\phi_1^{-1}(U_1 \cap U_2)$, $\phi_2^{-1}(U_1 \cap U_2)$ are open in $\mathbf{R}^1$, they are the a union of some non–intersecting finite or infinite intervals (connected components). First, we show that there are no finite intervals among the connected components of these sets. Assume the contrary. Let, for instance, the set $\phi_1^{-1}(U_1 \cap U_2)$ have a connected component (a, b). Since $\phi_2^{-1}\phi_1$ is a homeomorphism, the interval $(c, d) = \phi_2^{-1}\phi_1((a, b))$ is a connected component of the set $\phi_2^{-1}(U_1 \cap U_2)$. Note that therefore, none of the points $\phi_1(a)$ and $\phi_1(b)$ can coincide with the points $\phi_2(c)$, $\phi_2(d)$ (Fig. 91), as these pairs of points belong to the non–intersecting sets $U_1 \backslash (U_1 \cap U_2)$ and $U_2 \backslash (U_1 \cap U_2)$. Since the function $\phi_2^{-1}\phi_1$ homeomorphically maps (a, b) into (c, d), it increases or decreases monotonously. We consider the case of increasing $\phi_2^{-1}\phi_1$. Let $V(\phi_1(a))$, $\widetilde{V}(\phi_2(c))$ be arbitrary neighbourhoods of the points $\phi_1(a)$, $\phi_2(c)$ in M^1. Then $V_1(\phi_1(a)) = V(\phi_1(a)) \cap U_1$, $V_2(\phi_2(c)) = \widetilde{V}(\phi_2(c)) \cap U_2$ are neighbourhoods of the points $\phi_1(a)$, $\phi_2(c)$ in U_1, U_2, respectively, and $W_1(a) = \phi_1^{-1}(V_1(\phi_1(a)))$, $W_2(c) = \phi_2^{-1}(V_2(\phi_2(c)))$ are neighbourhoods of the points a and c in $\mathbf{R}^1$. For all sufficiently small $\epsilon > 0$, we have $(a, a + \epsilon) \subset W_1(a)$, $(c, c + \epsilon) \subset W_2(c)$, whence

(2)
$$\phi_1((a, a + \epsilon)) \subset \phi_1(W_1(a)) = V_1(\phi_1(a)),$$
$$\phi_2((c, c + \epsilon)) \subset \phi_2(W_2(c)) = V_2(\phi_2(c)).$$

Because $\phi_2^{-1}\phi_1$ is increasing, we have $\phi_2^{-1}\phi_1((a, a + \epsilon)) \cap (c, c + \epsilon) \neq \emptyset$ (Fig. 92), whence $\phi_1((a, a + \epsilon)) \cap \phi_2((c, c + \epsilon)) \neq \emptyset$. From the latter inequality and the inclusions (2) we obtain $V_1(\phi_1(a)) \cap V_2(\phi_2(c)) \neq \emptyset$. Thus, an intersection of neighbourhoods of the points $\phi_1(a)$, $\phi_2(c)$ is not empty. In a similar way, it is proved that an intersection of neighbourhoods of the points $\phi_1(b)$, $\phi_2(d)$ is not empty; the same is valid in the case of decreasing function $\phi_2^{-1}\phi_1$ for the points $\phi_1(a)$, $\phi_2(d)$, as well as for the points $\phi_1(b)$, $\phi_2(c)$. Since the fact that non–intersecting neighbourhoods for a pair of different points do not exist contradicts the Hausdorff property of the manifold M^1, it is, by the same token, proved that there are no finite intervals among the connected

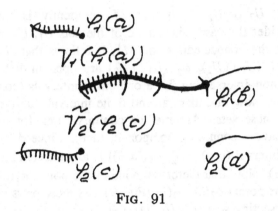

FIG. 91

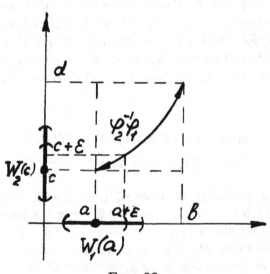

FIG. 92

components of sets $\phi_1^{-1}(U_1 \cap U_2)$, $\phi_2^{-1}(U_1 \cap U_2)$. Besides, the sets $\phi_1^{-1}(U_1 \cap U_2)$, $\phi_2^{-1}(U_1 \cap U_2)$ do not coincide with the whole line $\mathbf{R}^1$, since none of the sets U_1, U_2 are contained in each other. Therefore, only the following cases are logically possible:

(1)   each of the sets $\phi_1^{-1}(U_1 \cap U_2)$, $\phi_2^{-1}(U_1 \cap U_2)$ is an open half–line;

(2)   each of the sets $\phi_1^{-1}(U_1 \cap U_2)$, $\phi_2^{-1}(U_1 \cap U_2)$ consists of two open non–intersecting half–lines.

Consider the first case. Without loss of generality one may assume that the set $\phi_1^{-1}(U_1 \cap U_2)$ has the form $(-\infty, a)$, and the set $\phi_2^{-1}(U_1 \cap U_2)$ is of the form $(b, +\infty)$ (the latter can be obtained by considering, if necessary, the chart $(U_i, \phi_i(-x))$ instead of the chart $(U_i, \phi_i(x))$. Since the function $\phi_2^{-1}\phi_1$ homeomorphically maps $(-\infty, a)$ into $(b, +\infty)$, it is monotone. Note that $\phi_2^{-1}\phi_1$ increases, otherwise from arguments analogous to those presented above, it follows that the points $\phi_1^{-1}(a)$, $\phi_2^{-1}(b)$ have no intersecting neigh-bourhoods, and this contradicts the Hausdorff property of M^1. Let x_0 be any point from $U_1 \cap U_2$. Because $\phi_2^{-1}\phi_1$ is inrceasing (Fig. 93), we have $\phi_2^{-1}\phi_1((-\infty, \phi_1^{-1}(x_0))) = (b, \phi_2^{-1}(x_0))$, $\phi_2^{-1}\phi_1((\phi_1^{-1}(x_0), a)) = (\phi_2^{-1}(x_0), +\infty)$ (Fig. 94). Applying the homeomorphism ϕ_2 to both sides of the latter e-qualities, we obtain $\phi_1((-\infty, \phi_1^{-1}(x_0))) = \phi_2(b, \phi_2^{-1}(x_0))$, $\phi_1((\phi_1^{-1}(x_0), a)) = \phi_2((\phi_2^{-1}(x_0), +\infty))$. Hence $M^1 = \phi_1([\phi_1^{-1}(x_0), +\infty)) \cup \phi_2((-\infty, \phi_2^{-1}(x_0)])$ and, moreover, $\phi_1((\phi_1^{-1}(x_0), +\infty)) \cap \phi_2((-\infty, \phi_2^{-1}(x_0))) = \emptyset$. Therefore, if $\psi_1 : [0, +\infty) \to [\phi_1^{-1}(x_0), +\infty)$, $\psi_2 : (-\infty, 0] \to (-\infty, \phi_2^{-1}(x_0)]$ are certain ho-meomorphisms, then the homeomorphisms $\phi_1\psi_1 : [0, +\infty) \to \phi_1([\phi_1^{-1}(x_0), +\infty))$, $\phi_2\psi_2 : (-\infty, 0] \to \phi_2((-\infty, \phi_2^{-1}(x_0)])$ define a home-omorphism between $\mathbf{R}^1$ and M^1. Thus, in case (1), the manifold M^1 is homeomorphic to $\mathbf{R}^1$.

In the second case, $\phi_1^{-1}(U_1 \cap U_2) = (-\infty, a_1) \cup (a_2, +\infty)$, $\phi_2^{-1}(U_1 \cap U_2) = (-\infty, b_1) \cup (b_2, +\infty)$, where $a_1, a_2, b_1, b_2 \in \mathbf{R}$, $a_1 < a_2$, $b_1 < b_2$. Without restriction of generality, we may assume that $\phi_1((-\infty, a_1)) = \phi_2((-\infty, b_1))$, $\phi_1((a_2, +\infty)) = \phi_2((b_2, +\infty))$ (otherwise, instead of considering one of the charts $(U_1, \phi_i(x))$, $i = 1, 2$, the chart $(U_i, \phi_i(-x))$ should be considered). Let x_1 be a point from $\phi_1((-\infty, a_1))$, and x_2 a point from $\phi_2((a_2, +\infty))$. We show that

(3)
$$\phi_1((-\infty, \phi_1^{-1}(x_1))) = \phi_2((\phi_2^{-1}(x_1), b_1)),$$
$$\phi_1((\phi_1^{-1}(x_1), a_1)) = \phi_2((-\infty, \phi_2^{-1}(x_1))),$$
$$\phi_1((a_2, \phi_1^{-1}(x_2))) = \phi_2((\phi_2^{-1}(x_2), +\infty)),$$
$$\phi_1((\phi_1^{-1}(x_2), +\infty)) = \phi_2((b_2, \phi_2^{-1}(x_2))).$$

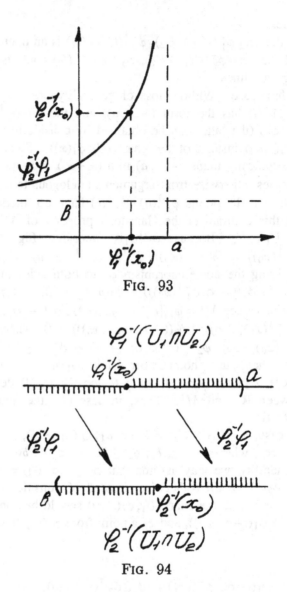

FIG. 93

FIG. 94

We shall verify the first of these equalities (the remaining ones can be verified similarly). Since the function $\phi_2^{-1}\phi_1$ homeomorphically maps $(-\infty, a_1)$ into $(-\infty, b_1)$, it is monotone; moreover, $\phi_2^{-1}\phi_1$ decreases, otherwise the points $\phi_1 1(a_1)$, $\phi_2^{-1}(b_1)$ would not have non–intersecting neighbourhoods and that would contradict the Hausdorff property of M^1. Because $\phi_2^{-1}\phi_1$ is decreasing

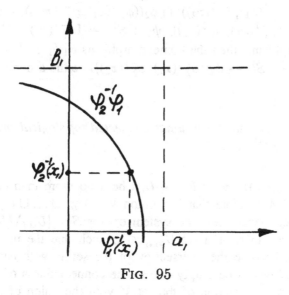

FIG. 95

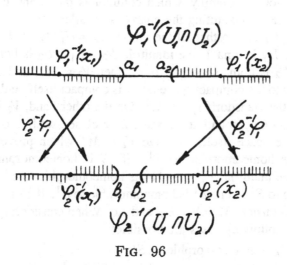

FIG. 96

(Fig. 95), we have $\phi_2^{-1}\phi_1((-\infty, \phi_1^{-1}(x_1))) = (\phi_2^{-1}(x_1), b_1)$ (Fig. 96). Applying the homeomorphism ϕ_2 to both sides of the latter equality, we obtain the required equality. Due to equalities (3), we have

$$M^1 = \phi_1([\phi_1^{-1}(x_1), \phi_1^{-1}(x_2)]) \cup \phi_2([\phi_2^{-1}(x_1), \phi_2^{-1}(x_2)]),$$

moreover, $\phi_1((\phi_1^{-1}(x_1), \phi_1^{-1}(x_2))) \cap \phi_2((\phi_2^{-1}(x_1), \phi_2^{-1}(x_2))) = \emptyset$. Therefore, if $\psi_1 : \bar{S}_+^1 \to [\phi_1^{-1}(x_1), \phi_1^{-1}(x_2)]$, $\psi_2 : \bar{S}_-^1 \to [\phi_2^{-1}(x_1), \phi_2^{-1}(x_2)]$ are certain homeomorphisms, then the homeomorphisms $\phi_1 \psi_1 : \bar{S}_+^1 \to \phi_1([\phi_1^{-1}(x_1), \phi_1^{-1}(x_2)])$, $\phi_2 \psi_2 : \bar{S}_-^1 \to \phi_2([\phi_2^{-1}(x_1), \phi_2^{-1}(x_2)])$ define a homeomorphism of S^1 into M^1. $\qquad\square$

Theorem 3. *Any one–dimensional connected topological manifold M^1 is homeomorphic to $\mathbf{R}^1$ or S^1.*

Proof. Let $\{(U_\alpha, \phi_\alpha)\}$ ($\phi_\alpha : \mathbf{R}^1 \to U_\alpha$) be a no more than countable atlas of the manifold M^1. Note that if a union $V = U_{\alpha_1} \cup \ldots \cup U_{\alpha_k}$ of any sets from $\{U_\alpha\}$ is connected, and the system of sets $S = \{U_\alpha\} \backslash \{U_{\alpha_1}, \ldots, U_{\alpha_k}\}$ is nonempty, then there is a set $U_{\alpha_{k+1}} \in S$ such that the union $V \cup U_{\alpha_{k+1}}$ is connected (otherwise the intersection of the set V with each of the sets of the system S would be empty due to the connectedness of sets U_α, and, consequently, the intersection of the set V with the union of all the sets of the system S would be empty, which contradicts the connectedness of M^1). Therefore, the sets constituting the charts of an atlas can be renumbered into a sequence $U_1, U_2, \ldots$ so that all the unions $V_k = U_1 \cup \ldots \cup U_k$, $k = 1, 2, \ldots$, are connected. By lemma 1, the manifold $V_2 = U_1 \cup U_2$ is homeomorphic to S^1 or $\mathbf{R}^1$. If V_2 is homeomorphic to S^1, then $V_2 = M^1$. Indeed, since V_2 is homeomorphic to the compact space S^1, it is compact itself, and, consequently, it is closed in the Hausdorff space M^1. On the other hand, V_2 is open in M^1. Since a nonempty, open and at the same time closed set in a connected space can only be the space itself, we have $V_2 = M^1$ which proves the theorem, provided V_2 is homeomorphic to S^1. If V_2 is homeomorphic to $\mathbf{R}^1$ then, analogously, the manifold $V_3 = V_2 \cup V_3$ is homeomorphic to S^1 or $\mathbf{R}^1$. If V_3 is homeomorphic to S^1, then M^1 is homeomorphic to S^1. If V_3 is homeomorphic to $\mathbf{R}^1$, then we consider $V_4 = V_3 \cup U_4$, and etc. When considering the manifolds $V_1, V_2, \ldots$, the following cases are possible:

(1) some V_k is homeomorphic to S^1;

(2) all V_k, $k = 1, 2, \ldots$ are homeomorphic to $\mathbf{R}^1$.

In the first case, M^1 is homeomorphic to S^1. We show that in case (2), M^1 is homeomorphic to $\mathbf{R}^1$. If the atlas $\{(U_\alpha, \phi_\alpha)\}$ is finite, then this is obvious. Let the atlas consist of an infinite set of charts. Construct inductively a sequence of finite intervals $(a_1, b_1) \subset (a_2, b_2) \subset \ldots$ and a sequence of homeomorphisms $\psi_k : (a_k, b_k) \to V_k$, $k = 1, 2, \ldots$, such that $\psi_{k+1}|_{(a_k, b_k)} = \psi_k$. Let (a_1, b_1) be some finite interval, and $\psi_1 : (a_1, b_1) \to V_1$ a homeomorphism. Suppose that the intervals $(a_1, b_1), \ldots, (a_k, b_k)$ and homeomorphisms $\psi_i : (a_i, b_i \to V_i$, $i = 1, \ldots, k$ are already constructed. We construct an interval (a_{k+1}, b_{k+1}) and a homeomorphism $\psi_{k+1} : (a_{k+1}, b_{k+1}) \to V_{k+1}$. Let $f_{k+1} : (a_k, b_k) \to V_{k+1}$ be any homeomorphism. Since $V_k \subset V_{k+1}$, $f_{k+1}^{-1}(V_k)$ is an interval (c_k, d_k) which is contained in (a_k, b_k). Let $g_{k+1} : \mathbf{R}^1 \to \mathbf{R}^1$ be a homeomorphism, for which $g_{k+1}((c_k, d_k)) = (a_k, b_k)$ (among these homeomorphisms there are both increasing and decreasing ones). Since the function $\psi_k^{-1} f_{k+1} g_{k+1}^{-1}|_{(a_k, b_k)}$ is a homeomorphism, it is monotone, and without restriction of generality, we may assume that this function is increasing (this can be guaranteed by choosing an increasing or decreasing homeomorphism g_{k+1} as necessary). Put $(a_{k+1}, b_{k+1}) = g_{k+1}((a_k, b_k))$. Define the mapping $\psi_{k+1} : (a_{k+1}, b_{k+1}) \to V_{k+1}$ by the equality

$$\psi_{k+1}(x) = \begin{cases} \psi_k(x), & \text{if } x \in (a_k, b_k); \\ f_{k+1} g_{k+1}^{-1}(x), & \text{if } x \in (a_{k+1}, b_{k+1}) \backslash (a_k, b_k). \end{cases}$$

We show that ψ_{k+1} is a homeomorphism. Indeed, the mapping $\Phi = \psi_{k+1}^{-1} f_{k+1} g_{k+1}^{-1}|_{(a_{k+1}, b_{k+1})}$ on the set $(a_{k+1}, b_{k+1}) \backslash (a_k, b_k)$ is the identity, and on the set, (a_k, b_k) it coincides with the homeomorphism $\psi_k^{-1} f_{k+1} g_{k+1}^{-1}$ which is increasing (Fig. 97); therefore, Φ is a homeomorphism, and, consequently, $\psi_{k+1} = f_{k+1} g_{k+1}^{-1} \Phi^{-1}$ is also a homeomorphism.

The constructed sequence of homeomorphisms $\psi_k : (a_k, b_k) \to V_k$, $k = 1, 2, \ldots$, defines a homeomorphism of the interval $\bigcup_{k=1}^{\infty} (a_k, b_k)$, and $M^1 = \bigcup_{k=1}^{\infty} V_k$ and, consequently, M^1 is homeomorphic to $\mathbf{R}^1$. $\square$

Theorem 4. *Any one–dimensional connected topological manifold M^1 with boundary $(\partial M^1 \neq \emptyset)$ is homeomorphic to $\mathbf{R}_+^1$ or $\bar{D}^1$.*

Proof. The double DM^1 of the manifold M^1 is a one–dimensional connected topological manifold (without boundary) and, consequently, it is homeomorphic to $\mathbf{R}^1$ or S^1, and M^1 is itself homeomorphic to a connected closed proper subset of $\mathbf{R}^1$ or S^1 which is not reduced to a point. Since any connected not one–point subset $\mathbf{R}^1$ contains, together with any two of its points, the interval joining these points, the connected not one–point sets in $\mathbf{R}^1$ have the form (a, b), $[a, b)$, $(a, b]$ or $[a, b]$, where $a < b$ (it is possible that $a = -\infty$, $b = +\infty$). Therefore, connected closed proper subsets in $\mathbf{R}^1$ which are not reduced to a point, have the form $(-\infty, b]$, $[a, +\infty)$ or $[a, b]$ and, consequently, they are

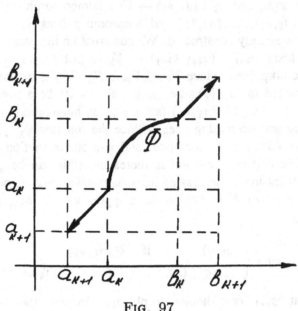

FIG. 97

homeomorphic to $\mathbf{R}^1_+$ or $\bar{D}^1$. In a similar manner, connected closed proper subsets of S^1 which are not reduced to a point, are homeomorphic to $\mathbf{R}^1_+$ or $\bar{D}^1$. Thus, M^1 is homeomorphic to $\mathbf{R}^1_+$ or $\bar{D}^1$ (a compact M^1 is homeomorphic to $\bar{D}^1$, and a non–compact one to $\mathbf{R}^1_+$).   □

The topological classification of compact two–dimesnional manifolds is carried out in Ch. 2, § 4, and Ch. 3, § 4. The classification of manifolds of dimension higher than 2 turns out to be a very difficult problem.

3. Regular and non–regular points of a smooth mapping. Let $f : M^n \to N^m$ be a C^r–mapping of C^r–manifolds, $r \geq 1$.

Definition 6. A point $x \in M^n$ is called a *regular (non–critical, non–singular)* point of a mapping f, if for a certain coordinate representation

$$\psi^{-1} f \phi : \phi^{-1}(W(x)) \to \psi^{-1}(V)$$

of the mapping f in a neighbourhood of the point x the point $\phi^{-1}(x)$ is regular. Otherwise the point x is said to be *non–regular (critical, singular)*.

Execise 11°. Show that the definition does not depend on the choice of a coordinate representation.

Definition 7. A point $y \in N^m$ is called a *regular (non–critical, non–singular) value* of a mapping f if its full preimage $f^{-1}(y)$ either consists only of regular points of the mapping f or is empty.

Otherwise the point y is called a *non–regular (critical, singular) value*.

Remark 2. The concept of a regular point of a mapping f can be extended to a case when M^n, N^m are manifolds with boundary. The definition of regularity for points that do not belong to the boundary ∂M^n, is the same. A point x of a boundary ∂M^n is called a *regular point* of the mapping f if for some coordinate representation $\psi^{-1} f \phi : \phi^{-1}(W(x)) \to \psi^{-1}(V)$ of the mapping f in a neighbourhood of the point x the following conditions are satisfied:

(1) in $\mathbf{R}^n$, there exists an open neighbourhood $\widetilde{W}$ of the point $\phi^{-1}(x)$ and a C^r–mapping $\Phi : \widetilde{W} \to \mathbf{R}^m$ coinciding with $\psi^{-1} f \phi$ on the set $\widetilde{W} \cap \phi^{-1}(W(x))$ for which $\phi^{-1}(x)$ is a regular point;

(2) the point x is the regular point of the restriction $f|_{\partial M^n}$ of the mapping f to the boundary ∂M^n.

The definitions of a non–regular point, regular and non–regular value are the same.

Note that when $n \leq m$, condition (2) follows from condition (1), and when $n > m$, condition (1) follows from condition (2); if $n > m$, the set of non–regular values of the mapping f coincides with the union of the sets of non–regular values of the mappings $f|_{M^n \setminus \partial M^n}$, $f|_{\partial M^n}$.

Exercise 12°. Show that the set of regular points of a mapping $f : M^n \to N^m$ of manifolds is open in M^n.

The set of regular values of a mapping of manifolds, contrary to the set of regular points, is not necessarily open (give examples!).

In topology, the fundamental theorem of analysis concerning the measure of the set of non–regular values of a smooth mapping is often applied.

First, recall that a set $A \subset \mathbf{R}^n$ has measure zero (this is denoted by mes $A = 0$), if for any $\epsilon > 0$ there exists a no more than countable covering A of closed parallelepipeda $U_1, U_2, \ldots$ such that $\sum_i \operatorname{Vol} U_i < \epsilon$ (here $\operatorname{Vol} U_i$ is the volume of the parallelepiped U_i in $\mathbf{R}^n$).

It is clear that a not more than countable union of sets of measure zero is a set of measure zero, and that if A has measure zero and $B \subset A$, then B also has measure zero.

Theorem 5. (Sard).[*] *Let U be an open set in $\mathbf{R}^n$, and $f : U \to \mathbf{R}^m$ a C^r–mapping, $r \geq \max(n - m, 0) + 1$. Then, if $n \geq m$, the set of non–regular values of the mapping f has measure zero; if $n < m$, then the set $f(U)$ of all the values of the mapping f has measure zero.*[**]

We will prove theorem 5 only for the case $n \leq m$, although later we will also apply it for $n > m$. The proof of the case $n > m$ is rather complicated, but the reader can find it in literature.

First, we prove the theorem for the case $n = m$.

Let Q be a closed n–dimensional cube in U with the side l. First we show that the set of non–regular values of a mapping $f : Q \to \mathbf{R}^n$ has measure zero. According to the mean value theorem, for any two points $y, z \in Q$ we have

$$(4) \qquad f_i(z) - f_i(y) = \sum_{j=1}^{n} \left(\frac{\partial f_i}{\partial x_j} \bigg|_{u^i} \right) (z_j - y_j), \quad i = 1, 2, \ldots, n,$$

where u^i is a certain point from the interval joining y and z. Since $f \in C^1$, the functions $\partial f_i / \partial x_j$, $i, j = 1, \ldots, n$ are continuous in Q and, consequently, bounded as continuous functions on a compact set. Therefore, from the equalities (4) the inequality

$$(5) \qquad \qquad \|f(y) - f(z)\| < c\|y - z\|,$$

---

[*]In the literature, this theorem is mostly known as the Sard theorem, although for the first time, this statement was presented by A. Brown in 1935. In 1939, A. P. Morse proved the theorem for the case of functions $f : \mathbf{R}^n \to \mathbf{R}^1$, referring to a weaker result in a non–published paper of M. Morse and A. Sard. Theorem 5 was published by A. Sard in 1942. In 1953, the result was discovered again by A.Ya. Dubovitsky and R. Thom. For the case of mappings on the plane $f : \mathbf{R}^2 \to \mathbf{R}^2$, the theorem was established by K. Knopp and R. Schmidt in 1926.

[**]As it was shown by H. Whitney and D. E. Men'shov, the proposition of smoothness can not be weakened.

where c is some constant, can be obtained.

Consider now the affine mapping

(6)
$$T_y(z) = (T_y^{(1)}(z), \ldots, T_y^{(n)}(z)) : \mathbf{R}^n \to \mathbf{R}^n,$$

$$T_y^{(i)} = f_i(y) + \sum_{j=1}^{n} \left(\left. \frac{\partial f_i}{\partial x_j} \right|_y \right) (z_j - y_j), \quad i = 1, \ldots n.$$

From the equalities (4), (6) we have

$$f_i(z) - T_y^{(i)}(z) =$$

$$= \sum_{j=1}^{n} \left(\left. \frac{\partial f_i}{\partial x_j} \right|_{u^i} - \left. \frac{\partial f_i}{\partial x_j} \right|_y \right) (z_j - y_j), \quad i = 1, \ldots, n.$$

Since the functions $\partial f_i / \partial x_j, i, j = 1, \ldots, n$ are uniformly continuous on Q (as continuous functions on a compact set), there exists a function $\lambda(\epsilon) : \mathbf{R}^1_+ \to \mathbf{R}^1_+$ such that $\lambda(\epsilon) \to 0$, for $\epsilon \to 0$ and

(7)
$$\|f(z) - T_y(y)\| \le \lambda(\|z - y\|) \cdot (\|z - y\|).$$

If y is a non–regular point of the mapping f, then rank $(\partial f / \partial x)|_y < n$, and, consequently, the image $T_y(\mathbf{R}^n)$ of the space $\mathbf{R}^n$ under the mapping T_y, is contained in some hyperplane $P_y \subset \mathbf{R}^n$.

If $\|z - y\| < \epsilon$, then due to (7) , we have $\|f(z) - T_y(z)\| < \epsilon\lambda(\epsilon)$, and, consequently, $f(z)$ lies in the "slice" P between hyperplanes parallel to P_y and at distance $\epsilon\lambda(\epsilon)$ from P_y (Fig. 98); on the other hand, from inequality (5) it follows that $f(z)$ is contained in the disc $D^n_{c\epsilon}(f(y))$ of radius $c\epsilon$ with centre at the point $f(y)$, and so it is contained in a closed cube with the edge $2c\epsilon$ and centre at $f(y)$, specifically in the cube $Q_{2c\epsilon}$ with one of its sides parallel to P_y. Thus, if y is a non–regular point of the mapping f, and $\|z - y\| < \epsilon$, then $f(z)$ belongs to the closed parallelepiped $Q_{2c\epsilon} \cap P$. If $\epsilon > 0$ is sufficiently small, then $\epsilon\lambda(\epsilon) < c\epsilon$, and $\text{Vol}(Q_{2c\epsilon} \cap P) = (2c\epsilon)^{n-1} 2\epsilon\lambda(\epsilon) = 2^n c^{n-1} \epsilon^n \lambda(\epsilon)$.

Now, divide the cube Q into p^n closed cubes $Q_i, i = 1, \ldots, p^n$, with edge length (l/p), where we take $p = [l\sqrt{n}/\epsilon + 1]$ (here the square brackets denote taking the integer part of a real number). Since $a < [a] + 1$, for any number

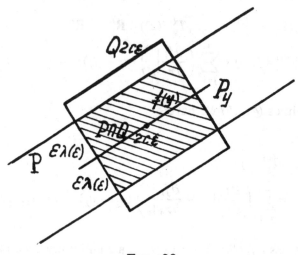

a, we have $l\sqrt{n}/\epsilon + 1 < p + 1$, and, consequently, the following inequality for
the length of the edge of cube Q_i is valid: $(l/p) < \epsilon/\sqrt{n}$. Since the distance
between two points of a cube with edge length r in $\mathbf{R}^n$ does not exceed $\sqrt{n}r$,
the distance between any of the points of each cube Q_i is less than ϵ. If a
certain cube Q_i contains a non–regular point y, then for any point $z \in Q_i$
we have $\|z - y\| < \epsilon$, therefore, the point $f(z)$, as has been shown above,
belongs to some closed parallelepiped of volume $2^n c^{n-1} \epsilon^n \lambda(\epsilon)$. The set of
non–regular values of the mapping f is contained in the union of the images
$f(Q_i)$ of the cubes Q_i containing non–regular points of the mapping f. Since
the number of cubes containing non–regular points of the mapping f, does not
exceed p^n, and the image of each such cube is contained in a parallelepiped of
volume $2^n c^{n-1} \epsilon^n \lambda(\epsilon)$, the sum of volumes of the parallelepipeda containing
non–regular values, does not exceed $p^n 2^n c^{n-1} \epsilon^n \lambda(\epsilon) = (l\sqrt{n} + \epsilon)^n 2^n c^{n-1} \lambda(\epsilon)$.
The latter expression tends to zero when $\epsilon \to 0$. Thus, the set of non–regular
values of the mapping $f : Q \to \mathbf{R}^n$ has measure zero.

Now, we show that the set of non–regular values of the mapping $f : U \to \mathbf{R}^n$
has measure zero.

Represent the set U in the form of a union $U = \bigcup_k Q'_k$ of not more than a
countable set of some n–dimensional closed cubes Q'_k, $k = 1, 2, \ldots$. Then the

set of non–regular values of the mapping $f : U \to \mathbf{R}^n$ is the union of sets of non–regular values of the mappings $f : Q'_k \to \mathbf{R}^n$, $k = 1, 2, \ldots$, each of which, as was shown above, has measure zero. Hence a set of non–regular values of the mapping $f : U \to \mathbf{R}^n$ has measure zero.

Now, we prove the theorem for $n < m$.

The points of the space $\mathbf{R}^m = \mathbf{R}^n \times \mathbf{R}^{m-n}$ are represented in the form (x, y), where $x = (x_1, \ldots, x_n) \in \mathbf{R}^n$, $y = (y_1, \ldots, y_{m-n}) \in \mathbf{R}^{m-n}$. Consider the C^1–mapping $F : U \times \mathbf{R}^{m-n} \to \mathbf{R}^m$ given by $F(x, y) = f(x)$. According to the proved case of this theorem, the set of non–regular values of the mapping $F : U \times \mathbf{R}^{m-n} \to \mathbf{R}^m$ has measure zero. Since $\operatorname{rank}_{(x,y)} F = \operatorname{rank}_x f < m$, the set of non–regular values of the mapping $F : U \times \mathbf{R}^{m-n} \to \mathbf{R}^m$ coincides with the image $F(U \times \mathbf{R}^{m-n})$. To conclude the proof, it remains to note that $F(U \times \mathbf{R}^{m-n}) = f(U)$. $\qquad \square$

Theorem 5 can easily be generalized to the case of mappings of manifolds.

Definition 8. It is said that a *subset A of a manifold M^n has measure zero* (mes $A = 0$), if A can be represented in the form of a union $A = \bigcup_k A_k$ of not more than a countable number of sets $A_1, A_2, \ldots$, where each A_k, $k = 1, 2, \ldots$ is contained in the set U_k of some chart (U_k, ϕ_k) from an atlas of a manifold M^n, and $\phi_k^{-1}(A_k)$ has measure zero in $\mathbf{R}^n$.

Theorem 6. *Let $f : M^n \to N^m$ be a C^r–mapping of C^r–manifolds, $r \geq \max(n-m, 0)+1$. If $n \geq m$, then the set of non–regular values of the mapping f has measure zero; if $n < m$, then the set $f(M^n)$ of all values of the mapping f is of measure zero.*

Proof. Let $\{(U_\alpha, \phi_\alpha)\}$, $\{(V_\beta, \psi_\beta)\}$ be atlases of the manifolds M^n, N^m, respectively, and let x be a point of M^n. Let (U_α, ϕ_α), (V_β, ψ_β) be charts of these atlases such that $x \in U_\alpha$, $f(x) \in V_\beta$, and let $W(x)$ be an open neighbourhood of the point x such that $W(x) \subset U_\alpha$, $f(W(x)) \subset V_\beta$. The mapping $\psi_\beta^{-1} f \phi_\alpha : \phi_\alpha^{-1}(W(x)) \to \psi_\beta^{-1}(V_\beta)$ is a coordinate representation of the mapping f in a neighbourhood of x (and of any point $y \in W(x)$). For each point $x \in M^n$, consider such a coordinate representation of the mapping f in a neighbourhood of x. The system of sets $\{W(x)\}_{x \in M^n}$ obtained under this consideration forms an open covering of M^n. Since the manifolds considered possess a countable base, by the Lindelöf theorem (see Ch. 2, § 11), the covering $\{W(x)\}_{x \in M^n}$ contains no more than a countable subcovering $\{W(x_k)\}_{x_k \in M^n}$. Let

(8) $$\psi_{\beta_k}^{-1} f \phi_{\alpha_k} : \phi_{\alpha_k}^{-1}(W(x_k)) \to \psi_{\beta_k}^{-1}(V_{\beta_k}), \quad k = 1, 2, \dots,$$

be the coordinate representations of the mapping f in neighbourhoods of the points x_k, $k = 1, 2, \dots$, respectively. Then the set C of non–regular values of the mapping f can be represented as a union $C = \cup_k C_k$ of the sets C_k of non–regular values of the mappings (8), and the set $f(M^n)$ as the union $f(M^n) = \bigcup_k B_k$ of the sets $B_k = f(W(x_k))$ of all values of the mappings (8). If $n \geq m$ then, by theorem 5, we have mes $C_k = 0$, $k = 1, 2, \dots$, therefore, mes $C = 0$; if $n < m$ then, by theorem 5, we have mes $B_k = 0$, $k = 1, 2, \dots$, therefore, mes $f(M^n) = 0$. $\qquad\square$

Corollary. *Let* $f : M^n \to N^m$ *be a* C^r*–mapping of* C^r*–manifolds,* $r \geq \max(n - m, 0) + 1$. *Then the set of regular values of a mapping* f *is everywhere dense in* N^m, *and is open in* N^m *if* M^n *is compact.*

Proof. Assume that the set of regular values of the mapping f is not everywhere dense in N^m. Then there exists a nonempty open set $U \subset N^m$ which does not contain regular values of the mapping f, i.e., U is contained in the set C of non–regular values of the mapping f. According to theorem 6, mes $C = 0$, therefore, mes $U = 0$, what contradicts the fact that a nonempty open set has measure not equal to zero.

Now, let M^n be compact. The set of regular points of the mapping f is open in M^n, therefore, the set A of non–regular points of the mapping f is closed and, consequently, is compact in the compact space M^n. Then the set $f(A)$ of non–regular values of the mapping f is also compact (as an image of a compact space under a continuous mapping) and, consequently, it is closed in the Hausdorff space N^m. Therefore, the set $N^m \backslash f(A)$ of regular values of the mapping f is open. $\qquad\square$

4. Immersions, submersions, imbeddings and submanifolds. For a C^r–mapping $f : M^n \to N^m$ of C^r–manifolds, $r \geq 1$, let each point $x \in M^n$ be regular. Such a mapping is called (1) a C^r*–immersion*, when $n \leq m$, (2) a C^r*–submersion*, when $n \geq m$.

A C^r–immersion is called a C^r*–imbedding*, if f is a homeomorphism of M^n onto the subspace $f(M^n)$ of the topological space N^m.

Example 1. The mapping $f : \mathbf{R}^2 \to \mathbf{R}^1$ given by the rule $f(x,y) = x$, is a C^∞–submersion.

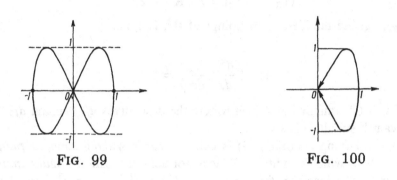

FIG. 99   FIG. 100

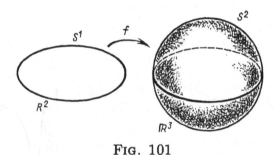

FIG. 101

Example 2. The mapping $f : \mathbf{R}^1 \to \mathbf{R}^2$ given by the rule $f(x) = (\sin x, \sin 2x)$ (Fig. 99) is a C^∞–immersion but not an imbedding, since it is not injective. The mapping $f|_{(0,2\pi)}$ also is not an imbedding, even though f is bijective on $(0, 2\pi)$. In the given case, the mapping f^{-1} is not continuous. Note, that the mapping $f|_{(0,\pi)}$ (Fig. 100) is a C^∞–imbedding.

Example 3. The mapping $f : S^1 \to S^2$ given by the rule $f(x,y) = (x,y,0)$ (Fig. 101) is a C^∞–imbedding.

Example 4. Let f_1, $f_2 : \mathbf{R}^1 \to \mathbf{R}^1$ be functions of class C^r.

The mapping $f = (f_1, f_2) : \mathbf{R}^1 \to \mathbf{R}^2$ (a curve of class C^r) can be considered as a C^r–mapping of manifolds $\mathbf{R}^1$, $\mathbf{R}^2$ with the natural C^∞–structure. We will specify the conditions under which the mapping f is an immersion. The condition of immersion means that every point $x \in \mathbf{R}^1$ is a regular point of the mapping

$$(1_{\mathbf{R}^2})^{-1} f 1_{\mathbf{R}^1} = f : \mathbf{R}^1 \to \mathbf{R}^2$$

(here $1_{\mathbf{R}^2}$, $1_{\mathbf{R}^1}$ are the identity mappings of $\mathbf{R}^2$, $\mathbf{R}^1$), i.e.,

$$(9) \qquad\qquad \text{rank} \left(\frac{df_1}{dx}, \frac{df_2}{dx} \right) = 1.$$

Thus, f is a C^r–immersion if everywhere the derivatives df_1/dx and df_2/dx do not vanish simultaneously.

A curve satisfying condition (9) is called a *curve without singular points*. Those points at which condition (9) does not hold, are called *singular points of the curve*. For instance, for the curve $f_1(x) = x^2$, $f_2(x) = x^3$ (Fig. 102), the point 0 is singular.

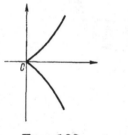

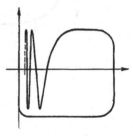

FIG. 102   FIG. 103

Example 5. The curve drawn in Fig. 103 (constructed with a help of the graph of the function $y = \sin(1/x)$), defines a C^∞–immersion, but not an imbedding of the half–line into the plane, although the mapping is bijective.

Another example of a similar kind is provided by the immersion $f : \mathbf{R}^1 \to \mathbf{C} \times \mathbf{C}$, which is given by the formula $f(x) = (e^{2\pi i \alpha_1 x}, e^{2\pi i \alpha_2 x})$, where α_1/α_2

is irrational. It is easy to verify that this is a bijective mapping (of rank 1), moreover, its image lies on the torus $S^1 \times S^1$ being an everywhere dense winding around the torus.

Note that the non–compactness of the line plays an essential role in these examples. Indeed, the following theorem holds.

Theorem 7. *If a manifold M^n is compact and $f : M^n \to N^m$ is an injective immersion, then f is an imbedding.*

Proof. This follows from the fact that an injective, continuous mapping $f : M \to N$ of a compact space M to the Hausdorff space N is a homeomorphism onto the subspace $f(M)$ (see Ch. 2, § 13).   □

Note that any C^r–immersion $f : M^n \to N^m$ is a C^r–imbedding on a certain neighbourhood of each point $x \in M^n$ (this follows from the theorem on rectification of a mapping, see § 1).

Example 6. A mapping $f : \mathbf{R}^n \to \mathbf{R}^N$, $N \geq n$, of class C^r, $r \geq 1$, determines an immersion on $\mathbf{R}^n$, if

$$\mathrm{rank}\left(\frac{\partial f}{\partial x}\right)\Big|_y = n$$

at any point $y \in \mathbf{R}^n$. Thus, f has no non–regular points and, according to the theorem on rectification of a mapping, it is a local homeomorphism between $\mathbf{R}^n$ and $f(\mathbf{R}^n)$. If, in addition, f is a homeomorphism of $\mathbf{R}^n$ onto $f(\mathbf{R}^n)$, then f is a C^r–imbedding.

Exercise 13°. Verify that the concept of a chart on a C^r–manifold in $\mathbf{R}^N$ (see § 2) is equivalent to that of a C^r–imbedding of $\mathbf{R}^n$ in $\mathbf{R}^N$.

Rather often manifolds lie in other ambient manifolds. However, it would be too general to call any such manifold a submanifold of the ambient manifold, just like in a topological space, a subset endowed with an arbitrary topology would not be called a subspace. It is necessary to impose reasonable restrictions in the form of the existence of a simple relation between the structures of an imbedded and ambient manifolds. In this setting, the following concept of imbedding turns out to be useful.

Definition 9. A *submanifold of a C^r-manifold N^m* is a subspace M_1 in N^m which is the image of a certain C^r-imbedding $f : M^n \to N^m$ with the C^r-structure induced by the homeomorphism f.

The structures of a submanifold and manifold are related in a simple way: for some atlas $\{(U_\alpha, \phi_\alpha)\}$ of a manifold N^m, the intersection $U_\alpha \cap M_1^n$ (provided it is nonempty) is the image of the subspace $\mathbf{R}^n \subset \mathbf{R}^m = \mathbf{R}^n \times \mathbf{R}^{m-n}$ under the homeomorphism ϕ_α; moreover, the restrictions $\phi_\alpha|_{\mathbf{R}^n} : \mathbf{R}^n \to U_\alpha \cap M_1^n$ define an atlas of the submanifold M_1^n.

Thus, locally, a submanifold in N^m is given in the corresponding local coordinates $\xi_1, \ldots, \xi_m$ on the manifold N^m by the equations $\xi_{n+1} = 0, \ldots, \xi_m = 0$.

Exercise 14°. Applying the theorem on rectification to the coordinate representation of the imbedding f, construct atlases on the manifolds N^m and M_1^n described above.

This interrelation of the structures of a manifold and a submanifold can be used as the basis for the concept of a submanifold.

Definiton 10. A subspace $M_1 \subset N^m$ is called an *n–dimensional submanifold of the C^r-manifold N^m*, $n \le m$, if in the given structure of the manifold N^m there exists a collection of charts $\{(U_\alpha, \phi_\alpha)\}$, $\phi_\alpha : \mathbf{R}^n \times \mathbf{R}^{m-n} \to U_\alpha$ such that $\phi_\alpha(\mathbf{R}^n) = U_\alpha \cap M_1$, when $U_\alpha \cap M_1 \ne \emptyset$, and $M_1 \subset \cup_\alpha U_\alpha$. And moreover, the mappings $\phi_\alpha|_{\mathbf{R}^n} : \mathbf{R}^n \to U_\alpha \cap M_1$ determine a C^r-atlas which specifies the structure of an n–dimensional C^r-manifold on M_1 (Fig. 104). Such a structure on the manifold $M_1 \subset N^m$ is called the structure compatible with the structure of the manifold N^m, or simply the *structure of the submanifold*.

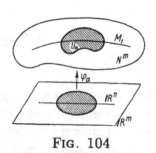

FIG. 104

The equivalence of definitions 9 and 10 is evident.

Exercise 15°. Verify that $\{(U_\alpha \cap M_1, \phi_\alpha|_{\mathbf{R}^n})\}$, $(U_\alpha \cap M_1 \ne \emptyset)$ is a C^r-atlas on M_1.

Exercise 16°. Let M^n be a C^r–submanifold in $\mathbf{R}^N = \mathbf{R}^n \times \mathbf{R}^{N-n}$ (see § 2), and $(U(x), \phi)$ a chart at a point $x \in M^n$. Show that (1) there exists a C^r–diffeomorphism $\phi : \mathbf{R}^N \to \widetilde{U}(x)$ from the space $\mathbf{R}^N$ on a certain, open in $\mathbf{R}^N$, neighbourhood $\widetilde{U}(x)$ of the point x such that

$$(10) \qquad \widetilde{\phi}|_{\mathbf{R}^n} = \phi,$$

and (2) that the set of charts

$$(11) \qquad \{(\widetilde{U}(x), \widetilde{\phi})\}_{x \in M^n} \cup (\mathbf{R}^N, 1_{\mathbf{R}^N})$$

forms a C^r–atlas on $\mathbf{R}^N$ (in the sense of definition 2, § 3).

From exercise 16° it follows that $\mathbf{R}^N$ with that atlas (11) is a C^r–manifold, and M^n, due to (10), is a submanifold of it. (This justifies the term a "C^r–submanifold in $\mathbf{R}^N$" used in § 2. However, to be more precise, instead of this term we should use the term a "submanifold of the C^r–manifold $\mathbf{R}^N$".)

Example 7. The equator of the sphere S^2 (see example 3) is a submanifold.

Exercise 17°. Show that the graph of the mapping $f(x) = |x|$, $x \in \mathbf{R}^1$, is not a submanifold of $\mathbf{R}^2$.

Submanifolds which are the Lie groups arise as images of homomorphisms of Lie groups.

Let $\phi : G_1 \to G_2$ be a homomorphism of Lie groups such that ϕ is an immersion and injection. Then the pair (G_1, ϕ) is called the *Lie subgroup of the Lie group* G_2. If, in addition, the image $\phi(G_1)$ is closed in G_2, then we speak of a *closed subgroup* (G_1, ϕ) *in* G_2.

In the case when $G_1 \subset G_2$ is an abstract subgroup of the Lie group G_2 and at the same time is an imbedded submanifold $G_1 \xrightarrow{i} G_2$, then the pair (G_1, i) with the induced smooth structure on G_1 is a *Lie subgroup (imbedded in G_2)*.

Exercise 18°. Show that under the conditions of the last sentence, G_1 is a Lie group (with the induced smooth structure from G_2).

Exercise 19°. Show that the Lie groups $O(n, \mathbf{R})$, $SO(n, \mathbf{R})$ are imbedded Lie subgroups in $GL(n, \mathbf{R})$. Verify that an everywhere dense winding on a torus is an example of a Lie subgroup that is not imbedded.

Manifolds often arise not as images under certain mappings, but as preimages. The following important theorem is useful not only when constructing new manifolds, but it also often facilitates the proof of the fact that the spaces investigated have the structure of manifolds.

Theorem 8. *Let $f : M^n \to N^m$ ($n \geq m$) be a C^r-mapping of C^r-manifolds ($r \geq 1$), and let N_1^k be a submanifold in N^m, consisting of only regular values of the mapping f. Then $M_1 = f^{-1}(N_1^k)$ is either empty, or a submanifold in M^n of dimension $n - m + k$.*

Proof. Assume that $M_1 \neq \emptyset$. Let x_0 be an arbitrary point in M_1. Since N_1^k is a submanifold, there exists a chart $(\widetilde{V}, \psi)(f(x_0) \in \widetilde{V})$ from the maximal atlas of the C^r-structure given on N^m such that the pair $(\widetilde{V} \cap N_1^k, \psi|_{\mathbf{R}^k})$ is a chart of an atlas for the C^r-structure on N_1^k. Let (U, ϕ) ($x_0 \in U$) be a chart of an atlas for the C^r-structure given on M^n such that $f(U) \subset \widetilde{V}$. Then, according to the condition, $\phi^{-1}(x_0)$ is a regular point of the mapping $\Phi = \psi^{-1} f \phi : \phi^{-1}(U) \to \mathbf{R}^m$ and, by the theorem on rectification of a mapping, there exist an open neighbourhood $V \subset \mathbf{R}^n$ of the point $\phi^{-1}(x_0)$, an open set $W \subset \mathbf{R}^n$ and a C^r-diffeomorphism $F : V \to W$ such that the mapping ΦF^{-1} on the set W is the standard projection of $\mathbf{R}^n$ on $\mathbf{R}^m$. Note that $\phi(V)$ is an open neighbourhood in M^n of the point x_0, and the pair $(\phi(V), \phi F^{-1})$ is a chart of the maximal atlas for the C^r-sructure given on M^n. Since ΦF^{-1} is the standard projection, and the set

$$\psi^{-1} f(\phi(V) \cap M_1) \subset \mathbf{R}^m$$

consists of the points of the form $(x_1, \ldots, x_k, 0, \ldots, 0)$, the set

$$F \phi^{-1}(\phi(V) \cap M_1) \subset \mathbf{R}^n$$

consists of points of the form $(x_1, \ldots, x_k, 0, \ldots, 0, x_{m+1}, \ldots, x_n) \in \mathbf{R}^{n-m+k}$. Thus, the chart $(\phi(V), \phi F^{-1})$ on M^n possesses the property

$$(\phi F^{-1})(\mathbf{R}^{n-m+k} \cap W) = \phi(V) \cap M_1.$$

Such a chart can be constructed for any point $x_0 \in M_1$. This proves that M_1 is a submanifold in M^n of dimension $n - m + k$. $\qquad \square$

Example 8. In particular, it follows from theorem 8 that the preimage of a regular value of the mapping $f : M^n \to N^m$ is either empty, or a submanifold in M^n of dimension $n - m$.

The following fundamental fact solves the main problem in the theory of manifolds.

Theorem 9. (Whitney). *For any C^r-manifold M^n, $r \geq 1$, there exists a C^r-imbedding of M^n in $\mathbf{R}^{2n+1}$.*

We shall prove the theorem only under the assumption that M^n is compact, and $r \geq 2$; the more general case requires a more detailed analysis.

At first, we consider a weaker statement.

Theorem 10. *For any compact C^r-manifold M^n, $r \geq 1$, there exists a C^r-imbedding M^n in the space $\mathbf{R}^N$ for a certain dimension N.*

Proof. Let $\{(U_\alpha, \phi_\alpha)\}_{\alpha \in I}$ be a C^r-atlas of the manifold M^n. By lemma 3, § 4, for any point $y \in U_\alpha$, there exist open neighbourhoods $V_{\alpha,1}(y)$, $V_{\alpha,2}(y)$ and a C^r-function $f_{\alpha,y} : M^n \to [0,1]$ such that

(1) $\bar{V}_{\alpha,1}(y) \subset V_{\alpha,2}(y) \subset U_\alpha,$

(2) $f_{\alpha,y}(x) = \begin{cases} 1, & \text{if } x \in \bar{V}_{\alpha,1}(y), \\ 0, & \text{if } x \in M^n \backslash V_{\alpha,2}(y), \end{cases}$

moreover, $f_{\alpha,y}(x) < 1$, if $x \notin \bar{V}_{\alpha,1}(y)$. For each point $y \in U_\alpha$ of every set U_α, $\alpha \in I$, consider the open neighbourhoods $V_{\alpha,1}(y)$, $V_{\alpha,2}(y)$ and a C^r-function $f_{\alpha,y}$ with the properties indicated. The system of sets $\{V_{\alpha,1}(y)\}_{\alpha \in I, y \in U_\alpha}$ forms an open covering of M^n. By the compactness of M^n, this covering contains a finite subcovering $\{V_{\alpha_1,1}(y_1), \ldots, V_{\alpha_k,1}(y_k)\}$.

Consider the mappings $\psi_i : M^n \to \mathbf{R}^n$, $i = 1, \ldots, k$:

$$\psi_i(x) = \begin{cases} f_{\alpha_i,y_i}(x) \cdot \phi_{\alpha_i}^{-1}(x), & \text{if } x \in U_{\alpha_i}, \\ 0, & \text{if } x \notin U_{\alpha_i}. \end{cases}$$

Obviously, $\psi_i \in C^r$, $i = 1, \ldots, k$. The mappings $\psi_1, \ldots, \psi_k; f_{\alpha_1,y_1}, \ldots, f_{\alpha_k,y_k}$ determine a C^r-mapping $\psi = (\psi_1, \ldots, \psi_k, f_{\alpha_1,y_1}, \ldots, f_{\alpha_k,y_k}) : M^n \to$

$\mathbf{R}^n \times \ldots \times \mathbf{R}^n \times \mathbf{R}^1 \times \ldots \times \mathbf{R}^1 = \mathbf{R}^{k(n+1)}$. We show that ψ is an immersion. Let z be some point in M^n. Then z belongs to a certain set $V_{\alpha_i,1}(y_i)$ of the subcovering. Since $f_{\alpha_i,y_i}(x) = 1$ for $x \in V_{\alpha_i,1}(y_i)$, the Jacobian matrix $\left(\frac{\partial(\psi\phi_{\alpha_i})}{\partial x}\right)\Big|_z$ of the mapping $\psi\phi_{\alpha_i}$ contains the identity matrix

$$\left(\frac{\partial((f_{\alpha_i,y_i} \cdot \phi_{\alpha_i}^{-1})\phi_{\alpha_i})}{\partial x}\right)\Big|_z = \begin{pmatrix} 1 & & 0 \\ & \ddots & \\ 0 & & 1 \end{pmatrix}$$

of dimension $n \times n$. Therefore, $\mathrm{rank}_z\,\psi\phi_{\alpha_i} = n$, and ψ is an immersion. Now we show that ψ is injective. Let $z' \in M^n$ and $z' \neq z$. If $z' \in \bar{V}_{\alpha_i,1}(y_i)$, then $\phi_{\alpha_i}(z) \neq \phi_{\alpha_i}(z')$ (since ϕ_{α_i} is a homeomorphism), and $f_{\alpha_i,y_i}(z) = f_{\alpha_i,y_i}(z') = 1$; therefore, $\psi(z) \neq \psi(z')$. If $z' \notin \bar{V}_{\alpha_i,1}(y_i)$, then $f_{\alpha_i,y_i}(z) \neq f_{\alpha_i,y_i}(z')$, and, so $\psi(z) \neq \psi(z')$. Thus, $\psi : M^n \to \mathbf{R}^{k(n+1)}$ is an injective immersion. Since M^n is compact, by theorem 7, ψ is an imbedding. $\square$

Now, we prove theorem 9.

By theorem 10, one may assume that the C^r–manifold M^n is a C^r–submanifold of a certain space $\mathbf{R}^N$. If $N \leq 2n + 1$, the statement is proved. Let $N > 2n+1$. We show that in this case there exists a projection $\mathrm{pr}_e : \mathbf{R}^N \to \mathbf{R}^{N-1}$ on the subspace $\mathbf{R}^{N-1} = \{(x_1, x_2, \ldots, x_{N-1}, 0)\} \subset \mathbf{R}^N$ parallel to some vector $e \in \mathbf{R}^N$, which gives an imbedding of M^n in $\mathbf{R}^{N-1}$. We shall search for e among the vectors of the unit sphere S^{N-1}. First, we prove that there exist projections which injectively map M^n into $\mathbf{R}^{N-1}$. When projecting M^n on $\mathbf{R}^{N-1}$ parallel to the vector e, the injectivity fails iff there exist $x, y \in M^n$, $x \neq y$, such that the vector $x - y$ is parallel to e. Thus, the injectivity of the projection $\mathrm{pr}_e : M^n \to \mathbf{R}^{N-1}$ will hold if the following condition is fulfilled:

$$(12) \qquad e \neq \frac{x - y}{\|x - y\|}, \quad x, y \in M^n, \quad x \neq y$$

We show that vectors e satisfying condition (12) exist. Consider the subset $K^{2n} = (M^n \times M^n) \backslash \Delta$ of the product $M^n \times M^n$ (here $\Delta = \{(x, x) : x \in M^n\}$ is the diagonal). Since M^n is Hausdorff, K^{2n} is open in $M^n \times M^n$ and, consequently, is a C^r–manifold of dimension $2n$. Consider the C^r–mapping $f : K^{2n} \to S^{N-1}$, $f(x, y) = (x - y)/\|x - y\|$. The vector $e \in S^{N-1}$ satisfies condition (12) iff it does not belong to the image $f(K^{2n})$. Since $2n < N - 1$, according to theorem 6, the set $f(K^{2n})$ has measure zero in S^{N-1}; therefore $S^{N-1} \backslash f(K^{2n}) \neq \emptyset$.

Now we show that there exist projections realizing an immersion of M^n in $\mathbf{R}^{N-1}$. Since M^n is compact, there exist C^r–charts $(U_1, \phi_1), \ldots, (U_s, \phi_s)$ in M^n such that $\bigcup_{i=1}^s U_i = M^n$. Let x be some point of M^n and (U_k, ϕ_k) a C^r–chart on M^n, $x \in U_k$. In order to have the projection $\mathrm{pr}_e : M^n \to \mathbf{R}^{N-1}$ being an immersion at the point x, it is necessary and sufficient for the subspace $(D_{\phi_k^{-1}(x)}\phi_k)(\mathbf{R}^n)$ (the image under the linear mapping $D_{\phi_k^{-1}(x)}\phi_k : \mathbf{R}^n \to \mathbf{R}^N$) to preserve its dimension under the projection pr_e. This is equivalent to the fact that $e \notin (D_{\phi_k^{-1}(x)}\phi_k)(\mathbf{R}^n)$. Thus, the projection $\mathrm{pr}_e : M^n \to \mathbf{R}^{N-1}$ is an immersion at every point $x \in U_k$, if the following condition is fulfilled

$$(13) \qquad e \notin \bigcup_{x \in U_k} (D_{\phi_k^{-1}(x)}\phi_k)(\mathbf{R}^n).$$

We shall show that there exist vectors e satisfying condition (13). Consider the subset $L = \mathbf{R}^n \times (\mathbf{R}^n \backslash 0)$ of the product $\mathbf{R}^n \times \mathbf{R}^n$. Since L is open in $\mathbf{R}^n \times \mathbf{R}^n$, it is a C^r–manifold of dimension $2n$. Consider the mapping $g_k(y, z) : L^{2n} \to \mathbf{R}^n$, $g_k(y, z) = (D_y\phi_k)(z)$. Since for any point $y \in \mathbf{R}^n$, $\mathrm{rank}_y\phi_k = n$, we have $\ker D_y\phi_k = \{0\}$, and, consequently, $g_k(y, z) \neq 0$. Therefore, one may consider the mapping

$$h_k(y, z) : L^{2n} \to S^{N-1}, \quad h_k(y, z) = g_k(y, z)/\|g_k(y, z)\|.$$

Evidently, $h_k \in C^{r-1}$. The vector $e \in S^{N-1}$ satisfies condition (13) iff it does not belong to the image $h_k(L^{2n})$. Since $2n < N - 1$, the set $h_k(L^{2n})$, by theorem 6, has measure zero in S^{N-1}. As $\bigcup_{i=1}^s U_i = M^n$, so $\mathrm{pr}_e M^n \to \mathbf{R}^{N-1}$ is an immersion at every point $x \in M^n$, provided $e \notin \bigcup_{i=1}^s h_i(L^{2n})$. Since the union of a finite number of sets of measure zero is a set of measure zero, $\mathrm{mes} \bigcup_{i=1}^s h_i(L^{2n}) = 0$, and, therefore, $S^{N-1} \backslash \bigcup_{i=1}^s h_i(L^{2n}) \neq \emptyset$.

Furthermore, the set $H = (f(K^{2n})) \cup \left(\bigcup_{i=1}^s h_i(L^{2n})\right)$ has measure zero in S^{N-1}, therefore $S^{N-1} \backslash H \neq \emptyset$. For any vector $e \in S^{N-1} \backslash H$ the projection $\mathrm{pr}_e : M^n \to \mathbf{R}^{N-1}$ is an injective C^r–immersion. Since M^n is compact, the projection pr_e, according to theorem 7, is a C^r–imbedding.

We have shown that if $N > 2n + 1$, then there exists a C^r–imbedding $\mathrm{pr}_e : M^n \to \mathbf{R}^{N-1}$. If, furthermore, $N - 1 > 2n + 1$, then there exists a projection realizing a C^r–imbedding of the image $\mathrm{pr}_e(M^n)$ in $\mathbf{R}^{N-2}$. The process of imbedding in a space of lower dimension can be continued up to the point when the dimension of the space is equal to $2n + 1$. The composition of these projections is the C^r–imbedding of M^n in $\mathbf{R}^{2n+1}$. $\quad\Box$

Remark 3. Under certain restrictions on manifolds and their dimensions, theorem 9 can be strengthened. In particular, when $n \geq 1$, the C^r–manifold M^n, $r \geq 1$, admits a C^r–imbedding in $\mathbf{R}^{2n}$ (Whitney) and even in $\mathbf{R}^{2n-1}$, if M^n is non–compact (Hirsch) or $n \neq 2^k$.

Theorem 9 can be formulated differently: *any C^r–manifold M^n is C^r–diffeomorphic to a submanifold of the space $\mathbf{R}^{2n+1}$.*

Since we agreed not to distinguish diffeomorphic manifolds, it follows from theorem 9 that the abstract concept of submanifold is not more general than the concept of a submanifold in Euclidean spaces, and we could limit ourselves by considering only them. However, this is not always appropriate. Many problems on manifolds are much simpler to solve without using imbeddings.

5. The degree modulo 2 of a mapping. Let $f : M^n \to N^n$ be a C^r–mapping of C^r–manifolds, $r \geq 1$. Let, in addition, M^n be compact. If $y \in N^n$ is a regular value of the mapping f, then $f^{-1}(y)$ is a submanifold in M^n of dimension zero or is empty. The submanifold $f^{-1}(y)$ is compact as a closed subset of the compact space M^n, and, consequently, it consists of a finite number of points $n(f^{-1}(y))$ (if $f^{-1}(y) = \emptyset$, then $n(f^{-1}(y) = 0)$. We show that if N^n is connected and $r \geq 2$, then the residue class mod 2 of the number $n(f^{-1}(y))$ does not depend on the choice of the regular value $y \in N^n$ of the mapping f. This residue class is called the *degree of the mapping f* mod 2 and denoted by $\deg_2 f$.

First, we introduce some important ideas.

Definition 11. Two C^r–mappings $f, g : M^n \to N^m$ of C^r–manifolds, $r \geq 0$, are called *C^r–homotopic* if there exists a C^r–mapping $F(x, t) : M^n \times [0, 1] \to N^m$ such that $F(x, 0) = f(x)$, $F(x, 1) = g(x)$ for all $x \in M^n$.

Exercise 20°. Verify that the product $M^n \times [0, 1]$, where M^n is a C^r–manifold, is a C^r–manifold with boundary, moreover, its boundary consists of two replicas $M^n \times 0$ and $M^n \times 1$ of the manifold M^n.

Exercise 21°. Show that a C^r–homotopy is an equivalence relation on the set $C^r(M^n, N^m)$ of all C^r–mappings of the C^r–manifold M^n to N^m.

Definition 12. Two C^r–diffeomorphisms $f, g : M^n \to N^n$ of C^r–manifolds, $r \geq 0$, are said to be C^r–*isotopic* if there exists a C^r–homotopy $F(x,t)$: $M^n \times [0 \times 1] \to N^n$ between f and g such that for each fixed $t \in [0,1]$ the mapping $F(x,t)$ is a C^r–diffeomorphism.

To start, we now prove some auxiliary statements (lemmas 2–5).

Lemma 2. *Let* y, z *be arbitrary diametrically opposite points of the circle* $S^1_{r/2}(0) \subset \mathbf{R}^2 \subset \mathbf{R}^n$, $n \geq 2$. *Then there exists a* C^∞–*diffeomorphism* h : $\mathbf{R}^n \to \mathbf{R}^n$, C^∞–*isotopic to the identity mapping* I, *such that* $h(y) = z$ *and* $h(x) = x$, *if* $x \in \mathbf{R}^n \backslash D^n_r(0)$.

Proof. By lemma 2, § 4, there exists a C^∞–function $g_r : \mathbf{R}^n \to [0,1]$ such that

$$g_r(x) = \begin{cases} 1, & \text{if } x \in \bar{D}^n_{r/2}(0), \\ 0, & \text{if } x \in \mathbf{R}^n \backslash D^n_r(0), \end{cases}$$

moreover, $g_r(x)$ is constant on the spheres $S^{n-1}_\rho(0)$. It is not difficult to see that the mapping $h : \mathbf{R}^n \to \mathbf{R}^n$ given by

$$h(x) = \left(\begin{array}{cc|ccc} \cos(\pi g_r(x)) & -\sin(\pi g_r(x)) & & 0 & \\ \sin(\pi g_r(x)) & \cos(\pi g_r(x)) & & & \\ \hline & & 1 & & 0 \\ & 0 & & \ddots & \\ & & 0 & & 1 \end{array} \right) \left(\begin{array}{c} x_1 \\ \vdots \\ x_n \end{array} \right),$$

is the required C^∞–diffeomorphism, and the C^∞–homotopy $F(x,t)$: $\mathbf{R}^n \times [0,1] \to \mathbf{R}^n$,

$$F(x,t) = \left(\begin{array}{cc|ccc} \cos(t\pi g_r(x)) & -\sin(t\pi g_r(x)) & & 0 & \\ \sin(t\pi g_r(x)) & \cos(t\pi g_r(x)) & & & \\ \hline & & 1 & & 0 \\ & 0 & & \ddots & \\ & & 0 & & 1 \end{array} \right) \left(\begin{array}{c} x_1 \\ \vdots \\ x_n \end{array} \right),$$

isotopically connects I and h (the mapping $F(x,t)$ for each fixed $t \in [0,1]$ rotates the disc $\bar{D}^n_{r/2}(0)$ over an angle πt parallel to the plane $\mathbf{R}^2$, and the spheres $S^{n-1}_\rho(0)$, $r/2 \leq \rho \leq r$ by the angle $t\pi g_r(x)$). $\qquad \square$

Lemma 3. *Let y, z be arbitrary points of $\mathbf{R}^n$, $n \geq 2$. Then there exists a C^∞–diffeomorphism $\widetilde{h} : \mathbf{R}^n \to \mathbf{R}^n$, C^∞–isotopic to the identity, such that $\widetilde{h}(y) = z$, and $\widetilde{h}(x) = x$ outside a certain disc $D^n_\nu(0)$.*

Proof. Let $\|y - z\| = r$. The transformation $A(x) = x - (y + z)/2$ transforms the points y, z into diametrically opposite points of the sphere $S^{n-1}_{r/2}(0)$. Let $B : \mathbf{R}^n \to \mathbf{R}^n$ be some rotation transforming the vector $(y - z)/2$ into the subspace $\mathbf{R}^2$. Then $BA(y)$, $BA(z)$ are diametrically opposite points of the circle $S^1_{r/2}(0)$. According to lemma 2, there exists a C^∞–diffeomorphism $h : \mathbf{R}^n \to \mathbf{R}^n$, C^∞–isotopic to the identity, such that $h(BA(y)) = BA(z)$ and $h(x) = x$, if $x \in \mathbf{R}^n \backslash D^n_r(0)$. Then the mapping $\widetilde{h} = (BA)^{-1}h(BA) : \mathbf{R}^n \to \mathbf{R}^n$ is the required diffeomeorphism, and if $H(x,t) : \mathbf{R}^n \times [0,1] \to \mathbf{R}^n$ is a C^∞–isotopy connecting I and h, then $\widetilde{H} = (BA)^{-1}H(BA) : \mathbf{R}^n \times [0,1] \to \mathbf{R}^n$ is a C^∞–isotopy connecting I and $\widetilde{h}$. $\square$

Lemma 4. *Let y, z be arbitrary points of a connected C^r–manifold N^n, $n > 0$, $r \geq 0$. Then there exists a C^r–diffeomorphism $d : N^n \to N^n$, C^r–isotopic to the identity, such that $d(y) = z$.*

Proof. First, assume that $n \geq 2$, and y, z belong to the set U of some chart (U, ϕ) of the atlas of the manifold N^n. According to lemma 3, there exists a C^∞–diffeomorphism $\widetilde{h} : \mathbf{R}^n \to \mathbf{R}^n$, C^∞–isotopic to the identity, such that $\widetilde{h}(\phi^{-1}(y) = \phi^{-1}(z)$, and $\widetilde{h}(x) = x$ outside some disc $D^n_\nu(0)$. Then, as it is easy to see, the mapping $d : N^n \to N^n$,

$$d(x) = \begin{cases} \phi\widetilde{h}\phi^{-1}(x), & \text{if } x \in U, \\ x, & \text{if } x \in N^n \backslash U, \end{cases}$$

is the required C^r–diffeomorphism, and if $\widetilde{H}(x,t) : \mathbf{R}^n \times [0,1] \to \mathbf{R}^n$ is a C^∞–isotopy connecting I and $\widetilde{h}$, then $F(x,t) : N^n \times [0,1] \to N^n$,

$$F(x,t) = \begin{cases} \phi\widetilde{H}\phi^{-1}, & \text{if } x \in U, \\ x, & \text{if } x \in N^n \backslash U, \end{cases}$$

is a C^r–isotopy connecting I and d.

Now, assume that y, z do not belong to the same chart. Then by the connectedness of N^n, there exist charts (U_i, ϕ_i), $i = 1, \ldots, k$ in N^n such that $y \in U_1$, $z \in U_k$ and $U_i \cap U_{i+1} \neq \emptyset$, $i = 1, \ldots, k - 1$. In each of the sets $U_i \cap U_{i+1}$, $i = 1, \ldots, k - 1$, select an arbitrary point u_i. Then the pairs of points $(y_1, u_1), (u_1, u_2), \ldots, (u_{k-2}, u_{k-2}), (u_{k-1}, z)$ belong to the sets $U_1, \ldots, U_k$, respectively, and, consequently, there exist C^r–diffeomorphisms $d_i : N^n \to N^n$, $i = 1, \ldots, k$, C^r–isotopic to the identity, such that $d_1(y) = u_1$, $d_2(u_1) = u_2, \ldots, d_{k-1}(u_{k-2}) = u_{k-1}, d_k(u_{k-1}) = z$. The diffeomorphism $d = d_k d_{k-1} \ldots d_1 : N^n \to N^n$, clearly, does what is required.

Thus, the lemma is proved for the case of $n \geq 2$. The proof for $n = 1$ is left to the reader. $\square$

Lemma 5. *Let $f : M^{m+1} \to N^n$ be a C^r–mapping of a compact C^r–manifold with boundary M^{n+1} into a C^r–manifold N^n, $r \geq 1$. If y_0 is a regular value of the mapping f, then $M_1 = f^{-1}(y_0)$ is either empty or a one–dimensional compact C^r–manifold with boundary; moreover, $\partial M_1 = M_1 \cap \partial M^{n+1}$.*

Proof. The compactness of M_1 follows from the fact that M_1 is closed (as the inverse image of a closed set under a continuous mapping) in the compact space M^{n+1}. We prove the main statement of the lemma. Since $L = M^{n+1} \backslash \partial M^{n+1}$ is an $(n + 1)$–dimensional C^r–manifold, then by theorem 8 for the mapping $f|_{L^{n+1}} : L^{n+1} \to N^n$, the set $(f|_{L^{n+1}})^{-1}(y_0) = M_1 \cap L^{n+1}$ is either empty or a one–dimensional submanifold in L^{n+1}. Therefore, if $M_1 \cap \partial M^{n+1} = \emptyset$, then the statement of lemma is proved (in this case, $\partial M_1 = \emptyset$). Consider the case $M_1 \cap \partial M^{n+1} \neq \emptyset$. For the points of the set $M_1 \cap \partial M^{n+1}$, we construct charts which are C^r–compatible with each other and with the charts of the manifold $M_1 \cap L^{n+1}$. According to the condition, y_0 is a regular value of the mapping $f|_{\partial M^{n+1}} : \partial M^{n+1} \to N^n$; in addition, ∂M^{n+1} is compact. Therefore, as it is shown at the beginning of this section, the set $(f|_{\partial M^{n+1}})^{-1}(y_0) = M_1 \cap \partial M^{n+1}$ is finite. Let x_0 be an arbitrary point from $M_1 \cap \partial M^{n+1}$, and (U, ϕ), (V, ψ) charts in M^{n+1} and N^n such that $x_0 \in U$, $y_0 \in V$ and $f(U) \subset V$. Moreover, choose U so that x_0 is the unique point of the set $M_1 \cap \partial M^{n+1}$ in U (this is possible, since the set $M_1 \cap \partial M^{n+1}$ is finite). According to the condition, $z_0 = \phi^{-1}(x)$ is a regular point of the mapping $\Phi = \psi^{-1} f \phi : \mathbf{R}_+^{n+1} \to \mathbf{R}^n$, i.e., in $\mathbf{R}^{n+1}$, there exists an open neighbourhood W_1 of the point z_0 and a C^r–mapping $\widetilde{\Phi} : W_1 \to \mathbf{R}^n$ coinciding with Φ on the set $W_1 \cap \mathbf{R}_+^{n+1}$ and such that z_0 is a regular point of the mappings $\widetilde{\Phi}$ and $\widetilde{\Phi}|_{W_1 \cap \mathbf{R}^n}$. According to the theorem on rectification of a mapping (see § 1), there exist an open neighbourhood $W_2 \subset W_1$ of the point z_0, an open set $W \subset \mathbf{R}^{n+1}$ and a C^r–diffeomorphism $F : W_2 \to W$ such that $\widetilde{\Phi} F^{-1}$, on the set W, is the standard projection of

$\mathbf{R}^{n+1}$ onto $\mathbf{R}^n$. Let $\widetilde{W} = D_\epsilon^{n+1}(u_0)$ be a suitable open disc with its centre at the point $u_0 = F(z_0)$ which is contained in W. Denote $W_2 = F^{-1}(\widetilde{W})$, $\widetilde{U} = \phi(\widetilde{W}_2 \cap \mathbf{R}_+^{n+1})$. We now consider the restrictions of the mappings F, $\widetilde{\Phi}$, ϕ to the sets $\widetilde{W}_2$, $\widetilde{W}_2$, $\widetilde{W}_2 \cap \mathbf{R}_+^{n+1}$, respectively, and keep the former notations F, $\widetilde{\Phi}$, ϕ for them. Since $\widetilde{\Phi} F^{-1}$, on the set $\widetilde{W}$, is the standard projection of $\mathbf{R}^{n+1}$ on $\mathbf{R}^n$, the preimage $\psi^{-1}(y_0)$ of the point y_0 under the mapping $\widetilde{\Phi} F^{-1}$ is the interval I obtained by intersecting the disc $\widetilde{W}$ with the line which passes through the point u_0 and is parallel to the vector $(0, \dots, 0, 1) \in \mathbf{R}^{n+1}$. The point u_0 divides the interval I into two semi–intervals I_1, I_2 with common end u_0. The set $F\phi^{-1}(\widetilde{U} \cap M_1)$, clearly, is contained in I; we show that it coincides with one of the semi–intervals I_1, I_2. To see this, we shall prove the equivalent statement that the set $\phi^{-1}(\widetilde{U} \cap M_1)$ coincides with one of the sets $F^{-1}(I_1)$, $F^{-1}(I_2)$. Since $\phi^{-1}(\widetilde{U} \cap M_1) \subset \mathbf{R}_+^{n+1} \cap F^{-1}(I)$, it is sufficient to show that the sets $F^{-1}(I_1)$, $F^{-1}(I_2)$ are contained in the two different semispaces $\mathbf{R}_+^{n+1}$, $\mathbf{R}_-^{n+1}$. First, let us show that each of the sets $F^{-1}(I_1)$, $F^{-1}(I_2)$ is contained in one of the semispaces $\mathbf{R}_+^{n+1}$, $\mathbf{R}_-^{n+1}$. By the choice of the neighbourhood U, we have $\widetilde{U} \cap M_1 \cap \partial M^{n+1} = \{x_0\}$. Consequently, $\widetilde{\Phi}^{-1}(\psi^{-1}(y_0)) = z_0 \in \mathbf{R}^n$. But $\widetilde{\Phi}^{-1}(\psi^{-1}(y_0)) = F^{-1}(u_0)$, thus, the intersection of each of the sets $F^{-1}(I_1 \backslash u_0)$, $F^{-1}(I_2 \backslash u_0)$ with the space $\mathbf{R}^n$ is empty. Therefore, if the sets $F^{-1}(I_1 \backslash u_0)$, $F^{-1}(I_2 \backslash u_0)$ had a nonempty intersection with each of the open semispaces $\mathbf{R}_+^{n+1} \backslash \mathbf{R}^n$, $\mathbf{R}_-^{n+1} \backslash \mathbf{R}^n$, then they would not be connected, contrary to the fact that the images of connected sets I_1, I_2 under the continuous mapping F^{-1} are connected (see Ch. 2, § 10). Now, we show that the sets $F^{-1}(I_1)$, $F^{-1}(I_2)$ are contained in different semispaces $\mathbf{R}_+^{n+1}$, $\mathbf{R}_-^{n+1}$. Consider $F^{-1}(I)$ as a smooth curve given by the diffeomorphism $F^{-1} : I \to \mathbf{R}^{n+1}$; denote it by $z = z(t)$, $t = u_{n+1}$, $t_0 = u_0$. Since $\widetilde{\Phi}^{-1}(\psi^{-1}(y_0)) = F^{-1}(I)$, we have $\widetilde{\Phi}(z(t)) \equiv \psi^{-1}(y_0)$ for $t \in I$. By differentiating the latter equality, we obtain $(D_{z_0}\widetilde{\Phi})\left(\frac{dz}{dt}\right)\big|_{t_0} = 0$, i.e. the vector $a = \left(\frac{dz}{dt}\right)\big|_{t_0}$ belongs to the kernel $\ker D_{z_0}\widetilde{\Phi}$. Since $a = \left(\frac{\partial F_1^{-1}}{\partial u_{n+1}}\big|_{u_0}, \dots, \frac{\partial F_{n+1}^{-1}}{\partial u_{n+1}}\big|_{u_0}\right)$ and the diffeomorphism F^{-1} has no singular points, we get $a \neq 0$. Obviously, $a \notin \mathbf{R}^n$, since, otherwise, $a \in \ker(D_{z_0}\widetilde{\Phi})|_{\mathbf{R}^n} = \ker D_{z_0}(\widetilde{\Phi}|_{\widetilde{W}_2 \cap \mathbf{R}^n})$, contrary to the fact that z_0 is a regular point of the mapping $\widetilde{\Phi}|_{\widetilde{W}_2 \cap \mathbf{R}^n}$. Thus, the tangent vector a for the curve $z = z(t)$ at the point z_0 is directed to one of semispaces $\mathbf{R}_+^{n+1}$, $\mathbf{R}_-^{n+1}$; therefore, the curve intersects $\mathbf{R}^n$ at the point z_0 passing from one semispace into other. Consequently, the set $F^{-1}(I) = F^{-1}(I_1) \cup F^{-1}(I_2)$ has a nonempty intersection with each of open semispaces $\mathbf{R}_+^{n+1}\backslash\mathbf{R}^n$, $\mathbf{R}_-^{n+1}\backslash\mathbf{R}^n$. Hence, taking into account, that each of the sets $F^{-1}(I_1)$, $F^{-1}(I_2)$ is contained in one of the semispaces $\mathbf{R}_+^{n+1}$, $\mathbf{R}_-^{n+1}$, we get that the sets $F^{-1}(I_1)$, $F^{-1}(I_2)$ are contained in different semispaces $\mathbf{R}_+^{n+1}$, $\mathbf{R}_-^{n+1}$.

Thus, the set $F\phi^{-1}(\widetilde{U} \cap M_1)$ coincides with I_1 or I_2. Let, for instance, $F\phi^{-1}(\widetilde{U} \cap M_1) = I_1$. So if $g : \mathbf{R}_+^1 \to I_1$ is some C^r–diffeomorphism, then $\phi F^{-1}g : \mathbf{R}_+^1 \to \widetilde{U} \cap M_1$ is a homeomorphism, and $\phi F^{-1}g(0) = x_0$; therefore, $(\widetilde{U} \cap M_1, \phi F^{-1}g)$ is a chart of the point x_0 on M_1. The charts constructed in this way for different points of the set $M_1 \cap \partial M^{n+1}$ are C^r–compatible with each other and with the charts of the submanifold $M_1 \cap L$, because their homeomorphisms are obtained from homeomorphisms of charts of the C^r–atlas on M^{n+1} by composition with C^r–diffeomorphisms. Thus, M_1 is a one–dimensional C^r–manifold with boundary, and $\partial M_1 = M_1 \cap \partial M^{n+1}$. $\square$

We now prove that the degree modulo 2 of a mapping does not depend on the choice of a regular value.

Lemma 6. *Let $f, g : M^n \to N^n$ be C^2–homotopic C^2–mappings of C^2–manifolds; moreover, let M^n be compact. If $y \in N^n$ is a regular value for f and g, then $n(f^{-1}(y)) \equiv n(g^{-1}(y)) \mod 2$.*

Proof. Let $F : M^n \times [0, 1] \to N^n$ be a C^2–homotopy connecting f and g. First, assume that y also is a regular value for F. If $f^{-1}(y) \cup g^{-1}(y) = \emptyset$, then the statement of lemma is evident, since $n(f^{-1}(y)) = n(g^{-1}(y)) = 0$. Consider the case of a nonempty set $f^{-1}(y) \cup g^{-1}(y)$. By lemma 5, the set $M_1 = F^{-1}(y)$ is a one–dimensional compact manifold with boundary; moreover, $\partial M_1 = M_1 \cap ((M^n \times 0) \cup (M^n \times 1)) = (f^{-1}(y) \times 0) \cup (g^{-1}(y) \times 1)$. Then the connected components of M_1 are compact (as closed sets of a compact space) and, consequently, they are homeomorphic to S^1 or $\bar{D}^1$ (see subsection 2). Thus, the boundary of each of the connected components of M_1 consists of an even number of points. Since the boundary $\partial M_1 = (f^{-1}(y) \times 0) \cup (g^{-1}(y) \times 1)$ is the union of the boundaries of the connected components, and the sets $f^{-1}(y)$, $g^{-1}(y)$ are finite (see the beginning of this section), the number $n(f^{-1}(y)) + n(g^{-1}(y))$ of points of the boundary ∂M_1 is even, therefore $n(f^{-1}(y)) \equiv n(g^{-1}(y)) \mod 2$.

Now, assume that y is not a regular value of the mapping F. We show that there exists an open neighbourhood of the point y and that for each of its points y' the equalities $n(f^{-1}(y')) = n(f^{-1}(y))$, $n(g^{-1}(y')) = n(g^{-1}(y))$ are valid. Let, for instance, $f^{-1}(y) = \{x_1, \ldots, x_k\}$. According to the inverse mapping theorem, there exist open neighbourhoods $U(x_i)$, $U_i(y)$, $i = 1, \ldots, k$ of the points $x_1, \ldots, x_k, y$ such that the mappings $f|_{U(x_i)} : U(x_i) \to U_i(y)$, $i = 1, \ldots, k$, are C^2–diffeomorphisms. Without loss of generality, one may assume that the neighbourhoods $U(x_1), \ldots, U(x_k)$ do not intersect. The set $M^n \backslash \bigcup_{i=1}^k U(x_i)$ is closed in M^n, and, thus, it is compact; consequently, its image $f\left(M^n \backslash \bigcup_{i=1}^k U(x_i)\right)$, under the continous mapping f, is compact in N^n, and therefore, it is closed in the Hausdorff space N^n. Then $U(y) = \bigcap_{i=1}^k U_i(y) \backslash f\left(M^n \backslash \bigcup_{i=1}^k U(x_i)\right)$ is an open neighbourhood of the point y, and $f^{-1}(y') \subset \bigcup_{i=1}^k U(x_i)$ for any point $y' \in U(y)$; consequently, $n(f^{-1}(y')) = n(f^{-1}(y))$. Similarly, there exists an open neighbourhood $V(y)$ of the point y such that for each of its points y' the equality $n(g^{-1}(y')) = n(g^{-1}(y))$ is valid. The neighbourhood $W(y) = U(y) \cap V(y)$ is, evidently, what is required. Note that all points of the neighbourhood $W(y)$ are regular points of the mappings f, g. Furthermore, the set of non–regular values of the mapping F coincides with the union of the sets of non–regular values of the mappings $F|_{(M^n \times [0,1]) \backslash \partial M^n \times [0,1])}$, $F|_{\partial(M^n \times [0,1])}$. Each of the latter sets, by theorem 6, has measure zero. Therefore, in the neighbourhood $W(y)$, a regular value y_0 of the mapping F exists. Thus, y_0 is a regular value of the mappings f, g, F and, as it is shown above, $n(f^{-1}(y_0)) \equiv n(g^{-1}(y_0))$ mod 2. But $n(f^{-1}(y_0)) = n(f^{-1}(y))$, $n(g^{-1}(y_0)) = n(g^{-1}(y))$, so $n(f^{-1}(y)) \equiv n(g^{-1}(y))$ mod 2.   □

Theorem 11. *Let* $f : M^n \to N^n$ *be a* C^2–*mapping of* C^2–*manifolds; moreover, let* M^n *be compact, and* N^n *connected. If* y, z *are regular values of the mapping* f, *then* $n(f^{-1}(y)) \equiv n(f^{-1}(z))$ mod 2.

Proof. By lemma 4, there exists a C^2–diffeomorphism $d : N^n \to N^n$ C^2–isotopic to the identity, such that $d(y) = z$. Then z is a regular value of the C^2–mapping $df : M^n \to N^n$, and df is C^2–homotopic to f. According to lemma 6, $n(f^{-1}(z)) \equiv n((df)^{-1}(z))$ mod 2. But $(df)^{-1}(z) = f^{-1}d^{-1}(z) = f^{-1}(y)$, therefore $n(f^{-1}(y)) \equiv n(f^{-1}(z))$ mod 2.   □

Thus, the degree modulo 2 of a mapping does not depend on the choice of a regular value.

Theorem 12. *Let $f, g : M^n \to N^n$ be C^2–homotopic C^2–mappings of C^2–manifolds; moreover, let M^n be compact and N^n connected. Then $\deg_2 f = \deg_2 g$.*

Proof. By theorem 6, the sets of non–regular values of the mappings f, g have measure zero, therefore, one may select a regular value y common to both f and g. In virtue of lemma 6, $n(f^{-1}(y)) \equiv n(g^{-1}(y)) \mod 2$, whence $\deg_2 f = \deg_2 g$. □

Example 9. If M^n is a compact, connected C^2–manifold, and $f : M^n \to M^n$ is a constant mapping, i.e., $f(x) = x_0$ for all $x \in M^n$, then for $n > 0$, $\deg_2 f = 0$, and for $n = 0$, $\deg_2 f = 1$.

Example 10. If M^n is a compact, connected C^2–manifold, $n \geq 0$, and $I : M^n \to M^n$ the identity mapping, then $\deg_2 I = 1$.

Remark 4. From exercises 9, 10 and theorem 12 it follows that the identity mapping of a compact, connected C^2–manifold M^n is not C^2–homotopic to a constant mapping for $n > 0$.

Remark 5. In the foregoing constructions above of the degree modulo 2 of a mapping, we used the compactness of the manifold M^n to provide, in particular, the compactness of the preimages of one–point sets under the mappings involved and under homotopies. In applications, it is often useful to eliminate compactness of M^n. To this end, we consider a class of mappings which possesses a priori the property of compactness for preimages.

Definition 13. A mapping $f : X \to Y$ of topological spaces is called *proper*, if the preimage $f^{-1}(A)$ is compact for any compact set $A \subset Y$.

It is easy to see that in the class of proper mappings of manifolds and proper homotopies, lemma 6, and theorems 11 and 12 are valid without the condition of compactness of the manifold M^n. Thus, for proper C^r–mappings of a C^r–manifold M^n into a connected C^r–manifold N^n, $r \geq 2$, the degree modulo 2 is well defined.

As an application of the degree modulo 2 of a mapping, we shall prove the classical Brouwer fixed point theorem.

Lemma 7. *There is no C^2–mapping $f : \bar{D}^n \to S^{n-1}$ under which every point of the sphere S^{n-1} is fixed (i.e., the sphere S^{n-1} is not a "smooth retract" of the ball $\bar{D}^n$).*

Proof. Assume the opposite; let $f : \bar{D}^n \to S^{n-1}$ be a C^2–mapping, and $f(x) = x$ for any point $x \in S^{n-1}$. Then the mapping $F : S^{n-1} \times [0, 1] \to S^{n-1}$ determined by the equality $F(x, t) = f(tx)$, is a C^2–homotopy connecting the identity mapping of the sphere with the constant mapping of the sphere into the point $f(0) \in S^{n-1}$, what is impossible for $n > 1$, by Remark 4. In the case of $n = 1$, the set $f(\bar{D}^1) = S^0$ is not connected, which contradicts the connectedness of an image under a constant mapping of topological spaces. $\square$

The non–existence of a retraction $\bar{D}^n \to S^{n-1}$ gives a fixed point principle (see Ch 3, § 4).

Lemma 8. *Any C^2–mapping $f : \bar{D}^n \to \bar{D}^n$ has a fixed point $x_* \in \bar{D}^n$, $f(x_*) = x_*$.*

Proof. Assume that some C^2–mapping $f : \bar{D}^n \to \bar{D}^n$ has no fixed points. Consider the mapping $g : \bar{D}^n \to \bar{D}^n$ which associates to each point $x \in \bar{D}^n$ the point $g(x)$ of intersection of the sphere S^{n-1} with the half line $tx + (1 - t)f(x)$, $t \geq 0$. The mapping g has the following analytic form:

$$g(x) = x + u(\sqrt{1 - (x, x) + (x, u)^2} - (x, u)),$$

where

$$u = (x - f(x))/\|x - f(x)\|, \quad (y, z) = \sum_{i=1}^{n} y_i z_i, \quad \|y\| = (y, y)^{1/2}.$$

Evidently, $g \in C^2$ and $g(x) = x$, if $x \in S^{n-1}$, which contradicts lemma 7. $\square$

Lemma 8 can be easily extended to the case of continuous mappings by applying the standard procedure of approximating continuous mappings by smooth ones.

Theorem 13. (Brouwer). *Any continuous mapping $f : \bar{D}^n \to \bar{D}^n$ has a fixed point.*

Proof. Assume the opposite, i.e., $f(x) \neq x$ for all $x \in \bar{D}^n$. Then the continuous function $\|f(x) - x\|$ attains its minimum $\mu > 0$ on the compact set $\bar{D}^n$. According to the Weierstrass approximation theorem, there exists a polynomial mapping $P_1 : \mathbf{R}^n \to \mathbf{R}^n$ such that $\|P_1(x) - f(x)\| < \mu/2$ for $x \in \bar{D}^n$. The mapping P_1, however, can take points from $\bar{D}^n$ into $\mathbf{R}^n \setminus \bar{D}_1^n$. We shall modify P_1 a little, and instead of P_1 consider the approximating mapping $P = P_1/(1 + \mu/2)$ that maps $\bar{D}^n$ into $\bar{D}^n$, and is such that $\|P(x) - f(x)\| < \mu/(1 + \mu/2)$, if $x \in \bar{D}^n$. Indeed, the inequality $\|P(x)\| \leq 1$, for $x \in \bar{D}^n$, follows from the inequality $\|P(x)\| = \|P_1(x) - f(x) + f(x)\|/(1 + \mu/2) \leq (\|P_1(x) - f(x)\| + \|f(x)\|)/(1 + \mu/2)$ and from the inequalities $\|f(x)\| \leq 1$, $\|P_1(x) - f(x)\| < \mu/2$ for $x \in \bar{D}^n$. The inequality $\|P(x) - f(x)\| < \mu/(1 + \mu/2)$ for $x \in \bar{D}^n$ follows from the inequality $\|P(x) - f(x)\| = \|P_1(x) - f(x) - \frac{\mu}{2}f(x)\|/(1 + \mu/2) \leq (\|P_1(x) - f(x)\| + \frac{\mu}{2}\|f(x)\|)/(1 + \mu/2)$ and from the inequalities $\|f(x)\| \leq 1$, $\|P_1(x) - f(x)\| < \mu/2$ for $x \in \bar{D}^n$. Then $\|P(x) - x\| = \|f(x) - x - f(x) + P(x)\| \geq \|f(x) - x\| - \|P(x) - f(x)\| > \mu - \mu/(1 + \mu/2) = \mu^2/(2 + \mu) > 0$ for $x \in \bar{D}^n$, and, thus, the C^∞–mapping P on $\bar{D}^n$ has no fixed points, contradicting lemma 8. $\square$

We shall indicate one more simple but useful application of the degree modulo 2 of a mapping.

Theorem 14. *Let* $f : M^n \to N^n$ *be a* C^2*-mapping of* C^2*-manifolds; moreover, let* M^n *be compact, and* N^n *connected. If* $\deg_2 f \neq 0$, *then the mapping* f *is surjective, i.e.* $f(M^n) = N^n$.

Proof. Assume the opposite. Let $y \in N^n$ and $y \notin f(M^n)$, i.e. $f^{-1}(y) = \emptyset$. Then y is a regular value of the mapping f and $n(f^{-1}(y)) = 0$; therefore, $\deg_2 f = 0$, contradicting an assumption in the theorem. $\square$

§ 6. Tangent bundle and tangent map

1. The idea of a tangent space. For further development of the analysis of smooth mappings it is necessary to construct an analogue of the differential of a function as it is widely used in analysis. The tangent map defined and studied in this section is the required generalization. First of all, the necessity to generalize the concept of tangent to a curve (or tangent plane to a surface) arises. Such a generalization is needed because of applications of manifolds in mechanics and physics. As it was mentioned in § 3, the configuration space of a mechanical system is, as a rule, a smooth manifold. Each point of this manifold represents a certain position of a mechanical system. Under the action of forces, the mechanical system changes its position. The point of the configuration space corresponding to it moves tracing out a certain trajectory, i.e. a path on a smooth manifold. An important characteristic of this motion is its velocity which, generally speaking, changes with time. The *state of a mechanical system* at each given moment is the pair (x, v), where x is the point of the manifold corresponding to the position of the system at the moment under consideration, and v is the velocity of change of the point x. The collection of all states of a mechanical system is called its *phase space*.

Naturally, the question arises, what mathematical concept can be associated with the physical concept of velocity and, moreover, how to describe the concept of phase space with mathematical precision. The solution of this question is suggested by the simplest physical examples. Thus, when a point mass moves along a curve, its velocity is interpreted as a certain vector which is tangent to that curve and directed along the path of motion. If a point mass moves across a two–dimensional surface, then its velocity is interpreted as a certain vector which is tangent to the given surface and the trajectory itself. The set of all possible velocities at a given point of a curve (surface) is, thus, a tangent line (tangent plane). The set of all velocities admissable at a given position of a mechanical system can be naturally interpreted in a similar way, i.e., as the tangent vector space at the corresponding point of the smooth manifold which is its configuration space.

2. The concept of tangent space to a manifold. Before defining this concept precisely, it should be noted that we shall now pay attention to whether n–dimensional space is to be considered as a metric space (under the Euclidean metric), or it is provided with the additional structure of a vector space. In the first case, it is said that the elements of $\mathbf{R}^n$ are points, and in the latter case, they are called vectors; we also call $\mathbf{R}^n$ a vector space (earlier, we did not distinguish these concepts). Thus, for instance, considering the derivative $D_{x_0}(f) : \mathbf{R}^n \to \mathbf{R}^m$ of a mapping $f : \mathbf{R}^n \to \mathbf{R}^m$ at a point x_0 (see Ch. 4, § 1), it should be noted that this derivative realizes a linear mapping of vector spaces. Let $\mathbf{R}^N$ be a subspace of $\mathbf{R}^N$. A pair $(x, v) \in \mathbf{R}^n \times \mathbf{R}^n$, where x is a point, and v is a vector, is called a *vector v at the point x* (a vector "attached to" the point x (or "marked off"), a vector with "origin" at the *point* x). Let x be a fixed (arbitrary) point in $\mathbf{R}^N$. The collection of all vectors $v \in \mathbf{R}^n$ "attached to" the point x is called the vextor space $\mathbf{R}^n$ at x; this collection possesses a natural structure of an n–dimensional vector space; it will be denoted by $\mathbf{R}_x^n$.

Now, consider a smooth manifold M^n in Euclidean space $\mathbf{R}^N$ (see § 2).

Definition 1. Let M^n be a C^r–submanifold in $\mathbf{R}^N$ ($r \geq 1$), $x \in M^n$ an arbitrary point. Let (U, ϕ) be a chart of M^n, $x \in U$. The *tangent space* $T_x M^n$ *to the manifold* M^n *at the point* x is the subspace of vectors at x, i.e. the image of the vector space $\mathbf{R}^n$ under the mapping $D_{\phi^{-1}(x)}\phi : \mathbf{R}^n \rightarrow \mathbf{R}^N$.

Recall that the linear mapping $D_{\phi^{-1}(x)}\phi$ is given by the Jacobian matrix $\left(\frac{\partial \phi}{\partial x}\right)\big|_{\phi^{-1}(x)}$. Since rank $\left(\frac{\partial \phi}{\partial x}\right)\big|_{\phi^{-1}(x)} = n$, the tangent space is of dimension n. We show that the definition of $T_x M^n$ does not depend on the choice of a chart. Let (V, ψ), $x \in V$, be another chart. The commutative diagram (on the left) generates the commutative diagram of linear mappings (on the right):

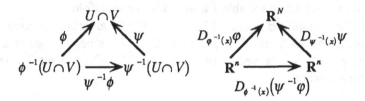

Since $\psi^{-1}\phi : \phi^{-1}(U \cap V) \rightarrow \psi^{-1}(U \cap V)$ is a diffeomorphism, $D_{\phi^{-1}(x)}(\psi^{-1}\phi) : \mathbf{R}^n \rightarrow \mathbf{R}^n$ is an isomorphism, and, consequently, denoting the image by Im, we have $\mathrm{Im}D_{\phi^{-1}(x)}(\psi^{-1}\phi) = \mathbf{R}^n$. Further, we obtain

$$\mathrm{Im}D_{\phi^{-1}(x)}\phi = \mathrm{Im}D_{\phi^{-1}(x)}[\psi(\psi^{-1}\phi)] =$$
$$= \mathrm{Im}\{[D_{\psi^{-1}(x)}\psi][D_{\phi^{-1}(x)}(\psi^{-1}\phi)]\} = \mathrm{Im}D_{\psi^{-1}(x)}\psi,$$

what proves the correctness of definition 1.

The elements of the space $T_x M^n$ are called the *tangent vectors to* M^n *at the point* x.

The tangent space at x for a manifold M^2 in $\mathbf{R}^3$ is a two–dimensional plane which passes through the point x and coincides with the tangent plane to the surface M^2 as it is usually considered in analysis.

Example 1. Let $U \subset \mathbf{R}^n$ be an open set considered as a submanifold in $\mathbf{R}^n$. Then for any point $x \in U$ we have $T_x U = \mathbf{R}_+^x$. $\Diamond$

We shall extend the concept of tangent space to arbitrary manifolds. In this case, generally speaking, it is impossible to talk about the derivative $D_{\phi^{-1}(x)}\phi$. However, from definition 1 of a tangent space, we can get another approach.

Let (U, ϕ) be a certain chart of the manifold M^n in $\mathbf{R}^N$, and $x \in U$ a point. The vector $(x, D_x \phi^{-1}(h))$ from $\mathbf{R}_x^n$ is called the *coordinate representation of the tangent vector* $(x, h) \in T_x M$ *in the chart* (U, ϕ). The question arises how the coordinate representations of a tangent vector h in various charts are related. Let (V, ψ) be another chart, $x \in V$. Differentiating the mapping $\phi^{-1} = (\phi^{-1}\psi)\psi^{-1}$, it is not difficult to see that the coordinate representations of the vector (x, h) in the charts (U, ϕ) and (V, ψ) are related by the equality:

$$(1) \qquad (x, D_x\phi^{-1}(h)) = (x, D_{\psi^{-1}(x)}(\phi^{-1}\psi)D_x\psi^{-1}(h)).$$

It is natural to identify a tangent vector with the set of all its coordinate representations. This remark can be used as a basis for a new definition of a tangent vector which is suitable for an arbitrary manifold.

Let M^n be a manifold of class C^r, $r \geq 1$. We fix an arbitrary point $x \in M^n$ and consider the set T of all triplets $(x, (U, \phi), h)$, where (U, ϕ) is a chart at the point x, and h is a vector of the space $\mathbf{R}^n$. On the set T, define the equivalence relation

$$(x, (U, \phi), h) \sim (x, (V, \psi), g) \Leftrightarrow h = D_{\psi^{-1}(x)}(\phi^{-1}\psi)(g).$$

Exercise 1°. Verify that this relation is an equivalence relation.

The equivalence class of $(x, (U, \phi), h)$ is called a *tangent vector* at the point x, and the triple $(x, (U, \phi), h)$ from the equivalence class the *representation of the tangent vector* in the chart (U, ϕ). Moreover, the vector h is called the *vector component of the representative* $(x, (U, \phi), h)$ and is denoted by h_ϕ.[*]

Consider the set of all tangent vectors at a point x. Denote it by $T_x M^n$. We fix a chart (U, ϕ), $x \in U$, and construct a mapping

$$(2) \qquad \tau_x : T_x M^n \to \mathbf{R}^n,$$

---

[*]Equality (1) indicates how the vector component varies under a change of chart.

by associating each tangent vector with the component h of its representative in the chart (U, ϕ). Clearly, τ_x is a bijection, and, consequently, the structure of the n–dimensional vector space $\mathbf{R}^n$ is naturally transferred to the set $T_x M^n$. More precisely, it means that in $T_x M^n$, the algebraic operations of addition and multiplication by a number are introduced in terms of the corresponding actions on the vector components of the representatives of the tangent vectors in the chosen chart (U, ϕ). If the representatives of the tangent vectors are defined in different charts, then they should first be replaced by equivalent representatives in the same chart. Thus, the algebraic operations on $T_x M^n$ are defined as follows:

(1) $\quad \{(x, (U, \phi), h)\} + \{(x, (V, \psi), g)\} =$

$$= \{((x, (U, \phi), h + D_{\psi^{-1}(x)}(\phi^{-1}\psi)(g)))\},$$

(2) $\quad \alpha\{x, (U, \phi), h)\} = \{(x, (U, \phi), \alpha h)\}.$

Exercise 2°. Prove that the algebraic operations given by (1) and (2) above are well–defined and verify the axioms of vector space.

Thus, to each point x of the manifold M^n there is associated a vector space which is called the *tangent space to M^n at the point x* and denoted by $T_x M^n$.

The dimension of a tangent space at each point is equal to n, i.e. it has the dimension of M^n. Indeed, this follows from the fact that the bijection τ_x above (with the given definition of algebraic operations) is an isomorphism of vector spaces.

We give another convenient definition of a tangent space. Let M^n be a smooth manifold, and $x \in M^n$ an arbitrary point. The *smooth curve χ in the manifold M^n* is a smooth mapping

$$\chi : (a, b) \longrightarrow M^n,$$

where (a, b) is a certain interval of the number axis considered as a manifold with the natural C^∞–structure.

Consider the set of smooth curves

$$\chi : (-a, a) \longrightarrow M^n, \quad \chi(0) = x.$$

Two such curves χ_1 and χ_2 are said to be *equivalent at the point* x if for any chart (U, ϕ) containing the point x, the curves $\phi^{-1}\chi_1$, $\phi^{-1}\chi_2$ in $\mathbf{R}^n$ possess the property

$$\frac{d}{dt}(\phi^{-1}\chi_1)(t)|_{t=0} = \frac{d}{dt}(\phi^{-1}\chi_2)(t)|_{t=0}.$$

Exercise 3°. Show that the definition of equivalence of the curves χ_1, χ_2 does not depend on the choice of a chart.

Exercise 4°. Show that equivalence of curves at a point is an equivalence relation on the set of curves on a manifold.

Definition 2. A *tangent vector to the manifold* M^n *at the point* x is an equivalence class of smooth curves passing through the point x.

Lemma 1. *The set of equivalence classes of smooth curves on a manifold* M^n *that pass through the point* x, *is an n–dimensional vector space.*

Indeed, by fixing the chart (U, ϕ), one may associate a class of equivalent curves at the point x with the n–dimensional vector $\alpha = \frac{d}{dt}(\phi^{-1}\chi)|_{t=0}$. Inversily, in the space $\mathbf{R}^n$, each vector α determines a line passing through the point $\phi^{-1}(x)$ with "direction coefficient" α, and the image of this line, under the mapping ϕ, will define a smooth curve χ in $\mathbf{R}^n$ passing through the point x such that $\alpha = \frac{d}{dt}(\phi^{-1}\chi)|_{t=0}$. Thus, we have a bijective correspondence between the equivalence classes of curves at a point x and vectors of the space $\mathbf{R}^n$.

Define the algebraic operations on the set of classes of curves which are equivalent at the point x in such a way that this bijection becomes an isomorphism of vector spaces:

(1) the *sum* $\{\chi_1\} + \{\chi_2\}$ of two classes is the class $\{\chi_3\}$ such that

$$\frac{d}{dt}(\phi^{-1}\chi_1)|_{t=0} + \frac{d}{dt}(\phi^{-1}\chi_2)|_{t=0} = \frac{d}{dt}(\phi^{-1}\chi_3)|_{t=0};$$

(2) the *product* $\lambda\{\chi\}$ of a class $\{\chi\}$ by a number λ is the class $\{\chi_\lambda\}$ such that

$$\frac{d}{dt}(\phi^{-1}\chi_\lambda)|_{t=0} = \lambda\frac{d}{dt}(\phi^{-1}\chi)|_{t=0}.$$

Exercise 5°. Show that the operations introduced above are well–defined, and verify the axioms of a vector space. □

The n–dimensional vector space of classes of curves equivalent at the point x on the manifold M^n, as is constructed above, is called the *tangent space to M^n at the point x*, and its elements are called *tangent vectors*. It is denoted by $T_x M^n$, as before. Note that for this definition of a tangent space, the isomorphism $\tau_x : T_x M^n \to \mathbf{R}^n$ corresponding to the chart (U, ϕ), $x \in U$, is given by the formula $\{\chi\} \to \frac{d}{dt}(\phi^{-1}\chi)|_{t=0}$.

It is natural to call the triple $\left(x, (U, \phi), \frac{d}{dt}(\phi^{-1}\chi)|_{t=0}\right)$ the *representative of the tangent vector $\{\chi\}$ in the chart (U, ϕ)*, and the vector $\frac{d}{dt}(\phi^{-1}\chi)|_{t=0}$ the *vector component of this representative*.

Example 2. We shall determine the tangent spaces of the groups $GL(n, \mathbf{R})$, $O(n, \mathbf{R})$, $SO(n, \mathbf{R})$.

The manifold $M^{n^2} = GL(n, \mathbf{R})$ is an open set in the vector space $L(n, \mathbf{R})$ of all real $n \times n$–matrices. The space $L(n, \mathbf{R})$ can be identified with the space $\mathbf{R}^{n^2}$. According to example 1, we have $T_x M^{n^2} = \mathbf{R}_x^{n^2}$; thus, the tangent space to $GL(n, \mathbf{R})$ at any point $x \in GL(n, \mathbf{R})$ coincides with $L(n, \mathbf{R})$.

The manifold $O(n, \mathbf{R})$ also lies in $L(n, \mathbf{R})$ and is distinguished by the property $x \cdot x^T = e$, where x^T is the transposed matrix of x and e is the identity matrix. We calculate the tangent space to $O(n, \mathbf{R})$ at the point $x = e$. Consider a smooth path $x = x(t)$, $x(0) = e$, satisfying the condition $x(t) \cdot x(t)^T = e$ identically in t. Differentiating with respect to t, we get $x'(t) \cdot x(t)^T + x(t) \cdot x'(t)^T = 0$; taking $t = 0$, we obtain $x'(0) + x'(0)^T = 0$. Thus, the tangent vector $x'(0)$ is a skew–symmetric $n \times n$–matrix. On the other hand, any skew–symmetric matrix c is the tangent vector at the point $t = 0$ to the curve $x(t) = e^{tc}$ (the exponential of the matrix tc), which lies in $O(n, \mathbf{R})$ and passes through the point e when $t = 0$. Consequently, $T_e O(n, \mathbf{R})$ coincides with a space of all skew–symmetric $n \times n$–matrices. Further, $SO(n, \mathbf{R})$ is an open set in $O(n, \mathbf{R})$. Consequently, $T_e SO(n, \mathbf{R}) = T_e O(n, \mathbf{R})$.

3. Tangent bundle. For each point x of a smooth manifold M^n the tangent space $T_x M^n$ can be defined. The next problem is to construct a topological space and even a smooth manifold consisting of all vectors from this family of vector spaces parametrized by the points $x \in M$.

Consider the disjoint union $TM^n = \sqcup T_x M^n$ of all the tangent spaces to a manifold M^n. Define the projection $\pi : TM^n \to M^n$ by mapping each vector from $T_x M^n$ into the point x. Then

$$\pi^{-1}(x) = T_x M^n;$$

we call this preimage the *fibre over the point* x.

Each chart (U, ϕ) of a manifold M^n determines a chart $(\pi^{-1}(U), \tau_\phi)$ in TM^n,

(3) $$\tau_\phi : \pi^{-1}(U) \to \mathbf{R}^n \times \mathbf{R}^n$$

in the following way: we associate to the tangent vector $a = \{x, (U, \phi), h)\}$ in the fibre over a point $x \in U$ the pair $(\phi^{-1}(x), \tau_x \alpha)$, where τ_x is defined earlier (see subsection 2), i.e.,

$$\tau_\phi \alpha = (\phi^{-1}(x), h_\phi),$$

where $(x, (U, \phi), h_\phi)$ is a representative of the vector α in the chart (U, ϕ). Clearly, τ_ϕ is bijective, therefore, the weakest topology can be introduced in $\pi^{-1}(U)$ for which τ_ϕ becomes a continuous mapping and even a homeomorphism (see Ch. 2, § 8). Since the set of all charts $\pi^{-1}(U)$ forms a covering of TM^n, a topology on TM^n can be defined by declaring the collection of all open sets in all charts $\pi^{-1}(U)$ to be the base for this topology. This turns TM^n into a topological space.

Remark. According to the formal definition, the pairs $(\pi^{-1}(U), \tau_\phi^{-1})$ should be called the charts on the space TM^n. For convenience, we reversed the homeomorphisms (verify that this change is inessential).

Thus, the chart (3) allows us to introduce local coordinates on the set $\pi^{-1}(U)$ by taking the coordinates of the pair $(\phi^{-1}(x), h_\phi)$ in $\mathbf{R}^n \times \mathbf{R}^n$. We shall call the pair $(\phi^{-1}(x), h_\phi)$ the coordinate representation of the tangent vector α in the chart (U, ϕ), and the vector h_ϕ the vector component (in the chart (U, ϕ)) of the tangent vector.

This terminology is justified by the following statement.

Lemma 2. *If M^n is a manifold of class C^r, $r \geq 1$, then the collection $\{\pi^{-1}(U), \tau_\phi)\}$ of all charts of the space TM^n is a C^{r-1}–atlas.*

Proof. Let (U, ϕ), $V, \psi)$ be two charts on a manifold M^n, $U \cap V \neq \emptyset$. Let $(\pi^{-1}(U), \tau_\phi)$, $(\pi^{-1}(V), \tau_\psi)$ be the corresponding charts on TM^n; then $(\pi^{-1}(U) \cap \pi^{-1}(V)) = \pi^{-1}(U \cap V) \neq \emptyset$. We have the commutative diagram

$$
\begin{array}{ccc}
 & \pi^{-1}(U \cap V) & \\
\tau_\phi \swarrow & & \searrow \tau_\psi \\
\mathbf{R}^n \times \mathbf{R}^n & \xrightarrow{\;\;\tau_\psi \tau_\phi^{-1}\;\;} & \mathbf{R}^n \times \mathbf{R}^n
\end{array}
$$

where the mapping

$$
\tau_\psi \tau_\phi^{-1} : \tau_\phi(\pi^{-1}(U \cap V)) \to \tau_\psi(\pi^{-1}(U \cap V))
$$

is a homeomorphism of open sets in $\mathbf{R}^n \times \mathbf{R}^n$ (the transition homeomorphism from one set of coordinates to another). It is sufficient to show that $\tau_\psi \tau_\phi^{-1} \in C^{r-1}$. Since $\tau_\phi \alpha = (\phi^{-1}(x), h_1)$, $\tau_\psi \alpha = (\psi^{-1}(x), h_2)$, $h_2 = D_{\phi^{-1}(x)}(\psi^{-1}\phi)h_1$, it is easy to see that

$$
(4) \qquad \tau_\psi \tau_\phi^{-1}(y, h) = ((\psi^{-1}\phi)(y), D_y(\psi^{-1}\phi)h),
$$

whence $\tau_\psi \tau_\phi^{-1} \in C^{r-1}$. $\qquad\qquad\square$

Note that transition transformation (4) has a special form: the coordinates of the point x are transformed with a help of the diffeomorphism $\psi^{-1}\phi$, and the vector component h of the tangent vector α by means of the linear transformation $D_{\phi^{-1}(x)}(\psi^{-1}\phi)$.

The structure of a smooth manifold of dimension $2n$ has now been defined on the topological space TM^n. Because of the special form of the transition representation (4) , this manifold is called the *tangent bundle of the manifold* M^n. The new term emphasizes the structure of the manifold TM^n which consists of fibres over each point in M^n that are tangent spaces. The structure of a smooth manifold defined by an atlas of charts of form (3), is called a *structure of a tangent bundle.*

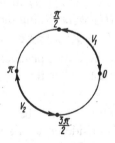

FIG. 105

Example 3. Let us define the structure of the tangent bundle TS^1 for the circle $S^1 \subset \mathbf{R}^2$. Consider S^1 as the set of points $e^{is}(s \in \mathbf{R}^1)$ and define on S^1 a C^∞–atlas consisting of two charts:

$$U_1 = \{e^{is} : s \in \left(0, \frac{3}{2}\pi\right)\},$$

$$\phi_1(s) = e^{is} : \left(0, \frac{3}{2}\pi\right) \to U_1,$$

$$U_2 = \{e^{is} : s \in \left(-\pi, \frac{\pi}{2}\right)\},$$

$$\phi_2(s) = e^{is} : \left(-\pi, \frac{\pi}{2}\right) \to U_2.$$

Indeed, the set $U_1 \cap U_2$ consists of two connected components V_1, V_2 (Fig. 105), and the transition homeomorphism on them is of the form:

$$\phi_2^{-1}\phi_1(s) = s : \phi_1^{-1}(V_1) \to \phi_2^{-1}(V_1),$$

$$\phi_2^{-1}\phi_1 s) = (s - 2\pi) : \phi_1^{-1}(V_2) \to \phi_2^{-1}(V_2),$$

and, consequently, it is a C^∞–diffeomorphism. Then the charts of the corresponding atlas on the tangent bundle are of the form

$$\tilde{U}_1 = \pi^{-1}(U_1), \quad \tau_{\phi_1} : \pi^{-1}(U_1) \to \left(0, \frac{3}{2}\pi\right) \times \mathbf{R}^1;$$

$$\tilde{U}_2 = \pi^{-1}(U_2), \quad \tau_{\phi_2} : \pi^{-1}(U_2) \to \left(-\pi, \frac{\pi}{2}\right) \times \mathbf{R}^1.$$

Remark. Since $D(\phi_2^{-1}\phi_1) = 1_{\mathbf{R}^1}$, the tangent vector given by the representative $(x, U_1, \phi_1), h)$ in the chart (U_1, ϕ_1) has the form $(x, (U_2, \phi_2), h)$ in the chart (U_2, ϕ_2). Gluing the manifold TS^1 from the direct products $(0, \frac{3}{2}\pi) \times \mathbf{R}^1$, $(-\pi, \frac{\pi}{2}) \times \mathbf{R}^1$ along the diffeomorphism $\tau_{\phi_2}\tau_{\phi_1}^{-1} = (\phi_2^{-1}\phi_1, 1_{\mathbf{R}^1})$, we obviously obtain the direct product $S^1 \times \mathbf{R}^1$. Thus, the tangent bundle TS^1 is homeomorphic to $S^1 \times \mathbf{R}^1$.

Exercise 6°. Let V be an open set in M^n. Show that the tangent bundle of the set V considered as a submanifold in M^n, coincides with $\pi^{-1}(V)$. Describe TV, $V \subset \mathbf{R}^n$.

Recall that in subsection 1, we were talking about the phase space of a system. The state of a system can be now characterized as an element from TM^n, i.e. as the tangent vector α at the point x. Moreover, x characterizes the position of the system in configuration space, and the vector α from $T_x M^n$ characterizes the velocity of the system.

4. Riemannian metric. The tangent bundle is related to the concept of a Riemannian metric on a manifold, which is important for geometrical problems. Consider a C^r–manifold M^n, $r \geq 1$, and its tangent bundle TM^n. Let a symmetric, positive–definite, bilinear form $A_x(u, v)$ depending, generally speaking, on x be given in each fibre of $T_x M^n$. We assume that this dependence is of class C^{r-1} in the sense that in the local coordinates on the chart $(\pi^{-1}(U), \tau_\phi)$ of the tangent bundle TM^n, the bilinear form $A_x(\tau_x^{-1} u, \tau_x^{-1} v)$ in a constant basis for the vector space $\mathbf{R}^n$ has a matrix $A_{ij}(x)$ whose elements are C^{r-1}–functions on U.

The form $A_x(u, v)$ is called a *Riemannian metric of class C^{r-1} on the manifold M^n*. It is often given in local coordinates on the tangent bundle as a bilinear form

$$\sum_{i,j} a_{ij}(x) u_i v_j, \quad x \in U,$$

where $u_1, \dots, u_n$, $v_1, \dots, v_n$ are the coordinates of the vectors u, v of the space $\mathbf{R}^n$. A Riemannian metric enables us to measure the lengths of vectors and the angles between them in tangent spaces, for example, if $v \in T_x M^n$ then the length $\|v\|_x$ of the vector v is determined by the equality $\|v\|_x^2 = A_x(v, v)$. The question arises whether the Riemannian metrics exist on smooth manifolds.

Theorem 1. *There exists a Riemannian metric of class C^{r-1} on any C^r–manifold M^n, $r \geq 1$.*

Proof. Consider an atlas $\{(U_\alpha, \phi_\alpha)\}$ on the manifold M^n. Let $\{V_\beta\}$ be a locally finite, open covering of M^n such that each V_β lies in some U_α (such a covering exists because of the paracompactness of M^n). Choose such an $\alpha = \alpha(\beta)$ for each β, and construct a C^r–partition of unity $\{g_\beta\}$ subordinate to the covering $\{V_\beta\}$. The idea to construct a Riemannian metric is that on each V_β (as a submanifold of M^n with the tangent bundle $\pi^{-1}(V_\beta)$) we construct its own Riemannian metric $A_x^\beta(u, v)$, and afterwards, by partition of unity, we "glue them together" to a "global" Riemannian metric

$$(5) \qquad A_x(u, v) = \sum_{\beta} g_{\beta}(x) A_x^{\beta}(u, v).$$

Exercise 7°. Verify that if $A_x^{\beta}(u, v)$ is a Riemannian metric on V_{β} (for each β), then formula (5) determines a Riemannian metric on M^n.

It remains to construct a Riemannian metric on V_{β}. According to the construction, $V_{\beta} \subset U_{\alpha(\beta)}$, therefore $\pi^{-1}(V_{\beta}) \subset \pi^{-1}(U_{\alpha(\beta)})$. Consequently, $\pi^{-1}(V_{\beta})$ belongs to the chart $(\pi^{-1}(U_{\alpha(\beta)}), \tau_{\phi_{\alpha(\beta)}})$ of the tangent bundle TM^n, and we get a mapping

$$(6) \qquad \tau^{\beta} = \tau_{\phi_{\alpha(\beta)}} : \pi^{-1}(V_{\beta}) \longrightarrow (\phi_{\alpha(\beta)}(V_{\beta})) \times \mathbf{R}^n.$$

In the local coordinates (6), we consider the bilinear form $B(u, v) = u_1 v_1 + \ldots + u_n v_n$ with the constant matrix $(a_{ij}(x)) = (\delta_{ij})$ (δ_{ij} being the Kronecker symbol). Now define a Riemannian metric on V_{β} by the equality

$$A_x^{\beta}(u, v) = B(\tau_x^{\beta} u, \tau_x^{\beta} v), \quad u, v \in T_x M^n, \quad x \in V_{\beta},$$

where τ_x^{β} is a restriction of τ^{β} to the fiber $T_x M^n$. $\qquad \Box$

5. The tangent mapping. When studying smooth mappings of surfaces (curves) in analysis and its applications, linearization is often used; it consists of replacing a surface (curve) by the tangent plane (line) in neighbourhoods of a point and its image; the mapping then is replaced by its differential, i.e., by a linear mapping. This method allows generalization to the case of mappings of smooth manifolds.

Let $f : M^n \to N^m$ be a smooth mapping of class C^r, $r \geq 1$, of smooth manifolds of the same class. Let $x \in M^n$ be an arbitrary point, and (U, ϕ), (V, ψ) charts in the manifolds M^n and N^m, respectively, such that $x \in U$, $f(x) \in V$; we assume also that $f(U) \subset V$. Consider the representation of the mapping f in the given charts:

$$\psi^{-1} f \phi : \phi^{-1}(U) \to \psi^{-1}(U)$$

and its derivative

(7) $$D_{\phi^{-1}(x)}(\psi^{-1}f\phi) : \mathbf{R}^n \to \mathbf{R}^m.$$

Definition 3. Let $a \in T_x M^n$ be an arbitrary tangent vector at a point x, and $(x, (U, \phi), h)$ its representative in the chart (U, ϕ). The linear mapping

$$T_x(f) : T_x M^n \to T_{f(x)} N^m,$$

under which the tangent vector a with representative $(x, (U, \phi), h)$ is transformed into a tangent vector b with representative $(f(x), (V, \psi), g)$ in the chart (V, ψ), where $g = D_{\phi^{-1}(x)}(\psi^{-1}f\phi)h$ is called the *tangent mapping of f at the point $x \in M^n$*.

Thus, under the tangent mapping, a point x is taken to the point $f(x)$ (as before) and the vector component h of the tangent vector (corresponding to the chosen chart) is transformed by the linear mapping (7).

Exercise 8°. Show that the tangent mapping $T_x(f)$ does not depend on the choice of charts.

The following basic properties of the tangent mapping are left to the readers, as a simple exercise, to be verified:

(1) to the identity mapping $1_{M^n} : M^n \to M^n$, there corresponds the identity mapping

$$T_x(1_{M^n}) = 1_{T_x M^n} : T_x M^n \to T_x M^n;$$

(2) the commutativity of the diagram of mappings

entails the commutativity of the diagram of tangent mappings

$$
\begin{array}{ccc}
& & T_yN^m \\
T_x(f) \nearrow & & \\
T_xM^n & & \downarrow T_y(g) \\
T_x(gf) \searrow & & \\
& & T_zp^k
\end{array}
$$

where $y = f(x)$, $z = g(y)$.

The collection of all tangent maps $\{T_x(f)\}_{x \in M^n}$ defines a mapping of tangent bundles

$$(8) \qquad\qquad T(f) : TM^n \to TM^n,$$

which is called the *mapping of manifolds tangent to f*.

Using the smooth structures on TM^n and TN^m, we can write the representation of the mapping $T(f)$ in the corresponding charts. Indeed, let (V, ψ) be a chart at the point $f(x)$, and (U, ϕ) a chart at the point x, and let moreover, $f(U) \subset V$. Consider the charts $(\pi^{-1}(U), \tau_\phi)$, $(\pi^{-1}(V), \tau_\psi)$ on the tangent bundles TM^n, TN^m, respectively. To the tangent vector $a \in \pi^{-1}(U)$ in the chart τ_ϕ there corresponds a pair $(\phi^{-1}(x), h)$; similarly, to the vector $b = T(f)a$ in the chart τ_ψ, there corresponds a pair $(\psi^{-1}f(x), D_{\phi^{-1}(x)}(\psi^{-1}f\phi)h)$. We have the following transition transformation:

$$(9) \qquad\qquad \tau_\psi T(f)\tau_\phi^{-1} : (y, h) \mapsto ((\psi^{-1}f\phi(y), D_y(\psi^{-1}f\phi)h),$$

which goes from the set $\tau_\phi(\pi^{-1}(U)) \subset \mathbf{R}^n \times \mathbf{R}^n$ to the set $\tau_\psi(\pi^{-1}(V)) \subset \mathbf{R}^m \times \mathbf{R}^m$. It is clear that mapping (9) belongs to the smoothness class C^{r-1}.

Thus, to each smooth mapping of manifolds of class C^r, $r \geq 1$ there is associated a smooth mapping (8) of class C^{r-1} of their tangent bundles. Properties (1) and (2) remain valid for the tangent mappings of tangent bundles.

In terms of the tangent mapping, we can reformulate the definition of a regular point of a smooth mapping of manifolds (see § 5, definition 6). Let $f : M^n \to N^m$ be a C^r–mapping ($r \geq 1$) of C^r–manifolds.

Definition 4. A point $x \in M^n$ is called a *regular point of a mapping* f if

$$\text{rank}\, T_x(f) = \min(n, m).$$

Exercise 9°. Verify that definition 4 is equivalent to definition 6 from § 5.

The advantage of definition 4 lies in the fact that it is given in invariant form, i.e. in a form independent of the choice of coordinate systems.

6. Orientation of manifolds. The concepts of tangent space and tangent bundle allows us to define the concept of orientability of smooth manifolds by generalizing the definition of an orientable surface, which is quite important in analysis.

Recall the concept of oriented vector space $\mathbf{R}^n$. It is said that two bases $(e_1, \ldots, e_n)$ and $(g_1, \ldots, g_n)$ in $\mathbf{R}^n$ have the same *orientation*, if the transformation from one basis to the other is a linear mapping with a positive determinant[*]

Exercise 10°. Show that orientation is an equivalence relation on the set of all bases in $\mathbf{R}^n$ and that the number of equivalence classes is equal to 2.

A space $\mathbf{R}^n$ is said to be *oriented* if one of the equivalence classes of bases is given.

Consider a C^r–submanifold M^n, $r \geq 1$, in the space $\mathbf{R}^N$. A submanifold is said to be *orientable* if orientations in each tangent space $T_x M^n$ and an atlas $\{(U_\alpha, \phi_\alpha)\}$ in M^n can be chosen in such a way that the corresponding diffeomorphisms $\phi_\alpha : \mathbf{R}^n \to U_\alpha$ preserve the orientations, i.e. for any point $x \in U_\alpha$, the tangent mapping $T_x \phi_\alpha^{-1} : T_x M^n \to \mathbf{R}^n$ tranforms the chosen orientation of the vector space $T_x M^n$ into a fixed given orientation of the vector space $\mathbf{R}^n$.

Otherwise, the submanifold is said to be *non–orientable*.

An atlas satisfying this condition is called an *orienting atlas*. Clearly, the diffeomorphisms $\phi_\alpha : \mathbf{R}^n \to U_\alpha$ for an orienting atlas are compatible with each other. The exact meaning of this compatibility is expressed by the following exercise.

---

[*] Orientation in the space $\mathbf{R}^0$ is naturally defined as the choice of the zero sign: $+0$ or -0.

Exercise 11°. Show that any two charts (U_α, ϕ_α), (U_β, ϕ_β) from an orienting atlas are positively compatible, i.e., possess the property that the determinant of the mapping $D_{\phi_\alpha^{-1}(x)}(\phi_\beta^{-1}\phi_\alpha) : \mathbf{R}^n \to \mathbf{R}^n$ is positive for any point $x \in \phi_\alpha^{-1}(U_\alpha \cap U_\beta)$; inversily, if any two charts of an atlas are positively compatible, then the atlas is orienting.

The property given in exercise 11° is used in defining an orientable manifold (not necessarily imbedded in $\mathbf{R}^N$).

On the set of orienting atlases of a manifold, we introduce an equivalence realation: two orienting atlases are *equivalent* if their union is an orienting atlas.

The choice of one of the equivalence classes is called an *orientation of the manifold*.

Exercise 12°. Verify that for any manifold, the number of the equivalence classes of orienting atlases is even, and in the case of a connected manifold it is equal to 0 or 2.

The simplest example of an orientable manifold is the space $\mathbf{R}^n$. In this case, the atlas consisting of one chart $(\mathbf{R}^n, 1_{\mathbf{R}^n})$ is orienting.

Exercise 13°. Show that any manifold possessing an atlas which consists of one chart, is orientable.

Exercise 13° suggests the following examples: an open set in $\mathbf{R}^n$ and, consequently, any open disc D^n is orientable.

The Cartesian product of orientable manifolds is another example of an orientable manifold. We leave this to the readers as an exercise to be verified.

Exercise 14°. Construct an orienting atlas on S^n.

Exercise 15°. Show that the manifold $G_k(\mathbf{R}^n)$ is orientable for even n, $0 < k < n$.

As to non–orientable manifolds, these are, for instance, the Möbius strip and the projective spaces $\mathbf{R}P^{n-1}$ for even $n - 1 > 0$. We do not give the proof here. If $n - 1$ is odd, then $\mathbf{R}P^{n-1}$ is orientable as follows from exercise 15°.

Remark. Notice that for $n = 0$, $n = 1$, any manifold M^n is orientable.

The concept of orientation allows us to improve the concept of the degree modulo 2 of a mapping as introduced in § 5. Considering a mapping of oriented manifolds, we count the number of points in the preimage of a regular value not mod 2 but algebraically, assuming that each point of the preimage has sign "+" or "−", depending on whether or not the tangent map at the given point preserves the orientation. As in the case of the degree modulo 2, it can be shown that this number does not depend on the choice of a regular value; it is called the *(oriented) degree of the mapping f* and denoted by $\deg f$. In the case of smooth mappings of spheres, the degree of a mapping so defined coincides with the degree of a mapping introduced in Ch. 3, § 4.

§ 7. Tangent vector as differential operator.
Differential of a function and cotangent bundle

1. A new definition of (tangent) vector. We continue our studies of the tangent vectors and present the definition in terms of the differentiation with respect to a vector. This enables us to give a new interpretation of the tangent bundle.

Consider Euclidean space $\mathbf{R}^n$ and a C^∞–function f defined in a neighbourhood of a point $x^0 \in \mathbf{R}^n$. Consider the vector space $\mathbf{R}^n_{x^0}$ of all n–dimensional vectors at the point x^0. If (x^0, v) is some vector from $\mathbf{R}^n_{x^0}$ then the derivative of the function f with respect to the vector v at the point x^0 is the derivative $\frac{d}{dt} f(x^0 + tv)|_{t=0}$, where $t \geq 0$ is a numerical parameter (in analysis a vector v of unit length is usually considered, and one speaks of the derivative in the direction v). We have the following formula in terms of the standard coordinate system

$$
\frac{d}{dt} f(x^0 + tv_1)|_{t=0} = \left(\frac{\partial f_1}{\partial x_1}\right)_{x^0} v_1 + \ldots + \left(\frac{\partial f_n}{\partial x_n}\right)_{x^0} v_n =
$$

(1)

$$
= (\operatorname{grad} f(x^0), v),
$$

where $x_1, \ldots, x_n$, $v_1, \ldots, v_n$ are the coordinates of the point x and vector v. Denote the derivative (1) by $f_v(x^0)$. For a given vector v and a point x^0

we have obtained a correspondence $f \mapsto f_v(x^0)$ which determines a certain function (functional) $l^v_{x^0}$ defined on smooth functions in neighbourhoods of the point x^0 with values in $\mathbf{R}^1$. Clearly, this functional is defined on the germs $\widehat{f}_{x^0}$ of smooth functions at the point x^0. Thus, we have a mapping

$$(2) \qquad\qquad l^v_{x^0} : \mathcal{O}(x^0) \to \mathbf{R}^1.$$

From the definition, the following properties of the functional (2) follow:

(1) $l^v_{x^0}(\widehat{fg}) = f(x^0)l^v_{x^0}(\widehat{g}) + g(x^0)l^v_{x^0}(\widehat{f})$ (formula for differentiating a product);

(2) $l^v_{x^0}(\widehat{f}) = 0$, if $f = \text{const}$ (the formula for differentiating a constant);

(3) $l^v_{x^0}(\alpha\widehat{f} + \beta\widehat{g}) = \alpha l^v_{x^0}(\widehat{f}) + \beta l^v_{x^0}(\widehat{g})$, $\alpha, \beta \in \mathbf{R}^1$ (linearity).

Consider the set $\{l\}_{x^0}$ of all functionals $l : \mathcal{O}(x^0) \to \mathbf{R}$ satisfying properties (1), (2), (3). Obviously, $\{l\}_{x^0}$ is a vector space and $l^v_{x^0} \in \{l\}_{x^0}$. Now, as the vector v "runs through" the space $\mathbf{R}^n_{x^0}$, one obtains a mapping

$$(3) \qquad\qquad \mathbf{R}^n_{x^0} \to \{l\}_{x^0}, \quad v \mapsto l = l^v_{x^0}.$$

Theorem 1. *The mapping (3) is an isomorphism of the vector spaces* $\mathbf{R}^n_{x^0}$ *and* $\{l\}_{x^0}$.

Proof. The linearity of the mapping (3) follows from formula (1). The mapping (3) is a monomorphism. Indeed, if $l^v_{x^0} = l^w_{x^0}$, then $(\text{grad } f(x^0), v) = (\text{grad } f(x^0), w)$ for any function f which is smooth in a neighbourhood of x^0; now take $f(x) = x_i$ (the coordinate of the point x), and obtain the equalities $v_i = w_i$, $i = 1, \ldots, n$, i.e. $v = w$.

We prove that mapping (3) is epimorphic. We have

$$(4) \qquad \begin{aligned} f(x) = f(x^0) &+ \sum_{i=1}^n A_i(x^0)(x_i - x_i^0) + \\ &+ \sum_{i,j=1}^n A_{ij}(x)(x_i - x_i^0)(x_i - x_i^0), \end{aligned}$$

where

(5)
$$A_i(x^0) = \frac{\partial f}{\partial x_i}(x^0), \quad i = 1, \dots, n,$$

and $A_{ij}(x)$ are the functions of class C^∞ (see § 1, exercise 4°).

Now, let $l : \mathcal{O}(x^0) \to \mathbf{R}^1$ be an arbitrary functional from $\{l\}_{x^0}$. Using axioms (1), (2), (3), we obtain from (4) the equalities

$$l(\widehat{f}) = \sum_{i=1}^{n} A_i(x^0) l(\widehat{x}_i - x_i^0) = \sum_{i=1}^{n} A_i(x^0) l(\widehat{x}_i),$$

where $l(\widehat{x}_i)$ is the value of l on the germ of the function x_i, i.e., on the coordinate of the point x.

Using (5), we finally obtain

(6)
$$l(\widehat{f}) = \left(\frac{\partial f}{\partial x_1}\right)\bigg|_{x^0} v_1 + \dots + \left(\frac{\partial f}{\partial x_n}\right)\bigg|_{x^0} v_n = l_{x^0}^v(f),$$

where $v_1 = l(\widehat{x}_1), \dots, v_n = l(\widehat{x}_n)$. $\qquad\qquad\square$

Using the isomorphism (3) we can identify the vector space $\mathbf{R}_{x^0}^n$ with the n–dimensional vector space $\{l\}_{x^0}$ of all functionals satsifying axioms (1), (2), (3). Using the standard coordinate system in $\mathbf{R}^n$ and equality (6) each functional l_{x_0} can be associated with the differential operator

(7)
$$\sum_{i=1}^{n} v_i \frac{\partial}{\partial x_i}\bigg|_{x^0},$$

which acts on smooth functions according to the formula

$$\left(\sum_{i=1}^{n} v_i \frac{\partial}{\partial x_i}\bigg|_{x^0}\right) f = \sum_{i=1}^{n} \left(\frac{\partial f}{\partial x_i}\right)\bigg|_{x^0} v_i.$$

Exercise 1°. Verify that the set of all differential operators (7) forms a vector space, and the indicated correspondence defines an isomorphism with the vector space $\{l\}_{x^0}$.

Thus, we have another isomorphism, viz. of the vector space $\mathbf{R}^n_{x^0}$ with the vector space of differential operators (7). Under this isomorphism, to the basis vector $e_i = (0, \ldots, 0, 1, 0, \ldots, 0)$ (where 1 is in the i–th place), there corresponds the differential operator $\frac{\partial}{\partial x_i}\big|_{x^0}$.

2. Tangent bundles. The interpretation of the vector space at a point x^0 given in subsection 1, suggests a corresponding generalization of this concept to smooth manifolds.

Let M^n be a manifold of class C^∞, and x^0 a point from M^n. Consider the algebra $\mathcal{O}(x^0)$ of germs of smooth functions at a point x^0 and functionals

$$(8) \qquad\qquad l_{x^0} : \mathcal{O}(x^0) \to \mathbf{R}^1.$$

Exercise 2°. Let (U, ϕ) be a chart at a point x^0 of the manifold M^n. Verify that the functional l_{x^0} defined by the equality

$$l_{x^0}(\widehat{f}) = l^v_{\phi^{-1}(x^0)}(f\phi), \quad \widehat{f} \in \mathcal{O}(x^0),$$

for any vector $v \in \mathbf{R}^n$ determines a functional (8) satisfying axioms (1), (2), (3).

Definition 1. A set of all functionals (8) satisfying properties (1), (2), (3) is called the *tangent space* $T_{x^0}M^n$ to the manifold M^n at the point x^0.

The tangent space $T_{x^0}M^n$ is a vector space with natural algebraic operations. An individual element l_{x^0} from $T_{x^0}M^n$ is called a tangent vector to the manifold M^n at the point x^0.

The correspondence $l^v_{\phi^{-1}(x^0)} \mapsto l_{x^0}$ (see exercise 2) turns out to be an isomorphism of the spaces $\mathbf{R}^n_{\phi^{-1}(x^0)}$ and $T_{x^0}M^n$. Indeed, the linearity of this mapping is obvious, and the inverse mapping is given by the formula

$$\widetilde{l}_{\phi^{-1}(x^0)}(\widehat{g}) = l_{x^0}(\widehat{g\phi^{-1}}), \quad g \in \mathcal{O}(\phi^{-1}(x^0)),$$

where $\mathcal{O}(\phi^{-1}(x^0))$, $\mathcal{O}(x^0)$ are the algebras of the germs at the points $\phi^{-1}(x^0) \in \mathbf{R}^n$, $x^0 \in M^n$, respectively. It is convenient to consider in future that the functional l_{x^0} is given not only on the germs $\widehat{g} \in \mathcal{O}(x^0)$, but also on the functions g defined in the neighbourhood of the point x^0 (assuming that $l_{x^0}(g) = l_{x^0}(\hat{g})$), and write l_{x^0} instead of $l_{x^0}(\hat{g})$.

Let $\Phi : M^n \to N^m$ be a smooth mapping of manifolds, and let $x^0 \in M^n$, $y^0 = \Phi(x^0) \in N^m$. The mapping Φ induces a mapping $\hat{\Phi} : \mathcal{O}(y^0) \to \mathcal{O}(x^0)$ between the algebras of germs according to the rule $\hat{g} \in \mathcal{O}(y^0)$, $\hat{g} \mapsto \hat{f}$, if $f = g\Phi$. This enables us to define the tangent mapping $T_{x^0}(\Phi) : T_{x^0}M^n \to T_{y^0}N^m$ by the rule $T_{x^0}(\Phi)l_{x^0} = l_{y^0}$, where $l_{y^0} = l_{x^0}\hat{\Phi}$.

The action of the mappings Φ and $\hat{\Phi}$ is shown in the diagrams

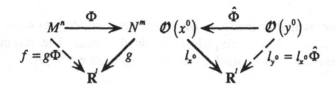

Exercise 3°. Verify that $l_{x^0}\hat{\Phi}$ is a tangent vector at the point y^0 of the manifold N^m, and that $T_{x^0}(\Phi)$ is a linear mapping.

The tangent mapping $T_{x^0}(\Phi)$ is often denoted by $(\Phi_*)_{x^0}$ (or $d_{x^0}\Phi$).

Exercise 4°. Show that $[1_{M^n})_*]_{x^0} = 1_{T_{x^0}M^n}$. If $\Phi : M^n \to N^m$, $\Psi : N^m \to P^k$ are smooth mappings of manifolds then $[(\Psi\Phi)_*]_{x^0} = (\Psi_*)_{\Phi(x^0)}(\Phi_*)_{x^0}$.

Exercise 5°. Prove that if Φ is a diffeomorphism, then $(\Phi_*)_{x^0}$ is an isomorphism of vector spaces (and, therefore, $m = n$).

We shall pass on to the construction of the tangent bundle. As in § 6, set $TM^n = \sqcup T_x M^n$ (disjoint union). The problem is to determine the structure of the tangent bundle on TM^n. We give the projection $\pi : TM^n \to M^n$ by associating to an element $l_x \in T_x M^n$ the point $x \in M^n$. Let (U, ϕ) be a chart of the point x. Consider the chart on the tangent space corresponding to the chart (U, ϕ),

$$(9) \qquad \qquad \tau_\phi : \pi^{-1}(U) \to \mathbf{R}^n \times \mathbf{R}^n.$$

Let $l_x \in T_x M^n$; then the tangent vector $\tilde{l}_{\phi^{-1}(x)}$ is defined on the algebra $\mathcal{O}(\phi^{-1}(x))$ by the rule

$$\tilde{l}_{\phi^{-1}(x)}(g) = l_x(g\phi^{-1}), \quad \hat{g} \in \mathcal{O}(\phi^{-1}(x)).$$

In virtue of the isomorphism of the space of differential operators and the tangent space (see exercise 1°), we have

$$(10) \qquad \tilde{l}_{\phi^{-1}(x)} = v_1 \frac{\partial}{\partial x_1} + \ldots + v_n \frac{\partial}{\partial x_n},$$

where the differential operators $\partial/\partial x_i$ act at the point $\phi^{-1}(x) = (x_1, \ldots, x_n)$, and $v = (v_1, \ldots, v_n)$ is a uniquely determined vector. The mapping (9) is given by a corresponence which is linear on each fibre $\pi^{-1}(x)$:

$$(11) \qquad l_x \mapsto (x_1, \ldots, x_n : v_1, \ldots, v_n).$$

The bijectivity of this mapping is evident; we define the topology on TM^n, as in § 6, by the condition of continuity of the mappings τ_ϕ for all charts from an atlas for the manifold M^n.

We show that mappings (9) and (11) define the structure of the tangent bundle. If (V, ψ) is another chart at the point x, then the tangent vector $\tilde{l}_{\psi^{-1}(x)}$ is determined similarly:

$$(12) \qquad \tilde{l}_{\psi^{-1}(x)} = w_1 \frac{\partial}{\partial y_1} + \ldots + w_n \frac{\partial}{\partial y_n},$$

where $w = (w_1, \ldots, w_n)$ is a vector at the point $\psi^{-1}(x) = (y_1, \ldots, y_n)$, and the mapping τ_ψ acts according to the rule $l_x \mapsto (y_1, \ldots, y_n; w_1, \ldots, w_n)$. Now calculate $\tau_\psi \tau_\phi^{-1}$. There is the mapping of class C^r:

$$(13) \qquad \begin{aligned} (y_1, \ldots, y_n) &= \psi^{-1}\phi(x_1, \ldots, x_n) = \\ &= ((\psi^{-1}\phi)_1(x_1, \ldots, x_n), \ldots, (\psi^{-1}\phi)_n(x_1, \ldots, x_n)). \end{aligned}$$

Now express $w_1, \ldots, w_n$ in terms of $v_1, \ldots, v_n$. Let $\hat{g} \in \mathcal{O}(\psi^{-1}(x))$, then we have the following equality from (12)

$$\tilde{l}_{\psi^{-1}(x)}(g) = w_1 \frac{\partial g}{\partial y_1} + \ldots + w_n \frac{\partial g}{\partial y_n},$$

but

$$\tilde{l}_{\psi^{-1}(x)}(g) = l_x(g\psi^{-1}) = l_x(g\psi^{-1}\phi\phi^{-1}) = \tilde{l}_{\phi^{-1}(x)}(g\psi^{-1}\phi).$$

Now use formula (10)

$$\tilde{l}_{\phi^{-1}(x)}(g) = v_1 \frac{\partial(g\psi^{-1}\phi)}{\partial x_1} + \ldots + v_n \frac{\partial(g\psi^{-1}\phi)}{\partial x_n}$$

and compare the two expressions for $\tilde{l}_{\psi^{-1}(x)}(g)$:

$$\sum_{i=1}^{n} w_i \frac{\partial g}{\partial y_i} = \sum_{j=1}^{n} v_j \frac{\partial(g\psi^{-1}\phi)}{\partial x_j}.$$

Since the germ $\hat{g} \in \mathcal{O}(\psi^{-1}(x))$ is arbitrary, we can take $g(y_1, \ldots, y_n) = y_i$; then from (13) we have

$$(g\psi^{-1}\phi)(x_1, \ldots, x_n) = (\psi^{-1}\phi)_i(x_1, \ldots, x_n),$$

and from the previous equality we obtain

(14) $$w_i = \sum_{j=1}^{n} v_j \frac{\partial(\psi^{-1}\phi)_i}{\partial x_j}(x_1, \ldots, x_n), \quad i = 1, \ldots, n.$$

The vector component of the tangent vector under the transformation (13) of coordinates, is transformed by the linear transformation (14) with the Jacobian matrix $\left(\frac{\partial(\psi^{-1}\phi)}{\partial x} \right)$, i.e., by the transformation $D_{\phi^{-1}(x)}(\psi^{-1}\phi)$. The transformations (13) and (14) smoothly depend on the point $\phi^{-1}(x)$ and, thus, show that the transformation $\tau_\psi \tau_\phi^{-1}$ is of class C^∞. The special form of this coordinate transformation means that the atlas $\{(\pi^{-1}(U), \tau_\phi)\}$ on TM^n determines the smooth structure of the tangent bundle.

Exercise 6°. Consider Euclidean space $\mathbf{R}^n$ with the structure given by the atlas consisting of one chart $(\mathbf{R}^n, 1_{\mathbf{R}^n})$. Verify that $T_x\mathbf{R}^n$ is isomorphic to $\mathbf{R}_x^n$.

Now we may consider $T\mathbf{R}^n$ to be the set of all (x, v), where $x \in \mathbf{R}^n$, and $v \in \mathbf{R}_x^n$. One may also assume that

$$TR^n = \left\{ \left(x_1, \ldots, x_n; v_1 \frac{\partial}{\partial x_1} + \ldots + v_n \frac{\partial}{\partial x_n} \right) \right\},$$

where the $x_1, \ldots, x_n$ are the coordinates of x, and $v_1, \ldots, v_n$ are the coordinates of v; the mapping defines the only chart of the corresponding atlas of the tangent bundle TR^n:

(15)
$$\left(x_1, \ldots, x_n; v_1 \frac{\partial}{\partial x_1} + \ldots + v_n \frac{\partial}{\partial x_n} \right) \rightarrow$$
$$\rightarrow (x_1, \ldots, x_n; v_1, \ldots, v_n).$$

By the same token, the structure of direct product is introduced on TR^n; the tangent bundle TR^n is a trivial bundle.

Exercise 7°. Show that the mapping (9) decomposes into a product $\tau_\phi = \tau_{1_{R^n}} (\phi^{-1})_*$ which acts according to the rule

$$l_x \overset{(\phi^{-1})_*}{\longmapsto} \left(x_1, \ldots, x_n; v_1 \frac{\partial}{\partial x_1} + \ldots + v_n \frac{\partial}{\partial x_n} \right) \overset{\tau_{1_{R^n}}}{\longmapsto}$$
$$\overset{\tau_{1_{R^n}}}{\longmapsto} (x_1, \ldots, x_n; v_1, \ldots, v_n).$$

It is often convenient to deal not with a coordinate representation of the vector l_x, but with its image $(\phi^{-1})_* l_x$; the vector component of the latter is $v_1 \frac{\partial}{\partial x_1} + \ldots + v_n \frac{\partial}{\partial x_n}$. When changing the chart (changing the coordinates), its coordinates $v_1, \ldots, v_n$ in the basis $\left\{ \frac{\partial}{\partial x_i} \right\}_{i=1}^n$ are transformed by formulae (14) to the coordinates $w_1, \ldots, w_n$ in the basis $\left\{ \frac{\partial}{\partial y_i} \right\}_{i=1}^n$.

3. Tangent maps. Let $\Phi : M^n \rightarrow N^m$ be a smooth mapping of manifolds. For each $x \in M^n$, we have a linear mapping $(\Phi_*)_x : T_x M^n \rightarrow T_y N^m$, where $y = \Phi(x)$. Thereby, the mapping $\Phi_* : TM^n \rightarrow TN^m$ is defined. Let us verify that Φ_* is a smooth mapping of tangent bundles.

Let (U, ϕ) be a chart at the point x, (V, ψ) a chart at the point y, $l_x \in T_x M^n$, and $l_y = (\Phi_*)_x l_x$. By (11), we have

(16)
$$l_x \overset{\tau_\phi}{\longmapsto} (x_1, \dots, x_n; v_1, \dots, v_n),$$
$$l_y \overset{\tau_\psi}{\longmapsto} (y_1, \dots, y_m; w_1, \dots, w_m),$$

and now it is necessary to find the transformation

(17)
$$\tau_\psi (\Phi_*) \tau_\phi^{-1} : (x_1, \dots, x_n; v_1, \dots, v_n) \longmapsto$$
$$\longmapsto (y_1, \dots, y_m; w_1, \dots, w_m).$$

Since $(\psi^{-1}\Phi\phi)(x_1, \dots, x_n) = (y_1, \dots, y_m)$, it remains to find the relation between the $\{v_i\}$ and $\{w_i\}$. If $\widehat{g} \in \mathcal{O}(\psi^{-1}(y))$, then the following equalities hold

$$\widetilde{l}_{\psi^{-1}(y)}(g) = l_y(g\psi^{-1}) = ((\Phi_*)_x l_x)(g\psi^{-1}) =$$
$$= l_x(g\psi^{-1}\Phi) = l_x(g\psi^{-1}\Phi\phi\phi^{-1}) = \widetilde{l}_{\phi^{-1}(x)}(g\psi^{-1}\Phi\phi).$$

But taking into account that the first and the last functionals are respectively equal to

(18)
$$\sum_{j=1}^m w_j \frac{\partial g}{\partial y_j}$$

and

(19)
$$\sum_{j=1}^n v_j \frac{\partial(g\psi^{-1}\Phi\phi)}{\partial x_j},$$

we obtain, just like in (14), by equalizing (18) to (19) and setting $g \equiv y_i$, that

(20)
$$w_i = \sum_{j=1}^n v_j \frac{\partial(\psi^{-1}\Phi\phi)_i}{\partial x_j}, \quad i = 1, \dots, m;$$

so that the mapping (17) is of class C^∞. Formula (20) confirms the fact established earlier that the vector component of a tangent vector is transformed by the linear transformation $D_{\phi^{-1}(x)}(\psi^{-1}\Phi\phi)$.

4. The differential of a function and the cotangent bundle. Consider the action of a vector $l_{x^0} \in T_{x^0} M^n$ on the function f, $\hat{f} \in \mathcal{O}(x^0)$. If we fix the function f, then a linear functional $l_{x^0} \mapsto l_{x^0}(f)$ on the space $T_{x^0} M^n$ arises. This functional is denoted by the symbol $(df)_{x^0}$ and is called the differential of the function f at the point x^0. According to the definition,

$$(df)_{x^0} l_{x^0} = l_{x^0}(f).$$

Thus, $(df)_{x^0}$ belongs to $(T_{x^0} M^n)^*$, i.e., the space dual to $T_{x^0} M^n$ which has a natural vector structure.

Let (U, ϕ) be a chart at the point x^0, and $\{x_i(x)\}_{i=1}^n$ the local coordinates of a point $x \in U$. Below, we identify tangent vectors with their corresponding differential operators. Let $\left\{ \frac{\partial}{\partial x_i}\big|_{x^0} \right\}_{i=1}^n$ be the basis for $T_{x^0} M^n$ (corresponding to (U, ϕ)), and $(dx_i)_{x^0}$ the differential of the function $x_i(x)$, then

$$(21) \qquad (dx_i)_{x^0} \left(\frac{\partial}{\partial x_j}\Big|_{x^0} \right) = \left(\frac{\partial}{\partial x_j} x_i \right)_{x^0} = \delta_{ij}$$

($\delta_{ij} = 0$, for $i \neq j$, $\delta_{ii} = 1$). Consequently, $\{(dx_i)_{x^0}\}_{i=1}^n$ is the dual basis to $\left\{ \frac{\partial}{\partial x_i}\big|_{x^0} \right\}_{i=1}^n$ in $(T_{x^0} M^n)^*$. Hence, it follows also that $(T_{x^0} M^n)^*$ consists of all possible linear combinations $\{a_1 (dx_1)_{x^0} + \ldots + a_n (dx_n)_{x^0}\}$ with real coefficients. For an arbitrary function f, $\hat{f} \in \mathcal{O}(x^0)$ and a vector $l_{x^0} = v_1 \frac{\partial}{\partial x_1}\big|_{x^0} + \ldots + v_n \frac{\partial}{\partial x_n}\big|_{x^0}$, we have the decomposition

$$(df)_{x^0} l_{x^0} = l_{x^0}(f) = v_1 \frac{\partial f}{\partial x_1}(x^0) + \ldots + v_n \frac{\partial f}{\partial x_n}(x^0).$$

With the help of (21), we obtain

$$(dx_i)_{x^0} l_{x^0} = (dx_i)_{x^0} \left(\sum_{j=1}^n v_j \frac{\partial}{\partial x_j}\Big|_{x^0} \right) = v_i$$

and, by substituting this in the previous equality, we find:

$$(df)_{x^0} l_{x^0} = \left(\sum_{i=1}^n \frac{\partial f}{\partial x_i}(x^0)(dx_i)_{x^0} \right) l_{x^0}.$$

This holds for all $l_{x^0} \in T_{x^0}M^n$, and so we have

$$(df)_{x^0} = \frac{\partial f}{\partial x_1}(x^0)(dx_1)_{x^0} + \ldots + \frac{\partial f}{\partial x_n}(x^0)(dx_n)_{x^0}.$$

Replacing x^0 by an arbitrary point $x \in U$, we may rewrite the last formula in a more convenient way:

$$(22) \qquad (df)_x = \frac{\partial f}{\partial x_1}(x)dx_1 + \ldots + \frac{\partial f}{\partial x_n}(x)dx_n,$$

here $\{dx_i\}_{i=1}^n$ are the basis differentials at the point x. Formula (22) justifies the term "differential" for $(df)_x$.

We shall deduce from (22) the relation between the differentials of the coordinates of various local coordinate systems at the point x. Let (V, ψ) be a chart which defines the coordinates $\{y_i(x)\}_{i=1}^n$, $x \in V$. If $x \in U \cap V$, then the coordinates $\{x_i(x)\}_{i=1}^n$ and $\{y_i(x)\}_{i=1}^n$ are related by the transformation $\psi^{-1}\phi$ (see (13)). From (22), we get the equalities

$$(dy_i)_x = \frac{\partial y_i}{\partial x_1}(x)dx_1 + \ldots + \frac{\partial y_i}{\partial x_n}(x)dx_n,$$

but

$$\frac{\partial}{\partial x_j}y_i(x) = \frac{\partial}{\partial x_j}(\psi^{-1}\phi)_i(x_1, \ldots, x_n),$$

therefore,

$$(23) \qquad (dy_i)_x = \sum_{j=1}^n \frac{\partial(\psi^{-1}\phi)_i}{\partial x_j}dx_j, \quad i = 1, \ldots, n.$$

Thus, under a transition from one system of local coordinates to another, the differentials of the coordinates considered as functions of a point on a manifold, are transformed by formulae (23), i.e., by the linear transformation which is defined by the Jacobian matrix $\left(\frac{\partial(\psi^{-1}\phi)_i}{\partial x_j}\right)$.

Consider the disjoint union $T^*M = \sqcup_{x \in M^n}(T_xM^n)^*$. We define a structure of a vector bundle on T^*M^n. A natural projection $p : T^*M^n \to M^n$ is already given. Let (U, ϕ) be a chart on M^n, $\{x_i(x)\}$ the local coordinates of the point $x \in U$. We define a chart on T^*M^n

$$(24) \qquad \sigma_\phi : p^{-1}(U) \to \mathbf{R}^n \times \mathbf{R}^n,$$

by the rule

$$(25) \qquad \sum_{i=1}^{n} a_i dx_i \mapsto (x_1, \dots, x_n; a_1, \dots, a_n),$$

for an element $\sum_{i=1}^{n} a_i dx_i$ from the fibre $p^{-1}(x) = (T_x M^n)^*$.

We show that $\{(p^{-1}(U), \sigma_\phi\}$ is an atlas for a C^∞–structure, if $\{(U, \phi)\}$ is the atlas of the manifold M^n. Let (V, ψ) be another chart at the point x, defining the local coordinates $\{y_i(x)\}_{i=1}^n$, and

$$(26) \qquad \sigma_\psi : p^{-1}(V) \to \mathbf{R}^n \times \mathbf{R}^n$$

the corresponding chart on $T^* M^n$. It is clear that $p^{-1}(U \cap V) = p^{-1}(U) \cap p^{-1}(V)$. If $x \in U \cap V$, then using the coordinates of chart (26) to the element $\sum_{i=1}^n a_i dx_i$ there is associated:

$$(27) \qquad \sum_{i=1}^{n} a_i dx_i \mapsto (y_1, \dots, y_n; b_1, \dots, b_n).$$

From (25) and (27), we know that in $(T_x M^n)^*$:

$$(28) \qquad \sum_{i=1}^{n} a_i (dx_i)_x = \sum_{i=1}^{n} b_i (dy_i)_x.$$

It is easy to obtain the relation between $\{a_i\}$ and $\{b_i\}$ from (28). Indeed, by substituting the expression $(dy_i)_x$ from (23) in (28), and afterwards identifying the coefficients of the differentials dx_i on both sides of the equality, we obtain

$$\begin{pmatrix} b_1 \\ \dots \\ b_n \end{pmatrix} = \left(\frac{\partial(\psi^{-1}\phi)_i}{\partial x_j} \right)^{*-1} \begin{pmatrix} a_1 \\ \dots \\ a_n \end{pmatrix},$$

where $*$ has the meaning of matrix trasnsposition. Hence, it is evident that the vector component of an element of the fibre $p^{-1}(x)$ in a coordinate representation under a change coordinates alters by a different rule than that for the vector component of the tangent vector, namely: it is transformed with the help of the Jacobian matrix $\left(\frac{\partial(\psi^{-1}\phi)}{\partial x} \right)^{*-1}$, while the vector component of the tangent vector is transformed by the Jacobian matrix $\left(\frac{\partial(\psi^{-1}\phi)}{\partial x}\right)$ itself. Variables changing under such a rule are called *covectors*. The elements of the set $(T_x M^n)^*$ are called the *covectors at the point x*.

Now, it is clear that having constructed charts (25) for all charts of a certain atlas on M^n, we have turned the set of all covectors, i.e., $T^* M^n$, into a smooth manifold; this manifold is called the *cotangent bundle*.

§ 8. Vector fields on smooth manifolds

The concepts presented in this section are important both for a great number of mathematical disciplines (differential equations, dynamic systems, topology of manifolds) and for applications to mechanics and physics; here, these connections are noted in the most elementary form. Just like in § 7, for the sake of simplicity in formulation, we consider all the objects to be of class C^∞ and will refer to them as smooth objects.

1. The tangent vector to a smooth path. Let M^n be a smooth manifold. Recall that a path in M^n is a continuous mapping $\chi : (a, b) \rightarrow M^n$ of an interval of the number line into the topological space M^n. Since (a, b) is a smooth submanifold in $\mathbf{R}^1$, one may consider smooth mappings χ and talk about a *smooth path*.

Let χ be a smooth path in M^n, $\chi(t)$ a point of this path, $t \in (a, b)$.

Definition 1. The *tangent vector to a path* χ at a point $\chi(t)$ is the tangent vector $l_{\chi(t)}$ to the maniffold M^n at the point $\chi(t)$ defined by the equality

$$(1) \qquad l_{\chi(t)}(f) = \frac{d}{dt} f(\chi(t))\Big|_t, \quad \hat{f} \in \mathcal{O}(\chi/(t)).$$

Exercise 1°. Verify that the right–hand side in (1) determines a tangent vector to the manifold M^n.

The tangent vector to the path $\chi(t)$ is usually denoted by $\chi'(t)$. Let us find the coordinate representation of the vector $\chi'(t)$. Let (U, ϕ) be a chart at the point $\chi(t)$. If $\hat{g} \in \mathcal{O}(\phi^{-1}(\chi(t)))$, then

$$l_{\chi(t)}(g\phi^{-1}) = \frac{d}{dt}(g\phi^{-1}\chi(t))\Big|_t =$$

$$= \sum_{i=1}^{n} \frac{dx_i(t)}{dt}\frac{\partial g}{\partial x_i}(x_1(t),\dots,x_n(t)),$$

where $\phi^{-1}\chi(t) = (x_1(t),\dots,x_n(t))$ is the corresponding path in $\mathbf{R}^n$. Hence, we obtain the coordinates of the vector $\chi'(t)$ in the chart (U,ϕ):

$$(2) \qquad \tau_\phi l_{\chi(t)} = \left(x_1(t),\dots,x_n(t); \frac{dx_1(t)}{dt},\dots,\frac{dx_n(t)}{dt}\right).$$

Here $x'(t) = (x_1'(t),\dots,x_n'(t))$, $x_i'(t) = \frac{dx_i(t)}{dt}$, is the vector component of the tangent vector.

If $A_x(u,v)$ is the Riemannian metric on M^n, then the length $\|\chi'(t)\|_{\chi(t)}$ of the tangent vector to the path and the length of the portion of the path for $t_1 \le t \le t_2$, are determined :

$$(3) \qquad S_{t_1}^{t_2} = \int_{t_1}^{t_2} \sqrt{A_{\chi(t)}(\chi'(t),\chi'(t))}dt = \int_{t_1}^{t_2} \|\chi'(t)\|_{\chi(t)}dt.$$

In local coordinates formula (3) has the form

$$S_{t_1}^{t_2} = \int_{t_1}^{t_2} \sqrt{g_{ij}(x(t))\frac{dx_i}{dt}\frac{dx_j}{dt}}dt = \int_{t_1}^{t_2} \sqrt{g_{ij}(x(t))dx_i dx_j},$$

where $g_{ij}(x)$ is the corresponding matrix of the bilinear form.

2. The dynamical group of a physical system and its infinitesimal generator.
The concepts of a smooth path and tangent vector to it are naturally applied to the mathematical investigations of physical systems.

We will speak of the set of all possible states of a physical system in a certain process and assume it to be a smooth manifold M^n called the *phase space* of the system. Then $x \in M^n$ denotes a possible state of the system, and the correspondence "point $x \mapsto$ state" is bijective. The state of the system varies with time according to the laws of physics; therefore, the point x corresponding to this state, changes it position with respect to time t. We will assume that the process is determinate, which means that the the state of the system is determined uniquely, in future and in past, by its present state. These processes are described by the dynamical group of a physical system defined in the following manner. If $x \in M^n$ is a point denoting the state of the system at the present moment ($t = 0$), then to the state of the system at a moment t there corresponds the point $\chi = \chi(t, x)$, $\chi(t, x) \in M^n$, $\chi(0, x) = x$. Thus, the point x describes a path $\chi = \chi(t, x)$, $-\infty < t < +\infty$, called the *phase trajectory (orbit)* of the point x. For each $t \in (-\infty, +\infty)$, a transformation $U_t : M^n \to M^n$ is determined by the rule $x \mapsto \chi(t, x)$. By the principle of causality, we have:

$$U_{t_1+t_2}(x) = U_{t_1}(U_{t_2}(x)), \quad t_1, t_2 \in (-\infty, +\infty);$$

and it follows that the family of the transformations $\{U_t\}$ is a group with inverse element $(U_t)^{-1} = U_{-t}$ and unit element $U_0 = 1_{M^n}$.

This group is called the *dynamical group of the physical system*. We assume that the mapping $\mathbf{R}^1 \times M^n \to M^n : (t, x) \mapsto U_t(x)$ is smooth; in this case, the group of the diffeomorphisms $\{U_t\}$ is said to *depend smoothly on t*. From the point of view of physics, the knowledge of the dynamical group means the complete description of the system in time. Such a description is not always possible.

The laws of physics are usually formulated much simpler in "infinitesimal form". That means the following: consider an orbit $\chi(t) = \chi(t, x)$ and its tangent vector $\chi'(0)$ (at a point x). For each point $x \in M^n$, we set $X(x) = \chi'(0)$. The collection of tangent vectors $\{X(x)\}$ is called a *vector field on the manifold M^n*; this field is also called the *infinitesimal generator of the dynamical group*. A law of physics is generally expressed by describing an infinitesimal generator. But then the problem of constructing (describing) the corresponding dynamical groups arises.

Below we consider in more detail the concept of vector fields.

3. Smooth vector fields. A *vector field on a manifold M^n* is a mapping

$$(4) \qquad\qquad X : M^n \to TM^n$$

such that $X(x) \in T_x M^n$ for any $x \in M^n$. The vector field is said to be *smooth* (of class C^∞), if the mapping (4) is smooth (of class C^∞). In local coordinates, a vector field is of the form

(5)
$$\left(x_1, \dots , x_n; X_1(x_1, \dots , x_n)\frac{\partial}{\partial x_1} + \dots \right.$$
$$\left. \dots + X_n(x_1, \dots , x_n)\frac{\partial}{\partial x_n} \right).$$

Exercise 2°. Show that the smoothness of a vector field is equivalent to the smoothness of the functions $X_i(x_1, \dots , x_n)$, $i = 1, \dots , n$.

Let U_t be a group of diffeomorphisms of the manifold M^n, depending smoothly on t, and let $\chi(t, x) = U_t(x)$ be the orbit of the point x.

Definition 2. A vector field $X(x)$ is called the *infinitesimal generator of the group* U_t, if for any orbit $\chi(t) = \chi(t, x)$ we have

(6) $\chi'(0) = X(x).$

Exercise 3°. Show that equality (6) is equivalent to the equality

(7) $\chi'(t) = X(\chi(t)), \quad t \in (-\infty, +\infty).$

Hint. Use the equality $\chi(t)=U_t(x)$ and the group property.

Exercise 4°. Show that the infinitesimal generator $X(x)$ is a smooth vector field.

Consider the problem of determining the group U_t for a given vector field $X(x)$. We search for the orbit $\chi(t) = \chi(t, x_0)$ using condition (7). In local coordinates, we have the system of differential equations

(8) $\dfrac{dx_i}{dt} = X_i(x_1(t), \dots , x_n(t)), \quad i = 1, \dots , n,$

where $(x_1(t), \dots , x_n(t))$ is the coordinate representation of the path χ, and $X_1, \dots , X_n$ are the coordinates of the vector component of the tangent vector

$X(x)$ (see (2) and (5)). The functions $X_i(x_1, \ldots, x_n)$ depend smoothly on $x_1, \ldots, x_n$. In order to determine the orbit (more precisely, that portion of it which lies in the chart), it is necessary to find a solution of the system of differential equations (8) which satisfies the condition $x_1(0) = x_1^0, \ldots x_n(0) = x_n^0$, where $(x_1^0, \ldots, x_n^0)$ are the coordinates of the point x^0. Using the theory of ordinary differential equations, we can find a unique solution $\chi(t) = \chi(t, x)$ for a sufficiently small interval $-\epsilon < t < +\epsilon$ and x from some neighbourhood of the point x^0; moreover, $\chi(t, x)$ depends smoothly on t, x. But in order to construct the (whole) group U_t, it is necessary to extend the solution $\chi(t, x)$ to the whole axis $-\infty < t < +\infty$ for any $x \in M^n$. This cannot always be done (for example, for the equation $y' = y^2$ on $\mathbf{R}^1$). However, if M^n is compact, then the required extension exists, which is not difficult to verify with the help of the theory of ordinary differential equations. This theory leads to the following theorem.

Theorem 1. *If M^n is a compact, smooth manifold, and X a smooth vector field on it, then it is the infinitesimal generator of a one–parameter group of diffeomorphisms which depends smoothly on the parameter.*

Note that the orbits are often called the integral curves of the vector field.

Example 1. Let $M^{2n} = TQ^n$ (the case considered in mechanics, where Q^n is the configuration space which we assume to be a smooth manifold). Let the local coordinates be $(q_1, \ldots, q_n)$ in Q^n, and $(q_1, \ldots, q_n; v_1, \ldots, v_n)$ in TQ^n.

A vector field on TQ^n of the form (the vector component only)

$$
L_{q,v} = v_1 \frac{\partial}{\partial q_1} + \ldots + v_n \frac{\partial}{\partial q_n} +
$$
$$
+ \alpha_1(q, v) \frac{\partial}{\partial v_1} + \ldots + \alpha_n(q, v) \frac{\partial}{\partial v_n}
$$

with smooth functions $\alpha_1(q, v), \ldots, \alpha_n(q, v)$, is said to be *special*. The integral curves of this field are described by the system of differential equations

$$
\frac{dq_i}{dt} = v_1, \qquad \frac{dv_i}{dt} = \alpha_i(q_1, \ldots, q_n; v_1, \ldots, v_n),
$$

which is equivalent to the second order system

$$(9) \qquad \frac{d^2 q_i}{dt^2} = \alpha_i \left(q_1, \ldots, q_n; \frac{dq_1}{dt}, \ldots, \frac{dq_n}{dt} \right),$$

$$i = 1, \ldots, n.$$

Classical mechanics deals with equations of the form (9).

Exercise 5°. Find $\pi_* L_{q,v}$, where $\pi : TQ^n \to Q^n$ is a projection.

4. The Lie algebra of vector fields. Let $C^\infty(M^n)$ be the set of all smooth mappings $f : M^n \to \mathbf{R}^1$. A smooth vector field X on M^n determines a mapping $C^\infty(M^n) \to C^\infty(M^n)$ by the rule $f \mapsto X(f)$, where $X(f)(x) = X(x)(f)$ for any point x from M^n. If (in local coordinates) $X(x) = \sum_{i=1}^n a_i(x)\frac{\partial}{\partial x_i}$, then $X(f)(x) = \sum_{i=1}^n a_i(x)\frac{\partial f}{\partial x_i}$. Clearly, it is a linear mapping of the vector space $C^\infty(M^n)$.

If X, Y are two vector fields, then their product, i.e., the commutator $[X, Y]$, is defined by the formula

$$[X, Y](f) = X(Y(f)) - Y(X(f)).$$

Exercise 6°. Verify that $[X, Y]$ is a vector field and calculate its local coordinates.

It is evident that the set of all vector fields on M^n forms a vector space (over the field $\mathbf{R}^1$) under the natural operations $X + Y$ and $\alpha \cdot X$. The commutator $[X, Y]$ depends linearly on the factors X, Y. Thus, the set of all vector fields on M^n is an algebra.

Exercise 7°. Show that for any vector fields X, Y, Z on M^n, the following equalities hold
   (1) $[X, Y] = -[Y, X]$ (antisymmetry);
   (2) $[[X, Y], Z] + [[Z, X], Y] + [[Y, Z], X] = 0$ (the Jacobi identity).

Thus the set of all vector fields on M^n is the Lie algebra.
   The set of all vector fields on a smooth manifold is not only a vector space, but also possesses the structure of the module over the ring of smooth functions

on M^n. In fact, the product $f \cdot X$, which is the vector field: $(f \cdot X)(x) = f(x) \cdot X(x)$, is defined for all $f \in C^\infty(M^n)$. It is clear that it is a smooth field depending linearly both on f and X.

The study of questions concerning Lie algebras of vector fields is one of the main trends in modern topology.

5. Covector fields. A smooth mapping $A : M^n \to T^*M^n$, for which $A(x)$ belongs to $T_x^*M^n$, i.e., the fibre over the point x, is called a *smooth covector field on M^n*.

In local coordinates, the field A is given in the form

$$A(x) = (x_1, \ldots, x_n; a_1(x)dx_1 + \ldots + a_n(x)dx_n),$$

where the $a_i(x)$ are smooth functions in the coordinates $(x_1, \ldots, x_n)$ of the point x.

If $f \in C^\infty(M^n)$ then $A(x) = (df)_x$ is a covector field on M^n which is smooth, since in local coordinates, the covector component $(df)_x$ is of the form

$$\frac{\partial f}{\partial x_1}dx_1 + \ldots + \frac{\partial f}{\partial x_n}dx_n,$$

and the coordinates $\frac{\partial f}{\partial x_i}$ of this covector are smooth functions.

Example 2. Let $M^{2n} = T^*Q^n$, where Q^n is the configuration space of a mechanical system. Let $(q_1, \ldots, q_n; p_1, \ldots, p_n)$ be local coordinates on T^*Q^n, where $(p_1, \ldots, p_n)$ are the coordinates of the covector at the point $(q_1, \ldots, q_n) \in Q^n$. If $H : T^*Q^n \to \mathbf{R}^1$ is a smooth function, then

$$dH = \frac{\partial H}{\partial q_1}dq_1 + \ldots + \frac{\partial H}{\partial q_n}dq_n +$$
$$+ \frac{\partial H}{\partial p_1}dp_1 + \ldots + \frac{\partial H}{\partial p_n}dp_n$$

is a covector field on the manifold T^*Q^n. We form a vector field on the same manifold

$$L = -\frac{\partial H}{\partial p_1}\frac{\partial}{\partial q_1} - \ldots - \frac{\partial H}{\partial p_n}\frac{\partial}{\partial q_n} +$$
$$+ \frac{\partial H}{\partial q_1}\frac{\partial}{\partial p_1} + \ldots + \frac{\partial H}{\partial q_n}\frac{\partial}{\partial p_n}.$$

The integral curves $(q_1(t), \ldots, q_n(t); p_1(t), \ldots, p_n(t))$ of the field L satisfy the system of differential equations

(10) $$\frac{dq_i}{dt} = -\frac{\partial H}{\partial p_i}, \quad \frac{dp_i}{dt} = \frac{\partial H}{\partial q_i}, \quad i = 1, \ldots, n.$$

In mechanics, one takes a function H which is equal to the sum of kinetic and potential energies, and the system (10) is then called a system of *equations of motion in Hamiltonian form*.

§ 9. Fibre bundles and coverings

1. Preliminary examples. In many problems, there naturally appear spaces that are "glued together" from direct products. Such a space represents a continuous family of spaces, i.e., the fibres are homeomorphic to each other and are indexed by the points of a space, viz. the base. We briefly mention only the very first concepts of the theory of such spaces. This theory has developed quite far. We consider some examples.

Let M^n be a smooth manifod, TM^n the tangent bundle, $\pi : TM^n \rightarrow M^n$ the projection of the tangent bundle onto its manifold. It is clear that for any point $x \in M^n$, the fibre $\pi^{-1}(x)$ is homeomorphic to the space $\mathbf{R}^n$ and, moreover, for a coordinate neighbourhood U of the point x, we have a homeomorphism $\pi^{-1}(U) \simeq U \times \mathbf{R}^n$ (see § 6). Sometimes, there exists a homeomorphism $TM^n \simeq M^n \times \mathbf{R}^n$ as, for instance, in the case when $M^n = S^1$. In general, however, this is not the case. (For example, for $M^n = S^2$.)

The tangent bundle is constructed "locally in x" as the direct product $U \times \mathbf{R}^n$ which is, of course, a corollary to its definition. The existence of similar structure for the sphere S^3 is more unexpected and related to the properties of complex numbers. We construct an example of a mapping of S^3 to S^2, for which the preimage of any point is homeomorphic to the circle. We consider S^3 as a sphere in $\mathbf{C}^2$, i.e.,

$$S^3 = \{(z_1, z_2) : |z_1|^2 + |z_2|^2 = 1\},$$

and the sphere S^2 as the extended complex plane (z–sphere). The formula $\pi(z_1, z_2) = z_1/z_2$ defines a mapping $\pi : S^3 \rightarrow S^2$. For $\lambda = e^{i\alpha}$, we have $\pi(\lambda z_1, \lambda z_2) = \pi(z_1, z_2)$, therefore $\pi^{-1}(z) \simeq S^1$ for any $z \in S^2$. Recall that the

z–sphere S^2 has the structure of C^∞–manifold with local coordinate z on the domain $U_1 = S^2 \backslash \infty$ and z^{-1} on the domain $U_2 = S^2 \backslash 0$ (see § 2).

Consider the direct products $U_1 \times S^1$, $U_2 \times S^1$. The sets $\pi^{-1}(U_1)$, $\pi^{-1}(U_2)$ turn out to be homeomorphic to $U_1 \times S^1$, $U_2 \times S^2$, respectively. To show this we define a mapping $\phi : U_1 \to S^3$ by the formula

$$\phi(z) = \left(\frac{z}{\sqrt{1 + |z|^2}}, \frac{1}{\sqrt{1 + |z|^2}} \right);$$

obviously, $\pi \phi = 1_{U_1}$, and the set $\pi^{-1}(z)$, $z \in U_1$, consists of points of the form $\lambda \phi(z)$, where $\lambda = e^{i\alpha}$. We define a mapping $\tilde{\phi} : U_1 \times S^1 \to S^3$ by the formula

$$\tilde{\phi}(z, \lambda) = \left(\frac{\lambda z}{\sqrt{1 + |z|^2}}, \frac{\lambda}{\sqrt{1 + |z|^2}} \right),$$

$$z \in U_1, \quad \lambda \in S^1.$$

It is clear that $\pi^{-1}(U_1) = \tilde{\phi}(U_1 \times S^1)$ and the diagram

$$
\begin{array}{ccc}
U_1 \times S^1 & \xrightarrow{\tilde{\phi}} & \pi^{-1}(U_1) \\
 & \searrow^{pr_1} \swarrow_{\pi} & \\
 & U_1 &
\end{array}
$$

where pr_1 is the projection of the direct product onto the first factor, is commutative. Analogously, we define a mapping $\tilde{\psi} : U_2 \times S^1 \to S^3$ by the formula

$$\tilde{\psi}(1/z, \lambda) = \left(\frac{\lambda \cdot (1/z)}{\sqrt{1 + |1/z|^2}}, \frac{\lambda \cdot 1}{\sqrt{1 + |1/z|^2}} \right)$$

$$1/z \in U_2, \quad \lambda \in S^1.$$

Then, $\pi^{-1}(U_2) = \tilde{\psi}(U_2 \times S^1)$. It is clear that the following diagram is commutative

$$
\begin{array}{ccc}
U_2 \times S^1 & \xrightarrow{\tilde{\psi}} & \pi^{-1}(U_2) \\
 & \searrow^{pr_1} \swarrow_{\pi} & \\
 & U_2 &
\end{array}
$$

Thus, the mapping π is arranged locally (over the coordinate neighbourhoods of S^2) as the projection of the direct product. However, the sphere S^3 is not homeomorphic to a direct product $S^2 \times S^1$ (the fundamental groups of these spaces are not isomorphic).

The mapping described is called the Hopf fibre bundle (Hopf fibration); it is remarkable in many respects. For example, the Hopf fibre bundle gives a generator of the group $\pi_3(S^2) \simeq \mathbf{Z}$. Note that for any two points $u, v \in S^2$, the circles $\pi^{-1}(u)$ and $\pi^{-1}(v)$ are linked in S^3 (see Fig. 106).

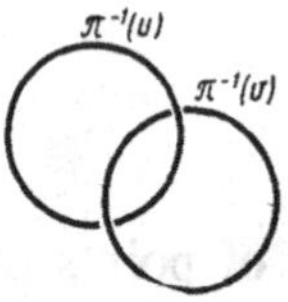

FIG. 106

2. The definition of a fibre bundle. The examples considered in subsection 1, naturally suggest the following definition.

Definition 1. The *locally trivial fibre bundle* is a quadruple (E, B, F, p), where E, B, F are topological spaces, p is a surjective mapping of E onto B, and, moreover, for any $x \in B$, there exists a neighbourhood U of the point x and a homeomorphism $\phi_U : p^{-1}(U) \to U \times F$ such that the diagram

(1)

$$p^{-1}(U) \xrightarrow{\quad \varphi_U \quad} U \times F$$
$$p \searrow \qquad \swarrow pr$$
$$U$$

is commutative. Here pr is the natural projection.

From the definition it follows that for any point x from U, the preimage $p^{-1}(x)$ is homeomorphic to the space F. It is called the *fibre* over the point x.

The spaces E, B, and F are called the (total) *space*, the *base* and the *fibre*, respectively, and the mapping p is the *projection* of the fibre bundle. The neighbourhoods U involved in definition 1 are called *coordinate neighbourhoods*, and the homeomorphisms ϕ_U are called *coordinate homeomorphisms* or *rectifying homeomorphisms*.

Although in topology wider classes of fibre spaces are considered than locally trivial fibre bundles, everywhere below, by a fibre bundle we will mean a locally trivial fibre bundle.

A fibre bundle is said to be *trivial* if there exists a homeomorphism ϕ_B : $E \to B \times F$ such that the diagram

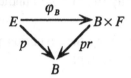

is commutative.

Note that the tangent bundle TM^n can be considered as the space of a locally trivial fibre bundle with base M^n, projection $p = \pi$, viz. the projection of the tangent spaces onto the corresponding points of the manifold M^n, and fibre $\mathbf{R}^n$. The coordinate neighbourhoods of the manifold M^n serve as neighbourhoods $U \subset M^n$.

The Hopf fibre bundle mapping considered above, is the projection of a locally trivial fibre bundle with space S^3, base S^2, and fibre S^1.

We present a number of examples of locally trivial fibre bundles.

Example 1. The Möbius strip M (the quotient space of the direct product $[0, 1] \times [-1, 1]$ with respect to the equivalence relation $(0, y) \sim (1, -y)$) is a fibre space with base S^1 (the "median") and fibre $[-1, 1]$.

The projection pr : $[0, 1] \times [-1, 1] \to [0, 1]$ given by rule $\mathrm{pr}(x, y) = x$, induces a quotient mapping $p : M \to S^1$, and that is the projection of this fibre bundle.

Example 2. The direct product $X \times Y$ of topological spaces X and Y forms a fibre space with the natural projection pr : $X \times Y \to X$, fibre Y, and base X.

Example 3. The sphere S^n is a fibre bundle space with base $\mathbf{R}P^n$, fibre consisting of two points (a disicrete set), and projection which associates a point $x \in S^n$ with its equivalence class $\{x, -x\} \in \mathbf{R}P^n$ (see Ch. 2, § 5).

The sphere S^{2n+1} is a fibre bundle space with the base $\mathbf{C}P^n$, fibre S^1, and projection associating the point $x \in S^{2n+1} \subset \mathbf{C}^{n+1}$ with its equivalence class in $\mathbf{C}P^n$ (see Ch 2, § 5).

Exercise 1°. Show that the tangent bundle of the manifold M^n is trivial iff there exists n (continuous) vector fields on M^n such that they are linearly independent at each point $x \in M^n$.

Exercise 2°. Show that a locally trivial fibre bundle over an interval is trivial.

A mapping $s : B \to E$ satisfying the condition $ps = 1_B$ is called a *cross–section* of the fibre bundle (E, B, F, p).

Exercise 3°. Show that the existence of a cross–section is a necessary condition for the triviality of a fibre bundle.

Exercise 4°. Do cross–sections of the Hopf fibre bundle exist? (Use the fact that $\pi_2(S^3) = 0$ and $s \cdot p = 1_{S^2}$ for any cross–section s)

Exercise 5°. Give an example of a non–trivial fibre bundle which possesses a cross–section.

We shall establish a certain relation between mappings to a fibre space and to its base.

Definition 2. A mapping $\Psi : X \to E$ is called the *lift of a mapping* $\Phi : X \to B$ if for any point $x \in X$, the equality $p\Psi(x) = \Phi(x)$ is fulfilled. It is also said that the mapping Ψ covers the mapping Φ.

The commutativity of the following diagram

(2)

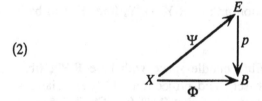

characterizes the relation just introduced.

Exercise 6°. Show that if a fibre bundle possesses a cross–section, then for any mapping to the base, there exists a lift of this mappping.

We present a necessary condition for the existence of a lift of a mapping in terms of the homotopy group functors (see Ch. 3, § 3).

Theorem 1. *Let (E, B, F, p) be a locally trivial fibre bundle with fibre F, bundle space E, and base B, where E and B are path–connected; let X be a topological space. Then for a mapping $\Phi : X \to B$ to have a lifting Ψ satisfying the condition $\Psi(x_0) = e_0$, where $x_o \in X$, $e_0 \in E$, $p(e_0) = b_0 = \Phi(x_0)$ (here x_0, e_0, and b_0 are fixed), it is necessary that*

(3) $$\Phi_n(\pi_n(X, x_0)) \subset p_n(\pi_n(E, e_0))$$

for all $n \geq 1$.

Proof. If such a lifting Ψ exists, then diagram (2) is commutative. Applying the homotopy group functors, we obtain the commutative diagrams (for all $n \geq 1$)

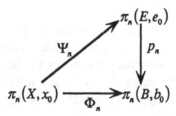

from which the required inclusions follow. □

Locally trivial fibre bundles possess the following important property.

The homotopy lifting property. Let (E, B, F, p) be a locally trivial fibre bundle with a Hausdorff and paracompact base B. Let X be an arbitrary topological space, $\Phi : X \times I \to B$ a homotopy; and let $f : X \to E$ be a lift of the mapping $\Phi|_{X \times 0}$, i.e., $pf = \Phi|_{X \times 0}$. Then there exists a lifting $\Psi : X \times I \to E$ of the homotopy Φ satisfying the condition $\Psi|_{X \times 0} = f$.

This statement will be proved for a special case in subsection 4.

3. Vector bundles. Let (E, B, F, p) be a locally trivial fibre bundle. Assume that U and V are coordinate neighbourhoods of the point $x \in B$. The homeomorphisms $g_V^U(x)$ of the space F can now be defined by the formula

$$g_V^U(x)h = \phi_V \phi_U^{-1}(x, h),$$
$$x \in U \cap V, \quad h \in F; \quad g_V^U(x) = 1_F.$$

If W is a third neighbourhood of the point x then the following equalities hold

$$g_W^U(x) = g_W^V(x)g_V^U(x).$$

Thus, for each point $x \in U \cap V$, a homeomorphism $g_V^U(x)$ is defined, i.e., there is a mapping $g_V^U : U \cap V \to H(F)$ of the set $U \cap V$ to the group $H(F)$ of homeomorphisms of the space F. The mappings g_V^U are called *coordinate transformations*. If F is a locally compact and the topology in $H(F)$ is induced by the imbedding of $H(F)$ into the space $C(F, F)$ with the compact open topology, then the coordinate transformation maps g_V^U, as can be easily seen, are continuous (see Ch. 3, § 1, exercise 11°).

Definition 3. A *vector bundle* is a locally trivial fibre bundle (E, B, F, p) whose fibre F is a finite–dimensional vector space and whose coordinate transformations g_V^U are continuous mappings to the group of invertible linear transformations of the space F (i.e., for constant U and V, $g_V^U(x)$ is a family of invertible linear operators continuously depending on $x \in U \cap V$).

Exercise 7°. Show that the tangent bundle is a vector bundle.

Definition 4. A *morphism of a locally trivial fibre bundle* (E, B, F, p) *to a locally trivial fibre bundle* (E', B', F', p') is a pair of continuous mappings $H : E \to E'$, $h : B \to B'$ such that $hp = p'H$.

The latter equality means that the diagram

$$
\begin{CD}
E @>H>> E' \\
@VpVV @VVp'V \\
B @>>> B'
\end{CD}
$$

is commutative (each fibre is taken into a fibre).

This definition turns the collection of locally trivial fibre bundles into a category.

Definition 5. Let (E, B, F, p), $E', B', F', p')$ be vector bundles whose fibres F and F' are vector spaces over the same field. Let (H, h) be a morphism of (E, B, F, p) to (E', B', F', p'). The morphism (H, h) is called a *morphism of vector bundles* if for any point $x \in B$ the composition

$$F \xrightarrow{\phi_x^{-1}} p^{-1}(x) \xrightarrow{H} (p')^{-1}(h(x)) \xrightarrow{\phi'_{h(x)}} F'$$

is a linear mapping, where ϕ_x, $\phi'_{h(x)}$ are the homeomorphisms of the fibres $p^{-1}(x)$, $(p')^{-1}(h(x))$ and the vector spaces F, F', respectively, which appear in the commutative diagram of definition 1.

Exercise 8°. Verify that vector bundles and their morphisms form a category.

Exercise 9°. Verify that by associating to a manifold its tangent bundle, and to a smooth mapping of manifolds the corresponding tangent mapping, we define a covariant functor from the category of smooth manifolds to the category of vector bundles (over the field **R**).

4. Coverings. Now we turn to a more precise study of one particular class of locally trivial fibre bundles.

Consider the circle $S^1 = \{z \in \mathbf{C} : |z| = 1\}$ and define the mapping p : $\mathbf{R}^1 \to S^1$ by the formula $p(t) = e^{2\pi i t}$. Since $p(t_1) = p(t_2)$ holds iff $t_1 - t_2 = k$, $k \in \mathbf{Z}$, the preimage $p^{-1}(z)$ of any point $z \in S^1$ is homeomorphic to the set of integers $\mathbf{Z}$ with the discrete topology. For any point $z \in S^1$, the mapping p homeomorphically maps each connected component of the set $p^{-1}(S^1 \backslash z) = \mathbf{R}^1 \backslash p^{-1}(z)$ onto $S^1 \backslash z$. The multi-valued mapping $p^{-1} : S^1 \backslash z \to \mathbf{R}^1 \backslash p^{-1}(z)$, $p^{-1}(u) = (1/2\pi i) \ln u$, has a countable number of single-valued branches, one of which we denote by ϕ.

Define the homeomorphism $\tilde{\phi} : (S^1 \backslash z) \times \mathbf{Z} \to \mathbf{R}^1 \backslash p^{-1}(z)$ by the formula $\tilde{\phi}(u, k) = \phi(u) + k$. Then we obtain a commutative diagram

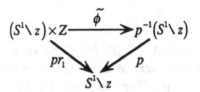

The system of sets $\{S^1 \setminus z\}_{z \in S^1}$ can be taken as a system of coordinate neigh-bourhoods in such a way that the quadruple $(\mathbf{R}^1, S^1, \mathbf{Z}, p)$ is a locally trivial fibre bundle whose fibre $\mathbf{Z}$ is discrete. Such fibre bundles appear in problems of analysis (especially function theory).

Definition 6. A locally trivial fibre bundle (E, B, F, p) is called a *covering* if the space E and the base B of the fibre bundle are path–connected, and the fibre F is a space with the discrete topology.

Instead of the quadruple (E, B, F, p) if no misunderstandings are likely, one writes $p : E \longrightarrow B$ and one calls the mapping p the covering. The fibre $p^{-1}(x)$ over each point of the covering is homeomorphic to the space F with the discrete topology, therefore, it is itself a discrete space.

In the definition of a covering (also of a locally trivial fibre bundle with a path–connected base), the requirements for the homeomorphism ϕ_U can be weakened by assuming that ϕ_U is only a homeomorphism on $U \times F_U$, where F_U is a space with the discrete topology (for a covering), dependent on the coordinate neighbourhood U. Under this definition, it is evident that $p^{-1}(x) \sim F_U$ (bijection or homeomorphism, respectively) for any $x \in U$. But it turns out that $F_U \sim F_V$ (bijection, homeomorphism) for any coordinate neighbourhoods U, V, and putting $F = p^{-1}(x_*)$, where x_* is a fixed point from B, we get back to definition 6 (or 1) (see the remark after the proof of lemma 1).

Example 4. The fibre bundle with as total space the sphere S^n over the projective space $\mathbf{R}P^n$ is a covering whose fibre consists of two points.

Example 5. The mapping $p : S^1 \rightarrow S^1$ (or $p : C \setminus 0 \rightarrow C \setminus 0$) given by the correspondence $z \mapsto z^n$, is a covering with a fibre consisting of n points.

A covering whose fibre consists of n points is called an n–sheeted covering.

Note that for a coordinate neighbourhood U of the covering (E, B, F, p), the preimage $p^{-1}(U)$ is homeomorphic to the product $U \times F$ which consists of non–intersecting "sheets", viz. the open sets $U \times \alpha$, $\alpha \in F$, and consequently, it itself consists of non–intersecting "sheets", viz. the open sets $W_\alpha = \phi_U^{-1}(U \times \alpha)$ which are homeomorphic to U; the $p_\alpha : W_\alpha \to U$, i.e., the restrictions of p to W_α, are homeomorphisms that establish this, as follows from the relation $p = \mathrm{pr}\phi_U$ expressing the commutativity of diagram 1.

Thus, the projection of the covering $p : E \to B$ is a local homeomorphism with the discrete preimage $p^{-1}(x)$ which is homeomorphic to the fibre F over each point $x \in B$.

However, the inverse is not true: a covering of the space B by the coordinate neighbourhoods U cannot be constructed for any local homeomorphism $p : E \to B$ (for instance, for the mapping $p : (a, b) \to S^1$ of a real interval onto the circle, given by the formula $p(t) = (\cos t, \sin t)$).

It is said that the points from the fibre $p^{-1}(x)$ "lie over" the point x, and the sheets W_α "lie over" U; the property indicated above of the projection of the covering $p : E \to B$ allows us to "lift" the subsets $A \subset U$ to the sheet W_α by considering the preimages $p_\alpha^{-1}(A)$, and also to "lift" mappings, paths, and homotopies into X. According to definition 2, a path $f : I \to E$ is called a lift of the path $g : I \to B$ (which then is said to cover the path g), if $pf = g$.

Lemma 1. *Let $p : E \to B$ be a covering. Then the following statements hold:*
(1) any path γ in B starting at a point $b_0 \in B$, has a unique covering path $\tilde\gamma$ in E which starts at an arbitrary fixed point $e_0 \in p^{-1}(b_0)$; (2) if $\gamma = \gamma_1 \cdot \gamma_2$ is the product of two paths γ_1, γ_2 in B, then the covering path is $\tilde\gamma = \tilde\gamma_1 \cdot \tilde\gamma_2$, where $\tilde\gamma_1$, $\tilde\gamma_2$ cover γ_1, γ_2, respectively and moreover, $\tilde\gamma_1(1) = \tilde\gamma_2(0)$; (3) if $\gamma = \gamma_1^{-1}$ is the path inverse to γ_1, then $\tilde\gamma = \tilde\gamma_1^{-1}$ and, moreover, $\tilde\gamma(0) = \tilde\gamma_1(1)$.

Proof. Let the path γ be given by a mapping $\gamma : I \to B$, $I = [0, 1]$, $\gamma(0) = b_0$. Each point of the path $\gamma(t)$ belongs to a certain coordinate neighbourhood U_t, and there exists a connected neighbourhood (i.e., an interval) Ω_t of each point $t \in I$ such that $\gamma(\Omega_t) \subset U_t$. We select a finite subcovering $\{\Omega_{s_i}\}_1^k$ of the interval I from its open covering $\{\Omega_t\}$. Let δ be the *Lebesgue number* of the covering $\{\Omega_{s_i}\}_1^k$, i.e. the supremum of all numbers S such that for each interval of length S there is an element of the covering $\{\Omega_1\}$ which wholly contains the interval. Divide the interval I into segments $\Delta_i = [t_{i-1}, t_i]$ of length less than δ by division points t_i, $i = 0, \dots, N$, $t_0 = 0$, $t_N = 1$.

Then $\gamma(\Delta_i)$ lies in a certain coordinate neighbourhood U_i, $i = 1, \ldots, N$. Consequently, each part γ_i of the path γ given by the mapping $\gamma_i : \Delta_i \to B$ permits a lift $\tilde{\gamma}_i$ to any sheet W_{α_i} which is given by the mapping $f_i = p_{\alpha_i}^{-1} \gamma_i$: $\Delta_i \to W_{\alpha_i}$, where $p_{\alpha_i} : W_{\alpha_i} \to U_i$ is a homeomorphism to the coordinate neighboourhood U_i. We choose a sheet containing the point $e_0 \in p^{-1}(b_0)$ as W_{α_1} and lift the part γ_1 of the path; then $\tilde{\gamma}_1$ starts at the point e_0. If $W_{\alpha_{i-1}}$ has already been chosen and the part γ_{i-1} of the path lifted, then as W_{α_i}, we choose the sheet containing the end point $f_{i-1}(t_{i-1})$ of the part of the path $\tilde{\gamma}_{i-1}$ which lies over the point $\gamma(t_{i-1})$. Then the part $\tilde{\gamma}_i$ of the lifted path starts at the point $f_{i-1}(t_{i-1})$. Thus, we can lift all the parts γ_i, $i = 1, \ldots, N$, of the path γ. Since the mappings $f_i : \Delta_i \to E$, $i = 1, \ldots, N$, are compatible at both ends of the adjacent intervals Δ_i, they can be combined into a mapping $f : I \to E$, $f(t) = f_i(t)$ for $t \in \Delta_i$. The mapping f gives a path $\tilde{\gamma}$ which covers the path γ. The uniqueness of the covering path follows from p being a local homeomorphism. This proves statement (1). Statements (2) and (3) are evident.   □

Remark. The construction of the coordinate neighbourhoods U_i, $i = 1, \ldots, N$, which cover the path $\gamma : I \to B$, allows us to prove the bijectivity of the fibres F_U over the coordinate neighbourhoods, where the homeomorphism ϕ_U goes from $p^{-1}(U)$ to $U \times F_U$ (see definition 6 of a covering and the subsequent discussion). It is evident that $F_U \sim F_V$ provided $U \cap V \neq \emptyset$, since for $x \in U \cap V$, we have $p^{-1}(x) \sim F_U$ and $p^{-1}(x) \sim F_V$. If $U \cap V = \emptyset$ then select points $x \in U$, $y \in V$, and join them with the path $\gamma : I \to B$, using the path–connectedness of B. For the indicated covering $\{U_i\}$, the fibres F_{U_i}, $F_{U_{i+1}}$ are homeomorphic to each other (in the discrete topology), whence $p^{-1}(x) \sim p^{-1}(y)$ and $F_U \sim F_V$.

Using the lemma just proved, it is easy to prove the homotopy lifting theorem, mentioned in subsection 2, for the case of coverings.

Theorem 2. (Homotopy lifting property for coverings). *Let (E, B, F, p) be a covering, X a topological space, $f : X \to E$ a mapping, and $\Phi : X \times I \to B$ a homotopy such that $pf = \Phi|_{X \times 0}$. Then there exists a unique lift Ψ of the homotopy Φ, i.e., a homotopy $\Psi : X \times I \to E$ such that $\Psi|_{X \times 0} = f$ and $p\Psi = \Phi$.*

Proof. The homotopy $\Phi : X \times I \to B$ for any $x \in X$, determines a path

$g_x : I \to B$, where $g_x(t) = \Phi(x, t)$, $t \in I$. The point $f(x)$ lies over the point $g_x(0) = \Phi(x, 0)$. According to lemma 1, the path g_x is uniquely lifted to E, $\widetilde{g}_x : I \to E$, under the condition $\widetilde{g}_x(0) = f(x)$. Set $\Psi(x, t) = \widetilde{g}_x(t)$. Thus, we have a mapping $\Psi : X \times I \to E$. It covers the mapping Φ, since

$$p\Psi(x, t) = p\widetilde{g}_x(t) = g_x(t) = \Phi(x, t), \quad (x, t) \in X \times I.$$

For $t = 0$, it is evident that $\Psi(x, 0) = f(x)$, i.e., $\Psi|_{X \times 0} = f$.

The following exercise concludes the proof of the theorem.

Exercise 10°. Prove that the mapping $\Psi : X \times I \to E$ is continuous. □

The lemma on lifting a path to a covering space and the homotopy lifting theorem for coverings allow us to investigate the relation between the fundamental groups of the base B and the covering space E. First, we formulate some immediate geometric corollaries.

Lemma 2. *Let $p : E \to B$ be a covering, and let $e_0 \in E$, $b_0 = p(e_0) \in B$ be any given points. Then the following statements hold:*

(1) *if α is a closed path in E with origin at the point e_0 and α is homotopic*) to a constant path, then $\beta = p\alpha$ is a closed path with origin at the point b_0 and it is also homotopic to a constant path;*

(2) *if α is a path in E with origin at the point e_0 covering a closed path β, then a homotopy of the path β lifts to a homotopy of the path α with fixed ends;*

(3) *if α is a path in E with origin at the point e_0 covering a closed path β which is homotopic to a constant path, then α is also closed and homotopic to a constant path.*

Proof. Statement (1) is evident because of the continuity of the mapping p. We prove statement (2). Let $\beta : I \to B$, $\beta(0) = \beta(1) = b_0$ be a closed path in B, and $f_t : I \to B$, $0 \leq t \leq 1$, $f_0 = \beta$ a homotopy. We denote by $\Psi_t : I \to E$, $0 \leq t \leq 1$, $\Psi_0 = \alpha$ the lift of the homotopy $f_t : p\Psi_t = f_t$, $0 \leq t \leq 1$ (cf Theorem 2 above). Since the ends of the path f_t are fixed, i.e., $f_t(0) = f_t(1) = b_0$ for all t, the ends $\Psi_t(0)$, $\Psi_t(1)$ of the path Ψ_t belong to $p^{-1}(b_0)$ for all t and depend continuously on t. Since the topology of the fibre $p^{-1}(b_0)$ is discrete,

---

*)Here, as in Ch. 3, we consider homotopies of paths with fixed ends.

the ends $\Psi_t(0)$, $\Psi_t(1)$ of the path are constant, i.e., the homotopy of the path α takes place while the ends $\alpha(0) = e_0$, $\alpha(1)$ are fixed. Finally, we prove statement (3). Let f_t be the presupposed homotopy of the path β, and Ψ_t a homotopy of the path α covering f_t. By assumption, f_1 is a constant mapping so that $f_1(1) = b_0$, whence $\Psi_1(I) \subset p^{-1}(b_0)$, i.e., Ψ_1 is a mapping to the fibre over b_0. The image $\Psi_1(I)$ of the segment I is a connected set, therefore by the discreteness of the topology of the fibre, $\Psi_1(I) = e_0' \in p^{-1}(b_0)$, in particular, $\Psi_1(0) = \Psi_1(1) = e_0'$. According to statement (2), the homotopy Ψ_t takes place with fixed ends, i.e., $\Psi_t(1) = \alpha(1)$, $\Psi_t(0) = \alpha(0)$, $0 \le t \le 1$. Consequently, $\alpha(1) = \alpha(0) = e_0'$, $\Psi_1(I) = e_0'$, i.e., the path α is closed and homotopic to a constant path.   □

We now study the relation between the fundamental groups of the space of a covering and its base.

The projection $p : E \to B$ induces a homomorphism of the fundamental groups $\pi_1(E)$ and $\pi_1(B)$ (see Ch. 3, § 3). The nature of this homomorphism is described by the following theorem.

Theorem 3. *The homomorphism $p_* : \pi_1(E) \to \pi_1(B)$ of the fundamental groups induced by the projection of a covering is a monomorphism.*

Proof. Let $x_0 \in E$, $b_0 \in B$ be given points such that $p(x_0) = b_0$; let $\pi_1(E, x_0)$, $\pi_1(B, b_0)$ be the corresponding fundamental groups, and let $p_* : \pi_1(E, x_0) \to \pi_1(B, b_0)$ be the homomorphism induced by the projection $p : E \to B$. Consider the preimage $p_*^{-1}(e)$ of the unit element of the group $\pi_1(B, b_0)$. It suffices to show that $p_*^{-1}(e) = e'$, where e' is the unit element of the group $\pi_1(E, x_0)$. If $[\alpha] \in p_*^{-1}(e)$ then α covers the path $\beta = p\alpha$ which is homotopic to a constant path in B. According to statement (3) of Lemma 2 above, the path α is also homotopic to a constant path (in E), therefore, $[\alpha] = e'$.   □

Thus, from theorem 3 it follows that the group $\pi_1(E)$ is isomorphic to a subgroup of the group $\pi_1(B)$ (namely, to the subgroup $N = p_*(\pi_1(E))$. Consider the cosets (for instance, right cosets) in the group $\pi_1(B)$ with respect to the subgroup N. The following important theorem holds.

Theorem 4. *For any covering $p : E \to B$, the fibre $p^{-1}(b_0)$ is in bijective correspondence with the set of cosets of the group $\pi_1(B)$ with respect to the subgroup N.*

Proof. We associate a point $x_\beta \in p^{-1}(b_0)$ to the homotopy class $[\beta] \in \pi_1(B, b_0)$ by the following rule: we lift the path β to the path α in E with origin at the point x_0 (see lemma 1 on lifting a path) and put $x_\beta = \alpha(1)$; according to lemma 2 (statement (2)), the end of the path α does not depend on the choice of a representative $\beta \in [\beta]$, and, therefore, the mapping $\pi_1(B, b_0) \to p^{-1}(b_0)$, $[\beta] \mapsto x_\beta$ is well defined. If $[\beta_1]$, $[\beta_2]$ belong to the same coset, then $[\beta_1] \cdot [\beta_2]^{-1} \in p_*(\pi_1(E, x_0))$; consequently, the loop $\beta_1 \beta_2^{-1}$ with origin at b_0 is homotopic to a certain loop $p\alpha$, where α is a loop in E with origin at x_0. Denote the lift of the loop $\beta_1 \cdot \beta_2^{-1}$ with origin at x_0 by α' and note that the loops α and α' are homotopic with fixed ends (statement (2) of lemma 2); consequently, α' is a closed loop which covers the loop $\beta_1 \cdot \beta_2^{-1}$. But now, by lemma 1 (statements (2), (3)), $\alpha' = \widetilde{\beta}_1 \cdot \widetilde{\beta}_2^{-1}$. The fact that the path $\widetilde{\beta}_1 \cdot \widetilde{\beta}_2^{-1}$ is closed, implies that the origin of path $\widetilde{\beta}_1$ coincides with that of path $\widetilde{\beta}_2$, and the same holds for their ends. Consequently, $x_{\beta_1} = x_{\beta_2}$. Thus, the mapping $[\beta] \mapsto x_\beta$ is constant on each coset. In addition, for different cosets different images result. In fact, assume the opposite, then there are a $[\beta_1]$ and $[\beta_2]$ from different cosets, but such that $x_{\beta_1} = x_{\beta_2}$; the latter means that the ends (and the origins) of the lifts $\widetilde{\beta}_1$, $\widetilde{\beta}_2$ coincide; therefore, $\widetilde{\beta}_1 \cdot \widetilde{\beta}_2^{-1}$ is a loop in E with origin at the point x_0, $p(\widetilde{\beta}_1 \cdot \widetilde{\beta}_2^{-1}) = \beta_1 \cdot \beta_2^{-1}$ is a loop (with origin at the point b_0), so that the homotopy class $[\beta_1 \cdot \beta_2^{-1}] = [\beta_1] \cdot [\beta_2]^{-1} = p_*[\widetilde{\beta}_1 \cdot \widetilde{\beta}_2^{-1}]$ of this loop belongs to $p_*(\pi_1(E, x_0))$, i.e., $[\beta_1]$, $[\beta_2]$ are from the same coset, which is contrary to the assumption. Finally, it remains to show that any point $\widetilde{x} \in p^{-1}(b_0)$ is the image x_β for a certain $[\beta]$. Take a path α in E joining the point x_0 with the point $\widetilde{x}$ (using the condition that E is path–connected), and set $\beta = p\alpha$; β is a closed path in B with the origin at the point b_0, the path α is its lift, so $x_\beta = \widetilde{x}$. $\qquad\square$

Corollary. *If the space of the covering $p : E \to B$ is 1–connected, i.e., $\pi_1(E) = 0$, then the fibre F and the fundamental group $\pi_1(B)$ are in bijective correspondence.*

Proof. Fix an $x_0 \in E$, $p(x_0) = b_0$ and consider $\pi_1(E, x_0) = e'$, $\pi_1(B, b_0)$. We have $p_*(\pi_1(E, x_0)) = e$, and consequently, the set of cosets coincides with the set $\pi_1(B, b_0)$. Thus, $\pi_1(B, b_0) \sim p^{-1}(b_0) \sim F$ (the equivalence is a bijection). $\qquad\square$

Definition 7. A covering (E, B, F, p) is said to be *universal* if the space E

is 1–connected, i.e., $\pi_1(E) = 0$. The space E is then called the *universal covering space* (of B).

Knowing a universal covering of a space is useful in calculating the fundamental group $\pi_1(B)$ (by the corollary above).

Example 6. The covering $p : \mathbf{R}^1 \to S^1$, $p(t) = e^{2\pi i t}$, $F = \mathbf{Z}$ is universal. We know already (see Ch. 3, § 3) that $\pi_1(\mathbf{R}^1) = 0$. Therefore, $\pi_1(S^1) \sim \mathbf{Z}$ (cf. also Ch 3, § 4).

Example 7. Consider the covering $p : S^n \to \mathbf{R}P^n$ with fibre $F \sim \mathbf{Z}_2$, $n \geq 2$. We have $\pi_1(S^n) = 0$, $n \geq 2$; hence $\pi_1(\mathbf{R}P^n) \sim \mathbf{Z}_2$.

However, the results obtained are not complete. By establishing a bijection of the group $\pi_1(B)$ with some other group, we cannot be sure that this bijection preserves group operations, i.e., is a homomorphism of groups. We shall strengthen theorem 4 and the corollary in this respect by assuming that on the covering space E is given, and that this an action of a group G is compatible with the covering structure.

We shall consider a group G which acts (from the left) on the space E, and identify, for brevity, an element $g \in G$ with the corresponding homeomorphism $h_g : E \to E$ (see Ch. 2, § 5).

Definition 8. It is said that a group G *acts discretely* (or that G is a *discrete group of transformations*), if the orbit O_y of any point $y \in E$ is a discrete subspace.

Definition 9. A group of transformations G which acts discretely on E, is said to be *properly discontinuous*, if for any point $y \in E$ there exists a neighbourhood $U(y)$ of the point y such that the images $g(U)$, $g \in G$ pairwise do not intersect. Such a neighbourhood will be called *elementary* below.

Definition 10. It is said that the group G *acts freely on E* (or *without fixed points*, if $g(y) \neq y$ for all $y \in E$, for any element $g \in G$, $g \neq e$.

Clearly, a properly discontinuous group of transformations G acts freely.
Let G be a properly discontinuous group of transformations of the space E.

Consider the space of orbits $E/G = B$ and the natural projection $p : E \to B$ (see Ch. 2, § 5).

Lemma 3. *Let E be a path–connected space, and G a properly discontinuous group of transformations in E. Then $p : E \to E/G = B$ is a covering with its fibre $p^{-1}(b)$, $b \in B$, equal to the orbit O_y of the point y, $p(y) = b$.*

Proof. By the definition of an orbit space, the projection p is a continuous mapping, and $p^{-1}(b) = O_y$, if $p(y) = b$; moreover, $O_y \sim G$. The path–connectedness of the space $p(E) = B$ follows from the path–connectedness of E and the continuity of p. It remains to construct coordinate neighbourhoods in B. Let $U(y)$ be an elementary neighbourhood of a point $y \in E$, $b = O_y$ an orbit of the point y, and $V(b)$ an open neighbourhood of the point $b \in B$ consisting of all orbits O_z, $z \in U_y$, which intersect the neighbourhood $U(y)$. For a properly discontinuous group G and an elementary neighbourhood $U(y)$, we have $p^{-1}(V) = \bigcup_{g \in G} g(U)$, moreover, the $g(U)$ are open in E and do not intersect. The image of $g(U)$ is a "sheet" W_g over V of the covering $p : E \to B$. Indeed, $p^{-1}(V) = \bigcup_{g \in G} W_g$, and, in addition, W_g is homeomorphic to V, since the restriction $p_g = p|_{W_g} : W_g \to V$ is a homeomorphism due to the fact that the mapping p_g is bijective and open. The neighbourhood $V(b)$ is a coordinate neighbourhood, since the homeomorphism $\phi_V : p^{-1}(V) \to V \times G$ (G is considered with the discrete topology) is given on the open non–intersecting sets W_g by the mappings $p_g : W_g \to V \times g$ for any $g \in G$. □

Example 8. The covering $p : S^{2n+1} \to L(k, k_1, \dots, k_n)$ of the sphere over a generalized lens space determined by the projection of the complex sphere $S_{\mathbf{C}}^n$ (homeomorphic to S^{2n+1}) to the quotient space $L(k, k_1, \dots, k_n)$ of orbits under the action of the group $\mathbf{Z}_k$ (see Ch. 2, § 5). A fibre of this covering coincides with an orbit of the group $\mathbf{Z}_k$, and hence consists of k elements. Since $\pi_1(S^{2n+1}) = 0$, $\pi_1(L) \sim \mathbf{Z}_k$.

Example 9. Consider $E = \mathbf{R}^n$ as an Abelian group; it contains the subgroup $\mathbf{Z}^n$ of all vectors with integer coordinates. The quotient group $\mathbf{R}^n / \mathbf{Z}^n$ equipped with the quotient topology, is called the n–dimensional torus T^n. The quotient mapping $p : \mathbf{R}^n \to T^n$ is a covering with the fibre $\mathbf{Z}^n$. Since $\pi_1(\mathbf{R}^n) = 0$, we conclude that $\pi_1(T^n) \sim \mathbf{Z}^n$.

For a covering $p : E \to E/G = B$, the following lemma holds.

Lemma 4. *Under the conditions of lemma 3, the subgroup $N = p_*(\pi_1(E, e_0))$ of the fundamental group $\pi_1(B, b_0)$, where $p(e_0) = b_0$, is a normal divisor.*

Proof. Let $[\beta] \in N$, $[\beta_1] \in \pi_1(B, b_0)$. To verify that $[\gamma] = [\beta_1]^{-1} \cdot [\beta] \cdot [\beta_1] \in N$, we lift the path $\gamma = \beta_1^{-1} \cdot \beta \cdot \beta_1$ to a path $\tilde{\gamma} = \tilde{\beta}_1^{-1} \cdot \tilde{\beta} \cdot \tilde{\beta}_1$, where $[\tilde{\beta}] \in \pi_1(E, e_0)$, $\tilde{\beta}_1$ runs from the point e_0 to a point e_1, $p(e_1) = b_0$, and $\tilde{\beta}_1^{-1}$ from the point e_1 to the point e_0. Consequently, $\tilde{\gamma}$ is a loop at the point e_1, and $p\tilde{\gamma} = \gamma$. Since the fibre $p^{-1}(b_0)$ is the orbit O_{e_1} of the group G, there exists an element $g_1 \in G$ such that $g_1(e_1) = e_0$. The homeomorphism g_1 maps the loop $\tilde{\gamma}$ to a loop $g_1\tilde{\gamma}$ at the point e_0 so that $[g_1\tilde{\gamma}] \in \pi_1(E, e_0)$. The path $g_1\tilde{\gamma}$ covers the path γ, since the mapping p is constant on the orbits of the group G; therefore, $p(g_1\tilde{\gamma}) = \gamma$ and $[\gamma] = p_*([g_1\tilde{\gamma}])$, i.e., $[\gamma] \in N$. $\square$

A covering for which the subgroup $N = p_*(\pi_1(E, e_0))$ is a normal divisor, is said to be *regular*.

For regular coverings, a set of cosets of the group $\pi_1(B, b_0)$ with respect to the subgroup N, is a quotient group.

Before proceeding with the calculation of $\pi_1(E/G)$, we introduce the important concept of the monodromy group of a covering.

Let $p : E \to B$ be a covering, and $b_0 \in B$ a fixed point of the base. We define an action of the group $\pi_1(B, b_0)$ on the fibre $p^{-1}(b_0) \sim F$. Let $[\beta] \in \pi_1(B, b_0)$, and $e_\alpha \in p^{-1}(b_0)$ be an arbitrary point of the fibre over b_0 whose subscript is an element $\alpha \in F$. Let $\tilde{\beta}$ be a lift of the path β starting at the point e_α; set $e_{\alpha'} = \tilde{\beta}(1)$, where α' is the element of F corresponding to $\tilde{\beta}(1)$. We know already that $\tilde{\beta}(1)$ does not depend on the choice of the path β from the class $[\beta]$, but depends on the class $[\beta]$ itself. Thus, the class $[\beta]$ determines a mapping $\sigma_\beta : p^{-1}(b_0) \to p^{-1}(b_0)$ by the rule $e_\alpha \mapsto e_{\alpha'}$ (or a mapping $\sigma_\beta : F \to F$ by the rule $\alpha \mapsto \alpha'$.)

It is easy to see that the mapping σ_β is a homeomorphism onto $p^{-1}(b_0)$.

The equalities $\sigma_{\beta_1 \cdot \beta_2} = \sigma_{\beta_2} \cdot \sigma_{\beta_1}$, $\sigma_\beta = 1_F$, if $[\beta] \in e$ (the unit element of $\pi_1(B, b_0)$), $\sigma_{\beta^{-1}} = \sigma_\beta^{-1}$ are easily derived from lemma 1 about lifting paths.

These equalities mean that the correspondence $\sigma : [\beta] \mapsto \sigma_\beta$ is a representation of the group $\pi_1(B, b_0)$ by "homeomorphisms", i.e., "permutations" of the discrete space $p^{-1}(b_0)$ (or F). This representation σ is called the *monodromy of the covering*, and the set of permutations $\{\sigma_\beta\}$, $[\beta] \in \pi_1(B, b_0)$, is called the *monodromy group of the covering.*

Thus, the monodromy σ is a homomorphism*) of the group $\pi_1(B, b_0)$ to the group of all permutations of the fibre.

It follows from theorem 3 that the point $e_\alpha \in p^{-1}(b_0)$ is fixed for those and only those permutations σ_β, for which $[\beta] \in p_*(\pi_1(E, e_\alpha))$. The group $p_*(\pi_1(E, e_\alpha))$ is called the isotropy subgroup of the point e_α in the group $\pi_1(B, b_0)$ acting on the fibre $p^{-1}(b_0)$. Moreover, $\sigma_\beta(e_\alpha) = \sigma_{\beta'}(e_\alpha)$ iff $[\beta']$ belongs to $p_*(\pi_1(E, e_\alpha))[\beta]$, i.e., to the coset containing the element $[\beta]$ (from which theorem 4 also follows immediately). For different points e_α, $e_{\alpha'} \in p^{-1}(b_0)$, the subgroups $p_*(\pi_1(E, e_\alpha))$, $p_*(\pi_1(E, e_{\alpha'}))$ are conjugate with respect to that element $[\beta] \in \pi_1(B, b_0)$ for which $\sigma_\beta(e_\alpha) = e_{\alpha'}$; indeed, if $\tilde{\beta}$ is a corresponding covering path, then the correspondence $\gamma \mapsto \gamma' = \tilde{\beta}^{-1} \cdot \gamma \cdot \tilde{\beta}$, where $[\gamma] \in \pi_1(E, e_0)$, establishes an isomorphism between $\pi_1(E, e_0)$ and $\pi_1(E, e_{\alpha'})$ transformed by the monomorphism p_* into an isomorphism $p_*(\pi_1(E, e_\alpha)) \rightarrow [\beta]^{-1}p_*(\pi_1(E, e_\alpha))[\beta] = p_*(\pi_1(E, e_{\alpha'}))$.

We are now going to calculate the monodromy group $\{\sigma_\beta\}$ for a covering $p : E \rightarrow E/G = B$ generated by a properly discontinuous group of transformations G.

Lemma 5. *The monodromy group of the covering* $p : E \rightarrow E/G = B$ *generated by a properly discontinuous group of transformations of a path-connected space E is isomorphic to G.*

Proof. Let $e_0 \in E$, $b_0 \in p(e_0)$ be distinguished points. We have $p^{-1}(b_0) = O_{e_0}$, where O_{e_0} is the orbit of the point e_0 of the group G, i.e., the set of points $\{g(e_0)\}$, $g \in G$. Let $[\beta] \in \pi_1(B, b_0)$ and σ_β be the corresponding monodromy transformation. Then there is $g_\beta \in G$ such that $\sigma_\beta(e_0) = g_\beta(e_0)$, and then $\sigma_\beta(g(e_0)) = g(\sigma_\beta(e_0))$ for $\forall g \in G$. The correspondence $\sigma_\beta \mapsto g_\beta$ determines a homomorphism of the monodromy group into the group G. Indeed, if $\sigma_{\beta_2} \cdot \sigma_{\beta_1}$ is the superposition of σ_{β_1} and σ_{β_2}, then $(\sigma_{\beta_2} \cdot \sigma_{\beta_1})e_0 = \sigma_{\beta_2}(g_{\beta_1}e_0) = g_{\beta_2}(g_{\beta_1}e_0) = (g_{\beta_1}g_{\beta_2})e_0$. Therefore, $\sigma_{\beta_2} \cdot \sigma_{\beta_1} \mapsto g_{\beta_1} \cdot g_{\beta_2}$.

Furthermore, the permutation $\sigma_\beta^{-1} = \sigma_{\beta^{-1}}$ corresponds to g_β^{-1}, and to the identity permutation $\sigma_\beta = 1_{O_{e_0}}$ ($[\beta] = e$) corresponds $g_\beta = e_G$ the unit element of the group G. We show that the homomorphism $\sigma_\beta \mapsto g_\beta$ is a monomor-

---

*)It would be more correct to call the monodromy an "antihomomorphism"; if we reverse the order of multiplication in $\pi_1(B, b_0)$ or in the group of all permutations of the fibre, then the monodromy becomes a "real" homomorphism.

phism of the monodromy group into the group G. In fact, if $g_\beta = e_G$, then $\sigma_\beta(ge_0) = ge_G e_0 = ge_0$ for any $g \in G$, and consequently, σ_β is the identity mapping of the fibre O_{e_0}.

The surjectivity of the homomorphism $\sigma_\beta \mapsto g_\beta$ follows from the fact that E is path–connected, which allows us to join, by some path α, the point e_0 with the point $g_* e_0$, where $g_* \in G$ is arbitrary, so that α is a lift of the loop $\beta_* = p\alpha$, and $\sigma_{\beta_*} e_0 = \alpha(1) = g_*(e_0)$; therefore, $\sigma_{\beta_*} \mapsto g_*$. Thus, the isomorphism of the monodromy group to the group G is established. $\qquad\square$

Now, it is not difficult to prove the following basic theorem.

Theorem 5. *For a covering $p : E \to E/G = B$ generated by a properly discontinuous group of transformations G of a path–connected space E, the quotient group of the group $\pi_1(B, b_0)$ with respect to the normal divisor $p_*(\pi_1(E, e_0))$, $p(e_0) = b_0$, is isomorphic to the group G.*

Proof. Consider the homomorphism $s : \pi_1(B, b_0) \to G$ given by the composition of the homomorphism of the group $\pi_1(B, b_0)$ into the monodromy group of the covering and the isomorphism of the monodromy group to the group G. i.e., the homomorphism is given by the correspondence $[\beta] \mapsto \sigma_\beta \mapsto g_\beta$. The preimage $s^{-1}(e_G)$ consists of those classes $[\beta]$ for which $g_\beta = e_G$, i.e., σ_β is the identity transformation of the fibre $p^{-1}(b_0)$. Consequently, $s^{-1}(e_G) = p_*(\pi_1(E, e_0))$, and the quotient homomorphism $\hat{s} : \pi_1(B, b_0)/p_*(\pi_1(E, e_0)) \to G$ is an isomorphism. $\qquad\square$

Corollary. *If the covering $p : E \to E/G = B$ is universal then the group $\pi_1(B)$ is isomorphic to the group G.*

We now go back to examples 6, 7, 8, 9.

The universal covering $p : \mathbf{R}^1 \to S^1$, $p(t) = e^{2\pi i t}$, is generated by a properly discontinuous group of transformations by the translation $t \mapsto t + n$, $n \in \mathbf{Z}$, of the axis $\mathbf{R}^1$. Consequently, $\pi_1(S^1) \simeq \mathbf{Z}$ (isomorphism). The monodromy group is also $\mathbf{Z}$ and acts on the fibre $F \sim \mathbf{Z}$ by translations $m \mapsto m + n$.

The universal covering $p : S^n \to \mathbf{R}P^n$, $n \geq 2$, is generated by a properly discontinuous group of transformations $\mathbf{Z}_2$ with the generator $a : S^n \to S^n$ acting by the rule $a(x) = -x$; consequently, $\pi_1(\mathbf{R}P^n) \simeq \mathbf{Z}_2$, $n \geq 2$. The monodromy group is $\mathbf{Z}_2$ and acts on the fibre $F = p^{-1}(b_0) = \{x_0, -x_0\}$,

$x_0 \in S^n$; for the generator σ, we have $\sigma(x_0) = -x_0$, $\sigma(-x_0) = +x_0$, i.e., σ permutes the points of the fibre. The generator of the group $\pi_1(\mathbf{R}P^n, b_0)$ corresponding to the element σ is formed by the homotopy class of the path $p\gamma$, where γ is a path on S^n joining the points x_0 and $-x_0$.

The universal covering $p : S^{2n+1} \to L(k, k_1, \ldots, k_n)$ is generated by a properly discontinuous action of the group $\mathbf{Z}_k$ with generator $a : S^{2n+1} \to S^{2n+1}$. Consequently, $\pi_1(L) \simeq \mathbf{Z}_k$, i.e. the monodromy group is also $\mathbf{Z}_k$ and acts on the fibre; its generator corresponds to the generator $[\gamma] \in \pi_1(L)$, where γ is the projection of the path in S^{2n+1} joining the point x_0 with the point $a(x_0)$. (Exercise: find a, $a(x_0)$ using subsection 3, § 5, Ch. 2).

The universal covering $p : \mathbf{R}^n \to T^n$ is generated by a properly discontinuous action of the group $\mathbf{Z}^n$ with generators a_i acting according to the rule

$$(x_1, \ldots, x_{i-1}, x_i, x_{i+1}, \ldots, x_n) \mapsto$$
$$\mapsto (x_1, \ldots, x_{i-1}, x_i + 1, x_{i+1}, \ldots, x_n),$$
$$i = 1, \ldots, n.$$

Consequently, $\pi_1(T^n) \simeq \mathbf{Z}^n$, and the generators $[\gamma_i]$, $i = 1, \ldots, n$, of the group $\pi_1(T^n)$ contain the loops γ_i which are obtained by the projection p from the paths in $\mathbf{R}^n$ joining the point 0 with the points $a_i(0)$. The monodromy group acts on the fibre $F \sim \mathbf{Z}^n$, and its generators σ_{a_i}, $i = 1, \ldots, n$, act on the integer vectors from $\mathbf{Z}^n$ by the rule:

$$(k_1, \ldots, k_n) \mapsto (k_1, \ldots, k_{i-1}, k_i + 1, k_{i+1}, \ldots, k_n).$$

For the further study of universal coverings, of the base of that covering, it is necessary to impose stronger conditions than path–connectedness. We introduce the following definitions.

Definition 11. A topological space X is said to be *locally path–connected* if for any point $x \in X$, there exists a base of open path–connected neighbourhoods. If the neighbourhoods of such a base possess additionally the property of 1–connectedness, then the space is said to be *locally 1–connected*.

It is easy to give examples of locally path–connected and locally 1–connected spaces (e.g., the Euclidean spaces $\mathbf{R}^n$ or manifolds). A locally 1–connected space need not necessarily be 1–connected, for instance, the circle S^1. In

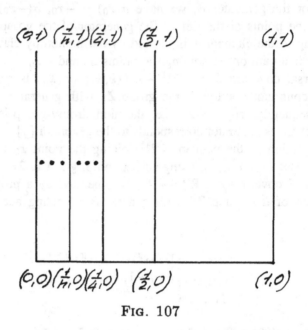

FIG. 107

fig. 107, a space is represented ("the comb space") which is path–connected, but does not possess the properties of local path–connectedness (and, consequently, local 1–connectedness). Fig. 108, depicting an infinite sequence of circles of radii $1/n$, $n = 1, 2, \ldots$, with a common point of tangency, illustrates a path–connected and locally path–connected, but not locally 1–connected space.

However, for the further constructions below, it is sufficient to assume a weaker condition than the local 1–connectedness. This condition is contained in the following definition.

Definition 12. A topological space X is said to be *semi–locally* 1–*connected* if for any point $x \in X$ there exists a neighbourhood where any two paths with common ends are homotopic within the whole space (or, which is equivalent, where any loop in that neighbourhood is contractible within the whole space).

It is not difficult to see that if the space X is locally path–connected and semi–locally 1–connected, then at every point $x \in X$ there exists a base of open path–connected neighbourhoods which possess the property that any two

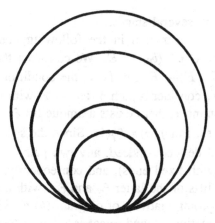

FIG. 108

paths with common ends in the neighbourhood of this base are homotopic in the whole space X.

An example of a semi–locally 1–connected, but not locally 1–connected space is the cone over the space drawn in fig. 108.

It should be also noted that a connected and locally path–connected space is path–connected.

The phrase "universal covering" refers to the fact that a 1–connected space covering B is a covering space over any other space which covers B. More precisely, the following statement is valid.

Theorem 6. *Let $(\widetilde{E}, B, \widetilde{F}, \widetilde{p})$ be a universal covering over a connected, locally path–connected space B. For any covering (E, B, F, p) over B, there exists a surjective mapping $f : \widetilde{E} \to E$ such that the diagram*

(4)
$$\widetilde{E} \xrightarrow{\ f\ } E$$
$$\widetilde{p} \searrow \quad \swarrow p$$
$$B$$

is commutative. Thus, the mapping f is the projection of a covering

$(\widetilde{E}, E, F', f)$ with fibre F' which is a discrete space being in a bijective correspondence with the group $\pi_1(E)$.

Proof. We prove this in several stages.

1. A mapping f is constructed in the following manner. Let $b_0 \in B$, $e_0 \in p^{-1}(b_0) \subset E$, $\widetilde{e}_0 \in \widetilde{p}^{-1}(b_0) \subset \widetilde{E}$. We construct the mapping f as the lift of the mapping $\widetilde{p} : \widetilde{E} \to B$, satisfying the condition $f(\widetilde{e}_0) = e_0$. For an arbitrary point $x \in \widetilde{E}$, consider a path $\gamma : I \to \widetilde{E}$ with origin at $\widetilde{e}_0$ and end at x. According to lemma 1, there exists a unique lift $\xi_\gamma : I \to E$ of the path $\widetilde{p}\gamma : I \to B$, $\xi_\gamma(0) = e_0$. Put $f(x) = \xi_\gamma(1)$. Since the space $\widetilde{E}$ is 1–connected, the mapping f is well defined. Indeed, any two paths γ and ω in $\widetilde{E}$ from x to y are homotopic (with fixed ends), and consequently, their projections $\widetilde{p}_\gamma$ and $\widetilde{p}_\omega$ in B and the lifts of the latter ξ_γ and ξ_ω (with a common origin) are homotopic in E. The commutativity of diagram (4) is evident.

The mapping f is continuous and, what is more, a local homeomorphism. This is evident for a sufficiently small neigbourhoods of the points $\widetilde{e}_o$ and e_0, namely, for the sheets $\widetilde{W}_\alpha$ and W_β lying in $\widetilde{E}$ and E, respectively, over a path–connected coordinate neighbourhood V. Indeed, for paths γ lying in a neighbourhood $\widetilde{W}_\alpha$, we obtain $\xi_\gamma = (p_\beta^{-1}\widetilde{p}_\alpha)\gamma$, and therefore, the mapping $f|_{\widetilde{W}_\alpha} = p_\beta^{-1}\widetilde{p}_\alpha$ is a local homeomorphism. In order to verify this fact, for any pair of points $x \in \widetilde{E}$, $y \in E$, where $f(x) = y$, it suffices to notice that x, y can be taken as new distinguished points $\widetilde{e}_0$ and e_0, and the mapping f under this assumption, is unaltered (the verifiation is left to the reader).

2. We show the surjectivity of f. Let y be an arbitrary point from E; consider the path $\gamma : I \to E$ with origin at e_0 and end at y. For the path $p\gamma : I \to B$, there exists a unique lift $\eta_\gamma : I \to \widetilde{E}$ with origin at $\widetilde{e}_0$ and end at a certain point $x = \eta_\gamma(1)$. Then the paths $f\eta_\gamma$ and γ have a common origin and cover the same path $p\gamma$ in B. Therefore, $f\eta_\gamma(1) = \gamma(1)$, i.e., $f(x) = y$, so that f is surjective.

3. We show that $f : \widetilde{E} \to E$ is the projection of a covering. For an arbitrary point $e \in E$, consider the intersection $\Omega = U \cap V$ of coordinate neighbourhoods U and V containing the point $p(e)$ for the coverings $(\widetilde{E}, B\widetilde{F}, \widetilde{p})$ and (E, B, F, p), respectively. Then Ω is a coordinate neighbourhood for both these coverings; without loss on generality, one may assume this neighbourhood to be path–connected. The following commutative diagram

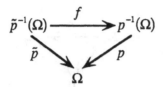

arises. The restriction of the mapping f to any sheet $\widetilde{W}_\alpha$ from $\tilde{p}^{-1}(\Omega)$ in this diagram is a homeomorphism

$$f|_{\widetilde{W}_\alpha} : \widetilde{W}_\alpha \to W_\beta, \quad f_{\widetilde{W}_\alpha} = p_\beta^{-1}\tilde{p}_\alpha,$$

where $W_\beta = f(\widetilde{W}_\alpha)$ is a sheet from $p^{-1}(\Omega)$.

We take a sheet W_β containing some point e. The set of those sheets $\widetilde{W}_\alpha$ of which the preimage $f^{-1}(W_\beta)$ consists, we denote by F_β'; the $\widetilde{W}_\alpha \in F_\beta'$ are the connected components of the preimage $f^{-1}(W_\beta)$. We provide the set F_β' with the discrete topology. Define the mapping

$$\psi_{W_\beta} : f^{-1}(W_\beta) \to W_\beta \times F_\beta'$$

by the formula

$$\psi_{W_\beta}(x) = (f(x), c(x)),$$

where $c(x)$ is the connected component containing the point $f(x)$ and which plays the role of "subscript of the sheet". Obviously, ψ_{W_β} is a local homeomorphism and a bijection, and, therefore, a homeomorphism.

By the same token, for an arbitrary point $e \in E$, a coordinate neighbourhood W_β and the coordinate homeomorphism ψ_{W_β} are constructed (the commutativity of the corresponding diagram is obvious). By the remark below Lemma 1 concerning the definition of a covering, the fibre F_β' does not depend on the choice of the point e and the coordinate neighbourhood $W_\beta \subset E$ up to bijection.

4. Thus, $f : \widetilde{E} \to E$ is a covering. Since it is universal ($\pi_1(\widetilde{E}) = 0$), its fibre F' is in bijective correspondence with the group $\pi_1(E)$.   □

Corollary. *Any two universal coverings* (E_1, B, F_1, p_1) *and* (E_2, B, F_2, p_2) *over a connected, locally path–connected space* B *are equivalent, i.e., there exists a homeomorphism* $f : E_1 \to E_2$ *such that the diagram*

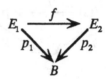

is commutative.

Proof. The local homeomorphism established by theorem 6 is a bijection by theorem 4.   □

We shall pass to the theorem on the existence of a universal covering.

Theorem 7. *Let X be a connected, locally path–connected, and semi–locally 1–connected space. Then there exists a universal covering over X.*

Proof. Note, first, that if in a base of a covering, one has a homotopy with fixed ends of a path, then the same holds for a covering path. Consequently, the points e of a 1–connected covering space correspond bijectively to the homotopy classes of paths in the base with origins at the distinguished point x_0 and the ends at the projections $p(e)$ of the points e. This property enables us to "invert the construction" and to conctruct a 1–connected covering space by means of homotopy classes of the paths in the base.

Thus, let x_0 be a fixed point in X. Consider a certain homotopy class $[\gamma_x]$ of paths γ_x in X with origin at the point x_0 and end at some point $x \in X$. The set $\Gamma(x)$ of all such clases, for fixed x, will be the fibre over the point x, and the union $E = \bigcup_{x \in X} \Gamma(x)$ of all fibres will be the space of the covering. The projection $p : E \to X$ is determined in a natural way: the class $[\gamma_x]$ is associated with the point x by the projection p. Evidently, $p^{-1}(x) = \Gamma(x)$.

First, we now construct the topology on E. For each point $[\gamma_x] \in E$, define a base of open neighbourhoods $\{\Omega_U([\gamma_x])\}$ as follows. Let U be an arbitrary,

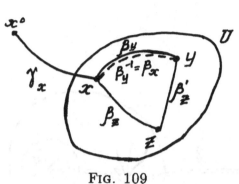

FIG. 109

open, path–connected neighbourhood of the point x. As a neighbourhood of the point $[\gamma_x]$, we take $\Omega_U([\gamma_x])$, i.e. the set of all homotopy classes $[\gamma_y]$ of those paths γ_y from x_0 to $y \in U$ which are products $\gamma_y = \gamma_x \cdot \beta_y$ of a certain path from the class $[\gamma_x]$ by a path β_y from x to y and lying in U; $[\gamma_y]$ depends only on $[\gamma_x]$ and the homotopy class $[\beta_y]$ of the path β_y. The neighbourhood $\Omega_U([\gamma_x])$ is "open", i.e., it is a neighbourhood of any of its points:

$\Omega_U([\gamma_y]) = \Omega_U([\gamma_x])$, if $[\gamma_y] \in \Omega_U([\gamma_x])$. Indeed, $\gamma_y = \gamma_x \cdot \beta_y$, $\gamma_y \cdot \beta_y^{-1} \sim \gamma_x \cdot (\beta_y \cdot \beta_y^{-1})$, and since $\beta_y \cdot \beta_y^{-1}$ is a loop at the point x homotopic to a constant path, we have $\gamma_x \sim \gamma_y \cdot \beta_y^{-1}$ (homotopy with fixed ends). Since $\beta_y^{-1} = \beta_x$ is a path in U from y to x, we obtain $\gamma_x \sim \gamma_y \cdot \beta_x$. If, now, $[\gamma_z] \in \Omega_U([\gamma_x])$, then $\gamma_z = \gamma_x \cdot \beta_z \sim \gamma_y \cdot (\beta_x \cdot \beta_z)$; if $[\gamma_z] \in \Omega_U([\gamma_y])$, then $\gamma_z = \gamma_y \cdot \beta_z' \sim \gamma_x \cdot (\beta_y \cdot \beta_z')$; hence we conclude that the neighourhoods $\Omega_U([\gamma_x])$ and $\Omega_U([\gamma_y])$ coincide. The discussion just given is illustrated in fig. 109.

Note that due to the semi–local 1–connectedness of the space X, there exists a path–connected open neighbourhood V of the point x for which the homotopy class of the product $\gamma_y = \gamma_x \cdot \beta_y$ does not depend on the choice of the path β_y from x to y. The neighbourhood V then serves as a coordinate neighbourhood of the covering under construction. The neighbourhoods Ω_V also form a base for the neigbourhoods of the point γ_y.

Now, check the continuity of the mapping p. It suffices to make certain that $p^{-1}(U)$ is open for any path–connected open neighbourhood U of the point x. Let $[\gamma_y] \in p^{-1}(U)$. Then $[\gamma_y]$ is contained in $p^{-1}(U)$ together with one of its neighbourhoods, namely, the neighbourhood $\Omega_U([\gamma_y])$, i.e., $p^{-1}(U)$ is open.

Further, we show that p is a local homeomorphism. For this, select a "coordinate neighbourhood" V of the point $x \in X$ and the corresponding neighbourhood $\Omega_V([\gamma_x])$ of some fixed point $[\gamma_x]$ of the fibre $\Gamma(x)$. If $[\gamma_y]$ is an arbitrary point from this neighbourhood, then $\gamma_y = \gamma_x \cdot \beta_y$, and all possible

paths β_y (from x to y in V)
fall into the unique homotopy
class $[\beta_y]$, due to the semi–
local 1–connectedness. Consequently, the correspondence
$[\gamma_y] \mapsto y$ defining the mapping $\Omega_V([\gamma_x]) \to V$, is bijective. Moreover, the mapping
$p|_{\Omega_V} : \Omega_V([\gamma_x]) \to V$ is a
homeomorphism, since it is
continuous (as a restriction of
the continuous mapping $p : E$
$\to X$ to an open set) and open
(as $p(\Omega_W([\gamma_y])) = W$ for any
"coordinate neighbourhood"
$W \subset V$ of the point y and

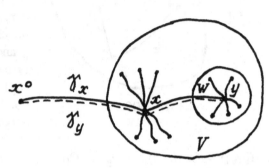

$[\gamma_y] \in \Omega_V([\gamma_x]))$ (see Fig. 110).

Thus, p is a local homeomorphism between E and X, and, moreover, for
a "coordinate neighbourhood" V of the point $x \in X$, we have $p^{-1}(V) =$
$\bigcup_{\alpha \in \Gamma(x)} W_\alpha$, where, for $\alpha = [\gamma_x]$, $W_\alpha = \Omega_V([\gamma_x])$ and $p_\alpha = p|_{W_\alpha} : W_\alpha \to V$
are homeomorphisms; each W_α is open in E and path–connected. In addition,
for $\alpha_1 \neq \alpha_2$, $W_{\alpha_1} \cap W_{\alpha_2} = \emptyset$. In fact, assuming the contrary, there must be
certain non–homotopic paths γ_x^1 and γ_x^2, and a path γ_z, $z \in V$, such that $[\gamma_z]$
lies is the intersection of the neighbourhoods $\Omega_V([\gamma_x^1])$, $\Omega_V([\gamma_x^2])$. By what
has gone before, $\gamma_z \sim \gamma_x^1 \cdot \beta_z$, $\gamma_z \sim \gamma_x^2 \cdot \beta_z'$ (homotopy with fixed ends).
Since $\beta_z' \sim \beta_z$, because of the semi–local 1–connectedness of X, we have
$\gamma_z \sim \gamma_x^1 \cdot \beta_z$, $\gamma_z \sim \gamma_x^2 \cdot \beta_z$, i.e., $\gamma_x^1 \cdot \beta_z \sim \gamma_x^2 \cdot \beta_z$; by multiplying both sides
of the last relation by β_z^{-1} and taking into account that the loop $\beta_z \cdot \beta_z^{-1}$ is
homotopic to a constant loop, we obtain $\gamma_x^1 \sim \gamma_x^2$, contradicting the original
assumption.

Thus, $p^{-1}(V)$ decomposes into the union of non–intersecting sheets W_α
which are open and path–connected in E (and homeomorphic to V), where α
runs through the fibre $\Gamma(x)$.

Now, it is natural to define the coordinate homeomorphism $\phi_V : p^{-1}(V) \to$
$V \times \Gamma(x)$ by taking as coordinates of a point $[\gamma_y] \in p^{-1}(V)$ the "number"
of the sheet W_α in which it lies and the point $y \in V$, i.e., the projection of
the point $[\gamma_y]$ under the homeomorphism $p_\alpha = p|_{W_\alpha} : W_\alpha \to V$; thus, we set
$\phi_V([\gamma_y]) = (y, [\gamma_x])$, if $[\gamma_y] \in \Omega_V([\gamma_x])$. It is obvious from the above that

the definition of the mapping ϕ_V is correct. It remains only to show that ϕ_V is a homeomorphism of an open set $p^{-1}(V)$ in E and the topological product $V \times \Gamma(x)$, where $\Gamma(x)$ is considered with the discrete topology.

The bijectivity of ϕ_V is obvious from the constructions presented above. The continuity of ϕ_V follows from the continuity of the two mappings p : $p^{-1}(V) \rightarrow V$, $p([\gamma_y]) = y$, and q_V : $p^{-1}(V) \rightarrow \Gamma(x)$, $q_V([\gamma_y]) = [\gamma_x]$ which occur in the definition of ϕ_V. The continuity of p was establised earlier, and the continuity of q_V follows from the fact that q_V is locally constant (on every sheet $W_\alpha = \Omega_V([\gamma_x])$). The continuity of ϕ_V^{-1} is a consequence of the discreteness of the topology of the fibre $\Gamma(x)$ and of p_α : $W_\alpha \rightarrow V$ being a homeomorphism. Indeed, a base of open neighbourhoods of the point $\{y \times \alpha\} \in V \times \Gamma(x)$ is formed by the sets $S(y) \times \alpha$, where $S(y) \subset V$ is a path–connected, open neighbourhood of the point y, and the preimage $\phi_V^{-1}(S(y) \times \alpha)$ is equal to $p_\alpha^{-1}(S(y))$ which is an open subset in W_α.

Thus, ϕ_V is a homeomorphism. The commutativity of the diagram which is required in the definition of a coordinate homeomorphism, is evident.

Now, we verify that the space E is path–connected. It sufficies to show that an arbitrary point $[\gamma_x]$ from E can be connected by a path in E with the point $[C_{x_0}]$, which is the homotopy class of the constant path (at the point x_0). Let γ_x : $I \rightarrow X$ be a representative of the class $[\gamma_x]$. Define the path $\zeta^s(t)$: $I \rightarrow X$ for fixed s, $0 \leq s \leq 1$, by the formula $\zeta^s(t) = \gamma_x(st)$. By associating the number s with the homotopy class $[\zeta^s]$ of the path ζ^s, we obtain a mapping ω : $I \rightarrow E$ satisfying the conditions $\omega(0) = [C_{x_0}]$, $\omega(1) = [\gamma_x]$. The continuity of the mapping ω can be easily established in sufficiently small segments of $[0, 1]$ whose images fall into the coordinate neighbourhoods $\Omega_V([\gamma_z])$, where $z = \gamma_x(s)$. Consequently, ω is a path in E with the origin at $[C_{x_0}]$ and the end at $[\gamma_x]$, whence the path–connectedness of E.

Thus, $(E, X, \Gamma(x), p)$ is a covering. In order to complete the proof of the theorem, we esablish the 1–connectedness of the space E. Consider a loop ϕ of the space E at the point e_0, where e_0 is the homotopy class of the constant mapping C_{x_0}. We show that the loop $\gamma = p\phi$: $I \rightarrow X$ (at the point x_0) is homotopic to a constant one. Note that by the construction of the space E, for an arbitrary path ξ : $I \rightarrow X$ with the origin at x_0 and its unique covering path η : $I \rightarrow E$ with origin at e_0, the end $\eta(1)$ of the path η is the homotopy class of the path ξ (in the class of paths with fixed ends). Since ϕ is a unique path with the origin at e_0 covering the path γ, we obtain that $\phi(1) = [\gamma] = [C_{x_0}] = e_0$, i.e., the paths γ and C_{x_0} are homotopic with fixed ends and that implies the contractibility of the loop $\gamma = p\phi$. By the fact that the projection p induces the monomorphism of fundamental groups, the loop ϕ is homotopic to a constant one.

Consequently, $\pi_1(E, e_0) = 0$, completing the proof of the theorem. $\square$

Note that condition (3) $(n = 1)$

$$f_1(\pi_1(X, x_0)) \subset p_*(\pi_1(E, e_0)),$$

which is necessary for lifting the mapping $f : X \to B$, is also sufficient for a connected, locally path–connected space X. In this case, the construction of a lift is based on lifting paths of the form $f\alpha$, where α is a path in X with the origin at x_0 and the end at an arbitrary point x. The correctness of this construction is verified with a help of the homotopy lifting property. The relation between the homotopy groups of a covering space and that of the base of the covering is very simple.

Theorem 8. *Let $p : E \to B$ be a covering. Then for $n \geq 2$ the homomorphism of homotopy groups*

$$p_n : \pi_n(E) \to \pi_n(B),$$

induced by the projection of the covering, is an isomorphism.

Proof. We split the proof of this theorem into three not complicated statements which are left to the readers as exercises.

Exercise 11°. Prove that if $p : E \to B$ is a covering, X a 1–connected, locally 1–connected space (with distinguished points e_0, b_0, x_0, respectively, $p(e_0) = b_0$), $f : X \to B$ a mapping such that $f(x_0) = b_0$, then there exists a unique mapping $F : X \to E$ such that $F(x_0) = e_0$ and $pF = f$.

Hint. To construct the mapping $F(x)$, consider a path α in X joining x_0 and x; then, construct a path β in E which covers the path $f\alpha$ in B; put $F(x)=\beta(1)$. In order to prove the uniqueness of F, use the 1–connectedness of X and for the proof of the continuity of F, the local 1–connectedness of X.

Exercise 12°. Prove that if $p : E \to B$ is a covering, then $p_n : \pi_n(E) \to \pi_n(B)$, for $n \geq 2$, is an epimorphism.

Hint. Show that by exercise 11°, any spheroid $\phi:(S^n,s_0)\to(B,b_0)$ can be covered by a spheroid $\Phi:(S^n,s_0)\to(E,e_0)$.

Exercise 13°. Prove that if $p : E \to B$ is a covering, then $p_n : \pi_n(E) \to \pi_n(B)$, for $n \geq 2$, is a monomoprhism.

Hint. Show that by exercise 11°, any homotopy $\psi:(S^n\times I,s_0\times I)\to(B,b_0)$ of spheroids in B can be covered by a homotopy $\Psi:(S^n\times I,s_0\times I)\to(E,e_0)$ of spheroids in E.

From theorem 8 and the result presented in Ch. 3, § 4, that $\pi_1(S^n) = \pi_2(S^n) = \ldots = \pi_{n-1}(S^n) = 0$, $\pi_n(S^n) \simeq \mathbf{Z}$ $(n \geq 2)$, we obtain

Corollary. *Let* $n \geq 2$. *Then* $\pi_k(\mathbf{R}P^n) = 0$, *for* $1 < k < n$, *and* $\pi_n(\mathbf{R}P^n) \simeq \mathbf{Z}$.

As it has been shown above, $\pi_1(\mathbf{R}P^n) \simeq \mathbf{Z}_2$ for $n \geq 2$.

5. Ramified coverings. To conclude this section, we dwell on the concept of a ramified covering. An example of a ramified covering (see the example in Ch. 1, § 4, of the Riemann surface of the function $w=\sqrt{z}$) is the mapping of the z–sphere S^2 into itself determined by the formula $f(z)=z^2$. Evidently, the quadruple

$$(S^2\setminus\{0,\infty\},S^2\setminus\{0,\infty\},\mathbf{Z}_2,f)$$

(where $\mathbf{Z}_2$ is a two–point space with the discrete topology) is a covering.

Definition 13. A quadruple $(\widetilde{M},M,\mathbf{Z}_n,p)$, where $p:\widetilde{M}\to M$ is called a *ramified covering* if (1) $\widetilde{M}$ and M are two–dimensional manifolds; $\mathbf{Z}_n$ is a space with the discrete topology consisting of n points; (2) for some finite set $T\subset\widetilde{M}$, the quadruple $(\widetilde{M}\setminus T,M\setminus p(T),\mathbf{Z}_n,p)$ is an n–sheeted covering; (3) for any point $y\in M$ and a sufficiently small neighbourhood $V(y)$ of it which is homeomorphic to a disc, the connected components of the set $p^{-1}(V(y))$ are homeomorphic to a disc.

The points $x\in T$ are called the singular points of the ramified covering.

Exercise 14°. Show that the Riemann surface p defined by an algebraic function

$$w^n+a_1(z)w^{n-1}+\ldots+a_{n-1}(z)w+a_n(z)=0,$$

where the $a_i(z)$, $i=1,\ldots,n$, are polynomials (see Ch 1, § 4), is a ramified covering $(P,S^2,\mathbf{Z}_n,p)$. Indicate the singular points of this covering. For $n=2$, compare it with the results obtained in Ch. 1, § 4.

Consider an open neighbourhood $V(p(x^i))$ of the image of a singular point x^i, which is homeomorphic to a disc, such that for all other singular points x^j it follows from the condition $p(x^j) \in \bar{V}(p(x^i))$ that $p(x^i) = p(x^j)$. The preimage of the boundary $\partial V(p(x^i))$ of this neighbourhood decomposes into several closed curves, viz. the circles which bound the connected components of the set $p^{-1}(V(p(x^i)))$ that are homeomorphic to open discs. Let $U(x^i)$ be the connected component of $p^{-1}(V(p(x^i)))$ containing the point x^i. The degree of the mapping (see Ch. 3, § 4)

$$p|_{\partial U(x^i)} : \partial U(x^i) \to \partial V(p(x^i))$$

is called the multiplicity of the branch point x^i; we denote it by k_i. It is evident that the multiplicity of the ramification can be defined for non–singular points, as well. If $p|_{U(x^i)} : U(x^i) \to V(p(x^i))$ is a homeomorphism, then, obviously, $\deg p|_{\partial U(x^i)} = \pm 1$. In the general case, the generators of $\pi_1(\partial U(x^i))$ and $\pi_1(\partial V(p(x^i)))$ are chosen arbitrarily and so is the sign of k_i. However, in a number of cases, the sign k_i is determined in a natural way. Thus, for a ramified covering $(S^2, S^2, \mathbf{Z}_2, z^2)$, the multiplicities of the points 0 and ∞ are equal to 2, and the multiplicity of any other point is equal to 1, For the ramified covering $S^2, S^2, \mathbf{Z}_2, \bar{z}^2)$, the multiplicities of the points 0 and ∞ are equal to -2, and that of any other point is -1,

Exercise 15°. Calculate the multiplicity of the ramification for singular points of the ramified covering from exercise 14°.

Let us establish the following important formula:

(1) $$\chi(\widetilde{M}) = n \cdot \chi(M) - \sum_i (|k_i| - 1),$$

which connects the multiplicities of the singular points with the Euler characteristics of the space of a ramified covering and its base.

We assume that the spaces $\widetilde{M}$ and M are compact and triangulable, i.e., closed surfaces. For any singular point $x^i \in T$, we select a neighbourhood $V(p(x^i))$, as it was done above. Consider now the quadruple

$$\left(\widetilde{M} \setminus \bigcup_i U(x^i), \; M \setminus \bigcup_i V(p(x^i)), \; \mathbf{Z}_n, p \right),$$

where x^i runs through the whole set T. This is clearly an n–sheeted covering (not ramified) whose space and base may be considered triangulable. These triangulations may be chosen sufficiently small and compatible so that the full preimage of a vertex, an edge and a triangle from the base are sets of n vertices, edges and triangles, respectively. Therefore, the following equality holds

(2) $$\chi\left(\widetilde{M} \setminus \bigcup_i U(x^i) \right) = n \cdot \chi\left(M \setminus \bigcup_i V(p(x^i)) \right).$$

Let the full preimage $p^{-1}(p(x^j))$ consist of m points $x^{j_1}, x^{j_2}, \ldots, x^{j_m}$. Then the full preimage $p^{-1}(V(p(x^j)))$ consists of m discs $U(x^{j_s})$. Since the boundary $\partial U(x^{j_s})$ is mapped onto $\partial V(p(x^j))$ locally homeomorphically with degree k_{j_s}, $s=1,\ldots,m$, the set $p^{-1}(y) \cap \partial U(x^{j_s})$, for every point $y \in \partial V(p(x^j))$, consists precisely of $|k_{j_s}|$ points. Consequently, for every singular point x^j, and the points $x^{j_s} \in p^{-1}(p(x^j))$, we have

$$\sum_{s=1}^{m} |k_{j_s}| = n, \tag{3}$$

and the number m of connected components of the set $p^{-1}(V(p(x^j)))$ satisfies the relation

$$n - \sum_{s=1}^{m} (|k_{j_s}| - 1) = m. \tag{4}$$

We glue the discs $\bar{U}(x^{j_s})$ which lie over the disc $V(p(x^j))$ to the space $\widetilde{M} \setminus \bigcup_i U(x^i)$. Denote the space obtained by $\widetilde{M}'$. Since the Euler characteristic of the disc is equal to 1 and that of its boundary is equal to 0, we obtain

$$\chi(\widetilde{M}') = \chi\left(\widetilde{M} \setminus \bigcup_i U(x^i)\right) + \sum_{s=1}^{m} \chi\left(U(x^{j_s})\right) -$$

$$- \sum_{s=1}^{m} \chi\left(\bar{U}(x^{j_s}) \cap \left(\widetilde{M} \setminus \bigcup_i U(x^i)\right)\right) =$$

$$= \chi\left(\widetilde{M} \setminus \bigcup_i U(x^i)\right) + m =$$

$$= \chi\left(\widetilde{M} \setminus \bigcup_i U(x^i)\right) + n - \sum_{s=1}^{m} (|k_{j_s}| - 1). \tag{5}$$

Gluing in one by one the other discs which are situated over the remaining points $p(x^i) \in M$, i.e., the projections of the singular points $x^i \in T \subset \widetilde{M}$, we get

$$\chi(\widetilde{M}) = \chi\left(\widetilde{M} \setminus \bigcup_i U(x^i)\right) + l \cdot n - \sum_i (|k_i| - 1), \tag{6}$$

where l is the number of different images $p(x^i)$ of the singular points x^i. Recall that $\chi\left(\widetilde{M} \setminus \bigcup_i U(x^i)\right)$ and $\chi\left(M \setminus \bigcup_i V(p(x^i))\right)$ are related by equality (6). Now note that

$$\chi(M) = \chi\left(M \setminus \bigcup_i V(p(x^i))\right) + l, \tag{7}$$

since M may be obtained from $M \setminus \bigcup_i V(p(x^i))$ by gluing in l discs. Thus, from (6),(10), and (11) we get

$$\chi(\widetilde{M}){=}n{\cdot}\chi\Big(M\backslash\bigcup_i V(p(x^i))\Big){+}n{\cdot}l{-}\sum_i(|k_i|{-}1){=}$$

$$=n{\cdot}\chi(M){-}\sum_i(|k_i|{-}1).$$

$\square$

Recalling the expression for the Euler characteristic in terms of the genus of a closed surface, it is easy to derive from formula (5) a formula for the sum of the multiplicities of the singular points on $\widetilde{M}$ in terms of the genus of M and the genus of $\widetilde{M}$. For instance, in the case of orientable M and $\widetilde{M}$ of genus $p,\bar{p}$, respectively, we have

$$\sum_i(|k_i|{-}1){=}2(\bar{p}{+}n(1{-}p){-}1).$$

This formula is the combinatorial analogue of the well–known Riemann–Hurwitz formula in the theory of Riemann surfaces.

Exercise 16°. Compare the last formula with that for the number of branch points for an algebraic function which was established at the end of § 4, Ch. 1.

§ 10. Smooth functions on a manifold and the cellular structure of a manifold (example)

1. Example of a function on a torus.
A manifold is a topological space which locally looks like an Euclidean space. However, the manifold as a whole, can be quite complicated. The study of the properties of manifolds presents considerable difficulties. How should non–diffeomorphic, non–homeomorphic or non–homotopy equivalent manifolds be distinguished? For example, are a complex projective space and a sphere of the same dimension homeomorphic? The most coarse of the equivalences mentioned is homotopy equivalence. Therefore, first of all, the homotopy type of a manifold should be investigated.

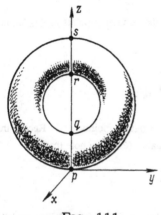

FIG. 111

An eminently useful tool in investigating the homotopy type of a manifold and solving many other problems in topology of manifolds is the theory of critical points of functions on manifolds. We illustrate this method by a simple example.

Consider the two–dimensional torus $M \subset \mathbf{R}^3$ tangent to the plane xy as in Fig. 111. Consider the function f on the torus whose value at the points of the torus with coordinates (x, y, z) is equal to z, i.e., the height of the point over the plane xy.

Exercise 1°. Verify that the function f so defined is a smooth function on the torus.

Denote the tangent point of the torus and the plane by p, and the points of the torus that lie over it on the perpendicular to that plane, by q, r and s in order of increasing height.

When studying functions on a manifold, we shall use the concepts of the *Lebesgue set* $(\phi \leq c) = \{x \in X : \phi(x) \leq c\}$ of a function $\phi : X \to \mathbf{R}$ and the *level set* $(\phi = c) = \{x \in X : \phi(x) = c\}$ of a function ϕ. These sets will be used essentially in the analysis of functions on manifolds in § 12.

Evidently, the level line $(f = c)$ of the function $f = z$ is the intersection of the torus with the plane $z = c$. The set of all points of the torus which lie not higher than the plane $z = c$, is the set $(f \leq c)$; we denote it by M^c. The set M^c is empty for $c < 0$; when $c = 0$, M^c consists of one point, i.e., is a zero–dimensional manifold; for $0 < c < f(q)$, M^c is homeomorphic to a plane closed disc; for $f(q) < c < f(r)$, this set is homeomorphic to a cylinder; for $f(r) < c < f(s)$, the set M^c is a torus with the cap cut–off (the cap being homeomorphic to an open plane disc); for $c \geq f(s)$, $M^c = M$. All these cases are illustrated in Fig. 112.

Exercise 2°. Describe the sets $M^{f(q)}$ and $M^{f(r)}$.

Exercise 3°. Show that when $0 < c < f(s)$ and $c \neq f(q), f(r)$, the set M^c is a two–dimensional manifold with boundary $(f = c)$.

Intuitively, the sets drawn in Fig. 112 are not only non–homeomorphic, but homotopy non–equivalent as well. This homotopy type of the set M^c alters when c passes through the values of the function f at the points p, q, r, s

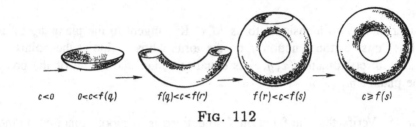

$$c<0 \qquad 0<c<f(q) \qquad f(q)<c<f(r) \qquad f(r)<c<f(s) \qquad c\geq f(s)$$

FIG. 112

FIG. 113 ~ • ~ FIG. 114

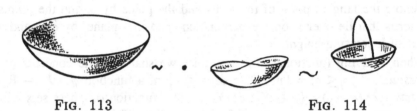

FIG. 115

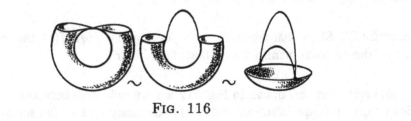

FIG. 116

which we singled out. Let us investigate this alteration in more detail.

It is obvious that the set M^c when $0 \leq c < f(q)$, is homotopy equivalent to the zero–dimensional disc (Fig. 113); for $c = f(q)$, the set M^c is homotopy equivalent to a disc with an arc glued to it (Fig. 114). As c moves, for $f(q) \leq c < f(r)$, this homotopy type is preserved (Fig. 115). For $c = f(r)$, the set M^c is a sylinder with the boundaries glued together at one point, therefore, M^c is homotopy equivalent to a cylinder with an arc glued to it or disc with two arcs (Fig. 116). When $f(r) < c < f(s)$, the set M^c is a torus with a cap cut–off; clearly, its homotopy type is the same as in the previous case (Fig. 117). Finally, for $c \geq f(s)$, the set M^c is the whole torus obtained from the previous type by gluing to it the cut–off cap which is homeomorphic to a plane disc.

FIG. 117

Thus, using the smooth function f we have "constructed" the torus from discs of different dimensions by sequentially gluing in discs and by converting to homotopy equivalent figures.

Below, it will be shown that any smooth manifold can be obtained in this way.

2. Cell complex. We shall analyze the operations of gluing in discs in the example under discussion and describe more presicely the meaning of "to glue in a disc".

Let X be a Hausdorff space and let $\bar{D}^n$ be the closed disc with radius 1 and centre at the origin of coordinates in the space $\mathbf{R}^n$, and S^{n-1} its boundary. Let $g : S^{n-1} \to X$ be a continuous mapping. The result of gluing the disc $\bar{D}^n$ to X by the mapping g is the quotient space $X \cup_g \bar{D}^n = (X \cup \bar{D}^n)/R$ of the disconnected (non–intersecting) union of X and $\bar{D}^n$ with respect to the equivalence R under which $u \sim g(u)$, $u \in S^{n-1}$.

A *cell* is the image e^n of the set $\text{int}(\bar{D}^n)$ in $X \cup_g \bar{D}^n$ with respect to the quotient mapping.

Thus, the disc is glued along its boundary to X with the help of the given continuous mapping g.

In case $n = 0$, the disc $\bar{D}^0$ is a point, its boundary the empty set; the result of gluing $\bar{D}^0$ to X is the disconnected union of X and $\bar{D}^0 = e^0$, i.e., the space X with a separate point. Another example is the gluing the disc $\bar{D}^n$ to the disc $\bar{D}^0$ which produces the sphere S^n.

One may glue the discs sequentially by taking the cell e^0 as an original space. Thus, we shall adopt the rule to glue the boundary of the disc $\bar{D}^k$ to a finite set of cells of dimension not higher than $(k - 1)$. A space which can thus be represented as the result of a sequential gluing in of discs (of different dimensions), is called a cell complex.

For example, the sphere S^n is a cell complex consisting of two cells e^n and e^0, where e^n is glued along its boundary to e^0.

We present a definition of a cell complex which is independent of the previous arguments, does not use sequential gluing of cells. However it is equivalent to the one obove.

Definition 1. A *cell complex* is a Hausdorff space K which can be represented as a union $\bigcup_{n=0}^{\infty} \left(\bigcup_{i \in I_n} e_i^n \right)$ of pairwise disjoint sets e_i^n called cells, and moreover, for each cell e_i^n, a mapping $g_i^n : \bar{D}^n \to K$ of a closed ball to the space K is given, called a *characteristic mapping*; the restriction of g_i^n to $\text{int}\,\bar{D}^n = D^n$ is a homeomorphism onto e_i^n. And, in addition, the following two axioms should be satisfied:

(1) the boundary $\partial e_i^n = \bar{e}_i^n \backslash e_i^n$ for each cell e_i^n is contained in the union of a finite number of cells of lower dimensions;

(2) the topology of K is such that a set $A \subset K$ is closed iff for each cell e_i^n, the full preimage $(g_i^n)^{-1}(A \cap \bar{e}_i^n) \subset \bar{D}^n$ is closed in $\bar{D}^n$.

A cell complex is said to be finite if it consists of a finite number of cells.

It should be noted that a space can often be decomposed into cells in different ways. A way of decomposing a cell complex into cells is called a cellular decomposition.

Exercise 4°. Show that the two–dimensional torus is a cell complex.

Exercise 5°. Show that the sets homotopy equivalent to the Lebesgue sets $(f \leq c)$ in the example of subsection 1, drawn in Fig. 113–117 on the right, are cell complexes, and compare their cellular decompositions. Keep in mind the changes in homotopy type of these complexes.

Consider a closed subspace L of a cell complex K. If L is a cell complex whose cells are also cells of the cell complex K with the same characteristic mappings, then L is called a *subcomplex of the complex* K.

Exercise 6°. Let K be a cell complex, L a subcomplex of it, and X a topological space. Let there be mappings $F : K \to X$ and $f : L \times I \to X$ be such that $f|_{L \times 0} = F|_L$. Show that there exists a mapping $\widetilde{F} : K \times I \to X$ such that $\widetilde{F}|_{L \times I} = f$ and $\widetilde{F}|_{K \times 0} = F$ (the Borsuk homotopy extension theorem).

Hint. Extend the homotopy to every 0–dimensional cell, then to each 1–dimensional one, etc.

Exercise 7°. Let K be a cell complex, L a contractible subcomplex of it. Show that the spaces K and K/L are homotopy equivalent.

Exercise 8°. Prove that any cell complex is a normal space.

Attention should be paid to the fact that in the example considered in subsection 1, the homotopy type of the set M^c changed while passing through the values $f(p)$, $f(q)$, $f(r)$ and $f(s)$. The points p, q, r, and s differ from the other points of the torus in the following way if in a neighbourhood of any of these points, say of the point p, we choose a local system of coordinates ξ, η on the torus, then both partial derivatives $\partial f/\partial \xi$ and $\partial f/\partial \eta$ at the point p (or at q, r, s, respectively) will vanish. These points are called critical points of the function f; the values of the function at these points are called the critical values of the function f.

Exercise 9°. Using the local coordinates of the plane x, y as local coordinates in neighbourhoods of the points p, q, r, s, show that $\partial f/\partial x = \partial f/\partial y = 0$ at any of these points. Expand the function at these points into a power series in x and y up to terms of second order inclusively. Pay attention to the fact that the number of minuses in the second degree terms is exactly the dimension of the cell which should be glued to M^a in order to obtain to M^b, if the critical value corresponding to the critical point under consideration, lies between the numbers a and b.

§ 11. Nondegenerate critical point and their indices

1. Nondegenerate critical points. Let M^n be a manifold of class C^∞, and $f : M^n \to \mathbf{R}$ a function of clas C^∞.

A point $p \in M^n$ is called a *critical point of the function f* if the equality $\partial f/\partial x_1 = \ldots = \partial f/\partial x_n = 0$ holds in local coordinates $x_1, \ldots, x_n$. T The number $f(p)$ is called a *critical value of the function f*. All the remaining points of the manifold M^n are called *non–critical points* of the function f. All numbers which are not critical values of the function f are called *non–critical values* of this function.

Exercise $1°$. Compare the concepts of critical and non–critical values of a function with those of regular and non–regular values of a smooth mapping (see § 5).

A critical point is said to be *isolated* if there exists a neighbourhood of it such that it has no other critical points in that neighbourhood. A critical point is said to be *nondegenerate* if the matrix of the second partial derivatives $A = \left(\frac{\partial^2 f}{\partial x_i \partial x_j} \right)\Big|_p$ is non–singular. Otherwise, the critical point is said to be *degenerate*.

Consider the quadratic form (Ax, x), where $x \in \mathbf{R}^n$. It is called the *Hessian* of the function f at the point p. The matrix A is symmetric; the quadratic form (Ax, x) can be reduced to the canonical form

$$(Ax, x) = -y_1^2 - y_2^2 - \ldots - y_\lambda^2 + y_{\lambda+1}^2 + \ldots + y_h^2$$

by a suitable choice of the coordinates $y_1, \ldots, y_h$, $h \le n$; if the matrix A is non–singular, then $h = n$.

The number λ is called the *index of the function f* at the point p, and the number $(n - h)$ the *degree of degeneracy of the function f* at the point p.

Example. We define a function on $\mathbf{R}^2$ by the formula $f(x, y) = x^3 - 3xy^2$. Clearly, the partial derivatives

$$\frac{\partial f}{\partial x}(x, y) = 3x^2 - 3y^2, \quad \text{and} \quad \frac{\partial f}{\partial y}(x, y) = -6xy$$

vanish simultaneously only at the point $(0, 0)$ which is thus an isolated critical point. All the second partial derivatives

$$\frac{\partial^2 f}{\partial x^2}(x, y) = 6x, \qquad \frac{\partial^2 f}{\partial x \partial y}(x, y) = -6y, \qquad \frac{\partial^2 f}{\partial y^2}(x, y) = -6x$$

are equal to zero at the point $(0, 0)$. Consequently, the matrix of the second partial derivatives of the function f at the point $(0, 0)$ is zero, and the Hessian of the function f at the point $(0, 0)$ is a quadratic form identically equal to zero. Thus, the critical point $(0, 0)$ is degenerate; the degree of degeneracy of the function f at the point $(0, 0)$ equals 2, and the index of f is zero.

Exercise 2°. Demonstrate the well–definedness (i.e., independence of the choice of a system of local coordinates) of the definitions of a critical point, nondegenerate critical point, degree of degeneracy and index of a function at a critical point.

Exercise 3°. Investigate the critical points of the following functions on $\mathbf{R}^1$ and $\mathbf{R}^2$:

(a) $f(x) = x^2$, (b) $f(x, y) = x^3$, (c) $f(x, y) = x^2 y^3$; investigate the critical points on the torus (see § 10, subsection 1).

2. The Morse lemma. A remarkable fact in the theory of critical points is the possibility to represent a function in a neighourhoood of a nondegenerate critical point as a quadratic form, and to describe the behavior of a function by its index.

Theorem 1. (**The Morse lemma**). *Let* $f : M^n \rightarrow \mathbf{R}$, *and let* p *be a nondegenerate critical point of the function* f. *Then in some neighbourhood* U *of the point* p *there exists a local system of coordinates* $y_1, \ldots, y_n$ *such that* $y_i(p) = 0$, $i = 1, \ldots, n$, *and the following identity holds in* U:

(1) $$f(u) = f(p) - y_1^2 - \ldots - y_\lambda^2 + y_{\lambda+1}^2 + \ldots + y_n^2,$$

where $y_1, \ldots, y_n$ *are the coordinates of the point* u, *and* λ *is the index of the function* f *at the point* p.

Proof. We shall show that if there exists a system of coordinates in which the function f is of form (1), then λ is the index of the function f at the point

p. Indeed, if such a system of coordinates exists, then the matrix of partial derivatives $\left(\frac{\partial^2 f}{\partial y_i \partial y_j}\right)\Big|_p$ is diagonal. The numbers on the diagonal are ± 2, and the number of negative eigen–values is, on one hand, equal to the number λ in representaion (1), and, on the other hand, is the index of f at the point p by definition.

Now, we prove that such a representation (1) for the function f exists. Let $x_1, \dots, x_n$ be a local system of coordinates such that the point p has the coordinates $(0, \dots, 0)$. In a certain neighbourhood U of the point p, lemma 1, § 1, can be applied to the function $f(u)-f(p)$. These gives an equality

$$f(x_1, \dots, x_n) - f(0, \dots, 0) = \sum_{i=1}^{n} x_i g_i(x_1, \dots, x_n),$$

moreover,

$$g_i(0, \dots, 0) = \frac{\partial f}{\partial x_i}(0, \dots, 0) = 0,$$

since p is a critical point of f.

We again apply lemma 1, § 1, but now to the functions g_i. We get

$$g_i(x_1, \dots, x_n) = \sum_{j=1}^{n} x_i h_{ij}(x_1, \dots, x_n),$$

and, consequently,

$$(2) \qquad f(x_1, \dots, x_n) - f(0, \dots, 0) = \sum_{i,j=1}^{n} x_i x_j h_{ij}(x_1, \dots, x_n).$$

Writing $\bar{h}_{ij} = \frac{1}{2}(h_{ij} + h_{ji})$, we obtain $\bar{h}_{ij} = \bar{h}_{ji}$ and

$$f(x_1, \dots, x_n) - f(0, \dots, 0) = \sum_{i,j=1}^{n} x_i x_j \bar{h}_{ij}(x_1, \dots, x_n).$$

Since $\bar{h}_{ij}(0, \dots, 0) = \frac{1}{2}\frac{\partial^2 f}{\partial x_i \partial x_j}(0, \dots, 0)$, the matrix $(\bar{h}_{ij}(0, \dots, 0))$ is non–singular. Thus, without loss of generality, we may assume that the matrix (h_{ij}) in (2) is symmetric.

If the functions h_{ij} were constant, then for the proof of the theorem it would be sufficient to reduce the quadratic form of $f(x_1, \ldots, x_n)$ to a canonical one. In the general case, the arguments need to be slightly modified.

In further investigations, it is convenient to assume additonally, that $\frac{\partial^2 f}{\partial x_1^2}(0, \ldots, 0) \neq 0$. This assumption does not restrict the generality, since this can be achieved by changing the local coordinates linearly (i.e., changing the chart). In fact, the quadratic form $\sum_{i,j=1}^{n} h_{ij}(0, \ldots, 0)x_i x_j$ can be reduced, by a linear non–singular change of the coordinates, to the form where the element a_{11} of its matrix is $\neq 0$. After this coordinate change in formula (2), we obtain for f (in the new coordinates $x_1', \ldots, x_n'$) a similar representation

$$f(x_1', \ldots, x_n') - f(0, \ldots, 0) = \sum_{i,j=1}^{n} x_i' x_j' h_{ij}'(x_1, \ldots, x_n),$$

but now, $h_{11}'(0, \ldots, 0) \neq 0$.

Thus, assuming that $h_{11}(0 \ldots, 0) \neq 0$, we can write (in some neigbourhood of the point $(0, \ldots, 0)$:

$$f(x_1, \ldots, x_n) - f(0, \ldots, 0) = \sum_{i,j=1}^{n} h_{ij} x_i x_j =$$

$$= h_{11} x_1^2 + 2 \sum_{i>1}^{n} h_{i1} x_i x_1 + \sum_{i,j>1}^{n} h_{ij} x_i x_j =$$

$$= \text{sign } h_{11}(0, \ldots, 0) \left(\sqrt{|h_{11}|} x_1 + \sum_{i>1}^{n} \frac{h_{i1}}{\text{sign } h_{11}(0, \ldots, 0) \sqrt{|h_{11}|}} x_i \right)^2 -$$

$$- \frac{1}{|h_{11}|} \sum_{i,j>1}^{n} h_{i1} h_{1j} x_i x_j + \sum_{i,j>1} h_{ij} x_i x_j =$$

$$= \text{sign } h_{11}(0, \ldots, 0) y_1^2 + \sum_{i,j>1} \left(h_{ij} - \frac{h_{i1} \cdot h_{j1}}{|h_{11}|} \right) x_i x_j,$$

where the new coordinate y_1 depends smoothly on $x_1, \ldots, x_n$:

$$y_1 = \sqrt{|h_{11}(x_1, \ldots, x_n)|} x_1 + \sum_{i>1}^{n} \frac{h_{i1}(x_1, \ldots, x_n) x_i}{\text{sign } h_{11}(0, \ldots 0) \sqrt{|h_{11}(x_1, \ldots, x_n)|}}.$$

Applying the inverse mapping theorem (see § 1), we see that the transformation $(x_1, x_2, \ldots, x_n) \mapsto (y_1, x_2, \ldots x_n)$ is a diffeomorphism in a neighbourhood of the point $(0, \ldots, 0)$.

Furthermore, note that the matrix

$$\left(h_{ij} - \frac{h_{i1} h_{j1}}{|h_{11}|} \right), \quad 1 < i, j \leq n,$$

is non–singular at the point $(0, \ldots, 0)$ and symmetric (verify!). Consequently, we can apply the arguments presented above to the function

$$\sum_{i,j>1}^{n} \left(h_{ij} - \frac{h_{i1} h_{j1}}{|h_{11}|} \right) x_i x_j$$

and so on, as is done in the classical Lagrange algorithm for reducing a quadratic form to the canonical one. As a result, we get an expression of form (1) for the function f.   □

Exercise 4°. Prove that any nondegenerate critical point is isolated.

Exercise 5°. Find representations (1) as specified by the Morse lemma for the height function on the torus (see § 10) at the critical points.

Exercise 6°. Prove that the points on a manifold (without boundary) where a smooth function attains its maximum or minimum are critical. Calculate the indices at these points if these points are known to be nondegenerate.

3. The gradient field. Let $A_x(u, v)$ be a Riemannian metric on M^n. Let us select a vector $y_x \in T_x M^n$ for each point $x \in M^n$, so that the following condition is fulfilled: for an arbitrary vector $l_x \in T_x M^n$ the following equality holds

(3) $$A_x(y_x, l_x) = (df)_x(l_x),$$

where $(df)_x(l_x)$ is the value of the differential of the function f at the point x on the vector l_x.

The field y_x obtained is called the *gradient field* of the function and denoted by grad $f(x)$.

Exercise 7°. Show that in local coordinates, the gradient field is of the form

$$\text{grad } f(x) = \left(x_1, \dots, x_n, \left(\sum_{i=1}^{n} a^{i1}(x) \frac{\partial f}{\partial x_i} \right) \frac{\partial}{\partial x_1}, \dots \right.$$

$$\left. \dots, \left(\sum_{i=1}^{n} a^{in}(x) \frac{\partial f}{\partial x_i} \right) \frac{\partial}{\partial x_n} \right),$$

where $a^{ij}(x)$ are the coefficients of the matrix which is inverse to the matrix $(a_{ij}(x))$ of the form $A_x(u, v)$.

Exercise 8°. Prove that the gradient field for a function of class C^∞, is a smooth vector field.

Exercise 9°. Prove that grad $f(x^0) = 0$ iff x^0 is a critical point of the function f.

§ 12. Critical points and homotopy type of manifold

In this section the homotopy type of a manifold is described by means of critical points of a smooth function on the manifold. This approach was first given by M. Morse. It will be shown that a compact manifold is homotopy equivalent to a cell complex. The details of the proofs (in a number of cases they are rather subtle) will be omitted.

1. Structure of the Lebesgue sets of smooth functions. Let M be a compact n–dimensional C^∞– manifold, f a function of class C^∞ on M all of whose critical points are non–degenerate. For any number a, the set $(f < a)$ is an open subset in M, and, therefore an open submanifold in M. Now, we assume that a is a non–critical value of f, and $f^{-1}(a) \neq \emptyset$. We show that the set $M^a = (f \leq a)$ is a manifold with boundary $(f = a)$. Let $u \in f^{-1}(a)$. According to the theorem on the rectification of a mapping (see § 1), the function f in a certain neighbourhood of the point u, can be represented in local coordinates as the projection π of the space $\mathbf{R}^n$ on the line $\mathbf{R}^1$ (Fig. 118). The preimage of the point a under this projection, is the subspace $\mathbf{R}^{n-1}$, i.e., the boundary of the half–space $\mathbf{R}^n_-$. At the points of this half–space, the function f takes values not greater than a. This indicates that in M^a, there exists a neighbourhood of the point u homeomorphic to the half–space $\mathbf{R}^n_-$.

Consequently, $M^a = (f \leq a)$ is an n–dimensional manifold with its boundary being the $(n-1)$–dimensional manifold $(f = a)$.

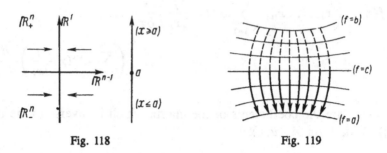

Fig. 118   Fig. 119

2. Conditions for homotopy equivalence of Lebesgue sets. Let a and b be non-critical values of a function f, and let the interval $[a,b]$ contain no critical values. We move the set $(f=c)$ to the set $(f=a)$ along the curves orthogonal to the level manifolds $(f=c)$, $a \leq c \leq b$ (Fig. 119). Thus, we define a deformation $\phi_a^b(t)$, $a \leq t \leq b$, of the manifold M^b to the manifold M^a. Therefore, M^a is a strong deformation retract of M^b, and M^a and M^b are homotopy equivalent.

A strict proof that the mapping ϕ_a^b exists should include the construction of the curves orthogonal to the level manifolds. They can be determined as integral curves of the vector field $X(u)$, where the vector $X(u) \in T_u M$ is defined by the condition $\langle X(u), h \rangle = 0$ for all $h \in T_u (f=c=f(u))$, i.e., from the condition of orthogonality of the vector $X(u)$ to the tangent space at u of the level manifold $(f=c)$. The symbol $\langle , \rangle$ denotes a Riemaniann metric which always exists on a manifold (see § 6). In order to have the level manifold transformed to the level submanifold at any moment t, we specify the vector field $X(u)$ by the formula

$$X(u) = \rho(u) \operatorname{grad} f(u),$$

where $\rho(u)$ is a smooth function on M equal to $1/\langle \operatorname{grad} f(u), \operatorname{grad} f(u) \rangle$ on $M^b \backslash M^a$, and to zero outside a certain neighbourhood $M^b \backslash M^a$ which does not contain critical points.

The deformation M^b to M^a can be well-defined also in the case when a is a critical value of f.

3. The change of homotopy type when passing through a critical value. Thus, the homotopy type of the set M^c does not change as long as the number c, while increasing (or decreasing), does not pass through a critical value c_0. Let us now see what happens when c does pass through a critical value.

We suggest the readers to prove the following useful statement.

Exercise $1°$. Prove that a smooth function on a compact manifold all of whose critical points are nondegenerate, has a finite number of critical points and critical values.

Consider a critical value c_0 and assume that a unique critical point p, $f(p)=c_0$ corresponds to it. Choose a neighbourhood U of the point p and local coordinates as specified by the Morse lemma (see § 11), in which the function f is represented in these coordinates $y_1,\dots,y_n$ in the form

$$f(u)=c_0-y_1^2-\dots-y_\lambda^2+y_{\lambda+1}^2+\dots+y_n^2.$$

Choose ϵ such that $[c_0-\epsilon,c_0+\epsilon]$ does not contain other critical points and a point with the local coordinates $(y_1,\dots,y_n)$, $\sum_{i=1}^n y_i^2 \leq 2\epsilon$, belongs to U.

We construct a smooth function F on M which differs from f only in U, and such that, moreover, the sets $(f \leq c_0+\epsilon)$ and $(F \leq c_0-\epsilon)$ are homotopy equivalent. Having arranged this, we compare the sets $(f \leq c_0-\epsilon)$ and $F \leq c_0-\epsilon$. We find this to be more convenient than the direct comparison of the sets $(f \leq c_0-\epsilon)$ and $(f \leq c_0+\epsilon)$. To construct the function F, it is necessary to have a smooth function μ on $\mathbf{R}^1$ such that possesses the following properties:

$$\mu(0)>\epsilon,\ \mu(x)=0\ \text{for}\ x>2\epsilon,\ -1<\mu'(x)\leq 0\ \text{for}\ -\infty<x<\infty.$$

The form of the graph of such a function μ is shown in Fig. 120.

Exercise 2°. Give an example of a function μ satisfying the properties indicated above.

Let us define the smooth function F by the formulae

$$F(v)=\begin{cases} f(v) & \text{for } v\notin U, \\ f(v)-\mu\left(\sum_{i=1}^\lambda y_i^2+2\sum_{i=\lambda+1}^n y_i^2\right) & \text{for } v\in U. \end{cases}$$

It is not difficult to see that the critical points of the function F coincide with those of the function f (although $f(p)\neq F(p)$).

To the critical point p of the function F, there corresponds the critical value $F(p)=c_0-\mu(0)<c_0-\epsilon$ Since the values of the function F coincide with those of the function f at all other critical points, the interval $[c_0-\epsilon,c_0+\epsilon]$ has no critical values of F. Therefore, the set $(F \leq c_0-\epsilon)$ is a strong deformation retract of the set $(F \leq c_0+\epsilon)$. But $(F \leq c_0+\epsilon)=(f \leq c_0+\epsilon)$. Therefore, $(F \leq c_0-\epsilon)$ is a strong deformation retract of the set $(f \leq c_0+\epsilon)$. Thus, these sets are homotopy equivalent.

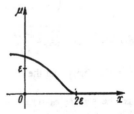

Fig. 120

Further, we shall compare the homotopy types of the sets $(f \leq c_0-\epsilon)$ and $(F \leq c_0-\epsilon)$ (instead of comparing the homotopy types of the sets $(f \leq c_0-\epsilon)$ and $(f \leq c_0+\epsilon)$). Let H be the closure of the set $(F \leq c_0-\epsilon)\backslash(f \leq c_0-\epsilon)$. Consider the cell e^λ consisting of those points $u\in U$ whose coordinates $y_1,\dots,y_n$ satisfy the conditions

$$\sum_{i=1}^\lambda y_i^2<\epsilon,\quad \sum_{i=\lambda+1}^n y_i^2=0.$$

The cell e^λ lies inside H; it is glued to the set $(f \leq c_0 - \epsilon)$ along the set of all points u for which $\sum_{i=1}^{\lambda} y_i^2 = \epsilon$.

Fig. 121 presents a neighbourhood of a critical point of index 1 on a two–dimensional manifold (e.g., of the point q from the example of § 10); the set $M^{c_0 - \epsilon} = (f \leq c_0 - \epsilon)$ is shaded, the set H is cross hatched, the cell e^λ is denoted by a thick line.

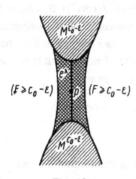

Fig. 121

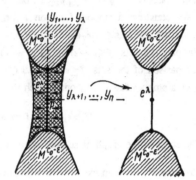

Fig. 122

We define a deformation Γ_t of the set $(F \leq c_0 - \epsilon) = M^{c_0 - \epsilon} \cup H$ on the set $M^{c_0 - \epsilon} \cup e^\lambda$ as follows: Γ_t is the identity mapping on $M^{c_0 - \epsilon}$, and defined on H by the formula

$$\Gamma_t(y_1, \ldots, y_n) =$$
$$= \begin{cases} y_1, \ldots, y_\lambda, t y_{\lambda+1}, \ldots, t y_n & \text{for } \sum_{i=1}^{\lambda} y_i^2 \leq \epsilon, \\ y_1, \ldots, y_\lambda, s_t y_{\lambda+1}, \ldots, s_t y_n & \text{for } \epsilon < \sum_{i=1}^{\lambda} y_i^2 < \sum_{i=\lambda+1}^{n} y_i^2 + \epsilon, \end{cases}$$

where $s_t = t + (1-t) \sqrt{\dfrac{\sum_{i=1}^{\lambda} y_i^2 - \epsilon}{\sum_{i=\lambda+1}^{n} y_i^2}}$, $0 \leq t \leq 1$. This deformation is shown by arrows in Fig. 122.

Exercise 3°. Verify that the deformation Γ_t is well–defined.

Thus, the set $(f \leq c_0 - \epsilon) \cup e^\lambda$ is a strong deformation retract of the set $(F \leq c_0 - \epsilon)$, and, consequently, of the set $(f \leq c_0 + \epsilon) = M^{c_0 + \epsilon}$. Thus, $M^{c_0 + \epsilon}$ is of the same homotopy type as the set $M^{c_0 - \epsilon} \cup e^\lambda$, i.e., with the set $M^{c_0 - \epsilon}$ with a cell[*] glued in in a certain way and of dimension equal to the index of the critical point that corresponds to the value c_0.

We considered the case when to a critical value of a function there corresponds a unique critical point. Now consider the general case.

---

[*] Here and below, we omit the gluing mappings in the notation.

Exercise 4°. Construct a smooth function on a two–dimensional manifold with all its critical points nondegenerate and such that several critical values correspond to one critical value.

Let $k > 1$ critical points correspond to one critical value c_0. All the constructions described above can be carried out simultaneously in a neighbourhood of each critical point. The set $M^{c_0+\epsilon}$ has the homotopy type of the set $M^{c_0-\epsilon} \cup e^{\lambda_1} \cup ... \cup e^{\lambda_k}$, i.e., of the set $M^{c_0-\epsilon}$ with the cells e^{λ_i} glued to it in a certain way, and, besides, the dimension λ_i is equal to the index of the i–th critical point corresponding to c_0.

Let c' be the least of the critical values that are greater than c_0, and assume that there are no other critical values in the ϵ–neighbourhoods of c_0 and c'. Let to the value c' there correspond k' critical points with the indices $\lambda'_1, ..., \lambda'_{k'}$. The set $M^{c_0-\epsilon} \cup e^{\lambda_1} \cup ... \cup e^{\lambda_k}$ is homotopy equivalent to the set M^a for $c_0 \leq a < c'$. The set $M^{c'}$ is, in turn, homotopy equivalent to the set $M^a \cup e^{\lambda'_1} \cup ... \cup e^{\lambda'_{k'}}$.

We establish the homotopy equivalence of the sets $M^{c'}$ and

$$\left(M^{c_0-\epsilon} \cup e^{\lambda_1} \cup ... \cup e^{\lambda_k} \right) \cup e^{\lambda'_1} \cup ... \cup e^{\lambda'_{k'}}.$$

For this we deform the set $M^a \cup e^{\lambda'_1} \cup ... \cup e^{\lambda'_{k'}}$ onto the set

$$\left(M^{c_0-\epsilon} \cup e^{\lambda_1} \cup ... \cup e^{\lambda_k} \right) \cup e^{\lambda'_1} \cup ... \cup e^{\lambda'_{k'}},$$

by using the deformation of M^a onto $M^{c_0-\epsilon} \cup e^{\lambda_1} \cup ... \cup e^{\lambda_k}$ contructed above.

Exercise 5°. Explain how the cells $e^{\lambda'_1}, ..., e^{\lambda'_{k'}}$ are glued to the set $M^{c_0-\epsilon} \cup e^{\lambda_1} \cup ... \cup e^{\lambda_k}$.

We emphasize that the gluing in of a cell is carried out, at each stage, not in an arbitrary, but in a strictly definite manner (up to the homotopy class of the mapping of the sphere, i.e., the boundary of the cell to the corresponding space). Therefore, the gluing in of a cell is defined by an element from the homotopy group of the corresponding space; the dimension k of this homotopy group π_k is equal to that of the cell reduced by one.

4. The homotopy type of a manifold. We shall outline the construction of a cell complex which is homotopy equivalent to a manifold M, as it was done for the torus in § 10.

Let c_1 be the least of the critical values of a function f. Evidently, for $a < c_1$, the set $(f \leq a)$ is empty. Since c_1 is the least critical value, all critical points corresponding to c_1 are the points where f assumes its minimum; their indices are equal to zero. The set $(f \leq c_1)$ consists of a finite number of points and can be regarded as a result of gluing several cells of dimension 0 to the empty set.

Let c_2 be another critical value next in size. For $c_1 < c < c_2$, the set $(f \leq c)$ is obtained by "inflating" the points from $(f \leq c_1)$; it consists of a finite number of sets homeomorphic to the n–dimensional disc and homotopy equivalent to the set $(f \leq c_1)$. The set $(f \leq c_2)$ is homotopy equivalent to the set $(f \leq c_1)$ with cells glued to it of different (in general, of any from 0 to n) dimensions equal to the indices of critical points corresponding to c_2. The latter set, obviously, is a cell complex.

By taking a critical value c_3 next in size, we obtain that $(f \leq c_3)$ is homotopy equivalent to the result of a subsequent gluing to $(f \leq c_1)$ of cells corresponding to the critical points with critical value c_2, and then the cells corresponding to the critical points with critical value c_3. This space can be made a cell complex by adjusting the mapping on the boundaries of the cells glued in.

Exercise 6°. Prove that each mapping of the sphere S^m to a cell complex K is homotopic to a mapping of the sphere to a subspace K^m of the space K consisting of cells of dimension not higher than m.

In the general case, the set $M^a = (f \leq a)$ when $a \geq \max_{u \in M} f(u)$, is homotopy eqivalent to a space which is a cell complex obtained from the empty set by sequentially gluing in cells which correspond to the critical points with the critical values c_i, in increasing order $-\infty < c_i \leq a$.

Note that if c_r is the greatest critical value, then the critical points at which the value of the function f is equal to c_r, are the points where f assumes its maximum, and, consequently, their indices are equal to the dimension of the manifold M.

Let us formulate the final statement.

Theorem 1. *Each smooth function f on a compact manifold M having only nondegenerate critical points, defines a homotopy eqivalence of the manifold M with a certain finite cell complex whose cells are in one–to–one correspondence with critical points of the function f; moreover, the dimension of a cell is equal to the index of the corresponding critical point.*

Now we dwell on the existence of a smooth function on a compact manifold possesing only nondegenerate critical points. Such a function can be constructed in the following way. Consider an imbedding of the manifold M into an Euclidean space $\mathbf{R}^l$ of a suffciently large dimension l. Define the function f by the formula $f(p) = (u - p, u - p)$, where $(,)$ is the standard scalar product on $\mathbf{R}^l$, u a constant vector in $\mathbf{R}^l$, and $p \in M \subset \mathbf{R}^l$. Applying the Sard theorem (see § 5), one can show that there exists a vector $u \in \mathbf{R}^l$ such that the function f possesses only nondegenerate critical points.

This result enables us to make the following important statement.

Theorem 2. *Any compact, smooth manifold has the homotopy type of a finite cell complex.*

It should be noted that theorem 1 does not permit to recover completely the homotopy type of a manifold using only information on critical points of the function. In the general case, we can determine the number of cells and their dimensions, but, generally speaking, do not know how to glue these cells to each other. Thus, it is in general impossible to recover the cell complex completely from the information on critical points.

5. The concept of the exact sequence of a fibre bundle. (Supplement to § 9.) The results on the calculation of the fundamental group of a base of a covering presented in § 9 can be generalized to locally trivial fibre bundles. Let (E,B,F,p) be a locally trivial fibre bundle, and e_0, b_0 distinguished points in $F=p^{-1}(b_0)$, b, respectively. The projection mapping p and the imbedding $i:F \to E$ induce homomorphisms of homotopy groups $\pi_n(E,e_0) \xrightarrow{p_n} \pi_n(B,b_0)$, $\pi_n(F,e_0) \xrightarrow{i_n} \pi_n(E,e_0)$, $n \geq 1$. We extend these mappings to the case $n=0$ by taking $\pi_0(F,e_0)$, $\pi_0(E,e_0)$, $\pi_0(B,b_0)$ to be the sets of path–connected components of the spaces F,E,B. The component containing the distinguished point $e_0(b_0)$ is called the "zero" element θ. Let us define another homomorphism $\partial_n : \pi_n(B,b_0) \to \pi_{n-1}(F,e_0)$, $n \geq 1$. Let $\phi :(I^n, \partial I^n) \to (B,b_0)$ be a spheroid. By the homotopy lifting property, there exists a lift $\Phi : I^n \to E$ of ϕ. Since $I^n = I^{n-1} \times I$, and $\phi_t : I^{n-1} \times I \to B$ is a homotopy of the mapping $\phi_0 : I^{n-1} \times \{0\} \to B$ covered by the mapping $\phi_0 : I^{n-1} \times \{0\} \to e_0$. Moreover, $p\Phi(y \times \{t\}) = \phi(y \times \{t\})$, $y \in I^{n-1}$, $t \in I$, and $p\Phi(\partial I^{n-1} \times \{t\}) = \phi(\partial I^{n-1} \times \{t\}) = b_0$, whence $\Phi(\partial I^n) \in F$. Thus, $\Phi : \partial I^n \to F$ is an $(n-1)$–dimensional spheroid determining a class $\alpha \in \pi_{n-1}(F,e_0)$. Put $\partial_n[\phi] = \alpha$. That ∂_n, $n \geq 1$, is well–defined; it is not difficult to see that ∂_n is a homomorphism. When $n=0$, we set $\partial_0 = 0$.

The *homotopy sequence of a locally trivial fibre bundle* (E,B,F,p) is the sequence infinite from the left

$$\cdots \xrightarrow{\partial_{n+1}} \pi_n(F,e_0) \xrightarrow{i_n} \pi_n(E,e_0) \xrightarrow{p_n} \pi_n(B,b_0) \xrightarrow{\partial_n}$$

$$\xrightarrow{\partial_n} \pi_{n-1}(F,e_0) \xrightarrow{i_{n-1}} \cdots \xrightarrow{p_0} \pi_0(B,b_0) \xrightarrow{\partial_0} 0.$$

This sequence possesses the following property: for every term of the sequence, the image of the "entering" homomorphism is equal to the kernel of the "leaving" one, i.e., $\mathrm{Im}\, \partial_{n+1} = \mathrm{Ker}\, i_n$, $\mathrm{Im}\, i_n = \mathrm{Ker}\, p_n$, $\mathrm{Im}\, p_n = \mathrm{Ker}\, \partial_n$, $n = 0,1,\dots$ This property is called the "exactness" of the homotopy sequence of the fibre bundle.

Example 1. Consider the covering $p : \mathbb{R}^1 \to S^1$, $p = e^{2\pi i t}$ (see § 9, subsection 4, example 6). The fibre of the covering is the group $\mathbb{Z}$ (with the discrete topology). It is clear that $\pi_k(\mathbb{Z}) \simeq \mathbb{Z}$, $k \geq 1$; $\pi_k(\mathbb{R}^1) = 0$, $k \geq 0$. Consequently, $\cdots \to 0 \to 0 \to \pi_k(S^1) \to 0 \to \cdots \to 0 \to 0 \to \pi_1(S^1) \to \mathbb{Z} \to 0 \to 0$ is the exact homotopy sequence of the fibre bundle $\mathbb{R}^1, S^1, \mathbb{Z}, p)$. Hence, $\pi_1(S^1) \simeq \mathbb{Z}$, $\pi_k(S^1) = 0$, $k \geq 0$.

Example 2. The Hopf fibre bundle (S^3, S^2, S^1, π) (see § 9, subsection 1). The exact homotopy sequence looks like $\to 0 \to \mathbb{Z} \to \pi_3(S^2) \to 0 \to 0 \to \pi_2(S^2) \to \mathbb{Z} \to 0 \to 0$, because $\pi_3(S^1) = \pi_2(S^1) = \pi_2(S^3) = \pi_1(S^3) = \pi_1(S^2) = 0$ (see example 1). By using the exactness of this sequence, we immediately obtain $\pi_3(S^2) \simeq \pi_2(S^2) \simeq \mathbb{Z}$.

The summary information on homotopy groups given at the end of Ch. 3, is obtained in a similar manner with the help of exact homotopy sequences for some special fibre bundles.

REVIEW OF THE RECOMMENDED LITERATURE

Systematic texts on the theory of manifolds and fibre bundles are [36,25, 56, 58, 60, 61, 64, 69, 70, 73, 74, 83, 81, 85, 87, 40, 42, 43, 17]

Visual illustrations of the basic concepts of the theory of manifolds and fibre bundles are contained in [14, 57].

A fundamental outline of the development of the ideas of modern topology is given in [62].

Problem books on the material of Ch. 4 are [59, 63]. The information from analysis used in Ch. 4, is contained in [22, 49, 61].

The classification of one–dimensional and two–dimensional manifolds can be found in [14, 88, 57, 74]. The classification of 3–dimensional and 4–dimensional manifolds is presented in an accessible way in [29, 30] (see also [88]).

The Sard theorem and degree theory for smooth mappings can be found in [36, 25, 57, 60, 64, 71, 73, 74, 83, 40].

For embedding theorems for manifolds one can consult [56, 57, 60, 61, 69, 73, 74, 83, 40].

Lie groups theory is dealt with in [25, 68, 87, 29, 17].

The theory of covering spaces can be found in [54, 80, 31].

The theory of critical points of smooth functions on manifolds is discussed in [55, 57, 70, 29, 30, 74].

Vector fields on smooth manifolds, dynamical systems, hamiltonian formalism are explained in [9, 11, 25, 60, 64, 73, 83, 87, 29].

In this chapter, we define the homology groups of any topological space. The idea of constructing homology groups, as was already mentioned, can be traced back to A. Poincaré. For the first time in history, the useful idea of algebraization of topological problems was carried out by constructing homology groups and fundamental groups. Homology theory still maintains a central position. In general, all topological invariants from homotopy groups to fibre bundles are expressed, in the end, through homological invariants. Given this circumstance, it is fortunate that, although the definition of homology groups is somewhat more complicated than, say, that of homotopy groups, they are far easier to calculate (as a rule).

Simplicial complexes and cell complexes of topological spaces enable us to study (using methods of algebraic topology) their most profound geometrical and topological properties. This composition is meant to invite the viewer to meditate upon the wonderful magic of human hands, which "operate" on topological spaces, reducing them to cell complexes and "with the help of algebra, verifying their harmony".

CHAPTER V

HOMOLOGY THEORY

In this chapter, the so–called homology groups will be defined for any topological space. An idea of constructing homology groups, as was already mentioned, goes back to Poincaré. The useful idea of "algebraization" of topological problems was carried out for the first time, by constructing homology groups and fundamental group. Homology theory still remains in a central position. In many cases, topological invariants are finally expressed in terms of homology groups and cohomology groups. This comes about because of a better computability of homology groups and cohomology groups, although their definitions are somewhat more complicated than, for example, the definition of homotopy groups.

§ 1. Introductory remarks

Here we illustrate some of the arguments that led to the concept of homology.

When studying spaces that are rather simple, geometric intuition often helps to distinguish spaces from the topological point of view, i.e., to see that they are not homeomorphic to each other. As a rule, it is not difficult to determine whether different rather simple concrete subsets of a line, plane, and sometimes three–dimensional space are homeomorphic (in the induced topologies) or not. The definition of manifold implies directly that a manifold X is not homeomorphic to a space Y which is not a C^0–manifold. However, when investigating manifolds with dimensions higher than 1 or 2, geometric intuition proves less effective.

In order to distinguish non–homeomorphic manifolds of high dimension, one can employ the following idea. Let M_1^n and M_2^n be two n–dimensional manifolds. We shall consider compact C^0–submanifolds in M_1^n, M_2^n.

If any q–dimensional submanifold ($q < n$) in M_1^n is the boundary of a $(q + 1)$–dimensional submanifold in M_1^n, and there is a q–dimensional submanifold in M_2^n which is not the boundary of a submanifold in M_2^n, then the manifolds in M_1^n, M_2^n are definitely non–homeomorphic. Thus, any 1–dimensional (compact) submanifold of the sphere S^2 is a boundary while on the torus $T^2 = S^1 \times S^1$, it is easy to indicate circles which are not boundaries of any two–dimensional submanifold in T^2 (Fig. 123).

FIG. 123

If, however, submanifolds that are not boundaries, exist both in M_1^n and M_2^n, then one may try to compare the "number" of such manifolds in M_1^n and M_2^n. Consider the set $\{V_\alpha^q\}$ of all q–dimensional cycles, i.e., of q–dimensional submanifolds (without boundary) of a manifold M^n. Let W^{q+1} be a submanifold of M^n with boundary consisting of connected manifolds $V_1^q, \ldots, V_m^q$, $V_i^q \in \{V_\alpha^q\}$. We shall then say that the cycle $V_1^q + \ldots + V_m^q$ is homologous to zero. Thus, an equivalence relation is introduced on the set $\{V_\alpha^q\}$: two cycles are equivalent (homologous), if they differ by a cycle homologous to zero; the equivalence classes of q–dimensional cycles are called the *q–dimensional homologies of the manifold M^n*.

If there exists an integer q such that the quantity of q–dimensional homologies of the manifold M_1^n is greater than that of the manifold M_2^n, then this means that M_1^n and M_2^n are non–homeomorphic.

The concept of a sum of two non–intersecting cycles was introduced for the set $\{V_\alpha^q\}$, but this does not yet mean that a group structure is introduced on it. Therefore, we cannot yet assume that homologies form a group. The intuitive definition of homology given above is inconvenient for calculations. It is more efficient to consider cycles (manifolds) as composed of certain elementary manifolds with boundaries. We shall show in next example how this might be possible.

Let Π^2 be the surface of a tetrahedron (Fig. 124); clearly, Π^2 is homeomorphic to the sphere S^2. We shall consider the zero–dimensional manifolds

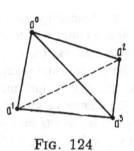

FIG. 124

which consist of vertices of the tetrahedron, the one–dimensional manifolds made up from its edges, and two–dimensional manifolds consisting of its faces (aasuming that 1–dimensional and 2–dimensional manifolds may have boundaries); it is only natural to treat the union of two manifolds as a sum. To use this remark for the algebraization of the objects under investigation, we consider the group of formal linear combinations[*] with integral coefficients of the vertices (the group of zero–dimensional chains), of the edges (the group of one–dimensional chains), and faces (the group of two–dimensional chains). Moreover, for every edge, we fix an order of vertices (a^i, a^j) and identify $(-1)(a^i, a^j)$ with (a^j, a^i); for each face we fix a direction of passing through the vertices (a^i, a^j, a^k) and identify $(-1)(a^i, a^j, a^k)$ with (a^j, a^i, a^k) and (a^i, a^k, a^j).

Define the boundary of the edge (a^i, a^j) as the sum $a^j + (-1)a^i$, and the boundary of the face (a^i, a^j, a^k) as the sum of the edges that bound this face (with the same orientation as was given for the face), i.e., $(a^i, a^j) + (a^j, a^k) + (a^k, a^i)$; we put the boundary of a vertex equal to zero. The operations of taking the boundary thus defined are extended to the groups of chains by linearity. A cycle is a chain with a boundary equal to zero; thus, a cycle is an algebraic analogue of a closed manifold (without boundary).

Since we are interested in homology, i.e., in classes of equivalent cycles differing from one another by a boundary, we consider the cosets of q–dimensional cycles with respect to the subgroup of boundaries of $(q + 1)$–dimensional chains ($q = 0, 1, 2$). These cosets form a group called the q–dimensional group of integral homology of the surface Π^2. The homology groups of Π^2 are easy to calculate; they are isomorphic to $\mathbf{Z}$, 0, and $\mathbf{Z}$ for dimensions 0, 1, and 2, respectively. Such constructions can be generalized to higher dimensions by using decomposition into tetrahedra and their (higher dimensional) analogues (simplices). Knowing how to decompose a given space into simplices, one can calculate its homology groups. In practice, however, the definition itself is rarely used for calculating homology groups. Instead various special techniques are applied (exact sequences, spectral sequences,

---

[*] The group of formal linear combinations of elements τ_α from a certain set $\{\tau_\alpha\}_{\alpha \in A}$ with coefficients in an Abelian group G is the direct sum $\oplus_{\alpha \in A} G_\alpha$ of groups $G_\alpha = \{g \cdot \tau_\alpha\}, g \in G$ isomorphic to G, where the isomorphism is given by the rule $g\tau_\alpha \mapsto g$.

etc.).

Homology groups are topological invariants, i.e., the homology groups of homeomorphic spaces coincide (are isomorphic). Other examples of topological invariants are the number of components of connectedness (or of path–connectedness) of a space, the Euler characteristic, the fundamental group and the higher homotopy groups (see Ch. 3). It is remarkable that the fundamental group and homotopy groups are closely related to the homology groups (see § 4 of this chapter), and that the remaining topological invariants mentioned can be calculated by means of the homology groups (see § 4 and § 8 of this chapter). Note also that the homology groups of homotopy equivalent spaces are isomorphic, and that the other invariants indicated above are also homotopy invariants.

In a number of cases geometric intuition helps not only to distinguish non–homeomorphic spaces but even to prove that they are non–homeomorphic.

Exercise 1°. Prove (without using homology groups) that the spaces drawn as the symbols O, Π, X, P, i, 8, are non–homeomorphic.

Exercise 2°. Is the orientation of a surface a topological invariant? Are S^2 and RP^2 homeomorphic?

§ 2. Homology of chain complexes

We shall start by investigating abstract algebraic objects.

A sequence (infinite)

$$(1) \qquad \cdots \xrightarrow{\partial_{k+1}} C_k \xrightarrow{\partial_k} C_{k-1} \xrightarrow{\partial_{k-1}} \cdots \xrightarrow{\partial_1} C_0 \xrightarrow{\partial_0} 0$$

of Abelian groups C_k and their homomorphisms ∂_k satisfying the condition $\partial_{k-1}\partial_k = 0$ for every $k \geq 1$, is called a *chain complex*. We denote it by C_*; the groups C_k are called *chain groups*, and homomorphisms ∂_k *differentials* or *boundary homomorphisms*.

The set $\operatorname{Ker} \partial_k = \{c \in C_k : \partial_k c = 0\}$ forms a subgroup in C_k which is called the *group of k–dimensional cycles*; its elements are called *k–dimensional cycles*. The set $\operatorname{Im} \partial_{k+1} = \{c \in C_k : c = \partial_{k+1}u\}$ also forms a subgroup in C_k called the *group of k–dimensional boundaries*; its elements are called *k–dimensional boundaries*.

A *homomorphism* ϕ_* *of a chain complex* C_* to a chain complex C'_* is a sequence of homomorphisms $\phi_k : C_k \to C'_k$ such that the diagram

$$\cdots \longrightarrow C_k \overset{\partial_k}{\longrightarrow} C_{k-1} \longrightarrow \cdots \longrightarrow C_1 \overset{\partial_1}{\longrightarrow} C_0 \longrightarrow 0$$

(2) $\qquad\quad \downarrow \phi_k \qquad\qquad \downarrow \phi_{k-1} \qquad\qquad\qquad\quad \downarrow \phi_1 \qquad\quad \downarrow \phi_0$

$$\cdots \longrightarrow C'_k \overset{\partial'_k}{\longrightarrow} C'_{k-1} \longrightarrow \cdots \longrightarrow C'_1 \overset{\partial'_1}{\longrightarrow} C'_0 \longrightarrow 0$$

is commutative, i.e. $\phi_{k-1}\partial_k = \partial'_k\phi_k$, for any k.

We introduce one of the most important concepts of algebraic topology, the idea of a homology group. Consider a chain complex C_*. Because $\partial_k \partial_{k+1} = 0$, there is the inclusion $\operatorname{Im} \partial_{k+1} \subset \operatorname{Ker} \partial_k$. The quotient group of cycles with respect to the group of boundaries $\operatorname{Ker} \partial_k / \operatorname{Im} \partial_{k+1}$ is called the *k–th homology group of the complex* C_* and denoted by $H_k(C_*)$. Two cycles c_1, c_2 from the same coset are said to be *homologous* and this is denoted by $c_1 \sim c_2$.

Let $\phi_* : C_* \to C'_*$ be a homomorphim of chain complexes. From the commutativity of diagram (2), it immediately follows that

$$\phi_k(\operatorname{Ker} \partial_k) \subset \operatorname{Ker} \partial'_k \quad \text{and} \quad \phi_k(\operatorname{Im} \partial_{k+1}) \subset \operatorname{Im} \partial'_{k+1}.$$

Therefore ϕ_* induces a *homomorphism of homology groups*:

$$\phi_{*k} : H_k(C_*) \to H_k(C'_*).$$

We shall continue the investigation of chain complexes and their homology groups. Let C_* and C^0_* be chain complexes such that the groups C^0_k are subgroups of the groups C_k, and the differentials ∂^0_k of the complex C^0_* are obtained by restricting ∂_k to C^0_k. In this case, the complex C^0_* is called a *subcomplex of the complex* C_*. This defines a homomorphism of chain complexes $i_* : C^0_* \to C_*$, where $i_k : C^0_k \to C_k$ is the given inclusion; i_* is called a *monomorphism of chain complexes*.

Consider the sequence of quotient groups $\widehat{C}_k = C_k / C^0_k$. The homomorphisms ∂_k induce homomorphisms $\widehat{\partial}_k : \widehat{C}_k \to \widehat{C}_{k-1}$.

Exercise 1°. Show that the groups $\widehat{C}_k$ and the homomorphisms $\widehat{\partial}_k$ form a chain complex $\widehat{C}_*$, and that the quotient epimorphisms $j_k : C_k \to \widehat{C}_k$ form a homomorphism of chain complexes $j_* : C_* \to \widehat{C}_*$ (epimorphims of chain complexes).

A sequence

$$\ldots \to A_{k+1} \xrightarrow{\psi_{k+1}} A_k \xrightarrow{\psi_k} A_{k-1} \to \ldots$$

of groups A_k and homomorphisms ψ_k is said to be *exact* if for every k, the image of the homomorphism ψ_{k+1} coincides with the kernel of the homomorphism ψ_k, i.e., $\operatorname{Im} \psi_{k+1} = \operatorname{Ker} \psi_k$.

Exercise 2°. Show that the sequence

$$0 \to C_k^0 \xrightarrow{i_k} C_k \xrightarrow{j_k} \widehat{C}_k \to 0$$

is exact for all k.

A sequence of chain complexes and their homomorphisms

(3) $0 \to C_*^0 \xrightarrow{i_*} C_* \xrightarrow{j_*} \widehat{C}_* \to 0,$

is exact if the corresponding sequence $0 \to C_k^0 \xrightarrow{i_k} C_k \xrightarrow{j_k} \widehat{C}_k \to 0$ is exact for all k, i.e., if C_*^0 is a subcomplex of C_k and $\widehat{C}_k$ is (isomorphic to) the corresponding quotient complex.

By the general definition, one can take the homology groups of a quotient complex $\widehat{C}_*$, i.e., the groups $H_k(\widehat{C}_*)$. The new groups turn out to be related to the groups $H_k(C_*)$ and $H_k(C_*^0)$ by a certain exact sequence.

Let us construct this sequence. The homomorphisms i_* and j_* induce homomorphisms

$$i_{*k} : H_k(C_*^0) \to H_k(C_*), \quad j_{*k} : H_k(C_*) \to H_k(\widehat{C}_*).$$

We obtain the short sequences:

$$H_k(C_*^0) \xrightarrow{i_{*k}} H_k(C_*) \xrightarrow{j_{*k}} H_k(\widehat{C}_*)$$
$$\xleftarrow{\quad} \overset{\delta_k}{\text{-----}}$$
$$H_{k-1}(C_*^0) \xrightarrow{i_{*k-1}} H_{k-1}(C_*) \xrightarrow{j_{*k-1}} H_{k-1}(\widehat{C}_*)$$

We shall show that there exist homomorphisms

$$\delta_k : H_k(\widehat{C}_*) \to H_{k-1}(C_*^0),$$

combining these short sequences into a long exact sequence

$$\ldots \to H_{k+1}(\widehat{C}_*) \xrightarrow{\partial_{k+1}} H_k(C^0_*) \xrightarrow{i_{*k}} H_k(C_*) \xrightarrow{j_{*k}}$$

(4)
$$\to H_k(\widehat{C}_*) \to \ldots \to H_0(\widehat{C}_*) \to 0.$$

We shall describe the homomorphisms δ_k. Consider the sequence (3). Let $\widehat{\alpha} \in H_k(\widehat{C}_*)$, $k > 0$, i.e., $\widehat{\alpha}$ is the coset of a certain element $\alpha \in \operatorname{Ker} \widehat{\partial}_k$ with respect to the subgroup $\operatorname{Im} \widehat{\partial}_{k+1}$. In turn, $\alpha \in \widehat{C}_k$, and it can be considered as the coset of a certain element $d \in C_k$ with respect to the subgroup C^0_k. From $\widehat{\partial}_k \alpha = 0$ it follows that $\partial_k d \in C^0_{k-1}$, and from $\partial_{k-1} \partial_k = 0$ that $\partial_k d \in \operatorname{Ker} \partial_{k-1}$.

Exercise 3°. Show that the coset $[\partial_k \delta]^0$ of the element $\partial_k d$ in $H_{k-1}(C^0_*)$ does not depend on the choice of the elements α and d from the corresponding cosets.

To each element $\widehat{\alpha}$ from $H_k(\widehat{C}_*)$ we associated an element $[\partial_k d]^0$ from $H_{k-1}(C^0_*)$ thereby defining a mapping which we denote by

$$\delta_k : H_k(\widehat{C}_*) \to H_{k-1}(C^0_*)$$

and call the *connecting homomorphism*.

Exercise 4°. Show that δ_k is, in fact, a homomorphism.

The construction of a connecting homomorphism can be extended by taking $\delta_0 : H_0(\widehat{C}_*) \to 0$.

Lemma 1. *The sequence* (4) *is exact.*

The proof is reduced to the direct checking of the relations

$$\operatorname{Im} \delta_{k+1} = \operatorname{Ker} i_{*k}, \quad \operatorname{Im} i_{*k} = \operatorname{Ker} j_{*k}, \quad \operatorname{Im} j_{*k} = \operatorname{Ker} \delta_k$$

and is left to the reader.

Exercise 5°. (S. Lang). Take any book on homological algebra and prove all the theorems without consulting the proofs given in that book.

§ 3. Homology groups of simplicial complexes

We shall apply the algebraic technique developed in § 2 to the construction of homology groups of geometric objects.

1. Simplicial complexes and polyhedra.

First, we give the necessary definitions.

Definition 1. The *standard k–dimensional simplex* σ^k, $k \geq 0$, is the convex closure of the $k + 1$ points in $\mathbf{R}^{k+1}$ with coordinates $(1, 0, 0, \dots, 0, 0)$, $(0, 1, 0, \dots, 0, 0), \dots, (0, 0, \dots, 0, 1)$, i.e., the collection of points with coordinates $(t_0, \dots, t_k)$ such that $t_i \geq 0$ for all i and $\sum_{i=0}^{k} t_i = 1$.

Definition 2. A *simplex of dimension k* or a *k–dimensional simplex* $\tau^k = (a^0, a^1, \dots, a^k)$ is the convex closure of $k+1$ points $a^0, \dots, a^k$ of the Euclidean space $\mathbf{R}^n$, $k \leq n$, lying in general position (which means that they do not lie in any m–plane of dimension lower than k), i.e., the collection of points of the form $x = \sum_{i=0}^{k} t_i a^i$, where $t_i \geq 0$, for all i, $\sum_{i=0}^{k} t_i = 1$.

The points a^i are called the *vertices of the simplex* $(a^0, \dots, a^k)$, and the numbers t_i *the barycentric coordinates of the point* $x \in (a^0, \dots, a^k)$.

The concept of a face of a simplex is defined in a natural way.

Definition 3. The convex closure of a subset of $s+1$ vertices of the simplex τ^k is called a *face of dimension s* or an *s–dimensional face* of the *k–dimensional simplex* τ^k, where $0 \leq s \leq k$.

The faces of dimension $s < k$ of the simplex τ^k of dimension k are called *proper*.

It is obvious that an s–dimensional face of a simplex is an s–dimensional simplex. In particular, the faces of a standard simplex (and the standard simplex itself) are simplices. It is easy to verify that a k–dimensional simplex is affinely homeomorphic to the standard simplex of the same dimension; the interior (in the carrier k–plane) of a simplex τ^k can be considered as a special case of a k–dimensional cell.

Thus, it is possible to construct cell complexes from simplices of different dimensions. The existence of the faces in a simplex allows us to combine simplices in a more ordered manner than cells in a general cell complex.

Definition 4. A *simplicial complex* K is a set $\{\tau_j^k\}$ of simplices in $\mathbf{R}^n$ satisfying the following conditions: (1) together with each simplex τ_j^k in K each of its faces is included in K; (2) two simplices can intersect only in a common face.

A simplicial complex is said to be *finite* if it consists of a finite number of simplices.

Consider the set–theoretic union $|K| \subset \mathbf{R}^n$ of all simplices from K. We introduce a topology on $|K|$ that is the strongest of all those in which the imbedding mapping of each simplex into $|K|$ is continuous. In other words, the set $A \subset |K|$ is closed iff $A \cap \tau_i^k$ is closed in τ_i^k for any $\tau_i^k \in K$. If the simplicial complex K is finite, then this topology coincides with the topology induced by the metric on $\mathbf{R}^n$.

Definition 5. A space $|K|$ and, more generally, any topological space X which is homeomorphic to $|K|$, is called a *polyhedron*.

Definition 6. A *triangulation of a polyhedron* X is a simplicial complex K such that the space $|K|$ is homeomorphic to X.

Examples of polyhedra are the closed surfaces (see Ch. 2, § 4). Their triangulation is given by partitioning a surface into topological triangles, their edges and vertices.

Consider a finite simplicial complex K. In the space $|K| \subset \mathbf{R}^n$ take the metric induced from $\mathbf{R}^n$. It is evident that there exist different triangulations of the space $|K|$. Let K' be a certain triangulation $|K|$. The *mesh of the triangulation* K' is the greatest of the lengths of the one–dimensional simplices included in K'.

Exercise 1°. Prove that a polyhedron is a) a normal space; b) a cell complex.

Exercise 2°. Prove that if K is a finite simplicial complex, then the space $|K|$ is a) a compact space; b) a finite cell complex.

Exercise 3°. Prove that a simplicial complex K is finite iff the polyhedron $|K|$ is compact.

Everywhere below, we consider finite simplicial complexes and compact polyhedra unless explicitly mentioned otherwise. It is easy to see that a compact polyhedron is a metrizable space.

Let X be a polyhedron, K a simplicial complex, and $\phi : |K| \to X$ a homeomorphism. The homeomorphism ϕ generates a decomposition (triangulation) of the space X into the sets $\Sigma_i^k = \phi(\tau_i^k)$, $\tau_i^k \in K$, which are called *curvilinear simplices*; the images of the vertices of the simplex τ_i^k are called the *vertices of the curvilinear simplex* Σ_i^k.

Exercise 4°. Show that the following are simplicial complexes: a) the set $\{\tau^n\}$, i.e., the collection of the simplex τ^n and all its faces, $|\{\tau^n\}| = \tau^n$, b) the set $\{\partial\tau^n\}$, i.e., the collection of proper faces of the simplex τ^n, where $|\{\partial\tau^n\}|$ coincides with the boundary $\partial\tau^n$ of the set τ^n in the carrier n–plane.

Exercise 5°. Show that the closed disc $\bar{D}^n$ and the sphere S^{n-1} are polyhedra, and specify a decomposition of them into curvilinear simplices.

Hint. Consider a homeomorphism of a simplex τ^n onto $\bar{D}^n$ and use the result of the previous exercise. The homeomorphism can be defined as follows: consider the simplex $\tau^n = (a^0, a^1, \dots, a^n) \in \mathbf{R}^n$, where

$$a^1 = (1,0,0,\dots,0,0), \qquad a^2 = (0,1,0,\dots,0,0), \dots$$
$$a^n = (0,0,0,\dots,0,1), \qquad a^0 = (-1,-1,-1,\dots,-1,-1).$$

Let $\Theta \in \mathbf{R}^n$ and $x \neq \Theta$. Put $\lambda(x) = \sup\{\alpha : \alpha \frac{x}{\|x\|} \in \tau^n\}$. Then for $x \in \bar{D}^n$, we define the mapping $\psi : \bar{D}^n \to \tau^n$ as follows

$$\psi(x) = \begin{cases} 0, & \text{when } x = \Theta, \\ \lambda(x) \cdot x, & \text{when } x \neq \Theta. \end{cases}$$

It is not difficult to verify that ψ is a homeomorphism; the required decomposition into curvilinear simplices is defined by its an inverse homeomorphism.

2. Homology of simplicial complexes and polyhedra.

Now, we associate to a simplicial complex K a certain chain complex. Enumerate the vertices of each simplex $\tau_i^k \in K$ by the numbers $0, 1, \dots, k$ in some order $a^{i_0}, a^{i_1}, \dots, a^{i_k}$. There are $(k+1)!$ such enumerations. Two enumerations are said to be equivalent if one of them can be obtained from the other by an even permutation of indices. The set of all enumerations splits into two equivalence classes denoted by Λ_i^+ and Λ_i^-, respectively.

Definition 7. A simplex τ^k with one of the classes Λ^+, Λ^- specified, i.e., a pair (τ^k, Λ^+) (τ^k, Λ^-), is called an *oriented simplex*; the corresponding class is its *orientation*.

It is more convenient to write an oriented simplex (τ_i^k, Λ_i^+) in a different way, i.e., to select some numeration $a^{i_0}, a^{i_1}, \ldots, a^{i_k}$ from the orientation class and denote it as follows:

$$(\tau_i^k, \Lambda_i^+) = [a^{i_0}, a^{i_1}, a^{i_2}, \ldots, a^{i_k}];$$

then

$$(\tau_i^k, \Lambda_i^-) = [a^{i_1}, a^{i_0}, a^{i_2}, \ldots, a^{i_k}].$$

Definition 8. A *group of k–dimensional chains* $C_k(K; G)$ of a simplicial complex K with coefficients in the Abelian group G is a quotient group of the group of formal linear combinations (finite) of the form $\sum_i g_i \cdot (\tau_i^k, \Lambda_i)$, $g_i \in G$, $\Lambda_i = \Lambda_i^+$ or $\Lambda_i = \Lambda_i^-$, with respect to the subgroup of elements of the form

(1)
$$g \cdot (\tau_i^k, \Lambda_i^+) + g \cdot (\tau_i^k, \Lambda_i^-)$$

and their linear combinations.

In other words, we identify the elements $g \cdot (\tau^k, \Lambda_i^-)$, $-g \cdot (\tau_i^k, \Lambda_i^+)$ in the group of formal linear combinations of oriented simplices.

We define the differential

$$\partial_k : C_k(K; G) \rightarrow C_{k-1}(K; G)$$

by the formula

$$\partial_k(g \cdot [a^{i_0}, a^{i_1}, \ldots, a^{i_k}]) =$$

(2)
$$= \sum_{j=0}^{k} (-1)^j g \cdot [a^{i_0}, \ldots, a^{i_{j-1}}, a^{i_{j+1}}, \ldots a^{i_k}]$$

for each oriented simplex, and extend it to the whole group $C_k(K; G)$ by additivity.

For $k = 0$, we put $\partial_0 : C_0(K; G) \to 0$.

Proposition 1. *For all $k \geq 1$, the equality $\partial_{k-1}\partial_k = 0$ is valid.*

Proof. Indeed, the sum $\partial_{k-1}\partial_k(g \cdot [a^{i_0}, \dots, a^{i_k}])$ (for $k \geq 2$) includes both the summands

$$(-1)^p(-1)^{q-1}g \cdot [a^{i_0}, \dots, a^{i_{p-1}}, a^{i_{p+1}}, \dots, a^{i_{q-1}}, a^{i_{q+1}}, \dots, a^{i_k}]$$

and

$$(-1)^p(-1)^q g \cdot [a^{i_0}, \dots, a^{i_{p-1}}, a^{i_{p+1}}, \dots, a^{i_{q-1}}, a^{i_{q+1}}, \dots, a^{i_k}],$$

which cancel each other.   □

Thus, the groups $C_k(K; G)$ and the differentials ∂_k form a chain complex which is denoted by $C_*(K; G)$. For G we may take, for instance, the group **Z** of integers.

Definition 9. The homology groups of a chain complex $C_*(K; G)$ are called the *homology groups of the simplicial complex* K with coefficients in the Abelian group G and they are denoted by $H_k(K; G)$.

Definition 10. The *homology groups $H_k(X; G)$ of a polyhedron X* with coefficients in an Abelian group G are the homology groups with coefficients in G of a triangulation K of the polyhedron X.

To prove that is well-defined, i.e. independent of the choice of triangulation, is technically complicated; we will discuss these questions in § 5.

3. Calculation of the homology of concrete polyhedra. We shall calculate the homology groups $H_k(\tau^n; G)$ of the polyhedron τ^n. It is evident that for τ^0, the space consisting of one point, we have

$$C_k(\{\tau^0\}; G) = \text{Ker}\,\partial_k = \text{Im}\,\partial_k = 0 \quad \text{when } k > 0,$$
$$C_0(\{\tau^0\}; G) = \text{Ker}\,\partial_0 \simeq G.$$

Hence, we obtain the homology groups

(3) $$H_k(\tau^0; G) = 0 \quad \text{for } k > 0; \quad H_0(\tau^0; G) \simeq G.$$

Before calculating $H_k(\tau^n; G)$, $n > 0$, we shall solve a more general problem. Consider a simplicial complex K lying in the hyperplane $\Pi^m \subset \mathbf{R}^{m+1}$ and a point $a \in \mathbf{R}^{m+1} \setminus \Pi^m$. The *cone* aK *over the complex* K *with vertex* a is the collection of simplices consisting of simplices $\tau_i^k \in K$, the simplex a and all the simplices of the form (a, τ_i^k), i.e., the simplices $(a, a^{i_0}, \dots, a^{i_k})$ such that $\tau_i^k = (a^{i_0}, \dots, a^{i_k})$ is a certain simplex in K.

Exercise 6°. Show that aK is a simplicial complex.

Preposition 2. *Let* aK *be the cone with vertex* a *over a simplicial complex* K. *Then*

(4) $$H_k(aK; G) = 0 \quad \text{when} \quad k > o; \quad H_0(aK; G) \simeq G.$$

Proof. Consider an arbitrary 0–dimensional chain $g \cdot a + \sum_i g_i a^i$ from $C_0(aK; G) = \text{Ker } \partial_0$. We have

$$g \cdot a + \sum_i g_i a^i = \left(g + \sum_i g_i\right) \cdot a + \sum_i (g_i \cdot a_i - g_i \cdot a).$$

Now

$$\sum_i (g_i \cdot a^i - g_i \cdot a) = \partial_1\left(\sum_i g_i \cdot [a, a^i]\right),$$

so that an arbitrary cycle $g \cdot a + \sum_i g_i \cdot a^i$ from $\text{Ker } \partial_0$ is homologous to a cycle $g' \cdot a = (g + \sum_i g_i) \cdot a$ which is not homologous to zero for $g' \neq 0$ in the group $C_0(aK; G)$. This gives an isomorphism $H_0(aK; G) \simeq G$.

Consider now an arbitrary k–dimensional cycle in $C_k(aK; G)$

$$z_k = \sum_i g_i \cdot [\tau_i^k] + \sum_j h_j \cdot [a, \tau_j^{k-1}] \in \text{Ker } \partial_k,$$

where $i \in I_k$, $j \in I_{k-1}$, $g_i, h_j \in G$, and the $[\tau_i^k]$, $[a, \tau_j^{k-1}]$ are oriented simplices. We have

$$\sum_i g_i \cdot [\tau_i^k] \sim \sum_i (g_i \cdot [\tau_i^k] - \partial_{k+1}(g_i \cdot [a, \tau_i^k])) = \sum_j g_j' \cdot [a \cdot \tau_j^{k-1}].$$

Therefore the cycle z_k is homologous to a cycle of the form

$$z'_k = \sum_j h'_j \cdot [a, \tau_j^{k-1}] = \sum_j (g'_j + h_j) \cdot [a, \tau_j^{k-1}].$$

The simplex $[\tau_j^{k-1}]$ is included into the sum $\partial_k(\sum_j h'_j \cdot [a, \tau_j^{k-1}])$ with the coefficient h'_j (only once!). Therefore, $\sum_j h'_j \cdot [a, \tau_j^{k-1}]$ is a cycle iff $h'_j = 0$ for all j.

Thus, we have established that any cycle from $\text{Ker} \, \partial_k$ in $C_*(aK; G)$ is homologous to zero in $C_k(aK; G)$ when $k > 0$. Consequently, $H_k(aK; G) = 0$ for $k > 0$. $\qquad \square$

Note that the complex $\{\tau^n\}$ corresponding to the simplex $\tau^n = (a^0, \ldots, a^n)$, is the cone $a^0\{\tau^{n-1}\}$ with the vertex a^0 over the complex $\{\tau^{n-1}\}$ which corresponds to the simplex $\tau^{n-1} = (a^1, \ldots, a^n)$. Therefore, from (3) and (4), we obtain the homology groups of an n–dimensional simplex:

$$(5) \qquad H_k(\tau^n; G) \simeq \begin{cases} 0 & \text{for } k > 0, \\ G & \text{for } k = 0 \end{cases}$$

for all $n \geq 0$.

We now turn to the calculation of the homology groups $H_k(|\{\partial \tau^n\}|; G)$ of a polyhedron $|\{\partial \tau^n\}|$ whose triangulation $\{\partial \tau^n\}$ consists of all proper faces of a simplex τ^n. Consider the case $n > 1$. Then for $k < n$, we have $C_k(\{\partial \tau^n\}; G) = C_k(\{\tau^n\}; G)$, and the differentials of the chain complexes $C_*(\{\partial \tau^n\}; G)$ and $C_*(\{\tau^n\}; G)$ coincide when $k < n$. Therefore, for $k < n - 1$

$$(6) \qquad H_k(\{\partial \tau^n\}; G) \simeq H_k(\{\tau^n\}; G).$$

Evidently, when $k > n - 1$,

$$(7) \qquad H_k(\{\partial \tau^n\}; G) = 0.$$

Since $H_{n-1}(\{\tau^n\}; G) = 0$, any cycle $z_{n-1} \in C_{n-1}(\{\tau^n\}; G)$ is the boundary $\partial_n(g \cdot [\tau^n])$ of a chain $g \cdot [\tau^n] \in C_n(\{\tau^n\}; G)$, and, thus, we have $\text{Ker} \, \partial_{n-1} = \text{Im} \, \partial_n \simeq G$ in the complex $C_*(\{\tau^n\}; G)$. The differentials in the complexes $C_*(\{\tau^n\}; G)$ and $C_*(\{\partial \tau^n\}; G)$ coincide on the chain groups $C_{n-1}(\{\tau^n\}; G) =$

$C_{n-1}(\{\partial\tau^n\}; G)$. Thus, the group $\mathrm{Ker}\,\partial_{n-1}$ is isomorphic to the group G in $C_*(\{\partial\tau^n\}; G)$, whereas $\mathrm{Im}\,\partial_n = \partial_n(C_n(\{\partial\tau^n\}; G)) = 0$. Consequently,

$$
(8) \qquad\qquad H_{n-1}(\{\partial\tau^n\}; G) \simeq G.
$$

Thus, when $n > 1$, the homology of the boundary of an n–dimensional simplex are calculated:

$$
(9) \qquad H_k(|\{\partial\tau^n\}|; G) \simeq \begin{cases} 0 & \text{for } k \neq 0, n-1, \\ G & \text{for } k = 0, n-1. \end{cases}
$$

Exercise 7°. Prove that

$$
(10) \qquad H_k(|\{\partial\tau^1\}|; G) \simeq \begin{cases} 0 & \text{for } k > 0, \\ G \oplus G & \text{for } k = 0. \end{cases}
$$

We shall dwell on a visual geometric interpretation of the homology groups of a simplicial complex. A cycle from $C_k(K; \mathbf{Z})$ is a set of k–dimensional simplices from K; each of them is taken a certain number of times; this set is closed in the sense that each $(k-1)$–dimensional simplex is included in the boundary of the k–dimensional cycle the same number of times with opposite orientations. Two k–dimensional cycles are equivalent (homologous), if their difference is the boundary of a $(k+1)$–dimensional chain, i.e., bounds a certain set of $(k+1)$–dimensional simplices; the group $H_k(|K|; \mathbf{Z})$ is the group of equivalence classes of such k–dimensional cycles. Roughly speaking, $H_k(|K|; \mathbf{Z})$ consists of those closed collections of k–dimensional simplices which cannot be "filled in" with a collection of $(k+1)$–dimensional simplices. Thus, intuitively, the group $H_k(|K|; \mathbf{Z})$ corresponds to the group generated by the $(k+1)$–dimensional "holes" in the space $|K|$.

Definition 11. A subset L of simplices from K which is a simplicial complex, is called a *subcomplex* of a simplicial complex.

Let L be a subcomplex of a simplicial complex K. Obviously, $C_*(L; G)$ is then a subcomplex of the chain complex $C_*(K; G)$. Therefore, a quotient complex is defined

$$
C_*(K, L; G) = C_*(K; G)/C_*(L; G).
$$

Denoting the homology groups of this quotient chain complex by $H_k(K, L; G)$, we obtain from the exact sequence of chain complexes

$$0 \to C_*(L; G) \xrightarrow{i_*} C_*(K; G) \xrightarrow{j_*} C_*(K, L; G) \to 0$$

a long exact sequence of homology groups

$$\ldots \to H_{k+1}(K, L; G) \xrightarrow{\delta_{k+1}} H_k(L; G) \xrightarrow{i_{*k}}$$
$$\to H_k(K; G) \xrightarrow{i_{*k}} H_k(K, L; G) \xrightarrow{\delta_k} H_{k-1}(L; G) \to \ldots$$

It is called the *long exact sequence of the pair* (K, L), the groups $H_k(K, L; G)$ are called *relative homology groups* or *homology groups of the pair* (K, L).

It is useful to "decode" the definition of relative homology groups.

Since the chain $\widehat{\gamma}_k$ from $C_k(K, L; G)$ is a coset of the group $C_k(K; G)$ with respect to the subgroup $i_k(C_k(L; G)) \simeq C_k(L; G)$, there exists in the coset $\widehat{\gamma}_k$, a unique representative, the chain γ_k from $C_k(K; G)$, which includes only those oriented simplices with non–zero coefficients of the complex K that are not oriented simplices of the subcomplex L. From the definition of a boundary homomorphism in a quotient complex it follows that the boundary homomorphism $\widehat{\partial}_k : C_k(K, L; G) \to C_{k-1}(K, L; G)$ transforms the chain $\widehat{\gamma}_k$ into a chain $\widehat{\gamma}_{k-1}$ which is a coset of the group $C_{k-1}(K; G)$ with respect to the subgroup $i_{k-1}(C_{k-1}(L; G)) \simeq C_{k-1}(L; G)$ with a representative $\partial_k \gamma_k \in C_{k-1}(K; G)$. In the chain $\partial_k \gamma_k$, we discard all summands $g_m[\tau_m^{k-1}]$ for which τ_m^{k-1} is a simplex in L. The chain γ_{k-1} obtained defines a coset $\widehat{\gamma}_{k-1}$ which is obviously the same as $\widehat{\partial}_k \widehat{\gamma}_k$.

It is clear that the chain complex $C_*(K, L; G)$ is isomorphic to a chain complex $\widetilde{C}_*$ that is defined as follows: the chains are formal linear combinations of oriented simplices (in the sense of definition 8) from $K \backslash L$, and the boundary homomorphism associates to the k–dimensional chain γ_k the chain of dimension $k - 1$ which is obtained by calculating the value of the boundary homomorphism ∂_k on γ_k (in the chain complex $C_*(K; G)$) and by deleting all (superflous) summands $g_m \cdot [\tau_m^{k-1}]$ for which τ_m^{k-1} belongs to L. Since an isomorphism of chain complexes induces an isomorphism of homology groups, $H_k(\widetilde{C}_*) \simeq H_k(K, L; G)$, $k = 0, 1, 2, \ldots$.

Thus, we have arrived at a clearer definition of the homology groups of a pair. Note that the chains, cycles and boundaries of the complex $\widetilde{C}_*$ are called relative (for the pair (K, L)).

Now, we explain the geometric meaning of the connecting homomorphism

$$\delta_k : H_k(K, L; G) \to H_{k-1}(L; G).$$

Let $\tilde{h}_k \in H_k(K, L; G)$ be the homology class of the relative cycle $z_k \in \tilde{C}_k$. Consider z_k as a chain in $C_*(K; G)$ and calculate its boundary $\partial_k z_k$ in it. According to the definition of a relative cycle, the chain $\partial_k z_k$, after collecting similar terms, will include with non–zero coefficients only oriented simplices from L. Therefore, $\partial_k z_k$ can be considered as a chain in $C_*(L; G)$. It can be easily checked that $\partial_k z_k$ is a cycle whose homology class $h_{k-1} \in H_{k-1}(L; G)$ does not depend on the choice of the representative z_k of the class $\tilde{h}_k$. By the general construction of the connecting homomorphism (§ 2), $\partial_k \tilde{h}_k = h_{k-1}$. If a relative cycle is realized as a manifold with boundary lying in L and consisting of k–dimensional oriented simplices, then $\partial_k z_k$ is just that boundary with the corresponding orientations of $(k-1)$–dimensional simplices.

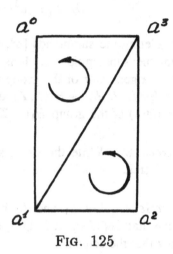

FIG. 125

Example. (Fig. 125). Let a simplicial complex consist of simplices

$$a^0, a^1, a^2, a^3,$$
$$(a^0, a^1), (a^1, a^2), (a^2, a^3), (a^3, a^0), (a^1, a^3),$$
$$(a^0, a^1, a^3), (a^1, a^2, a^3),$$

and let its subcomplex L consist of the same simplices with the exception of

$$(a^1, a^3), (a^0, a^1, a^3), (a^1, a^2, a^3).$$

Thus, $|K|$ is a rectangle (with "interior"), and $|L|$ its boundary. It is clear that the chain $\gamma_2 \in C_2(K; \mathbf{Z})$,

$$\gamma_2 = [a^0, a^1, a^3] + [a^1, a^2, a^3],$$

is a relative cycle of the pair (K, L). Indeed, its boundary

$$\partial_2 \gamma_2 = [a^3, a^0] + [a^0, a^1] + [a^1, a^2] + [a^2, a^3]$$

includes with non–zero coefficients only oriented simplices from the subcomplex L. The chain $\gamma_1 = [a^1, a^3]$ from $C_1(K; \mathbf{Z})$ is simultaneously both a relative cycle (verify!) and a relative boundary, since it can be obtained from

$$\partial_2[a^0, a^1, a^3] = [a^1, a^3] + [a^3, a^0] + [a^0, a^1]$$

by deleting the summands $[a^3, a^0]$ and $[a^0, a^1]$, which are oriented simplices from the subcomplex L. It is easy to see that the relative cycle γ_2 determines a generator of the group $H_2(K, L; \mathbf{Z})$. The connecting homomorphism $\delta_2 : H_2(K, L; \mathbf{Z}) \rightarrow H_1(L; \mathbf{Z})$ associates to this generator the element (also a generator) of the group $H_1(L; \mathbf{Z})$ consisting of the cycle $\partial_2 \gamma_2$.

Exercise 8°. Write the exact sequence of the pair (K, L) for the example considered.

Exercise 9°. Let L_1 and L_2 be subcomplexes of a simplicial complex K. Prove that $L_1 \cap L_2$ and $L_1 \cup L_2$ are also subcomplexes of the complex K, and show that the sequence

$$0 \rightarrow C_*(L_1 \cap L_2; G) \xrightarrow{I_*} C_*(L_1; G) \oplus C_*(L_2; G) \rightarrow C_*(L_1 \cup L_2; G) \rightarrow 0,$$

is exact, where $I_k(\sum_i g_i \cdot [\tau_i^k]) = (\sum_i g_i \cdot [\tau_i^k], -\sum_i g_i \cdot [\tau_i^k])$. Derive from this the following exact sequence

$$\dots \rightarrow H_{k+1}(L_1 \cup L_2; G) \rightarrow$$
$$\rightarrow H_k(L_1 \cap L_2; G) \rightarrow H_k(L_1; G) \oplus H_k(L_2; G) \rightarrow$$
(11)$\qquad \rightarrow H_k(L_1 \cup L_2; G) \rightarrow H_{k-1}(L_1 \cap L_2; G) \rightarrow \dots$

This sequence is called the *Mayer–Vietoris exact sequence*.

Exact sequnece (11) helps us to compute homology groups of complicated simplicial complexes.

Exercise 10°. Using (5), (9), (10), and (11) calculate the homology groups of the complex consisting of simplices of dimension 0 and 1, which is drawn in Fig. 126.

Hint. Consider the complex as a successive union of subcomplexes.

Exercise 11°. Show that for an orientable surface M_p of genus p we have the isomorphism $H_2(M_p; \mathbf{Z}) \simeq \mathbf{Z}$.

Hint. Show that any two–dimensional cycle is a multiple of a cycle which is equal to the sum of all curvilinear 2–simplices of a triangulation M_p taken with compatible orientations.

Note that for any path–connected polyhedron X, one has $H_0(X; G) \simeq G$. Indeed, if a is a vertex of a triangulation K of a polyhedron X, then any cycle from $C_0(K; G)$ is homologous to a cycle of the form $g \cdot a$ which is not homologous to zero when $g \neq 0$; for a cycle of the form $\sum_i g_i a^i$, this fact is easily established by a sequence of one–dimensional simplices "going" from the points a^i to the point a. More generally, for any polyhedron X, the group

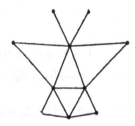

FIG. 126

$H_0(X; G)$ is isomorphic to the direct sum of copies from the group G the number of which is equal to the number of connectedness components of the polyhedron (note that for polyhedra, the concepts of connectedness and path–connectedness are equivalent, i.e., a connected polyhedron is path–connected).

Using the definition of homology groups of a polyhedron, we can perform a number of interesting calculations. First, it should be noted that by definitions 5 and 6, the homology groups of homeomorphic polyhedra are identical (isomorphic).

Therefore, the homology groups of a closed n–dimensional disc (and an n–dimensional cube) coincide with the homology groups of the simplex τ^n; the homology groups of an $(n-1)$–dimensional sphere (and the boundaries of an n–dimensional cube) coincide with the homology groups of the polyhedron $|\{\partial \tau^n\}|$, i.e., with that of the boundary of the simplex (see Exercises 4° and 5°, and formulae (5), (9), and (10)).

Using the exact sequence (11) it is not difficult to calculate the homology groups of the polyhedron $C = S^1 \times [0; 1]$, which is a cylinder over a circle. Cutting along two generating intervals I_1 and I_2 gives two bent rectangles P_1 and P_2 with a common boundary $I_1 \cup I_2$. Now write down the exact sequence (with coefficients in $\mathbf{Z}$)

$$0 \to C_*(I_1 \cup I_2; \mathbf{Z}) \to C_*(P_1; \mathbf{Z}) \oplus C_*(P_2; \mathbf{Z}) \to C_*(P_1 \cup P_2; \mathbf{Z}) \to 0.$$

For the homology groups, this gives the exact sequence (Mayer–Vietoris)

$$
\begin{aligned}
0 \to H_2(I_1 \cup I_2; \mathbf{Z}) &\to H_2(P_1; \mathbf{Z}) \oplus H_2(P_2; \mathbf{Z}) \to \\
&\to H_2(C; \mathbf{Z}) \to H_1(I_1 \cup I_2; \mathbf{Z}) \to H_1(P_1; \mathbf{Z}) \oplus H_1(P_2; \mathbf{Z}) \to \\
&\to H_1(C; \mathbf{Z}) \to H_0(I_1 \cup I_2; \mathbf{Z}) \to \\
&\to H_0(P_1; \mathbf{Z}) \oplus H_0(P_2; \mathbf{Z}) \to H_0(C; \mathbf{Z}) \to 0
\end{aligned}
$$

or

$$
\begin{aligned}
0 \to 0 \oplus 0 \to 0 \oplus 0 &\to H_2(C; \mathbf{Z}) \to 0 \oplus 0 \to 0 \oplus 0 \to \\
&\to H_1(C; \mathbf{Z}) \to \mathbf{Z} \oplus \mathbf{Z} \to \mathbf{Z} \oplus \mathbf{Z} \to \mathbf{Z} \to 0.
\end{aligned}
$$

It follows that $H_2(C; \mathbf{Z}) = 0$ and $H_1(C; \mathbf{Z}) \simeq \mathbf{Z}$.

Thus,

$$H_i(C; \mathbf{Z}) \simeq \begin{cases} \mathbf{Z} & \text{when } i = 0, 1, \\ 0 & \text{when } i = 2. \end{cases}$$

Now, it is easy to calculate the homology groups of the torus $T^2 \simeq S^1 \times S^1$. We decompose the torus into two bent cylinders C_1 and C_2 and obtain the exact sequence

$$0 \to C_*(S_1^1 \cup S_2^1; \mathbf{Z}) \to C_*(C_1; \mathbf{Z}) \oplus C_*(C_2; \mathbf{Z}) \to C_*(T^2; \mathbf{Z}) \to 0.$$

For the homology groups, we have the exact sequence

$$
\begin{aligned}
0 \to H_2(S_1^1 \cup S_2^1; \mathbf{Z}) &\to H_2(C_1; \mathbf{Z}) \oplus H_2(C_2; \mathbf{Z}) \to \\
&\to H_2(T^2; \mathbf{Z}) \to H_1(S_1^1 \cup S_2^1; \mathbf{Z}) \to H_1(C_1; \mathbf{Z}) \oplus H_1(C_2; \mathbf{Z}) \to \\
&\to H_1(T^2; \mathbf{Z}) \to H_0(S_1^1 \cup S_2^1; \mathbf{Z}) \to \\
&\to H_0(C_1; \mathbf{Z}) \oplus H_0(C_2; \mathbf{Z}) \to H_0(T^2; \mathbf{Z}) \to 0.
\end{aligned}
$$

Since the torus T^2 is path–connected, we have $H_0(T^2; \mathbf{Z}) \simeq \mathbf{Z}$, and by exercise $11°$, $H_2(T^2; \mathbf{Z}) \simeq \mathbf{Z}$; therefore we obtain an exact sequence

$$0 \to 0 \to 0 \oplus 0 \to \mathbf{Z} \to \mathbf{Z} \oplus \mathbf{Z} \to \mathbf{Z} \oplus \mathbf{Z} \to H_1(T^2; \mathbf{Z}) \to$$
$$\to \mathbf{Z} \oplus \mathbf{Z} \to \mathbf{Z} \oplus \mathbf{Z} \to \mathbf{Z} \to 0,$$

and it follows that $H_1(T^2; \mathbf{Z}) \simeq \mathbf{Z} \oplus \mathbf{Z}$. Thus,

$$H_i(T^2; \mathbf{Z}) \simeq \begin{cases} 0 & \text{when } i > 2, \\ \mathbf{Z} & \text{when } i = 0, 2, \\ \mathbf{Z} \oplus \mathbf{Z} & \text{when } i = 1. \end{cases}$$

Exercise $12°$. Show that for an orientable surface M_p of genus p, we have

$$H_1(M_p; \mathbf{Z}) \simeq \underbrace{\mathbf{Z} \oplus \ldots \oplus \mathbf{Z}}_{2p}.$$

Hint. Consider M_p as a result of gluing p handles ("bent cylinders") to the sphere with $2p$ holes and apply the exact Mayer–Vietoris sequence.

Exercise $13°$. Show that

$$H_k(\mathbf{R}P^2; \mathbf{Z}) \simeq \begin{cases} \mathbf{Z}, & k = 0, \\ \mathbf{Z}_2, & k = 1, \\ 0, & k > 1 \end{cases}$$

and

$$H_k(\mathbf{R}P^2; \mathbf{Z}_2) \simeq \begin{cases} \mathbf{Z}_2, & k = 0, 1, 2, \\ 0 & k > 2. \end{cases}$$

Hint. Use a simplicial decomposition of $\mathbf{R}P^2$.

Exercise $14°$. Show that for a non–orientable surface N_q of genus q

$$H_k(N_q; \mathbf{Z}) \simeq \begin{cases} \mathbf{Z}, & k = 0, \\ \underbrace{\mathbf{Z} \oplus \ldots \oplus \mathbf{Z}}_{q-1} \oplus \mathbf{Z}_2, & k = 1, \\ 0, & k > 1 \end{cases}$$

Exercise 15°. Show that

$$H_k(M_p; \mathbf{Z}_2) \simeq \begin{cases} \mathbf{Z}_2, & k = 0, 2, \\ \underbrace{\mathbf{Z}_2 \oplus \ldots \oplus \mathbf{Z}_2}_{2p}, & k = 1, \\ 0, & k > 2 \end{cases}$$

and

$$H_k(N_q; \mathbf{Z}_2) \simeq \begin{cases} \mathbf{Z}_2, & k = 0, 2, \\ \underbrace{\mathbf{Z}_2 \oplus \ldots \oplus \mathbf{Z}_2}_{q}, & k = 1, \\ 0, & k > 2. \end{cases}$$

4. Barycentric subdivisions. Simplicial mappings.

Let $\tau^k = (a^0, \ldots, a^k)$ be a k–dimensional simplex. The point with barycentric coordinates $1/(k+1), \ldots, 1/(k+1)$ is called the *barycentre of the simplex* τ^k. We denote this point by $b^{0,1,\ldots,k}$; in a more general case, denote by $b^{i_0,\ldots,i_p}$ the point whose barycentric coordinates t_i are defined as follows:

$$t_i = \begin{cases} \frac{1}{p+1}, & i = i_0, \ldots, i_p, \\ 0 & \text{otherwise.} \end{cases}$$

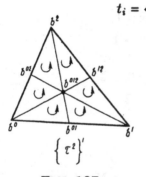

$\{\tau^2\}'$

FIG. 127

For all possible sets $a^{i_0}, \ldots, a^{i_p}$ of $p+1$ vertices ($0 \leq p \leq k$), the points $b^{i_0,\ldots,i_p}$ corresponding to them are the barycentres of the p–dimensional faces $(a^{i_0}, \ldots, a^{i_p})$ of the simplex τ^k (the 0–dimensional faces are the vertices a^i, and the k–dimensional face is the simplex τ^k itself). Consider all possible simplices of the form

$$(b^{i_0,i_1,\ldots,i_p}, b^{i_0,i_1,\ldots,i_{p-1}}, \ldots, b^{i_0,i_1}, b^{i_0}), \quad 0 \leq p \leq k.$$

The collection of all these simplices and their faces forms a simplicial complex called a *barycentric subdivision of the simplex* τ^k (Fig. 127).

Let K be a simplicial complex. The barycentric subdivisions of all its simplices form a simplicial complex K' which is called the *barycentric subdivision of the complex K*. We shall also consider the simplicial complexes $K^{(2)} = (K')', \ldots, K^{(r)} = (K^{r-1})'$.

The barycentric subdivision of a complex defines a chain homomorphism

$$\Theta_* : C_*(K; G) \to C_*(K'; G).$$

The homomorphism Θ_0 is defined on the vertices a^i by the formula

(12)
$$\Theta_0(g \cdot a^i) = g \cdot a^i,$$

and on simplices of greater dimension, it can be defined inductively by the formula

(13)
$$\Theta_p(g \cdot [a^{i_0}, \ldots, a^{i_p}]) =$$
$$= [b^{i_0, \ldots, i_p}, \Theta_{p-1} \partial_p (g \cdot [a^{i_0}, \ldots, a^{i_p}])],$$

which means that if

$$\Theta_{p-1} \partial_p (g \cdot [a^{i_0}, \ldots, a^{i_p}]) = \sum_k g_k \cdot [c_k^{j_0}, \ldots, c_k^{j_{p-1}}],$$

holds, then we have

$$\Theta_p(g \cdot [a^{i_0}, \ldots, a^{i_p}]) = \sum_k g_k \cdot [b^{i_0, \ldots, i_p}, c_k^{j_0}, \ldots, c_k^{j_{p-1}}].$$

By linearity, Θ_p can be extended to the whole group $C_p(K; G)$. It is easy to verify that Θ_* is a chain homomorphism.

Repeating this one obtains natural the homomorphisms

$$\Theta_*^{(r)} : C_*(K; G) \to C_*(K^{(r)}; G).$$

Let K and L be simplicial complexes. A mapping $f : |K| \to |L|$ is said to be *simplicial* if the image of each simplex τ^k from K is a certain simplex from L, and the mapping $f|_{\tau^k}$ is linear in barycentric coordinates:

$$f(t_0 a^{i_0} + \ldots + t_k a^{i_k}) = t_0 f(a^{i_0}) + \ldots + t_k f(a^{i_k}).$$

The concepts of barycentric subdivision and simplicial mapping have meaning also when concerning polyhedra consisting of curvilinear simplices, because barycentric coordinates can be transferred to curvilinear simplices by a triangulation homeomorphism.

Let $f : |K| \to |L|$ be a simplicial mapping. Define the homomorphisms

$$\widehat{f}_p : C_p(K; G) \to C_p(L; G)$$

as follows: for each simplex $(a^{i_0}, \dots, a^{i_p}) \in K$ we put

$$\widehat{f}_p(g \cdot [a^{i_0}, \dots, a^{i_p}]) = \begin{cases} g \cdot [fa^{i_0}, \dots, fa^{i_p}], & \text{if} \\ \quad (fa^{i_0}, \dots, fa^{i_p}) \text{ is a simplex of dimension } p, \\ 0, & \text{if } (fa^{i_0}, \dots, fa^{i_p}) \\ \quad \text{is a simplex of dimension less than } p, \end{cases}$$

and we extend $\widehat{f}_p$ to $C_p(K; G)$ by linearity.

Exercise 16°. Show that the collection of homomorphisms $\{\widehat{f}_p\}$ is a homomorphism of chain complexes

$$\widehat{f}_* : C_*(K; G) \to C_*(L; G)$$

and, consequently, induces homomorphisms

$$f_{*p} : H_p(K; G) \to H_p(L; G).$$

Exercise 17°. Show that simplicial mappings and simplicial complexes form a category and that the assignements

$$K \mapsto H_p(K; G),$$
$$(f : K \to L) \mapsto (f_{*p} : H_p(K; G) \to H_p(L; G))$$

is a covariant functor from this category to the category of Abelian groups.

Exercise 18°. Show that the assignments that takes an abelian group G to the homology group $H_p(K; G)$ of a simplicial (fixed) complex K with coefficients in G is a covariant functor from the category of Abelian groups to itself.

§ 4. Singular homology theory

1. Singular homology groups. In this section, we construct another functor from the category of homotopy types of spaces to the category of Abelian groups, i.e., a homology functor. In order to apply the algebraic constructions of § 2 to the study of topological spaces, it is necessary to work out methods of constructing chain complexes for a given space X. In algebraic topology, there are several methods for doing this, all of them requiring some or other properties of the space X; here we present one of the most general methods.

A continuous mapping $f^k : \sigma^k \to X$ of the standard simplex σ^k to a topological space X is called a *singular k–dimensional simplex* of the topological space X.

Let G be a ring with unit element[*], for instance, the ring of integers $\mathbf{Z}$. A formal linear combination $\sum_i g_i f_i^k$ of singular k–dimensional simplices of the space X with coefficients g_i from G, only a finite number of which differ from zero, is called a *k–dimensional singular chain* of the space X. The set of all k–dimensional singular chains X with coefficients in G is denoted by $C_k^s(X; G)$. It is an Abelian group under the addition of chains as linear combinations. If $G = \mathbf{Z}$, then the group $C_k^s(X; \mathbf{Z})$ is a free Abelian group and its generators are all possible singular k–dimensional simplices.

Let us define the differential

$$\partial_k^s : C_k^s(X; G) \to C_{k-1}^s(X; G).$$

To this end, we consider standard the $(k-1)$–dimensional and k–dimensional simplices σ^{k-1} and σ^k. Associate to the point

$$(t_0, \ldots t_{i-1}, t_i, \ldots, t_{k-1}) \in \sigma^{k-1}$$

the point

$$(t_0, \ldots t_{i-1}, 0, t_i, \ldots, t_{k-1}) \in \sigma^k.$$

This correspondence determines a mapping $\triangle_i^{k-1} : \sigma^{k-1} \to \sigma^k$ from σ^{k-1} to the i–th $(k-1)$–dimensional face of the simplex σ^k. If f^k is a k–dimensional singular simplex, then the composition $f^k \triangle_i^{k-1}$, clearly, is a

---

[*]We can take for G, as in § 3, an arbitrary Abelian group. However, for the convenience of decomposition, it is useful to have a coefficient 1 available, in order not to have to write it in explicitly in many defining formulas (in § 3, we did not use this).

$(k-1)$–dimensional singular simplex. For any simplex f^k, $k \geq 1$, we put

$$\partial_k^s f^k = \sum_{i=0}^{k} (-1)^i \cdot (f^k \Delta_i^{k-1})$$

and define the homomorphism ∂_k^s on the whole group $C_k^s(X; G)$ by linearity:

$$\partial_k^s \left(\sum_i g_i f_i^k \right) = \sum_i g_i \partial_k^s f_i^k.$$

If $k = 0$, then it is natural to set $\partial_0^s f^0 = 0$ and, in argument with the previous, to extend ∂_0^s to $C_0^s(X; G)$ as the zero mapping.

Exercise 1°. Verify that $\partial_k^s \partial_{k+1}^s = 0$.

Hint. It suffices to verify this equality on an arbitrary simplex f^{k+1}.

As has been seen, the sequence of groups $C_k^s(X; G)$ and the homomorphisms ∂_k^s forms a chain complex which we denote by $C_*^s(X; G)$. It is called a *singular chain complex* of the space X.

Let $\phi : X \to Y$ be a continuous mapping. For any k–dimensional singular simplex $f^k : \sigma^k \to X$ of the space X, the composition ϕf^k is a k–dimensional singular simplex of the space Y. Obviously, ϕ induces a homomorphism $\phi_k : C_k^s(X; G) \to C_k^s(Y; G)$.

Exercise 2°. Prove that the system of homomorphisms ϕ_k forms a homomorphism of chain complexes

$$\phi_* : C_*^s(X; G) \to C_*^s(Y; G),$$

i.e., for $k \geq 1$, the equalities $\bar{\partial}_k^s \phi_k = \phi_{k-1} \partial_k^s$ hold, where ∂_k^s, $\bar{\partial}_k^s$ are the differentials of the complexes $C_*^s(X; G)$, $C_*^s(Y; G)$, are fulfilled.

Definition 1. The homology groups of the complex $C_*(X; G)$ are called the *singular homology groups of the space* X with coefficients in G; the k–th homology group is denoted by $H_k^s(X; G)$, and the collection of groups $\{H_k^s(X; G)\}_{k \geq 0}$ is denoted by $H_*^s(X; G)$.

Example. We calculate the homology groups of a point $*$. It is obvious that $C_k^s(*; G) \simeq G$, because there exists only one singular simplex $f^k : \sigma^k \to *$ for any K. The value of the differential on it when $k \geq 1$ is calculated by the formula

$$\partial_k^s f^k = \sum_{i=1}^{k}(-1)^i \cdot f^k \Delta_i^{k-1} = \sum_{i=0}^{k}(-1)^i \cdot f^{k-1} =$$

$$= \begin{cases} 0, & \text{when } k \text{ is odd,} \\ f^{k-1}, & \text{when } k \text{ is even.} \end{cases}$$

Recall that $\partial_k^s = 0$ for $k = 0$. Hence, we obtain that if k is odd, then

$$\text{Im } \partial_{k+1}^s = C_k^s(*; G) = \text{Ker } \partial_k^s \simeq G,$$

if k is even and not equal to zero, then

$$\text{Im } \partial_{k+1}^s = \text{Ker } \partial_k^s = 0.$$

Finally, $\text{Im } \partial_1^s = 0$, $\text{Ker } \partial_0^s \simeq G$; consequently,

(1) $$H_0^s(*; G) \simeq G; \quad H_i^s(*; G) \simeq 0, \quad i > 0.$$

$\Diamond$

Since a continuous mapping $\phi : X \to Y$ induces a homomorphism $\phi_* : C_*^s(X; G) \to C_*^s(Y; G)$ of singular chain complexes of the spaces X and Y, it induces *homomorphisms of singular homology groups*

$$\phi_{*k} : H_k^s(X; G) \to H_k^s(Y; G).$$

Exercise 3°. Show that if $\phi : X \to Y$, $\psi : Y \to Z$ are continuous mappings, then $(\psi\phi)_{*k} = \psi_{*k}\phi_{*k}$. Show that the identity mapping X gives rise to the identity mapping of the homology groups, i.e., $(1_X)_{*k} = 1_{H_k^s(X; G)}$. From this, derive that the homology groups of homeomorphic spaces coincide (the theorem on topological invariance of homology groups).

Exercise 4°. Show that a constant mapping $X \to Y$, i.e., a mapping transferring X to the point $y_0 \in Y$, induces the trivial (zero) homomorphism in the homology groups of higher dimensions, $k > 0$.

2. Properties of singular homology groups. In subsection 1, we have constructed a covariant functor, or more precisely, a collection of functors $H_*^s = \{H_k^s(\cdot; G)\}_{k \geq 0}$ from the category of topological spaces to the category of Abelian groups. We shall investigate the most important properties of this functor.

Theorem 1. *Let two mappings* ϕ, $\psi : X \to Y$ *be homotopic. Then the induced homomorphisms on the homology groups coincide.*

First, we shall prove the following statement.

Lemma 1. *Let B be a convex set of a Euclidean space; then*

$$(2) \qquad\qquad\qquad H_*^s(B; G) \simeq H_*^s(*; G).$$

Proof. Let $f^k : \sigma^k \to B$ be a singular simplex. Define the singular simplex $D_k f^k : \sigma^{k+1} \to B$ by the equality

$$(3) \quad \begin{aligned} &D_k f(t_0, \dots, t_{k+1}) = \\ &\quad = \begin{cases} t_0 w + (1 - t_0) f^k \left(\frac{t_1}{1-t_0}, \dots, \frac{t_{k+1}}{1-t_0} \right), & \text{when } t_0 \neq 1, \\ w, & \text{when } t_0 = 1, \end{cases} \end{aligned}$$

where w is a point from B, and the t_i are the barycentric coordinates of a point from σ^{k+1}.

Extending D_k by linearity to the whole group $C_k^s(B; G)$, we obtain a homomorphism

$$D_k : C_k^s(B; G) \to C_{k+1}^s(B; G).$$

It follows from formula (3) that the homomorphisms D_k and the differentials ∂_k^s are related as follows

$$(4) \quad \begin{aligned} \partial_{k+1}^s D_k &= 1_{C_k^s(B;G)} - D_{k-1} \partial_k^s, \quad \text{for } k > 0, \\ \partial_1^s D_0 f^0 &= f^0 - h^0, \end{aligned}$$

where the singular simplex h^0 maps σ^0 to the point w from B.

Let $z_k \in \operatorname{Ker} \partial_k^s$, $k > 0$. Then, by (4), we have $\partial_{k+1}^s D_k z_k = z_k$, whence $z_k \in \operatorname{Im} \partial_{k+1}^s$. Thus, $H_k^s(B; G) = 0$, for $k > 0$. Similarly, the 0–dimensional cycle f^0 is homologous to the cycle h^0, and, consequently, $H_0^s(B; G) \simeq G$. □

The construction applied in the proof of lemma 1 is quite useful. We present the following definition.

Let C_*, C_*' be chain complexes, and ϕ_*, $\psi_* : C_* \to C_*'$ homomorphisms. A *chain homotopy* connecting ϕ_* with ψ_*, is a system of homomorphisms $\{D_k\}$,

$$D_k : C_k \to C_{k+1}'$$

such that the relation

(5) $$\partial_{k+1}' D_k + D_{k-1} \partial_k = \psi_k - \phi_k, \quad D_{-1} \overset{\text{def}}{\equiv} 0$$

holds.

The homomorphisms of this relation are shown in the following diagram

The homomorphisms ϕ_* and ψ_* are said to be *chain–homotopic*. If $\{D_k\}$ is a chain homotopy connecting ϕ_* and ψ_*, then for $z_k \in \operatorname{Ker} \partial_k$, we have

$$(\psi_k - \phi_k) z_k = \partial_{k+1}' D_k z_k \in \operatorname{Im} \partial_{k+1}'.$$

Hence, the homomorphisms of homology groups induced by the chain homomorphism ϕ_* and ψ_* coincide.

Exercise 5°. Let the chain homomorphisms ϕ_*, $\psi_* : C_* \to C_*'$ and the systems of homomorphisms $\{D_k^1\}$, $\{D_k^2\}$, $D_k^i : C_k \to C_{k+1}'$, $i = 1, 2$ be such that $\partial_{k+1}' D_k^1 + D_{k-1}^2 \partial_k = \psi_k - \phi_k$. Show that the homomorphisms of the homology groups induced by the homomorphisms ϕ_* and ψ_*, coincide.

We show that homotopic mappings of topological spaces induce chain–homotopic homomorphisms of chain complexes. To this end, we apply the following construction. Let X be a topological space, $X \times I$ the cylinder over it; it is natural to call the mappings α^X, $\beta^X : X \to X \times I$ defined by the formulae

$$\alpha^X(x) = (x, 0), \quad \beta^X(x) = (x, 1),$$

the top and bottom of the cylinder. It is clear that α^X and β^X are homotopic.

Lemma 2. *For any space X, there exists a chain homotopy $\{D_k^X\}$ connecting α_*^X with β_*^X, i.e.,*

(6)
$$\beta_k^X - \alpha_k^X = D_{k-1}^X \partial_k^s + \partial_{k+1}^s D_k^X.$$

Proof. We construct a chain homotopy $\{D_k^X : C_k^s(X; G) \to G_{k+1}^s(X \times I, G)\}$ by induction on k.

For $k = 0$, we put $D_0^X f^0 = f^0 \times 1_I$, where the singular simplex $f^0 \times 1_I$ is defined by the formula

$$f^0 \times 1_I(t_0, t_1) = (f^0(1), t_1),$$

and extend $D_0^X f_0$ to $C_0^s(X; G)$ by linearity.

For $k > 0$, we assume that the homomorphisms D_m^X are already defined when $m < k$ for any X, and that they are functorial.

Consider the chain

$$c_k \in C_k^s(\sigma^k \times I; G),$$
$$c_k = \beta_k^{\sigma^k}(1_{\sigma^k}) - \alpha_k^{\sigma^k}(1_{\sigma^k}) - D_{k-1}^{\sigma^k}\partial_k^s(1_{\sigma^k}),$$

where 1_{σ^k} is regarded as a singular simplex. By the induction assumption,

$$\partial_k^s c_k = (\beta_{k-1}^{\sigma^k} - \alpha_{k-1}^{\sigma^k} - \partial_k^s D_{k-1}^{\sigma^k})\partial_k^s(1_{\sigma^k}) =$$
$$= D_{k-2}^{\sigma^k}\partial_{k-1}^s\partial_k^s(1_{\sigma^k}) = 0;$$

consequently, $c_k \in \operatorname{Ker} \partial_k^s \subset C_k^s(\sigma^k \times I; G)$. But $\sigma^k \times I$ is a convex subset of Euclidean space; by lemma 1, $H_k^s(\sigma^k \times I; G) = 0$. Therefore, $c_k \in \operatorname{Im} \partial_{k+1}^s$, i.e., there exists a chain $u_{k+1} \in C_{k+1}^s(\sigma^k \times I; G)$ such that $\partial_{k+1}^s u_{k+1} = c_k$.

Put $D_k^{\sigma^k}(1_{\sigma^k}) = u_{k+1}$. Now, let $f^k : \sigma^k \to X$ be a singular simplex of the space X. We define the chain $D_k^X f^k$ by the formula

$$D_k^X f^k = (f^k \times 1_I)_{k+1} D_k^{\sigma^k} 1_{\sigma^k} = (f^k \times 1_I)_{k+1} u_{k+1},$$

where $(f^k \times 1_I)(x, t) = (f^k(x), t)$, $x \in \sigma^k$, $t \in I$. Since f_k and ∂_k commute, and the D_{k-1}^X are functorial, we obtain

$$\begin{aligned}
\partial_{k+1}^s D_k^X f^k &= \partial_{k+1}^s (f^k \times 1_I)_{k+1} u_{k+1} = \\
&= (f^k \times 1_I)_k \partial_{k+1}^s u_{k+1} = (f^k \times 1_I)_k c_k = \\
&= (f^k \times 1_I)_k (\beta_k^{\sigma^k} - \alpha_k^{\sigma^k} - D_{k-1}^{\sigma^k} \partial_k^s)(1_{\sigma^k}) = \\
&= \beta_k^X f^k - \alpha_k^X f^k - D_{k-1}^X \partial_k^s f^k .
\end{aligned}$$

Extending D_k^X by linearity to $C_k^s(X; G)$, we obtain the required homomorphism D_k^X. $\qquad\square$

It should be stressed that the construction of $\{D_k^X\}$ is functorial, i.e., for any continuous mapping $\phi : X \to Y$ the following diagram

$$\begin{array}{ccc}
C_k^s(X; G) & \xrightarrow{D_k^X} & C_{k+1}^s(X \times I; G) \\
\downarrow \phi_k & & \downarrow (\phi \times 1_I)_{k+1} \\
C_k^s(Y; G) & \xrightarrow{D_k^Y} & C_{k+1}^s(Y \times I; G)
\end{array}$$

is commutative.

Proof of theorem 1. Let $F : X \times I \to Y$ be a homotopy connecting ϕ and ψ. Define a chain homotopy $\{D_k : C_k^s(X; G) \to C_{k+1}^s(Y; G)\}$, connecting ϕ_* with ψ_* as the compositions $\{D_k = F_{k+1} D_k^X\}$ of the homomorphisms of the sequence

$$C_k^s(X; G) \xrightarrow{D_k^X} C_{k+1}^s(X \times I; G) \xrightarrow{F_{k+1}} C_{k+1}^s(Y; G).$$

The statement of the theorem follows from the fact that chain–homotopic homomorphisms of chain complexes induce identical homomorphisms of the homology groups. $\qquad\square$

Corollary. *A homotopy equivalence induces a homology group isomorphism.*

Thus, homotopy equivalent spaces, in particular homeomorphic spaces, have identical (isomorphic) homology groups.

Exercise 6°. Show that if X is a contractible space, then $H_0^s(X; G) \simeq G$, $H_k^s(X; G) = 0$ for $k > 0$.

Exercise 7°. Show that for the homology of a disjoint union $X \cup Y$ there is an isomorphism

$$H_k^s(X \cup Y; G) \simeq H_k^s(X; G) \oplus H_k^s(Y; G).$$

Show that $H_0^s(S^0; G) \simeq G \oplus G$, $H_k^s(S^0; G) = 0$ for $k > 0$.

Exercise 8°. Show that if X and Y are path–connected (see Ch 2, § 10), then $H_0^s(X; G) \simeq G \simeq H_0^s(Y; G)$, and any continuous mapping $\phi : X \to Y$ induces an isomorphism

$$\phi_{*0} : H_0^s(X; G) \simeq H_0^s(Y; G).$$

Note that the result of exercise 6° give the homology groups of the open and closed discs D^n and $\bar{D}^n$.

Let a space X be such that each connected component is path–connected (such spaces are polyhedra, finite cell complexes, manifolds, manifolds with boundary, etc.). Then from exercises 7° and 8°, it immediately follows that the group $H_0^s(X; G)$ is isomorphic to the direct sum of as many copies of the group G as there are (path) connected components in the space X.

Let X_0 be a subspace of X, $i : X_0 \to X$ an imbedding. Setting

$$C_k^s(X, X_0; G) = C_k^s(X; G)/C_k^s(X_0; G),$$

we have, by § 2, an exact sequence of chain complexes

$$0 \to C_*^s(X_0; G) \xrightarrow{i_*} C_*^s(X; G) \xrightarrow{j_*} C_*^s(X, X_0; G) \to 0.$$

The homology groups of the complex $C_*^s(X, X_0; G)$ are called the *singular homology groups of the pair* (X, X_0) and denoted by

$$H_*^s(X, X_0; G) = \{H_k^s(X, X_0; G)\}_{k \geq 0}.$$

It follows immediately from the lemma of § 2 that there is a long exact homology sequence

$$\ldots \to H_{k+1}^s(X, X_0; G) \xrightarrow{\delta_{k+1}} H_k^s(X_0; G) \xrightarrow{i_{*k}}$$

$$(7) \qquad \to H_k^s(X; G) \xrightarrow{j_{*k}} H_k^s(X, X_0; G) \xrightarrow{\delta_k} \ldots \to H_0^s(X, X_0; G) \to 0.$$

Long exact homology sequences are a basic tool in homology theory.

We shall present the following more visual description of the homology groups of a pair following from the definitions introduced above. Let all "similar" summands be collected in a chain $\gamma = \sum_i g_i \cdot f_i^k$, i.e., all the singular simplices f_i^k occurring in the sum are pairwise different and all coefficients g_i are not equal to zero; the carrier (or support) of the chain γ is the subset of the space X equal to the union of images of all the mappings f_i^k involved in γ with nonzero coefficients. According to the definition of a quotient complex (see § 2), an element $\hat{z}_k$ of the kernel Ker $\hat{\partial}_k$ of the boundary homomorphism $\hat{\partial}_k : C_k^s(X, X_0; G) \to C_{k-1}^s(X, X_0; G)$ is a coset consisting of all chains $C_k^s(X, X_0; G)$ such that 1) two different representatives of the element $\hat{z}_k$ differ ($\hat{z}_k$ is a coset!) only by summands from $i_k C_k^s(X_0; G) \subset C_k^s(X; G)$, i.e., by a chain with its carrier in the subspace X_0 (i_k is the chain homomorphism of dimension k, induced by the imbedding $i : X_0 \to X$); 2) the carrier of the boundary $\partial_k^s z_k$ of any representative z_k of the element $\hat{z}_k$ is contained in X_0 ($\hat{z}_k$ is a cycle "modulo X_0"). Similarly, the element $\hat{b}_k$ of the image of the homomorphism $\hat{\partial}_{k+1}$ is a coset consisting of chains b_k in $C_k^s(X; G)$ such that 1) two representatives differ by a chain with carrier in X_0; 2) each representative b_k of the coset $\hat{b}_k$ can be writtten as $\partial_{k+1}^s \gamma_{k+1} + \gamma_k^0$, where γ_{k+1} is a certain chain from $C_{k+1}^s(X; G)$, and γ_k^0 is any chain with carrier in X_0. Thus, elements of the homology groups are relative ("modulo the subspace X_0") cycles $\hat{z}_k$ considered up to relative ("modulo X_0") boundaries $\hat{b}_k$. In Fig. 128, an example of a two–dimensional relative cycle is presented.

The connecting homomorphism δ_k associates to the element $\hat{h}_k$ from the homology groups of the pair $H_k^s(X, X_0; G)$ the element h_{k-1}^0 from the homology group $H_{k-1}^s(X_0; G)$ of the subspace X_0 by the following rule derived from the definition of a connecting homomorphism for chain complexes (see the end of § 2). Let a relative cycle $\hat{z}_k \in C_k^s(X, X_0; G)$ be a representative of the element $\hat{h}_k$, and the cycle $z_k \in C_k^s(X; G)$ a representative of the element $\hat{z}_k$ regarded as a coset. Consider the chain $\partial_k^s z_k \in C_{k-1}^s(X; G)$. It is clear that 1) the carrier of the chain $\partial_k^s z_k$ is contained in X_0, and therefore the chain $\partial_k^s z_k$ can be

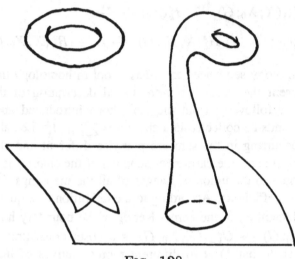

FIG. 128

regarded as a chain from $C^s_{k-1}(X_0; G)$; 2) since $\partial^s_{k-1}\partial^s_k z_k = 0$, the chain $\partial^s_k z_k$ is a cycle in $C^s_*(X_0; G)$. The cycle $\partial^s_k z_k$ is not, generally speaking, a boundary in $C^s_*(X_0; G)$, because z_k may not belong to $i_k C^s_k(X_0; G) \simeq C^s_k(X_0; G)$, i.e., the carrier of the chain z_k is not necessarily in X_0. That is the homology class h^0_{k-1} of the cycle $\partial^s_k z_k$ in $H^s_{k-1}(X_0; G)$ is the image $\delta_k \widehat{h}_k$ of the homology class $\widehat{h}_k$. Obviously, an arbitrary choice of the representatives $\widehat{z}_k$ and z_k leads only to the fact that for two different chains u_k and v_k which determine the same class $\widehat{h}_k$, the difference $\partial^s_k u_k - \partial^s_k v_k$ is a boundary in $C^s_*(X_0; G)$, and not only in $C^s_*(X; G)$. Therefore the homology class h^0_{k-1} is well defined.

Exercise 9°. Let $*$ be a point from X. Show that $H^s_k(X; G) \simeq H^s_k(X, *; G)$ for $k \geq 1$.

Exercise 10°. Let the imbedding $i : X_0 \to X$ be a homotopy equivalence. Show that $H^s_k(X, X_0; G) = 0$ for all k.

It should be noted that for singular homology the statement that the Mayer–Vietoris sequence is exact (see § 3), generally speaking, is incorrect. (Why? Try to construct a counterexample). However, under additional assumptions, this sequence is exact. For instance, if K_1 and K_2 are subcomplexes of a simplicial complex K, then for the singular homology of the spaces $|K_1|$ and $|K_2|$ the Mayer–Vietoris sequence is exact.

For the development of applications of singular homology (see § 6), we shall need the knowledge how to refine singular simplices. An exact formulation of the required result is given in exercise 12° below. This result can be established by the barycentric subdivision of singular simplices.

Consider the barycentric subdivision of the standard simplex σ^k. We denote by $\langle c^{i_0}, c^{i_1}, \dots, c^{i_q} \rangle$ the composition of the linear (in barycentric coordinates) mapping of a standard simplex σ^k to the simplex $(c^{i_0}, c^{i_1}, \dots, c^{i_q})$ from the barycentric subdivision of σ^k which transforms the j–th vertex of the standard simplex into the j–th vertex c^{i_j} from the set $\{c^{i_0}, c^{i_1}, \dots, c^{i_q}\}$, and the imbedding mapping of the simplex $(c^{i_0}, c^{i_1}, \dots, c^{i_q})$ into the simplex σ^k.

Note that the identity mapping 1_{σ^k} of the simplex σ^k can be regarded as an element of the group $C_k^s(\sigma^k; G)$. Its boundary is $\partial_k^s 1_{\sigma^k} = \sum_{i=0}^k (-1)^i \Delta_i^{k-1}$ (see subsection 1).

Now, let X be an arbitrary topological space. Define the homomorphisms

$$\Omega_k : C_k^s(X; G) \to C_k^s(X; G), \quad k = 0, 1, \dots,$$

inductively. Put

(8a) $$\Omega_0 = 1_{C_0^s(X;G)}.$$

Assume that the homomorphisms Ω_{k-1} are already defined for an arbitrary topological space X, and, moreover, that for the singular complex 1_{σ^k} of the space σ^k, the chain $\Omega_{k-1}(\partial_k^s 1_{\sigma^k})$ can be given in the form

(8b) $$\Omega_{k-1}(\partial_k^s 1_{\sigma^k}) = \sum_j g_j \langle c_j^0, c_j^1, \dots, c_j^{k-1} \rangle,$$

where c_j^r are vertices of the barycentric subdivisions of the $(k-1)$–dimensional faces of the simplex σ^k. It is obvious that the requirement is fulfilled for $k - 1 = 0$. Now, put

(8c) $$\Omega_k(1_{\sigma^k}) = \sum_j g_j \cdot \langle b^{0,1,\dots,k}, c_j^0, c_j^1, \dots, c_j^{k-1} \rangle$$

where $b^{0,1,\dots,k}$ is the barycentre of the simplex σ^k, and g_j and c_j^r are the same as in (8b).

Now define the homomorphism Ω_k on the singular simplex $f^k : \sigma^k \to X$ for an arbitrary space X as follows.

Let $f^k_* : C^s_*(\sigma^k; G) \to C^s_*(X; G)$ be the homomorphism of chain complexes induced by the mapping $f^k : \sigma^k \to X$. Put

(8d) $\Omega_k(f^k) = f^k_k \Omega_k(1_{\sigma^k}).$

Extending Ω_k to $C^s_k(X; G)$ by linearity:

(8e) $\Omega_k \left(\sum_i g_i f^k_i \right) = \sum_i g_i \Omega_k(f^k_i),$

completes the definition of Ω_k. It is clear that the chain $\Omega_k(\partial^s_{k+1} 1_{\sigma^{k+1}})$ has a representation similar to (8b). These inductive construction of Ω_k is complete.

The homomorphisms are functorial and commute with the differentials $\partial^s_k \Omega_k = \Omega_k \partial^s_k$ (verify!). The collection of homomorphisms Ω_k forms a homomorphism $\Omega_* : C^s_*(X; G) \to C^s_*(X; G)$ of the complex $C^s_*(X; G)$ into itself.

Exercise 11°. Show that the homomorphism Ω_* and $1_{C^s_*(X;G)}$ are chain–homotopic.

Exercise 12°. Let A, B be closed, non–intersecting spaces of a normal Hausdorff space X. Show that for any cycle $z_k \in C^s_k(X; G)$ there exists a cycle $(\Omega_k)^r z_k$ homologous to it such that for each singular simplex f^k involved in the cycle $(\Omega_k)^r z_k$, its image does not intersect both A and B at the same time.

3. Homology and homotopy. It is natural to try to determine a relation between the singular homology groups of and the homotopy groups of a space. It turns out that this problem is quite complicated; only partial results are known. Thus, the one–dimensional homology group of a path–connected space is defined completely by its fundamental group.

Theorem 2. *Let X be a path–connected space with a distinguished point x_0; then*

(9) $H^s_1(X;Z) \simeq \pi_1(X,x_0)/[\pi_1(X,x_0),\pi_1(X,x_0)],$

where $[\pi_1(X,x_0),\pi_1(X,x_0)]$ *is the commutator subgroup*) of the group* $\pi_1(X,x_0)$.

---

*)Recall that the *commutator subgroup* $[\pi, \pi]$ of a group π is the subgroup generated by all commutators, i.e., the elements of the form $g_1 \cdot g_2 \cdot g_1^{-1} \cdot g_2^{-1}$, where $g_1, g_2 \in \pi$. The commutator subgroup of a group is its normal divisor.

We outline the proof of formula (9) dwelling on geometric ideas only. In the first place, note that any loop of the space X (with the origin at the point x_0) is a singular cycle (the singular simplex $[0;1] \to [0;1]/0 \sim 1 \simeq S^1 \xrightarrow{\alpha} X$ is a cycle). Hence, there is a homomorphism of the group $\pi_1(X,x_0)$ into $H_1^*(X;Z)$, which we denote by Θ.

Secondly, it can be shown that Θ is an epimorphism. Indeed, each cycle in $H_1^*(X;Z)$ defines (ambiguously) several loops in the space X which, possibly, start at different points. These different loops can be transformed into one loop joining their origins with the point x_0 by a path which is passed through in both forward and reverse directions (Fig. 129). The complex loop obtained at the point x_0 is transformed by the homomorphism Θ into the original singular cycle. (More precisely, the class of this loop is transformed into the class of the original cycle).

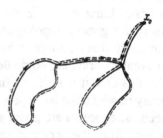

FIG. 129

Third, the commutator subgroup of the group $\pi_1(X,x_0)$ lies in the kernel of Θ. In fact, the loop $\alpha \cdot \beta \cdot \alpha^{-1} \cdot \beta^{-1}$ under the action of the homomorphism Θ is transformed, roughly speaking, into the cycle $\alpha + \beta + \alpha^{-1} + \beta^{-1}$; the group of singular cycles is commutative, and the cycles $\alpha + \alpha^{-1}$ and $\beta + \beta^{-1}$ are homologous to zero. Consequently, the loop $\alpha \cdot \beta \cdot \alpha^{-1} \cdot \beta^{-1}$ is transformed into a cycle homologous to zero.

It can be shown that the commutator subgroup makes up the whole kernel of the homomorphism Θ (in fact, when proving that Θ is epimorphic, one constructs a sort of "inverse" homomorphism of the group H_1^* into $\pi_1/[\pi_1,\pi_1]$).

Exercise 13°. Complete the proof of theorem 2 by the scheme outlined.

We present the following statement without proof.

Theorem 3. (Hurewicz). *Let X be a path–connected topological space such that $\pi_k(X)=0$ for $k<q$ and $\pi_q(X) \neq 0$, $q>1$. Then $H_k^*(X;Z)=0$ for $0<k<q$ and $H_q^*(X;Z) \simeq \pi_q(X)$, and for any mapping $f:X \to X$ the following diagram:*

$$
\begin{array}{ccc}
\pi_q(X) & \xrightarrow{\ f_q\ } & \pi_q(X) \\
\downarrow & & \downarrow \\
H_q(X;Z) & \xrightarrow{\ f_{*q}\ } & H_q(X;Z)
\end{array}
$$

is commutative.

§ 5. Axioms of homology theory. Cohomology

In the two previous sections, two homology theories have been considered, viz. simplicial and singular. In addition, several more homology theories arise in algebraic topology. Historically, simplicial homology theory was introduced earliest. Later on, different approaches to the construction of a homology theory for general topological spaces were developed (the Alexandrov–Čech homology theory, singular homology theory, etc.). The problem of when two different theories are equivalent turned out to be quite complicated.

To this end, it is quite useful to employ an axiomatic approach to homology theory in which the basic properties of correspondence between topological and algebraic concepts are given axiomatically, and all the remaining properties are derived from the accepted axioms. Such a system of axioms was developed by Eilenberg and Steenrod. Here we formulate their axioms.

A *homology theory* H_* with connecting homomorphisms δ_* is a collection of covariant functors $\{H_k\}$, $k = 0, 1, 2, \ldots$, from the category of pairs of topological spaces (X, A), $A \subset X$, into the category of Abelian groups, and the collection of functorial homomorphisms $\{\delta_k\}$, $k = 1, 2, \ldots$,

$$\delta_k(X, A) : H_k(X, A) \to H_{k-1}(A, \emptyset)$$

such that the following axioms hold.

1. Homotopy axiom. Let two mappings $f, g : X \to Y$ be homotopic, and $F : X \times I \to Y$ a homotopy connecting them. Let $A \subset X$, $B \subset Y$, and $F(A \times I) \subset B$. Then

$$H_*(f) = H_*(g) : H_*(X, A) \to H_*(Y, B)$$

for arbitrary X, Y, A, B, f, g.

2. Exactness axiom. For any pair (X, A) and corresponding imbeddings $i : (A, \emptyset) \to (X, \emptyset)$, $j : (X, \emptyset) \to (X, A)$, the following sequence

$$\ldots \xrightarrow{\delta_{k+1}(X, A)} H_k(A, \emptyset) \xrightarrow{H_k(i)} H_k(X, \emptyset) \xrightarrow{H_k(j)}$$

(1) $$\to H_k(X, A) \xrightarrow{\delta_k(X, A)} H_{k-1}(A, \emptyset) \to \ldots \to H_0(X, A) \to 0$$

is exact.

3. Excision axiom. Let (X, A) be an arbitrary pair, and let U be open in X and $\bar{U} \subset \text{Int}\, A$. Then the imbedding of pairs $j : (X \setminus U, A \setminus U) \to (X, A)$ induces the isomorphism

$$H_*(j) : H_*(X \setminus U, A \setminus U) \approx H_*(X, A).$$

4. Dimension axiom. For the space $*$ consisting of one point, $H_k(*, \emptyset) =$ when $k > 0$.

Exercise 1°. Verify that the axioms of homology theory hold for singular homology theory.

The axioms of homology theory are complete in the following sense.

The uniqueness theorem. *Let H_* and $\bar{H}_*$ be two homology theories. If there exists an isomorphism $h_0 : H_0(*, \emptyset) \simeq \bar{H}_0(*, \emptyset)$, then these theories are naturally isomorphic on the category of pairs of compact polyhedra, i.e.,*

    (1) *for any pair of compact polyhedra (X, A) such that the triangulation of A is a subset of the triangulation of X, and for each $k \geq 0$, a unique family of isomorphisms $h_k(X, A) : H_k(X, A) \simeq \bar{H}_k(X, A)$ is defined; moreover, $h_0(*, \emptyset) = h_0$.*

    (2) *for any mapping $f : (X, A) \to (Y, B)$ of pairs of compact polyhedra and each $k \geq 0$, the relations $H_k(f) = \bar{H}_k(f)$ hold; this means the commutativity of the diagram*

$$
\begin{array}{ccc}
H_k(X,A) & \xrightarrow{\;H_k(f)\;} & H_k(Y,B) \\
h_k(X,A) \Big\updownarrow & & \Big\updownarrow h_k(Y,B) \\
\bar{H}_k(X,A) & \xrightarrow{\;\bar{H}_k(f)\;} & \bar{H}_k(Y,B)
\end{array}
$$

    (3) *the diagrams*

$$
\begin{array}{ccc}
H_{k+1}(X,A) & \xrightarrow{\;\delta_{k+1}(X,A)\;} & H_k(A,\emptyset) \\
h_{k+1}(X,A) \Big\updownarrow & & \Big\updownarrow h_k(A,\emptyset) \\
\bar{H}_{k+1}(X,A) & \xrightarrow{\;\bar{\delta}_{k+1}(X,A)\;} & \bar{H}_k(A,\emptyset)
\end{array}
$$

arising from the isomorphism of the long exact sequences (1) *are commutative.*

The proof of this theorem is omitted because it falls outside the limits of an elementary course.

In particular, the singular and simplicial theories coincide on the category of pairs of compact polyhedra. Thus, for a compact polyhedron $|K|$, one has the isomorphism

$$H_*(|K|; G) \simeq H_*^s(|K|; G).$$

We shall apply this fact (the proof is not given here) in § 6 and § 8 to pass from one homology theory to another.

Note that the independence of simplicial homology of a compact polyhedron of the choice of a triangulation can be established both within the simplicial theory itself and by using singular homology theory. The latter method constructs an isomorphism between the homology groups of an arbitrary triangulation of a polyhedron and the groups of singular homology of this polyhedron. The details of these arguments are quite fairly complicated, and are not given here.

Exercise 2°. Using the uniqueness theorem, establish the validity of the exact Mayer–Vietoris sequence for singular homology theory:

$$\ldots \to H_k^s(|K_1| \cap |K_2|; G) \to H_k^s(|K_1|; G) \oplus H_k^s(|K_2|; G) \to$$
$$\to H_k^s(|K_1| \cup |K_2|; G) \to H_{k-1}^s(|K_1| \cap |K_2|; G) \to$$

(4) $\ldots \to H_0^s(|K_1| \cup |K_2|; G) \to 0,$

where K_1, K_2 are subcomplexes of a finite simplicial complex K.

Note that there exist homology theories satisfying axioms (1)–(3) but not the dimension axiom. These homology theories are called generalized homology theories, and their investigation is a central topic in modern algebraic topology.

In algebraic topology, besides homology groups, so-called cohomology groups are used. The main difference of cohomology theory from homology theory is the fact that a cohomology theory is the collection H^* of contravariant functors H^k, therefore most arrows change their directions compared to homology theory.

The fundamental object in cohomology is a cochain complex C^*, i.e. a sequence

$$0 \to C^0 \xrightarrow{d^0} C^1 \xrightarrow{d^1} \ldots \to C^{k-1} \xrightarrow{d^{k-1}} C^k \xrightarrow{d^k}$$
$$\xrightarrow{d^k} C^{k+1} \xrightarrow{d^{k+1}} C^{k+1} \to \ldots$$

of Abelian groups C^* (cochain groups) and their homomorphisms d^k (differentials or coboundary homomorphisms) such that $d^{k+1}d^k = 0$. The quotient groups $H^k(C^*) = \text{Ker } d^k / \text{Im } d^{k-1}$ are called the cohomology groups of a cochain complex. Cochain complexes are often obtained from chain complexes using the following approach. Let C_* be a chain complex, G an Abelian group. Put $C^k = \text{Hom}(C_k, G)$, i.e., a set of all homomorphisms of the group C_k into the group G. For $\psi^k \in \text{Hom}(C_k; G)$, we define the element $d^k \psi^k \in \text{Hom}(C_{k+1}, G)$ by the formula

$$(\delta^k \psi^k) \gamma_{k+1} = \psi^k(\partial_{k+1} \gamma_{k+1})$$

on an arbitrary element $\gamma_{k+1} \in C_{k+1}$. Thus, the boundary homomorphisms ∂_k of the chain complex C_k define the coboundary homomorphisms d^k of the cochain complex C^*. Obviously,

$$(d^{k+1} d^k \psi^k) \gamma_{k+2} = (d^k \psi^k)(\partial_{k+2} \gamma_{k+2}) =$$

$$= \psi^k(\partial_{k+1} \partial_{k+2} \gamma_{k+2}) = \psi^k(0) = 0,$$

so that C^* really is a cochain complex.

Applying this method to $C_* = C_*(K; Z)$, i.e., the chain complex of the simplicial complex K with integral coefficients, we obtain a cochain complex $C^*(K; G)$, where $C^k(K; G) = \text{Hom}(C_k(K; Z), G)$. The cohomology groups $H^k(C^*(K; G))$ are called the cohomology groups of the simplicial complex K (or the polyhedron $|K|$) with the coefficients in G. In the singular case, however, we have to overcome certain difficulties related to the infinite number of generators for the groups of singular chains. For cohomology theory, there is also uniqueness a system of axioms and a uniqueness theorem; they are similar to the axioms and the theorem for homology theory. An important advantage of cohomology theory compared to homology theories is that the cohomology groups of geometric objects form a ring with, generally speaking, a nontrivial multiplication. In many problems, homology and cohomology is applied simultaneously.

§ 6. Homology of spheres. Degree of a mapping

1. The homology groups of spheres. We shall calculate the singular homology groups of the spheres S^n. The knowledge of these groups enables us to introduce the concepts of the degree of a mapping, and the characteristic and index of a singular point of a vector field, which are quite useful in applications.

Let X be a cell complex, Y a finite subcomplex. We show that

(1) $$H_k^s(X, Y; G) \simeq H_k^s(X/Y; G)$$

when $k > 0$, and where X/Y is the quotient space of X with respect to Y.

Note, first, that the cell complex X/Y is homotopy equivalent to the complex $X \cup_i CY$, where CY is the cone[*] over Y with the vertex $*$, and $i : Y \to X$ is the given imbedding. Indeed, the complex X/Y coincides with the complex $(X \cup_i CY)/CY$. Since CY is a contractible subcomplex of the complex $X \cup_i CY$, the complexes $(X \cup_i CY)/CY$ and $X \cup_i CY$ are homotopy equivalent (see Ch 4, § 10, exercise 7°). Therefore,

$$H_k^s(X/Y; G) \simeq H_k^s(X \cup_i CY; G)$$

and when $k > 0$

$$H_k^s(X/Y; G) \simeq H_k^s(X \cup_i CY, *; G)$$

(see § 4, exercise 9°).

The cone CY is homotopy equivalent to the point $x \in CY$, and consequently,

$$H_k^s(X \cup_i CY, *; G) \simeq H_k^s(X \cup_i CY, CY; G).$$

Consider the imbedding of pairs $I : (X, Y) \to (X \cup_i CY, CY)$; it induces a homomorphism $I_* : H_*^s(X, Y; G) \to H_*^s(X \cup_i CY, CY; G)$.

Let us show that I_* is an isomorphism. We break the cone CY into two parts C^1Y and C^2Y, as shown in Fig. 130. It is clear that $H_*^s(X \cup_i C^2Y, C^2Y; G) \simeq H_*^s(X, Y; G)$.

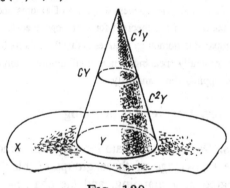

Each cycle $z_k \in C_k^s(X \cup_i CY, CY; G)$ can be replaced by the cycle $(\Omega_k)^r z_k$ homologous to it such that the image of each singular simplex from $(\Omega_k)^r z_k$ intersecting with X, will not intersect with C^1Y, and inverse, each singular simplex intersecting with C^1Y will not intersect with X (see § 4, exercise 12°). Deleting all simplices intersecting with C^1Y from the chain

FIG. 130

$(\Omega_k)^r z_k$, we obtain a cycle $z_k' \in C_k(X \cup_i CY, CY; G)$ which is homologous to the original one. On the other hand, z_k' can be regarded as a cycle in the group of chains $C_k^s(X \cup_i C^2Y, C^2Y; G)$; therefore, I_k is an epimorphism.

---

[*] Recall that for a topological space Y, the cone CY is defined as the quotient space $(Y \times I)/Y \times 0$.

Similarly, it can be shown that I_k is a monomorphism.

We shall give an application of formula (1) to the calculation of the homology of the sphere S^n. The homology of the disc $\bar{D}^n$ will be required. Since $\bar{D}^n$ contracts to a point, the homology groups of the disc are isomorphic to the homology groups of the point, namely,

$$H_k^s(\bar{D}^n; G) \simeq \begin{cases} G & \text{when } k = 0, \\ 0 & \text{when } k > 0 \end{cases}$$

(see § 4, exercise 6°). We start calculating with low dimensions n. Since S^0 is the disjoint union of two points, $H_0^s(S^0; G) \simeq G \oplus G$, $H_k^s(S^0; G) = 0$ when $k > 0$. Furthermore, due to the path–connectedness of S^n, when $n > 0$, we have

$$H_0^s(S^n; G) \simeq G, \quad n > 0.$$

Note that the sphere S^n is homeomorphic to the quotient space $\bar{D}^n/S^{n-1}$. Therefore, by (1),

$$H_k^s(\bar{D}^n, S^{n-1}; G) \simeq H_k^s(S^n; G) \quad \text{when } k > 0.$$

We shall use this fact.

Consider the long exact homology sequence of the pair $(\bar{D}^1, S^0)$ replacing the homology groups of the pair by the homology groups of the circle S^1 when $k > 0$:

$$\ldots \to H_k^s(S^0; G) \to H_k^s(\bar{D}^1; G) \to H_k^s(S^1; G) \to$$
$$\to H_{k-1}^s(S^0; G) \to \ldots \to H_1^s(S^0; G) \to H_1^s(\bar{D}^1; G) \to$$
(2) $$\to H_1^s(S^1; G) \to H_0^s(S^0; G) \to H_0^s(\bar{D}^1; G) \to H_0^s(\bar{D}^1, S^0; G).$$

Noting that $H_k^s(\bar{D}^1; G) = 0$, when $k \geq 1$, and $H_{k-1}^s(S^0; G) = 0$, for $k > 1$, we obtain from (2) a short exact sequence

$$0 \to H_k^s(S^1; G) \to 0, \quad k > 1,$$

and it follows that $H_k^s(S^1; G) = 0$, for $k > 1$. In addition, the homomorphism $H_0^s(S^0; G) \to H_0^s(\bar{D}^1; G)$ is epimorphic (verify by definition). Therefore, the exact sequence (2) leads to the short exact sequence

$$0 \to H_1^s(S^1; G) \to G \oplus G \xrightarrow{pr_1 + pr_2} G \to 0,$$

whence we obtain the isomorphism $H_1^s(S^1; G) \simeq G$.

Now we shall apply induction. Assume that when $1 \leq q \leq n - 1$, for the spheres S^q, the isomorphisms

$$H_k^s(S^q; G) \simeq \begin{cases} G, & k = 0, q, \\ 0, & k \neq 0, q \end{cases}$$

has been established.

Consider the long exact homology sequence of the pair $(\bar{D}^n, S^{n-1})$ while replacing, as above, the homology of a pair by the homology of the sphere S^n,:

$$\ldots \to H_k^s(\bar{D}^n; G) \to H_k^s(S^n; G) \to$$

(3)
$$\to H_{k-1}^s(S^{n-1}; G) \to H_{k-1}^s(\bar{D}^n; G) \to \ldots$$

For $k > 1$, we have $H_k^s(\bar{D}^n; G) = 0$, $H_{k-1}^s(\bar{D}^n; G) = 0$, therefore the portion (3) of the long exact sequence takes the form

$$0 \to H_k^s(S^n; G) \to H_{k-1}^s(S^{n-1}; G) \to 0,$$

so that $H_k^s(S^n; G) \simeq H_{k-1}^s(S^{n-1}; G)$, $k > 1$. Thus, for $n \geq 2$, we get

$$H_2^s(S^n; G) = 0, \ldots, H_{n-1}^s(S^n; G) = 0,$$
$$H_n^s(S^n; G) \simeq H_{n-1}^s(S^{n-1}; G) \simeq G, \quad H_{n+1}^s(S^n; G) = 0, \ldots$$

In order to calculate $H_1^s(S^n; G)$, we take $k = 1$ in (3):

$$\ldots \to H_1^s(\bar{D}^n; G) \to H_1^s(S^n; G) \to H_0^s(S^{n-1}; G) \xrightarrow{i_{*0}} H_0^s(\bar{D}^n; G) \to \ldots$$

Since S^{n-1}, $\bar{D}^n$ are path–connected, we have $H_0^s(S^{n-1}; G) \overset{i_{*0}}{\simeq} H_0^s(\bar{D}^n; G) \simeq G$ (see § 4, exercise 8°); hence, Ker $i_{*0} = 0$, and because of the exactness of (3), we obtain the short exact sequence

$$0 \to H_1^s(S^n; G) \to 0, \text{ i.e., } H_1^s(S^n; G) = 0, \quad n \geq 2.$$

The induction hypothesis has been extended to $q = n$. Therefore, finally we have

$$H_0^s(S^n; G) \simeq G; \quad H_j^s(S^n; G) = 0 \quad j \neq 0, \quad n \geq 1;$$
$$H_n^s(S^n; G) \simeq G, \quad n \geq 1; \quad H_0^s(S^0; G) \simeq G \oplus G;$$
(4)
$$H_j^s(S^0; G) = 0, \quad j \geq 1.$$

Thus, the homology groups S^n have been calculated.

In calculating the homology of S^n, we did not use the uniqueness theorem of homology theory (see § 5). It could be used is the following way. Since the sphere S^n is homeomorphic to the boundary $\partial \tau^{n+1}$ of the simplex τ^{n+1}, we have the isomorphism

$$(5) \qquad H_i(|\{\partial \tau^{n+1}\}|; G) \simeq H_i^s(S^n; G),$$

hence, using the results (see § 3, (9)) concerning $H_*(\{\partial \tau^{n+1}\}; G)$, we obtain the same result as in (4).

Note that in Ch. 3, § 4, the Brouwer fixed point theorem and the theorem that a retraction of the n–disc onto its boundary sphere is impossible, were based on the functoriality of homotopy groups and on the result $\pi_n(S^n) \simeq \mathbf{Z}$, which has not yet been proved. Now, using the fact that $H_n^s(S^n; \mathbf{Z}) \simeq \mathbf{Z}$ and homology functors, one may assume that both these important theorems are strictly proved. Indeed, their proofs only involved the axioms for a functor to the category of Abelian groups and on knowing the corresponding group for S^n.

Exercise $1°$. Deduce from theorem 3, § 4, that

$$\pi_k(S^n) = 0, \quad k < n; \quad \pi_n(S^n) \simeq \mathbf{Z}.$$

We shall discuss the question of topological invariance of the concept of dimension of a Euclidean space. From algebra, it is known that two Euclidean spaces of the same dimension are isomorphic, and, consequently, are homeomorphic. It is known also that the spaces $\mathbf{R}^m$ and $\mathbf{R}^n$ are not isomorphic (as vector spaces) for $m \neq n$. A question arises whether they are possibly homeomorphic. The following theorem gives a negative answer to this question and thereby states that the dimension of a Euclidean space is a topological invariant.

Theorem 1. *If $m \neq n$, then the spaces $\mathbf{R}^m$ and $\mathbf{R}^n$ are not homeomorphic.*

Proof. Consider the one–point compactifications $\tilde{\mathbf{R}}^m = \mathbf{R}^m \cup \xi^m$ and $\tilde{\mathbf{R}}^n = \mathbf{R}^n \cup \xi^n$ of the spaces $\mathbf{R}^m$ and $\mathbf{R}^n$ (see Ch. 2, § 14). A basis of neighbourhoods of the points ξ^m, ξ^n is given by the complements of closed balls with centres at the origin of coordinates on the spaces $\mathbf{R}^m$ and $\mathbf{R}^n$, respectively. It is easy to see that the one–point compactification of an Euclidean space is homeomorphic to the sphere of the same dimension.

Assume that there exists a homeomorphism $\Phi : \mathbf{R}^m \to \mathbf{R}^n$. It can be extended to a mapping $\widetilde{\Phi} : \widetilde{\mathbf{R}}^m \to \widetilde{\mathbf{R}}^n$, by putting $\widetilde{\Phi}(\xi^m) = \xi^n$. It is clear that the mapping $\widetilde{\Phi}$ is then also a homeomorphism. Hence, we obtain that the spheres S^m and S^n are also homeomorphic. Then, by the topological invariance of homology groups, we have

$$H_k^s(S^m; \mathbf{Z}) \simeq H_k^s(S^n; \mathbf{Z}) \quad \text{for all } k.$$

However, we know that this does not hold when $m \neq n$. Therefore, the assumption concerning the existence of a homeomorphism $\Phi : \mathbf{R}^m \to \mathbf{R}^n$, for $m \neq n$, is not correct. $\qquad\square$

2. The degree of a mapping. Now we pass to the investigation of homomorphisms of homology groups induced by mappings of n–dimensional spheres. It follows from the path–connectedness of spheres that if $\phi : S_1^n \to S_2^n$ is a mapping of one replica of the sphere to another, then the homomorphism $\phi_{*0} : H_0^s(S_1^n; G) \to H_0^s(S_2^n; G)$ is an isomorphism. The homomorphism

$$\phi_{*n} : H_n^s(S_1^n; G) \to H_n^s(S_2^n; G)$$

is not, generally speaking, an isomorphism. If we take the group of integers $\mathbf{Z}$ as the group of coefficients G and fix isomorphisms $H_n^s(S_i^n; \mathbf{Z}) \simeq \mathbf{Z}$, $i = 1, 2$, then the homomorphism ϕ_{*n} can be considered as a endomorphism $\pi_{*n} : \mathbf{Z} \to \mathbf{Z}$ of the group $\mathbf{Z}$. Such a homomorphism is defined uniquely by the value of ϕ_{*n} on the generating element $1 \in \mathbf{Z}$, because $\phi_{*n}(m) = m \cdot \phi_{*n}(1)$.

Definition 1. The number $\phi_{*n}(1)$ is called the *degree of a mapping* ϕ and denoted by $\deg \phi$.

Note that $\deg \phi$, in general, can take any integral values. The sign of $\deg \phi$ depends on the choice of the generating elements in the groups $H_n^s(S_1^n; \mathbf{Z})$, $H_n^s(S_2^n; \mathbf{Z})$, i.e., on the isomorphisms of these groups with the group $\mathbf{Z}$. If γ is a generating element of the group $H_n^s(S^n; \mathbf{Z})$, then $(-\gamma)$ is also a generating element; thus, the isomorphism $H_n^s(S^n; \mathbf{Z}) \simeq \mathbf{Z}$ can be established in two ways. If ϕ is a mapping of S^n into itself, then $\deg \phi$ does not depend on the choice of a generating element.

Note that from theorem 3, § 4, there follows that the concepts $\deg \phi$ of the degree of a mapping $\phi : S^n \to S^n$ as determined by homotopy groups (see Ch. 3,§ 4) or by homology groups, are identical.

Obviously, if $\phi, \psi : S^n \to S^n$ are homotopic mappings, then $\deg \phi = \deg \psi$. The opposite statement (the Hopf theorem) whose proof we do not present, also holds.

Exercise $2°$. Prove that for $\phi, \psi : S^n \to S^n$, the formula $\deg (\phi \psi) = \deg \phi \cdot \deg \psi$ is valid.

Hint. Use the functoriality of homology groups (see Ch. 5, § 4, subsection 8).

Exercise $3°$. Show that the degree of the constant mapping of the sphere S^n into itself is equal to zero.

Exercise $4°$. Let a mapping $\Phi : \mathbf{R}^{n+1} \to \mathbf{R}^{n+1}$ be such that $\Phi(x) \neq 0$ for $r \leq \|x\| \leq R$; mappings $\widetilde{\Phi}_\rho : S^n \to S^n$ are defined by the equalities

$$\widetilde{\Phi}_\rho(x) = \frac{\Phi(\rho x)}{\|\Phi(\rho x)\|}, \quad x \in S^n, \quad r \leq \rho \leq R.$$

Prove that $\deg \widetilde{\Phi}_r = \deg \widetilde{\Phi}_R$.

Hint. Construct a homotopy connecting the mappings $\widetilde{\Phi}_r$ and $\widetilde{\Phi}_R$.

The following two exercises can be easily solved on the basis of the isomorphism of singular and simplicial homologies.

Exercise $5°$. Let $A : \mathbf{R}^{n+1} \to \mathbf{R}^{n+1}$ be a non–singular linear operator. Define a mapping $\widetilde{A} : S^n \to S^n$ by the formula

$$\widetilde{A}(x) = \frac{A(x)}{\|A(x)\|}, \quad x \in S^n.$$

Let A be a diagonalizable operator with the eigenvalues $\lambda_1 = \ldots = \lambda_m = -1$ and $\lambda_{m+1} = \ldots = \lambda_{n+1} = 1$, where $m \leq n + 1$. Prove that for this operator, $\deg \widetilde{A} = (-1)^m$.

Hint. The operator A can be represented as a composition of m operators B_i each of which has one eigenvalue λ_i equal to -1, and the remaining n eigenvalues equal to 1. Then the mapping

$\tilde{A}$ is a composition of the mappings $\tilde{B}_i$ and $\deg \tilde{A} = \prod_{i=1}^{m} \deg \tilde{B}_i$. The degree of each $\tilde{B}_i$ is equal to (-1). To see this, construct a triangulation K of the sphere S^n invariant with respect to the mapping $\tilde{B}_i$; such a triangulation is obtained from the union of two cones over a triangulation of the "equator", i.e., the sphere S^{n-1} which lies in the eigenspace corresponding to $\lambda=1$, and the vertices of two cones, the "north" and "south" poles, lie in the eigenspace which corresponds to $\lambda_i=-1$. A generating element of the group $H_n(K;Z)$ consists of the cycle γ which is equal to the sum of all n–dimensional simplices of K with compatible orientation. The mapping $\tilde{B}_i$ transfers the cycle γ into the cycle $(-\gamma)$ (changing the orientation of all simplices); therefore $\deg \tilde{B}_i = -1$

Exercise 6°. Prove that for an arbitrary non–singular linear operator A : $\mathbf{R}^{n+1} \to \mathbf{R}^{n+1}$, the formula $\deg \tilde{A} = \operatorname{sign} |A|$ holds.

Hint. Show that A is homotopic in the class of non–singular linear operators to an operator A' whose matrix is diagonal and whose diagonal elements are equal to ± 1.

Consider a continuous mapping $\Phi : U \to \mathbf{R}^{n+1}$, where U is a domain in $\mathbf{R}^{n+1}$. Usually, while investigating the solutions of an equation

$$(6) \qquad\qquad \Phi(x) = 0,$$

the mapping Φ is called a vector field (vector $\Phi(x)$ is associated to the point x), and the solutions of equation (6) are called the singular points of the vector field Φ.

In practice, the mapping Φ is not necessarily continuous on the whole domain U. If it has a finite number of discontinuity points (or points in which the value is not determined), then these points are also said to be singular. Most subsequent results are also valid for such vector fields.

Let x^0 be an isolated singular point of a vector field Φ, i.e., $\Phi(x^0) = 0$, and let there be no other solutions of equation (6) in a neighbourhood of the point x^0. Then for a sufficiently small R, when $0 < r < R$, the degree of the mapping $\tilde{\Phi}_r : S^n \to S^n$ given by the equality

$$(7) \qquad\qquad \tilde{\Phi}_r(x) = \frac{\Phi(rx + x^0)}{\|\Phi(rx + x^0)\|},$$

is defined; this degree does not depend on the choice of r (compare with exercise 3°).

Definition 2. The degree, $\deg \tilde{\Phi}_r$, of the mapping $\tilde{\Phi}_r$ (for sufficiently small r) is called the *index of the isolated singular point x^0 of the vector field Φ*; we will denote it by $\operatorname{ind}(x^0, \Phi)$.

Let a field Φ have no singular points on the boundary $S_r^n(x^0)$ of the ball $D_r^{n+1}(x^0)$ of radius r and centre at the point x^0 (now we do not assume that x^0 is a singular point and r is small). It is obvious that in this case, formula (7) also defines a mapping $\widetilde{\Phi}_r : S^n \to S^n$.

Definition 3. The degree, $\deg \widetilde{\Phi}_r$, of the mapping $\widetilde{\Phi}_r$ is called the *characteristic of the vector field* Φ on the boundary of the ball $D_r^{n+1}(x^0)$. We will denote the characteristic by $\kappa(\Phi, S_r^n(x^0))$.

Along with the term "characteristic of a vector field", the term "rotation of a vector field" is often used, similarly to the 2–dimensional case, where for $\phi : S^1 \to S^1$, the degree $\deg \phi$ is the algebraic number of rotations of the vector $\phi(x)$, when x runs through the circle S^1 (in the positive direction).

Theorem 2. *Let a field Φ have no singular points in a closed ball $\bar{D}_r^{n+1}(x^0)$; then $\kappa(\Phi, S_r^n(x^0)) = 0$.*

Proof. The mapping $\widetilde{\Phi}_r$ is homotopic to the constant mapping Φ_0 of S^n to the point $\Phi(x^0)/\|\Phi(x^0)\| \in S^n$ with degree equal to zero. The corresponding homotopy is given, for example, by the formula

$$(8) \qquad \Phi(t, x) = \frac{\Phi(trx + x^0)}{\|\Phi(trx + x^0)\|}; \quad 0 \le t \le 1, \quad x \in S^n.$$

$\square$

Corallary. *If $\kappa(\Phi, S_r^n(x^0)) \ne 0$, then the field Φ has at least one singular point in the ball $D_r^{n+1}(x^0)$.*

Note that the characteristic $\kappa(\Phi, S_r^n(x^0))$ is defined even if the field Φ is given only on the boundary $S_r^n(x^0)$ of the ball $D_r^{n+1}(x^0)$.[*]

The following theorem immediately follows from theorem 2.

Theorem 3. *Let a field Φ be given on the sphere $S_r^n(x^0)$ and have no singular points. If $\kappa(\Phi, S_r^n(x^0)) \ne 0$, then Φ cannot be extended to the ball $\bar{D}_r^{n+1}(x^0)$ without singular points.*

---

[*] The field $\Phi : S_r^n(x^0) \to \mathbf{R}^{n+1}$ in this case, is not, evidently, a vector field on the manifold $S_r^n(x^0)$ in the sense of Ch. 4, § 8.

The theorem inverse to theorem 3 is also valid; it follows from the Hopf theorem which has already been mentioned.

Exercise 7°. Let x^0 be a singular point of a smooth vector field Φ on $U \subset \mathbf{R}^{n+1}$ and let the Jacobian matrix $\left(\frac{\partial \Phi}{\partial x}\right)$ of the mapping Φ at the point x^0 be non–singular (such points are said to be nondegenerate). Prove that x^0 is an isolated singular point of the field Φ and that

$$\text{ind}\,(x^0, \Phi) = \text{sign det} \left(\frac{\partial \Phi}{\partial x}\right)\Big|_{x^0}.$$

Hint. Construct a homotopy connecting the vector fields Φ_r and $\left(\frac{\partial \Phi}{\partial x}\right)\Big|_{x^0}$ on S^n.

Exercise 8°. Let the mapping $\Phi : \mathbf{C} \to \mathbf{C}$ be defined by the formula $\Phi(z) = z^n$, where $n > 0$ is an integer. Regarding Φ as a mapping $\Phi : \mathbf{R}^2 \to \mathbf{R}^2$, calculate the index of the zero singular point x^0 of the field Φ. Do the same for the mapping $\Psi(z) = (\bar{z})^n$.

Consider a vector field $X(x)$ on a manifold M^n. Let $x^0 \in M^n$ be an isolated singular point of the field $X(x)$, i.e., $X(x^0) = 0$, and let a neighbourhood $U(x^0) \subset M^n$ of the point x^0, in which $X(x) \neq 0$ for $x \neq x_0$, exist. The field $X(x)$, in local coordinates, has the form

$$\left(x_1, \dots, x_n; X_1(x_1, \dots, x_n)\frac{\partial}{\partial x_1} + \dots \right.$$
$$\left. \dots + X_n(x_1, \dots, x_n)\frac{\partial}{\partial x_n}\right).$$

The index $\text{ind}\,(x^0, X)$ of a singular point x^0 of a vector field $X(x^0)$ on a manifold can be defined as an index of the singular point $(x_1^0, \dots, x_n^0)$ (here, the x_i^0 are the coordinates of the point x^0) of the vector field $\Phi = \{X_1(x_1, \dots, x_n), \dots, \dots, X_n(x_1, \dots, x_n)\}$ in the space $\mathbf{R}^n$.

Exercise 9°. Prove that the index $\text{ind}\,(x^0, X)$ does not depend on the choice of local coordinates.

Exercise $10°$. Let f be a smooth function on a manifold, x^0 a nondegenerate critical point of index λ of the function f (see Ch. 4, § 11). Prove that $\text{ind}\,(x^0, \text{grad}\, f) = (-1)^\lambda$.

Hint. Use the results of exercise $7°$, § 12, Ch. 4, and those of exercise $6°$ of this section.

3. Rotation of a vector field.

We shall generalize the concept of the characteristic of a vector field to the case when of a boundary of a region of arbitrary form.

Let $U \subset \mathbf{R}^n$, $n \geq 2$, be a bounded region, $\bar{U}$ its closure, and ∂U its boundary. Consider a continuous vector field $\Phi : \bar{U} \to \mathbf{R}^n$, which has no singular points on the boundary; thus, a continuous mapping $\Phi : \partial U \to \mathbf{R}^n \backslash 0$ is defined, but singular points are permitted in $U = \text{int}\, \bar{U}$.

Developing the concepts of subsection 2 further, we shall define the global characteristic $\kappa(\Phi, \partial U)$ (or "rotation") of a vector field Φ on the boundary ∂U. In the first step of the construction, we pass to a smaller polyhedral (closed) region $P_\alpha \subset U$ which "well approximates" the closure $\bar{U} : P_\alpha \supset \bar{U} \backslash S_\alpha(\partial U)$, where $S_\alpha(\partial U)$ is the α–neighbourhood of the boundary ∂U for a sufficiently small $\alpha > 0$ (i.e., the union of the open balls of radius α with centre at points $x \in \partial U$). In order to construct such a polyhedron P_α, it suffices to decompose the space $\mathbf{R}^n$ into congruent cubes of a sufficiently small diameter $d = \sqrt{n}\rho$, where ρ is a side of a cube, with the help of planes parallel to the coordinate planes, and to require that $\rho < \alpha/\sqrt{n}$. Then P_α is the union of all the cubes intersecting with the closure $\overline{U \backslash S_\alpha(\partial U)}$ (which is nonempty if α is sufficiently small). Triangulating each cube from P_α in a standard way (i.e., on each face I^k of the cube I^n consequently perform a cone construction from its centre over the boundary ∂I^k for $k = 1, 2, \ldots, n$), we obtain a triangulation K of the polyhedron P_α. We orient all the n–dimensional simplices $\tau_i^n \in K$ identically; this means that if $[\tau_i^n] = [a_i^0, \ldots, a_i^n]$ is an oriented simplex, then the frames $e_i^1, \ldots, e_i^n$, where $e_i^k = \overrightarrow{a_i^0 a_i^k}$, $k = 1, 2, \ldots, n$, for any i defines an orientation of the space $\mathbf{R}^n$ coinciding with a pregiven one. We form the integral chain $x_n = \sum_i 1 \cdot [\tau_i^n]$ and call it the fundamental chain of P_α. Its boundary $\partial_n x_n = \sum_i \partial_n[\tau_i^n]$ possesses the following important property: it consists of $(n-1)$–dimensional simplices $\partial_n x_n = \sum_i 1 \cdot [t_j^{n-1}]$, where the simplices t_j^{n-1} belong to the set–theoretic boundary ∂P_α, and $\partial P_\alpha = \bigcup_i t_j^{n-1}$; and the orientation $[t_j^{n-1}]$ is induced by the orientation of the unique simplex $[\tau_i^n]$ whose boundary includes $[t_j^{n-1}]$ (i.e., $\partial_n[\tau_i^n] = [t_j^{n-1}] + \ldots$). This integral cycle $z_{n-1} = \partial_n x_n$ is called the *fundamental cycle of the boundary* ∂P_α.

It can be shown that the group $H_{n-1}(\partial P_\alpha; \mathbf{Z})$ is a free Abelian group possessing as many generators as there are connected components in the boundary ∂P_α; denote this number by L (generally speaking, L depends on $\alpha : L = L_\alpha$). In addition,

$$\partial_n x_n = \sum_i 1 \cdot [t_j^{n-1}] = \sum_{i=1}^{L} \left(\sum_{p=1}^{M_i} 1 \cdot [t_p^{n-1}] \right),$$

where the simplices t_p^{n-1} from every interior sum are contained in the same connected component ∂P_α; the homology class $[z_{n-1}^i]$ of each interior sum $\sum_{p=1}^{M_i} 1 \cdot [t_p^{n-1}] = z_{n-1}^i$, is a generator of the group $H_{n-1}(\partial P_\alpha; \mathbf{Z})$.

The second step of the construction is related to the vector field $\Phi : P_\alpha \rightarrow \mathbf{R}^n$ on the polyhedron P_α. Since the field Φ has no singular points on the boundary ∂P_α, there will be no singular points, because of the continuity in the neighbourhood $S_\alpha(\partial U)$ (or more precisely, on the intersection $S_\alpha(\partial U) \cap \bar{U}$) for a sufficiently small $\alpha > 0$. Let us determine such an α and construct the polyhedron P_α, as described above. Then the field Φ has no singular points on the boundary ∂P_α, and the mapping $\Phi : \partial P_\alpha \rightarrow \mathbf{R}^n \backslash 0$ is defined. This mapping induces a homomorphism of singular homology groups $\Phi_{*n-1} : H_{n-1}^s(\partial P_\alpha; \mathbf{Z}) \rightarrow H_{n-1}^s(\mathbf{R}^n \backslash 0; \mathbf{Z})$. Since $\mathbf{R}^n \backslash 0 \sim S^{n-1}$ (the homotopy equivalence), $H_{n-1}^s(\mathbf{R}^n \backslash 0; \mathbf{Z})$ and $H_{n-1}^s(S^{n-1}; \mathbf{Z})$ are isomorphic and free Abelian groups on one generator. By the uniqueness theorem, the homology groups $H_{n-1}(\partial P_\alpha; \mathbf{Z})$ and $H_{n-1}^s(\partial P_\alpha; \mathbf{Z})$ are isomorphic. Let $[z_{n-1}^i]^s$ be the image of the generator $[z_{n-1}^i]$ from $H_{n-1}(\partial P_\alpha; \mathbf{Z})$ under this isomorphism, and $\tilde{\gamma}_{n-1}^s$ a generator in $H_{n-1}^s(\mathbf{R}^n \backslash 0; \mathbf{Z})$. Then $\Phi_{*n-1}[z_{n-1}^i]^s = m_i \cdot \tilde{\gamma}_{n-1}^s$, where $m_i \in \mathbf{Z}$. Consider the homology class $[z_{n-1}]^s \in H_{n-1}^s(\partial P_\alpha; \mathbf{Z})$ corresponding to the class $[z_{n-1}] \in H_{n-1}(\partial P_\alpha; \mathbf{Z})$ of the cycle $z_{n-1} = \partial_n x_n = \sum_{i=1}^{L} z_{n-1}^i$. Then $\Phi_{*n-1}[z_{n-1}]^s = m \cdot \tilde{\gamma}_{n-1}^s$, where $m \in \mathbf{Z}$. The integer m is called a characteristic $\kappa(\Phi, \partial P_\alpha)$ (the "rotation") of the vector field Φ on the boundary ∂P_α. Note that by the uniqueness theorem of homology theory, $[z_{n-1}]^s = \sum_{i=1}^{L} [z_{n-1}^i]^s$, and, consequently, $m = \sum_{i=1}^{L} m_i$.

It is obvious that the sign of $\kappa(\Phi, \partial P_\alpha)$ depends on the choice of generator in the group $H_{n-1}^s(\mathbf{R}^n \backslash 0; \mathbf{Z})$. It is convenient to choose generators in the groups $H_{n-1}^s(\mathbf{R}^n \backslash 0; \mathbf{Z})$, $H_{n-1}^s(S^{n-1}; \mathbf{Z})$ in the following way. Take a simplex τ^n with a barycentre at the point 0 of the space $\mathbf{R}^n$. Then the central projection π gives a homeomorphism of the boundary $\partial \tau^n$ and S^{n-1}; since the boundary $\partial \tau^n$ is naturally triangulated, S^{n-1} also has a corresponding triangulation, and π becomes a simplicial mapping of the polyhedra $\partial \tau^n$, S^{n-1} and induces a chain mapping $\widehat{\pi}_*$ of their chain complexes (over $\mathbf{Z}$). Let $[\tau^n]$ be oriented according to the orientation of $\mathbf{R}^n$. Then $\partial_n [\tau^n] = q_{n-1}$ is a fundamental cycle of the polyhedron $\partial \tau^n$, as follows from formula (9), § 3, for the homology $H_*(\{\partial \tau^n\}; \mathbf{Z})$. Its image $\gamma_{n-1} = \widehat{\pi}_{n-1} q_{n-1}$ in $C_{n-1}(S^{n-1}; \mathbf{Z})$ is a fundamental cycle of the polyhedron $S^{n-1} = \partial D^n$, and the corresponding homology class $[\gamma_{n-1}]$ is a generator in $H_{n-1}(S^{n-1}; \mathbf{Z})$, which is the one we choose. Obviously, $[\gamma_{n-1}] = \pi_{*n-1}[q_{n-1}]$. The isomorphism between H_* and H_*^s homology defines a generator γ_{n-1}^s in $H_{n-1}^s(S^{n-1}; \mathbf{Z})$, and, moreover, $\gamma_{n-1}^s = \pi_{*n-1}[q_{n-1}]^s$. The generator $\widetilde{\gamma}_{n-1}^s \in H_{n-1}^s(\mathbf{R}^n \backslash 0; \mathbf{Z})$ is defined by specifying a mapping $r : \mathbf{R}^n \backslash 0 \to S^{n-1}$ and its homotopy inverse $\phi : S^{n-1} \to \mathbf{R}^n \backslash 0$, where $r(x) = x/\|x\|$ and $\phi(x) = x$; r and ϕ induce isomorphisms $\phi_{*n-1} = r_{*n-1}^{-1} : H_{n-1}^s(S^{n-1}; \mathbf{Z}) \to H_{n-1}^s(\mathbf{R}^n \backslash 0; \mathbf{Z})$, and $\widetilde{\gamma}_{n-1}^s = \phi_{*n-1}(\gamma_{n-1}^s)$.

Having thus specified generators for $H_{n-1}^s(\mathbf{R}^n \backslash 0; \mathbf{Z})$, $H_{n-1}^s(S^{n-1}; \mathbf{Z})$, we have, for the definition of the characteristic m, the equality $\Phi_{*n-1}[z_{n-1}]^s = m \cdot \widetilde{\gamma}_{n-1}^s$, or, as $\widetilde{\gamma}_{n-1}^s = \phi_{*n-1}(\gamma_{n-1}^s) = r_{*n-1}^{-1}(\gamma_{n-1}^s)$, the equality $r_{*n-1} \Phi_{*n-1}[z_{n-1}]^s = m\gamma_{n-1}^s$. Since $r_{*k} \Phi_{*k} = (r \cdot \Phi)_{*k}$, where $(r \cdot \Phi)(x) = \Phi(x)/\|\Phi(x)\|$, is the mapping $r \cdot \Phi = \widetilde{\Phi} : \partial P_\alpha \to S^{n-1}$, m can be interpreted as the degree of the mapping $\widetilde{\Phi} : \partial P_\alpha \to S^{n-1}$ as in definition 3, § 6.

The third step, i.e., the proof that the characteristic $\kappa(\Phi, \partial P_\alpha)$ is independent on the choice of the polyhedron P_α, is omitted because of the unwieldy elementary construction we have applied (here, one should make use of the invariance of simplicial homology and the uniqueness theorem).

Definition 4. The *characteristic (rotation)* $\kappa(\Phi, \partial U)$ *of a vector field* Φ *on the boundary* ∂U is the characteristic $\kappa(\Phi, \partial P_\alpha)$ on the boundary ∂P_α of some polyhedral region P_α of the indicated type (see the first step).

Theorem 4. *If* $\Phi : \bar{U} \times [0; 1] \to \mathbf{R}^n$ *is a homotopy of vector fields with no singular points on* ∂U *(i.e.,* $\Phi : (\partial U) \times [0; 1] \to \mathbf{R}^n \backslash 0$*), then the characteristic* $\kappa(\Phi_t, \partial U)$ *of the vector field* $\Phi_t = \Phi : (\partial U \times \{t\}) \to \mathbf{R}^n \backslash 0$ *does not change when changing* $t \in [0; 1]$*; in particular,* $\kappa(\Phi_0, \partial U) = \kappa(\Phi_1, \partial U)$.

Proof. By the continuity of the mapping $\Phi(x, t)$ in the variables (x, t), there can be found a $\alpha_0 > 0$ such that the field Φ_t in the α_0–neighbourhood $S_{\alpha_0}(\partial U)$

has no singular points for any $t \in [0; 1]$. Then $\Phi_t : \partial P_{\alpha_0} \to \mathbf{R}^n \setminus 0$ for any $t \in [0; 1]$, where P_{α_0} is constructed as in the first step above, is a continuous homotopy; therefore, the homomorphism $(\Phi_t)^s_{*n-1}$ of singular homology is constant with respect to t, giving the constancy of $\kappa(\Phi_t, \partial P_\alpha)$ and $\kappa(\Phi_t, \partial U)$. $\square$

Theorem 5. *If a continuous field $\Phi : \bar{U} \to \mathbf{R}^n$ has no singular points on a region $\bar{U}$, then $\kappa(\Phi, \partial U) = 0$.*

Proof. Let $z_{n-1} = \partial_n x_n$ be a fundamental cycle of a boundary ∂P_α. By the assumptions of the theorem, we have a commutative diagram

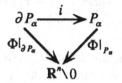

where i is the imbedding, $\Phi|_{\partial P_\alpha}$ and $\Phi|_{P_\alpha}$ are restrictions of the mapping $\Phi : \bar{U} \to \mathbf{R}^n \setminus 0$. Considering homomorphisms of the H^s_*–homology generated by these mappings, and also the natural isomorphisms h_{n-1} of the groups H_{*n-1} and H^s_{*n-1} (see the uniqueness theorem of homology theory), we obtain a commutative diagram

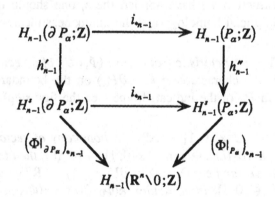

Since $i_{*n-1}[z_{n-1}] = 0$ in $H_{n-1}(P_\alpha; \mathbf{Z})$, because the cycle $z_{n-1} = \partial_n x_n$ is homologous to zero, $z_{n-1} \sim 0$, in $C_{n-1}(P_\alpha; \mathbf{Z})$, we have $h''_{n-1}(i_{*n-1}[z_{n-1}]) =$

0, and from the commutativity of the square we obtain $i_{*n-1}(h'_{n-1}[z_{n-1}]) = 0$; hence, from the equality $[z_{n-1}]^s = h'_{n-1}[z_{n-1}]$ there also follows that $i_{*n-1}[z_{n-1}]^s = 0$ in $H^s_{n-1}(P_\alpha; \mathbf{Z})$. This fact, together with the commutativity of the triangle, implies the following equalities:

$$(\Phi|_{\partial P_\alpha})_{*n-1}[z^s_{n-1}] = (\Phi|_{P_\alpha})_{*n-1} i_{*n-1}[z_{n-1}]^s =$$
$$= (\Phi|_{P_\alpha})_{*n-1}(i_{*n-1}[z_{n-1}]^s) = 0,$$

giving zero for the rotations

$$\kappa(\Phi, \partial P_\alpha) = \kappa(\Phi, \partial U).$$

$\square$

Corollary 1. *If for a continuous vector field $\Phi : \bar{U} \to \mathbf{R}^n$ without singular points on a boundary ∂U the characteristic $\kappa(\Phi, \partial U)$ differs from zero, then inside the region there is a point $x^* \in U$ with $\Phi(x^*) = 0$.*

Proof. Assume the opposite and immediately obtain a contradiction with the theorem above. $\square$

It was noted above that in applications, it is useful to regard as a singular point, not only zeros ($\Phi(x_*) = 0$), but also points of discontinuity or indeterminacy of a vector field in a more general way (see, for example, Ch. 1, § 6). The following statement is an insignificant modification of the previous.

Corollary 2. *If on some ϵ–neighbourhood $S_\epsilon(\partial U)$, $\epsilon > 0$, a vector field Φ is continuous, and on the boundary it has no zeros, and $\kappa(\Phi, \partial U) \neq 0$, then the field cannot be extended inside the region $\bar{U}$ without a singular point in the generalized sense.*

In fact, assuming the opposite, we can use the same arguments as above.

Remark. If $\bar{U}$ is a polyhedron consisting of a union of n–dimensional simplices (assuming connectedness of U), then in the last proposition, it is sufficient to require that Φ be continuous only on the boundary ∂U.

Let now $x_0 \in U$ be an isolated singular point (in the extended meaning) of a vector field Φ, i.e., there exists a disc $\bar{D}_r(x_0)$ which has no other singular point different from x_0. We have a continuous mapping

$$\Phi : (\bar{D}_r(x_0) \backslash x_0) \to \mathbf{R}^n \backslash 0.$$

Consider the polyhedra $\bar{D}_r(x_0)$, $\bar{D}_1(x_0)$ and their boundaries $S_r^{n-1}(x_0)$, $S_1^{n-1}(x_0)$, respectively; for the latter, we define the fundamental cycles γ_{n-1}^r, γ_{n-1}^1 by transferring the barycentre of the fixed simplex τ^n to the point x_0 and using homeomorphisms, i.e. the central projections $\pi^{(r)}$, $\pi^{(1)}$ from the point x_0 of the boundary $\partial \tau^n$ onto the spheres $S_r^{n-1}(x_0)$, $S_1^{n-1}(x_0)$. For the central projection $\pi : S_1^{n-1}(x_0) \to S_r^{n-1}(x_0)$ which, obviously, is simplicial, we have $\gamma_{n-1}^r = \widehat{\pi}_{n-1}\gamma_{n-1}^1$, since $\pi = \pi^{(r)}(\pi^{(1)})^{-1}$, and correspondingly, $\widehat{\pi}_{n-1} = \widehat{\pi}_{n-1}^{(r)} \cdot (\widehat{\pi}_{n-1}^{(1)})^{-1}$ for the chain complexes; hence, the equality $[\gamma_{n-1}^r] = \pi_{*n-1}[\gamma_{n-1}^1]$ of simplicial homology classes and for the corresoponding singular ones $[\gamma_{n-1}^r]^s = \pi_{*n-1}[\gamma_{n-1}^1]^s$ follows.

The mapping $\Phi : S_r^{n-1}(x_0) \to \mathbf{R}^n \backslash 0$ induces a homomorphism $\Phi_* : H_{n-1}^s(S_r^{n-1}(x_0); \mathbf{Z}) \to H_{n-1}^s(\mathbf{R}^n \backslash 0; \mathbf{Z})$; let $\Phi_{*n-1}[\gamma_{n-1}^r]^s = m \cdot \widetilde{\gamma}_{n-1}^s$. The number m is called the index $\mathrm{ind}\,(x_0, \Phi)$ of the singular point x_0 of the vector field Φ.

If we use the relation between the homology classes $[\gamma_{n-1}^r]^s$ and $[\gamma_{n-1}^1]^s$, then we obtain an equivalent equality $\Phi_{*n-1}\pi_{*n-1}[\gamma_{n-1}^1]^s = m\widetilde{\gamma}_{n-1}^s$; since $\Phi_{*n-1}\pi_{*n-1} = (\Phi\pi)_{*n-1}$, and taking into account that $\Phi\pi : S_1^{n-1}(x_0) \to \mathbf{R}^n \backslash 0$ is given by the formula $(\Phi\pi)(x) = \Phi(r(x-x_0)+x_0)$, $x \in S_1^{n-1}(x_0)$, we conclude that the present definition of $\mathrm{ind}\,(x_0, \Phi)$ is equivalent to the one given earlier (definition 2, subsection 2). The index $\mathrm{ind}\,(x_0, \Phi)$ does not depend on the radius of the sphere $S_r^{n-1}(x_0)$ (r can be taken arbitrarily small).

Theorem 6. (On the algebraic number of singular points). *Let a vector field Φ be continuous and have no singular points on the boundary of the region $\bar{U}$, and let it have a finite number of singular points (in the extended meaning) $\{x_i\}_{i=1}^k$ inside the region. Then the following equality holds*

$$(9) \qquad \kappa(\Phi, \partial U) = \sum_{i=1}^k \mathrm{ind}\,(x_i, \Phi),$$

where the sum on the right is called the algebraic number of singular points.

Proof. Since the mapping Φ is continuous on $\bar{U}\backslash\{x_1,\dots,x_k\}$, it is possible to construct a polyhedron P_α as indicated in the first step of the definition of the relation of a vector field. Some of the points $x_1,\dots,x_k$ can be (on one of the boundaries) of one of the n–dimensional simplices of the triangulation of the polyhedron P_α; by using sufficiently small shifts of a polyhedron that are given by a parallel transfer on a vector which is orthogonal to a plane containing such a bound, one can achieve that all singular points will appear inside one of the n–dimensional triangulation simplices $\{t_i^n\}_{i=1}^k$.

We assume that P_α also possesses this property, and that these special simplices t_i^n do not intersect with one another and with ∂P_α, i.e., $t_i^n \cap t_j^n = \emptyset$ for $i \neq j$, and $t_i^n \cap \partial P_\alpha = \emptyset$. We divide the fundamental chain x_n of the polyhedron P_α into two parts $x_n = x_n^1 + x_n^2$, where $x_n^1 = \sum_{i=1}^k 1 \cdot [t_i^n]$, i.e. x_1^n is the sum over those simplices which contain a singular point, and x_n^2 consists of the remaining simplices $\{\tau_j^n\}$: $x_n^2 = \sum 1 \cdot [\tau_j^n]$; it should be recalled that all n–dimensional simplices are oriented in the same way (given by the orientation of $\mathbf{R}^n$). The carrier of a chain $c_l = \sum_i g_i \tau_i^l$ in a complex $C_*(K; G)$ corresponding to the simplicial complex K, is the union of all simplices τ_i^l included with nonzero coefficients into the chain c_l. The carriers of the chains x_n^1 and x_n^2 we denote by Q_1, Q_2, respectively. For the fundamental cycle of the boundary ∂P_α, there is the equality $z_{n-1} = z_{n-1}^1 + z_{n-1}^2$, where $z_{n-1}^1 = \partial_n x_n^1$, $z_{n-1}^2 = \partial_n x_n^2$ are fundamental cycles of the polyhedra ∂Q_1 (Q_1 is not connected, if $k > 1$) and ∂Q_2; since the carrier of $|\partial_n x_n^1|$ is in $Q_1 \cap Q_2$, one can take z_{n-1}^1, z_{n-1}^2 to be cycles of the polyhedron ∂Q_2; moreover, the homology class $[z_{n-1}^2] = 0$ in $H_{n-1}(Q_2; \mathbf{Z})$; since $\partial P_\alpha \subset \partial Q_2$, one can see z_{n-1} as a cycle of the polyhedron ∂Q_2.

From the equality $z_{n-1}^2 = z_{n-1} - z_{n-1}^1$, we obtain $[z_{n-1}^2] = [z_{n-1}] - [z_{n-1}^1]$ in $H_{n-1}(\partial Q_2; \mathbf{Z})$, and $[z_{n-1}^2]^s = [z_{n-1}]^s - [z_{n-1}^1]^s$ in $H_{n-1}^s(\partial Q_2; \mathbf{Z})$, where the upper index (s) indicates that the corresponding class in singular homology is concerned. Now consider the commutative diagram

(10)

$$
\begin{array}{ccc}
H_{n-1}(\partial Q_2; \mathbf{Z}) & \xrightarrow{\,i_{*n-1}\,} & H_n(Q_2; \mathbf{Z}) \\
\downarrow{\scriptstyle h'_{n-1}} & & \downarrow{\scriptstyle h''_n} \\
H_{n-1}^S(\partial Q_2; \mathbf{Z}) & \xrightarrow{\,i_{*n-1}\,} & H_n^s(Q_2; \mathbf{Z})
\end{array}
$$

where the i_{*n-1} are homomorphisms generated by the imbedding $i : \partial Q_2 \to Q_2$, and h', h'' are isomorphisms of homology theories H_* and H_*^s. The class

$[z_{n-1}^2]$ from $H_{n-1}(\partial Q_2; \mathbf{Z})$ belongs to the kernel $\operatorname{Ker} i_{*n-1}$, since $[z_{n-1}^2] = 0$ in $H_{n-1}(Q_2; \mathbf{Z})$. From the commutativity of the diagram, we obtain, as in theorem 5, $i_{*n-1}[z_{n-1}^2]^s = 0$ in $H_{n-1}^s(Q_2; \mathbf{Z})$; moreover, the polyhedra ∂Q_2, Q_2 here play the role of the polyhedra ∂P_α, P_α in the proof of theorem 5; therefore, we also have $(\varPhi|_{\partial Q_2})_{*n-1}[z_{n-1}^2]^s = 0$ in $H_{n-1}^s(\mathbf{R}^n\backslash 0; \mathbf{Z})$. On the other hand, we have in $H_{n-1}^s(\mathbf{R}^n\backslash 0; \mathbf{Z})$

$$(\varPhi|_{\partial Q_2})_{*n-1}[z_{n-1}^2]^s =$$

$$= (\varPhi|_{\partial Q_2})_{*n-1}[z_{n-1}]^s - (\varPhi|_{\partial Q_2})_{*n-1}[z_{n-1}^1]^s,$$

which gives

(11)
$$(\varPhi|_{\partial Q_2})_{*n-1}[z_{n-1}]^s = (\varPhi|_{\partial Q_2})_{*n-1}[z_{n-1}^1]^s.$$

Taking into account the equality $[z_{n-1}^1]^s = \sum_i [\partial_n[t_i^n]]^s$, we rewrite (11) in the form

(12)
$$(\varPhi|_{\partial Q_2})_{*n-1}[z_{n-1}]^s = \sum_i (\varPhi|_{\partial Q_2})_{*n-1}[\partial_n[t_i^n]]^s.$$

Since $\partial Q_2 = \partial P_\alpha \cup (\bigcup_i \partial t_i^n)$ and the simplices t_i^n do not intersect, the groups C_{*}^s, H_{*}^s of singular chains and homology of the polyhedron ∂Q_2 decompose into the direct sum of groups corresponding to the simplices t_i^n in ∂Q_2 and ∂P_α. Identifying the classes $[z_{n-1}]^s$, $[\partial_n[t_i^n]]^s$ in $H_{n-1}^s(\partial Q_2, \mathbf{Z})$ with the classes $[z_{n-1}]^s$, $[\partial_n[t_i^n]]^s$ in $H_{n-1}^s(\partial P_\alpha; \mathbf{Z})$, $H_{n-1}^s(\partial t_i^n; \mathbf{Z})$ according to this decomposition, we rewrite the last relation (12) as follows

$$(\varPhi|_{\partial P_\alpha})_{*n-1}[z_{n-1}]^s = \sum_i (\varPhi|_{\partial t_i^n})_{*n-1}[\partial_n[t_i^n]]^s;$$

taking into account that

$$(\varPhi|_{\partial t_i^n})_{*n-1}[\partial_n[t_i^n]]^s = m_i \cdot \tilde{\gamma}_{n-1}^s,$$

where $m_i = \operatorname{ind}(x_i, \varPhi)$, we obtain $(\varPhi|_{\partial P_\alpha})_{*n-1}[z_{n-1}]^s = \sum_i m_i \tilde{\gamma}_{n-1}^s$, and, consequently, $\kappa(\varPhi, \partial P_\alpha) = \sum_i \operatorname{ind}(x_i, \varPhi)$, which completes the proof of the theorem. $\qquad\square$

Formula (9) is one of the most important in the theory of singular points of vector fields and fixed points of a mapping.

§ 7. Homology of a cell complex

We pass to the study of the homology of spaces which have the homotopy type of a cell complex. This class of spaces is interesting, first of all, because it is quite extensive (see Ch. 4, § 12), and, secondly, the homology of a cell complex can be calculated in a very simple and precise way.

Let X be a finite cell complex. We construct a chain complex $\widetilde{C}_*(X;G)$ in the following way. As the group $\widetilde{C}_k(X;G)$, we take the Abelian group of formal linear combinations $\sum_i g_i \cdot \tau_i^k$, where $g_i \in G$ are arbitrary elements, and the τ_i^k are the k–dimensional cells of the complex X; the summation is done over all k–dimensional cells. Consequently, the group $\widetilde{C}_k(X;G)$ is isomorphic to the direct sum of as many copies of the group G, as there are cells of dimension k in the cellular decomposition of X. Besides, we shall assume that each copy of G corresponds precisely to one of the k–dimensional cells.

Let us define the differential $\widetilde{\partial}_k : \widetilde{C}_k(X;G) \rightarrow \widetilde{C}_{k-1}(X;G)$. Let τ^k be a k–dimensional cell of X; its boundary is contained in the union of cells of dimensions not higher than $k-1$ (the $(k-1)$–dimensional skeleton of X denoted by X^{k-1}). According to the definition of a cell complex, the cell τ^k is given by a gluing mapping $f : S^{k-1} \rightarrow X^{k-1}$. Consider the composition $S^{k-1} \rightarrow X^{k-1} \rightarrow X^{k-1}/X^{k-2}$, where the last arrow is a quotient mapping. The space X^{k-1}/X^{k-2} is a cell complex; it consists of one cell of dimension zero, i.e., the point $*$ into which the space X^{k-2} is collapsed, and has as many cells of dimension $(k-1)$ glued along their boundaries to the point $*$, as there were in the skeleton X^{k-1}, i.e., in X. Such a space is called the wedge of $(k-1)$–dimensional spheres. We select a cell τ_j^{k-1} in X^{k-1}; in the wedge of spheres X^{k-1}/X^{k-2}, this cell corresponds to a certain sphere S_j^{k-1}. Consider the composition of mappings

$$ S^{k-1} \xrightarrow{f} X^{k-1} \rightarrow X^k/X^{k-2} \rightarrow S_j^{k-1}, $$

where the last arrow denotes the quotient mapping obtained by collapsing to $*$ the subspace of the space X^{k-1}/X^{k-2} consisting of all spheres except S_j^{k-1}. The degree of this composition is called the *incidence coefficient* of the cells τ^k and τ_j^k and denoted by $[\tau^k, \tau_j^{k-1}]$; the incidence coefficient shows how many times the boundary of the cell τ^k is "folded" onto the cell τ_j^{k-1} in gluing the cell τ^k to the skeleton X^{k-1}. We denote the set of cells of dimension $k-1$ in the cell complex X by Ω^{k-1}. For each cell τ^k, we define the differential

$\tilde{\partial}_k$ by the formula

$$\tilde{\partial}_k \tau^k = \sum_{\tau_j \in \Omega^{k-1}} [\tau^k, \tau_j^{k-1}] \cdot \tau_j^{k-1}$$

and extend $\tilde{\partial}_k$ to $\tilde{C}_k(X; G)$ by linearity.[*]

When $k = 1$, the incidence coefficient $[\tau^1, \tau_j^0]$ can be equal to 0, 1, or -1. If $f((0; 1))$ defines the gluing of the one–dimensional cell τ^1, then

$$[\tau^1, \tau_j^0] = \begin{cases} 0, & \text{if } [f(0) \cup f(1)] \cap \tau_j^0 = \emptyset, \\ & \text{or } f(0) = f(1) = \tau_j^0, \\ 1, & \text{if } f(1) = \tau_j^0 \text{ and } f(0) \neq \tau_j^0, \\ -1, & \text{if } f(0) = \tau_j^0 \text{ and } f(1) \neq \tau_j^0, \end{cases}$$

It can be shown that $\tilde{\partial}_{k-1} \cdot \tilde{\partial}_k = 0$.

Thus, a chain complex $\tilde{C}_*(X; G)$ has been constructed. It turns out that its homology coincides with the singular homology of the cell complex X. To proof of this fact involves the technique of exact sequences; since it is very long, we do not present it here.

The advantage of calculating homology with the help of the complex $\tilde{C}_*(X; G)$ is obvious: the groups $\tilde{C}_k(X, \mathbf{Z})$ have a finite number of generators contrary to the groups $C_k^s(X; \mathbf{Z})$. Consequently, the subgroups of k-dimensional cycles and boundaries also have a finite number of generators as does the quotient group $H_k^s(X; \mathbf{Z})$. The theory of Abelian groups implies that

$$H_k^s(X; \mathbf{Z}) \simeq \underbrace{(\mathbf{Z} \oplus \ldots \oplus \mathbf{Z})}_{\rho_k} \oplus \mathbf{Z}_{\rho_1^k} \oplus \ldots \oplus \mathbf{Z}_{\rho_{s_k}^k},$$

where $\mathbf{Z}_{\rho_i^k}$ is the finite cyclic group of order ρ_i^k; moreover, ρ_i^k is divisible by ρ_{i-1}^k. The number ρ_k is called the k–dimensional Betti number, and the numbers ρ_i^k are the k–dimensional torsion numbers of the space X.

In spite of some complexity of the proof, the method described appears to be quite convenient from the practical point of view, and enables us to calculate described homology of a whole range of concrete spaces in a simple way.

---

[*] We assume, just like in § 4, that G is a ring with identity.

Exercise 1°. By representing the sphere S^n in the form of the cell complex $S^n = e^n \cup e^o$, $n \geq 1$, calculate the homology of S^n. Show that $\rho_k = 0$, $k \neq 0, n$; $\rho_0 = \rho_n = 1$ and that each $\rho_i^k = 0$.

Now we calculate homology of the complex projective space CP^n. For this we represent CP^n in the form of a cell complex. A point from CP^n is given by a large circle $(e^{i\alpha}\xi_1, e^{i\alpha}\xi_2, \ldots, e^{i\alpha}\xi_{n+1})$, $0 \leq \alpha < 2\pi$, from S_C^n (i.e., $\xi_i \in C$, $|\xi_1|^2 + \ldots + |\xi_{n+1}|^2 = 1$). We define a cell τ^{2k}, where $0 \leq k \leq n$, by the characteristic mapping $g^{2k} : \bar{D}^{2k} \to CP^n$ which associates the point

$$(\xi_1, \ldots, \xi_k) \in \left\{ \xi \in C^k : \sum_{j=1}^{k} |\xi_j| \leq 1 \right\} \simeq \bar{D}^{2k}$$

with the point from CP^n given by the large circle

$$(e^{i\alpha}\xi_1, \ldots, e^{i\alpha}\xi_k, e^{i\alpha}\sqrt{1 - |\xi_1|^2 - \ldots - |\xi_k|^2}, 0, \ldots, 0).$$

When $k = 0$, this is the large circle (point in CP^n) $(e^{i\alpha} \cdot 1, 0, \ldots, 0)$.

Thus, the space CP^n is represented in the form of a cell complex (verify!) consisting of cells of every even dimension up to $2p$, one for each dimension, and no cells of other dimensions. Therefore,

$$\tilde{C}_k(CP^n; G) \simeq \begin{cases} G & \text{when } k = 2m \text{ and } k \leq 2n, \\ 0 & \text{when } k = 2m + 1 \text{ or } k > 2n. \end{cases}$$

Indeed, since one of the groups $\tilde{C}_{k-1}(CP^n; G)$, $\tilde{C}_k(CP^n; G)$ is trivial, the boundary homomorphism can only be trivial in the complex $\tilde{C}_*(CP^n; G)$ consisting of groups $\tilde{C}_k(CP^n; G)$. So $H_k^o(CP^n; G) \simeq \tilde{C}_k(CP^n; G)$, i.e.,

$$H_k^o(CP^n; G) \simeq \begin{cases} G & \text{when } k = 2i \leq 2n, \\ 0 & \text{otherwise.} \end{cases}$$

The homology of complex projective space CP^n can be calculated also in a different way. First, we define a smooth function f on the manifold CP^n all of whose critical points are nondegenerate, and then establish with its help the structure of a cell complex which is homotopy equivalent to CP^n, and calculate its homology groups.

We shall consider CP^n as the space of orbits of the group S^1 acting on S^{2n+1}. We define a function $\phi : C^{n+1} \to \mathbf{R}$ by putting $\phi(z_0, \dots, z_n) = \sum_{j=0}^{n} c_j |z_j|^2$, where the c_j are certain real numbers and, in addition, $c_j < c_{j+1}$. Let

$$(z_0, \dots, z_n) \in S^{2n+1} \subset C^{n+1}, \quad \text{i.e.,} \quad \sum_{j=0}^{n} |z_j|^2 = 1.$$

It is easy to see that for any complex number λ such that $|\lambda| = 1$, the equality $\phi(z_0, \dots, z_n) = \phi(\lambda z_0, \dots, \lambda z_n)$. Thus, ϕ defines a function on CP^n. Denote it by $f : CP^n \to \mathbf{R}$.

Now, construct on CP^n the following local coordinate system. Let U_j be a set of equivalence classes of points $(z_0, \dots, z_n) \in S^{2n+1}$ such that $z_j \neq 0$. Put $|z_j| \cdot \frac{z_k}{z_j} = x_{jk} + iy_{jk}$. The functions $x_{jk}(z_0, \dots, z_n)$, $y_{jk}(z_0, \dots, z_n)$, $k = 0, \dots, j-1, j+1, \dots, n$, define a diffeomorphism of the set U_j to the open unit ball in $\mathbf{R}^{2n}$.

Exercise $2°$. Verify that the sets U_j and the mappings given by the functions

$$x_{jk}, y_{jk}, \quad k = 0, \dots, j-1, j+1, \dots, n, \quad j = 0, 1, \dots, n,$$

form an atlas for the smooth manifold CP^n.

Since $|z_k|^2 = x_{jk}^2 + y_{jk}^2$ and $|z_j|^2 = 1 - \sum_{k \neq j}(x_{jk}^2 + y_{jk}^2)$, the function f can be represented in terms of the local coordinates on U_j in the form

$$f(\dots, x_{jk}, y_{jk}, \dots) = c_j + \sum_{k \neq j}(c_k - c_j)(x_{jk}^2 + y_{jk}^2).$$

The point $x_{jk} = y_{jk} = 0$, $k = 0, 1, \dots, j-1, j+1, \dots, n$, is the only critical point of the function f in U_j. This critical point is nondegenerate, and its index is equal to double the number of negative differences $c_k - c_j$, i.e., to double the number of c_k which are less than c_j. Therefore, the index of the critical point in U_0 is zero, that of the critical point in U_1 is equal to two, etc. In general, the index of the critical point in U_j equals to $2j$.

Thus, the function f has n critical points whose indices are equal to $2j$, $0 \leq j \leq n$. Therefore, (see Ch. 4, § 12) the space CP^n has the homotopy type of a cell complex K which consists of cells of even dimensions $2j$, $0 \leq j \leq n$, one for each dimension. For such a complex K, we have

$$\tilde{C}_k(K; G) \simeq \begin{cases} G & \text{when } k = 2j \leq 2n, \\ 0 & \text{otherwise.} \end{cases}$$

Since one of the groups $\tilde{C}_k(K;G)$, $\tilde{C}_{k-1}(K;G)$ is trivial, the differential can only be trivial in the complex $\tilde{C}_*(K;G)$ consisting of groups $\tilde{C}_k(K;G)$. We obtain isomorphisms

$$H_k^s(K;G) \simeq \tilde{C}_k(K;G).$$

Taking into account that the homology groups of homotopy equivalent spaces coincide, we get the final result

$$H_k^s(CP^n;G) \simeq \begin{cases} G & \text{when } k = 2j \leq 2n, \\ 0 & \text{otherwise.} \end{cases}$$

Now, consider real projective space RP^n. Its homology groups can also be calculated by representing RP^n in the form of a cell complex.

A point from RP^n is defined as a pair $\{x, -x\}$, where $x = (x_1, \ldots, x_{n+1}) \in S^n$, i.e., $\sum_{i=1}^{n+1} x_i^2 = 1$. We define the cell τ^k, where $0 \leq k \leq n$, by a characteristic mapping $g^k : \bar{D}^k \to RP^n$ which associates the point $(\xi_1, \ldots, \xi_k) \in \left\{ \xi \in R^k : \sum_{i=1}^{k} \xi_i \leq 1 \right\} = \bar{D}^k$ with the point from RP^n of the form $\{x, -x\}$, where

$$x = (\xi_1, \ldots, \xi_k, \sqrt{1 - \xi_1^2 - \ldots - \xi_k^2}, 0, \ldots, 0).$$

When $k = 0$, we have $x = (1, 0, \ldots, 0)$.

Thus, the space RP^n is represented in the form of a cell complex which consists of cells of dimensions from 0 to n, one for each dimension. Note that in the cell complex obtained, the k–dimensional skeleton is the space RP^k, where $0 \leq k \leq n$.

First, we calculate $H_*^s(RP^n; Z_2)$. In the corresponding chain complex,

$$\tilde{C}_k(RP^n; Z_2) \simeq Z_2, \quad k = 0, 1, \ldots, n.$$

In order to calculate the boundary homomorphisms, we consider the incidence coefficients $[\tau^k, \tau^{k-1}]$, i.e., the degrees of the mappings $\phi_k : S^{k-1} \to S^{k-1}$ arising as the compositions

$$S^{k-1} \to RP^{k-1} \to RP^{k-1}/RP^{k-2} \to S^{k-1},$$

where $2 \leq k \leq n$.

Each of these mappings ϕ_k can be represented differently in the form of a composition $\phi_k = \beta_k \alpha_k$. The mapping ϕ_k acts as follows: first, in the sphere

S^{k-1}, the equator S^{k-2} is contracted to the point, then, in the space obtained (a wedge of two spheres), the two spheres are glued together in such a way that each point is glued to the point which "earlier" (in the initial sphere) was centrally symmetric to the given one. Thus, α_k maps S^{k-1} to the wedge of two $(k-1)$-dimensional spheres and, moreover, the generator γ of the group $H^s_{k-1}(S^{k-1}; \mathbf{Z})$ becomes the sum of the two generators $\gamma_1 + \gamma_2$ of the $(k-1)$-dimensional homology group of that wedge. The mapping β_k glues two spheres together; at the homology level, the generators γ_1 and γ_2 become $\pm\gamma$, because on each of the spheres of the wedge, β_k is a homeomorphism. Thus, the homomorphism $(\phi_k)_{*k-1}$ can convert the generator γ either in $2\cdot\gamma = \gamma+\gamma$, or in $0\cdot\gamma = \gamma-\gamma$. In any case, $\deg\phi_k \equiv 0 \,(\mathrm{mod}\,2)$, i.e., $[\tau^k, \tau^{k-1}] \equiv 0 \,(\mathrm{mod}\,0)$ for $2 \le k \le n$. So the boundary homomorphism $\tilde{\partial}_k$ in $C_*(\mathbf{R}P^n; \mathbf{Z}_2)$ is trivial for $0 \le k \le n$. It is not difficult to see that when $k = 1$, $[\tau^1, \tau^0] = 0$. Consequently, the boundary homomorphism $\tilde{\partial}_k$ is trivial for all $k = 0, 1, \ldots, n$. Therefore, the homology groups $H^s_k(\mathbf{R}P^n; \mathbf{Z}_2)$ are isomorphic to the chain groups $C_k(\mathbf{R}P^n; \mathbf{Z}_2)$, i.e.,

$$H^s_k(\mathbf{R}P^n; \mathbf{Z}_2) \simeq \begin{cases} \mathbf{Z}_2 & \text{when } 0 \le k \le n, \\ 0 & \text{when } k > n. \end{cases}$$

We pass to the computation of the integral homology groups $H^s_*(\mathbf{R}P^n; \mathbf{Z})$. Obviously, $\tilde{C}_k(\mathbf{R}P^n; \mathbf{Z}) \simeq \mathbf{Z}$, for $0 \le k \le n$, and $\tilde{C}_k(\mathbf{R}P^n; \mathbf{Z}) = 0$ for $k > n$. To calculate $\tilde{\partial}_k$, we again consider the mappings $\phi_k = \beta_k \cdot \alpha_k$. It can be shown (using the result of exercise $5°$, § 6) that the homomorphism $(\beta_k)_{*k-1}$ for even $k - 1$ (i.e., odd k) transforms the generator γ_1 into γ, and γ_2 into $(-\gamma)$, and for odd $k - 1$ (even k), both generators γ_1 and γ_2 convert into γ. Therefore $(\phi_k)_{*k-1}\gamma = 0 \cdot \gamma$ for odd k, and $(\phi_k)_{*k-1}\gamma = 2 \cdot \gamma$, for even k. Consequently, $[\tau^k, \tau^{k-1}] = 0$, when k is odd, and $[\tau^k, \tau^{k-1}] = 2$, when k is even. Thus, the boundary homomorphism $\tilde{\partial}_k$ is trivial when k is odd and is multiplication by 2 for an even $k \le n$. Consequently, $H_0(\mathbf{R}P^n; \mathbf{Z}) \simeq \mathbf{Z}$, $H_1(\mathbf{R}P^n; \mathbf{Z}) \simeq \mathbf{Z}_2$, $H_2(\mathbf{R}P^n; \mathbf{Z}) \simeq 0$, $H_3(\mathbf{R}P^n; \mathbf{Z}) \simeq \mathbf{Z}_2$, etc. (for $n > 3$).

We obtain the following result: for $n = 2m$

$$H^s_k(\mathbf{R}P^n; \mathbf{Z}) \simeq \begin{cases} \mathbf{Z} & \text{for } k = 0, \\ \mathbf{Z}_2 & \text{for } k = 2p+1, \ 1 \le k < n, \\ 0 & \text{otherwise;} \end{cases}$$

for $n = 2m + 1$,

$$H_k^\circ(RP^n; Z) \simeq \begin{cases} Z & \text{for } k = 0, n, \\ Z_2 & \text{for } k = 2p + 1, \ 1 \leq k < n, \\ 0 & \text{otherwise;} \end{cases}$$

Solve the following exercise using the description of the homotopy type of a manifold (Ch. 4, § 12, subsection 4).

Exercise 3°. Show that if M^n is a compact, smooth manifold of dimension n, then $H_k^\circ(M^n; G) = 0$ when $k > n$.

§ 8. Euler characteristic and Lefschetz number

Quite important in applications is the question when a continuous mapping $f : X \to X$ of a topological space X into itself has a fixed point, i.e., when there exists a point $x \in X$ such that $f(x) = x$. Sufficient conditions for the existence of fixed points can be given in terms of homology groups and their homomorphisms. The present section deals with these questions. Everywhere below, we only consider topological spaces which are compact polyhedra.

1. The Lefschetz number of a simplicial mapping. Below, we shall assume the group of coefficients G to be a field. Consider a simplicial mapping $f : |K| \to |K|$, where, as in § 3, K is a finite simplicial complex. The induced homomorphism of the simplicial homology groups

$$f_{*p} : H_p(K; G) \to H_p(K; G)$$

is an endomorphism of the vector space $H_p(K; G)$. Choose a basis in $H_p(K; G)$ to obtain a matrix for this endomorphism; its trace, $\mathrm{Sp}(f_{*p})$, does not depend on the choice of this basis.

Definition 1. The *Lefschetz number of a simplicial mapping* $f : |K| \to |K|$ of a compact polyhedron $|K|$ into itself is the quantity

(1) $$\Lambda_f = \sum_{p=0}^{\infty} (-1)^p \mathrm{Sp}(f_{*p}).$$

Because the simplicial complex K is finite, the sum (1) is the sum of a finite number of summands (the finiteness of K is also required for $H_p(K; G)$ to be finite–dimensional vector spaces and for the traces $\mathrm{Sp}(f_{*p})$ to be well defined).

By $\mathrm{Sp}(\widehat{f}_p)$, we denote the trace of a matrix of the endomorphism $\widehat{f}_p$: $C_p(K; G) \to C_p(K; G)$ of the vector space $C_p(K; G)$.

Theorem 1. *For a simplicial mapping f,*

$$(2) \qquad \Lambda_f = \sum_{p=0}^{\infty} (-1)^p \mathrm{Sp}(\widehat{f}_p).$$

Theorem 1 states that the alternating sum of the traces of an endomorphism of a chain complex is equal to the alternating sum of the traces of the induced endomorphisms of the homology groups.

The proof of theorem 1 requires the following two lemmas. Both are easy exercises from linear algebra.

Lemma 1. *Let $A : E \to E$ be an endomorphism of a vector space E, E_0 a vector subspace of the space E, and $AE_0 \subset E_0$. Then A defines the endomorphism $\widetilde{A} : E/E_0 \to E/E_0$ and*

$$(3) \qquad \mathrm{Sp}(A) = \mathrm{Sp}(A|_{E_0}) + \mathrm{Sp}(\widetilde{A}).$$

Lemma 2. *Let $\Delta : E \to E$ be an isomorphism of vector spaces; let $A : E \to E$, $B : F \to F$ be operators such that the diagram*

$$
\begin{array}{ccc}
E & \xrightarrow{\Delta} & F \\
{\scriptstyle A}\downarrow & & \downarrow{\scriptstyle B} \\
E & \xrightarrow{\Delta} & F
\end{array}
$$

is commutative, i.e., $B\Delta = \Delta A$; then

$$(4) \qquad \mathrm{Sp}(A) = \mathrm{Sp}(B).$$

Proof of theorem 1. Since $\widehat{f}_* : C_*(K; G) \to C_*(K; G)$ is a homomorphism of chain complexes, we have

$$\widehat{f}_p(\operatorname{Ker} \partial_p) \subset \operatorname{Ker} \partial_p \quad \text{and} \quad \widehat{f}_p(\operatorname{Im} \partial_{p+1}) \subset \operatorname{Im} \partial_{p+1}.$$

Let us introduce the following notations:

$$\operatorname{Ker} \partial_p = Z_p, \quad \operatorname{Im} \partial_{p+1} = B_p, \quad C_p(K; G)/\operatorname{Ker} \partial_p = T_p, \quad Z_p/B_p = H_p.$$

By lemma 1, we have

$$\operatorname{Sp}(\widehat{f}_p) = \operatorname{Sp}(\widehat{f}_p|_{z_p}) + \operatorname{Sp}(\widetilde{\widehat{f}}_p|_{T_p}) =$$

$$(5) \qquad = \operatorname{Sp}(\widehat{f}_p|_{B_p}) + \operatorname{Sp}(\widetilde{\widehat{f}}_p|_{H_p}) + \operatorname{Sp}(\widetilde{\widehat{f}}_p|_{T_p}).$$

But the differential ∂_p induces a canonical isomorphism $\widetilde{\partial}_p : T_p \to B_{p-1}$, and, moreover, the diagram

$$
\begin{array}{ccc}
T_p & \xrightarrow{\widetilde{\partial}_p} & B_{p-1} \\
\widetilde{f}_p \downarrow & & \widetilde{f}_{p-1} \downarrow \\
T_p & \xrightarrow{\widetilde{\partial}_p} & B_{p-1}
\end{array}
$$

is commutative.
By lemma 2, we obtain

$$(6) \qquad \operatorname{Sp}(\widetilde{\widehat{f}}_p|_{T_p}) = \operatorname{Sp}(\widehat{f}_{p-1}|_{B_{p-1}}).$$

In view of the fact that $C_0(K; G) = \operatorname{Ker} \partial_0$, we have

$$(7) \qquad \operatorname{Sp}(\widetilde{\widehat{f}}_0|_{T_0}) = \operatorname{Sp}(\widetilde{\widehat{f}}|_0) = 0.$$

It is clear that by definition the homomorphism $\widetilde{\widehat{f}} : H_p \to H_p$ coincides with the homomorphism $f_{*p} : H_p(K; G) \to H_p(K; G)$. Thus, from equalities (5), (6), (7) we obtain

$$\operatorname{Sp}(\widehat{f}_p) = \operatorname{Sp}(\widehat{f}_p|_{B_p}) + \operatorname{Sp}(f_{*p}) + \operatorname{Sp}(\widehat{f}_{p-1}|_{B_{p-1}}),$$

whence

$$\sum_{p=0}^{\infty}(-1)^p \operatorname{Sp}(\widehat{f}_p) =$$

$$= \sum_{p=0}^{\infty}(-1)^p[\operatorname{Sp}(\widehat{f}_p|_{B_p}) + \operatorname{Sp}(f_{*p}) + +\operatorname{Sp}(\widehat{f}_{p-1}|_{B_{p-1}})] =$$

$$= \sum_{p=0}^{\infty}(-1)^p \operatorname{Sp}(f_{*p});$$

thus,

(8)
$$\sum_{p=0}^{\infty}(-1)^p \operatorname{Sp}(\widehat{f}_p) = \sum_{p=0}^{\infty}(-1)^p \operatorname{Sp}(f_{*p}).$$

$\square$

Corollary. *The Lefschetz number Λ_f over the field of coefficients of characteristic zero and, in particular, over the fields* $\mathbf{Q}$, $\mathbf{R}$ *or* $\mathbf{C}$ *is an integer.*

Indeed, by considering in $C_p(K; G)$ a basis consisting of oriented simplices, we obtain that $\operatorname{Sp}(\widehat{f}_p)$ is an integer, and, consequently, Λ_f is an integer because of (4). $\square$

The folowing theorem throws light on the number Λ_f.

Theorem 2. *Let $f : |K| \to |K|$ be a simplicial mapping and $\Lambda_f \neq 0$. Then there exists a fixed point of the mapping f, i.e., a point $x \in |K|$ such that $f(x) = x$.*

Proof. By (2), it follows from $\Lambda_f \neq 0$ that $\sum_{p=0}^{\infty}(-1)^p \operatorname{Sp}(\widehat{f}_p) \neq 0$, therefore, there is a p such that $\operatorname{Sp}(\widehat{f}_p) \neq 0$. In a basis consisting of oriented simplices, the matrix of the endomorphism $\widehat{f}_p$ consists of elements equal to 0, +1, −1 (because f is simplicial). Since $\operatorname{Sp}(\widehat{f}_p) \neq 0$, there exists $\tau_i^p \in K$ such that $f_p[\tau_i^p] = \pm[\tau_i^p]$. Consequently, $f|_{\tau_i^p}$ is a homeomorphism of τ_i^p onto itself which is linear in barycentric coordinates; hence, the barycentre of τ_i^p is a fixed point of the mapping f. $\square$

We shall discuss an important supplement to theorem 2. Let $f : |K^{(r)}| \to |K|$ be a simplicial mapping; evidently, it is a continuous mapping $f : |K| \to |K|$, but not necessarily simplicial. We introduce the composition $\Xi_*^{(r)} \widehat{f}_*$ (see § 3) of chain homomorphisms:

$$C_*(K^{(r)}; G) \xrightarrow{\widehat{f}_*} C_*(K; G) \xrightarrow{\Xi_*^{(r)}} C_*(K^{(r)}; G).$$

The chain homomorphism $\Xi_*^{(r)} \widehat{f}_*$ induces homomorphisms of homology groups $(\Xi_*^{(r)} \widehat{f}_*)_*: H_*(K^{(r)}; G) \to H_*(K^{(r)}; G)$. By definition, we set

(9)
$$\Lambda_f = \sum_{p=0}^{\infty} (-1)^p \operatorname{Sp} [\Xi_*^{(r)} \widehat{f}_*)_{*p}].$$

Exercise 1°. Prove that

(10)
$$\sum_{p=0}^{\infty} (-1)^p \operatorname{Sp} (\Xi_p^{(r)} \widehat{f}_p) = \sum_{p=0}^{\infty} (-1)^p \operatorname{Sp} [\Xi_*^{(r)} \widehat{f}_*)_{*p}]$$

and, that if $\Lambda_f \neq 0$, then there exist simplices $\tau_p \in K^{(r)}$ and $\mu^p \in K$ such that $\tau^p \subset \mu^p$ and $f(\tau^p) = \mu^p$.

Hint. Consider the chain homomorphism $\Xi_*^{(r)} \widehat{f}_*$ and use the arguments of theorems 1 and 2.

Consider now the example when $f = 1_K : |K| \to |K|$ is the identity mapping of the polyhedron $|K|$. We denote the dimension of the vector space $H_p(|K|; G)^{*)}$ by β_p, and the number of p–dimensional simplices in the simplicial complex K by d_p. Obviously,

$$\operatorname{Sp}((1_K)_{*p}) = \beta_p, \quad \operatorname{Sp}((1_K)_p) = \operatorname{Sp}(1_{C_p(K;G)}) = d_p.$$

Formula (8) acquires the form

(11)
$$\sum_{p=0}^{\infty} (-1)^p d_p = \sum_{p=0}^{\infty} (-1)^p \beta_p.$$

---

*)If G is a field of characteristic zero, then β_p coincides with the p–dimensional Betti number ρ_p of the space $|K|$ in the sense of the definition in § 7.

Formula (11) establishes a relation between the geometric and homological characteristics of a polyhedron.

Definition 2. The *Euler characteristic of a compact polyhedron* $|K|$, is the value

$$(12) \qquad \chi(|K|) \overset{\text{def}}{=} \sum_{p=0}^{\infty} (-1)^p d_p = \sum_{p=0}^{\infty} (-1)^p \beta_p.$$

It is clear that $\chi(|K|) = \Lambda_{1_{|K|}}$.

Exercise 2°. Show that the equality $\chi(S^n) = 1 + (-1)^n$ holds.

2. The Lefschetz number of a continuous mapping

In the previous argument, we only considered simplicial mappings. But the construction of a Lefschetz number and the statement of theorem 2 can be generalized to the case of arbitrary continuous mappings. For that, we shall use the uniqueness theorem of homology theory (see § 5). We shall use the approximation of a continuous mapping of a polyhedron by a simplicial mapping.

Theorem 3 (on simplicial approximation). *Let* $X = |L|$ *be a compact polyhedron,* $f : X \to X$ *a continuous mapping. Then for any* $\epsilon > 0$ *there are a triangulation* K *of the polyhedron* X, *and a number* r *such that for the* r-*th barycentric subdivision* $K^{(r)}$ *there is a simplicial mapping* $f_\epsilon : |K^{(r)}| \to |K|$ *such that for any point* $x \in X$, *the inequality* $\rho(f(x), f_\epsilon(x)) < \epsilon$ *holds.*

Proof. We select on the polyhedron K a triangulation K such that the mesh of the triangulation K is less than ϵ. The star, $\mathrm{St}\,a$, of a vertex $a \in K$ is the interior of the union of all simplices whose vertices involve a. It is obvious that the stars of all vertices of K form a covering X; a covering is also formed by the preimages $\{f^{-1}(\mathrm{St}\,a^p)\}_{a^p \in K}$.

Since X is compact, by the Lebesgue number lemma (see Ch. 2, § 13, theorem 13), there exists a number $\nu > 0$ such that any set of diameter $\partial < \nu$ is contained in one of the sets $f^{-1}(\mathrm{St}\,a^p)$. We choose r such that the mesh of $K^{(r)}$ is smaller than $\nu/2$. Then the mapping f takes any star $\mathrm{St}\,b^q$, $b^q \in K^{(r)}$, into a certain star $\mathrm{St}\,a^p$, $a^p \in K$. We define a simplicial mapping

$f_\epsilon : |K^{(r)}| \to |K|$ by the equalities

(13) $$f_\epsilon(b^q) = a^p.$$

Exercise 3°. Verify that formula (13) really determines a simplicial mapping.

Now, we calculate $\rho(f(x), f_\epsilon(x))$ for $x \in X$. If x is a vertex of $K^{(r)}$, then $f(\mathrm{St}\,x) \subset \mathrm{St}\,a^p$, $a^p \in K$, and, in particular, $f(x) \in \mathrm{St}\,a^p$, therefore

$$\rho(f(x), a^p) = \rho(f(x), f_\epsilon(x)) < \epsilon.$$

If $x \in \mathrm{Int}\,(b^0, \ldots, b^q)$, where $(b^0, \ldots, b^q) \in K^{(r)}$, then $x \in \bigcap_{i=0}^q \mathrm{St}\,b^i$. We have $f(x) = \bigcap_{a^i = f_\epsilon(b^i)} \mathrm{St}\,a^i$, so that $f(x)$ lies in the simplex determined by the vertices $a^i = f_\epsilon(b^i)$. Because of the fact that f_ϵ is a simplicial mapping, $f_\epsilon(x)$ falls into the same simplex from K. Thus, in this case also $\rho(f(x), f_\epsilon(x)) < \epsilon$.
□

Exercise 4°. Show that the mapping f_ϵ is homotopic to the mapping f.

Exercise 5°. Show that for a compact polyhedron X, there exists a positive number $\alpha = \alpha(x)$ such that if $\rho(f(x), g(x)) < \alpha$ for each $x \in X$ ($f, g : X \to X$ are continuous mappings), then the mappings f and g are homotopic.

By the uniqueness theorem of homology theory, there is an isomorphism $H_*(K; G) \simeq H_*^s(X; G)$ for the compact polyhedron $X = |K|$ and, consequently, $\dim_G \bigoplus_p H_p^s(X, G) < \infty$. Therefore, the following definition makes sense.

Definition 3. The *Lefschetz number of a continuous mapping* $f : X \to X$ of a compact polyhedron X into itself is the quantity

(14) $$\Lambda_f = \sum_{p=0}^{\infty} (-1)^p \mathrm{Sp}\,(f_{*p}^s),$$

where $f_{*p}^s : H_p^s(X; G) \to H_p^s(X; G)$.

By the uniqueness theorem of homology theory, for a simplicial mapping
$f : |K^{(r)}| \to |K|$, $r \geq 1$, the equality

(15)
$$\sum_{p=0}^{\infty}(-1)^p \mathrm{Sp}\,(f^s_{*p}) = \sum_{p=0}^{\infty}(-1)^p \mathrm{Sp}\,[\Xi^r_* \hat{f}_*)_{*p}]$$

is valid. Thus, definitions 1 and 3 are consistent.

Obviously, for homotopic continuous mappings $f, g : X \to X$, we have
$\Lambda_f = \Lambda_g$. Therefore, the Lefschetz number of a simplicial approximation
$f_\epsilon : |K^{(r)}| \to |K|$, where K is a triangulation of X is equal to that of f. It is
possible to define the Lefschetz number of continuous mapping as that of its
simplicial approximation without using singular homology theory.

The following theorem is quite useful for various applications. To prove it,
we use the uniqueness theorem of homology theory.

Theorem 4 (Lefschetz). *Let* $f : X \to X$ *be a continouus mapping of a
compact polyhedron* $X = |L|$ *into itself and* $\Lambda_f \neq 0$. *Then there exists a fixed
point of the mapping* f, *i.e., a point* $x \in X$ *such that* $f(x) = x$.

Proof. Assume that f has no fixed points. Then there is $\gamma > 0$ such that
$\rho(f(x), x) \geq \gamma$ for all $x \in X$.

Consider a triangulation K of mesh $\gamma/3$ of the polyhedron $X = |L|$ and a
simplicial approximation $f_{\gamma/3} : K^{(r)} \to K$ of the mapping f. For arbitrary
points x, y of any simplex $\tau^q \in K^{(r)}$, we have the inequalities $\rho(f_{\gamma/3}(x), y) \geq$
$\rho(f(x), x) - \rho(x, y) - \rho(f_{\gamma/3}(x), f(x)) \geq \gamma/3$. This means that an inclusion
$\tau^q \subset f_{\gamma/3}(\tau^q)$ is imposssible. On the other hand, because $\Lambda_{f_{\gamma/3}} = \Lambda_f \neq 0$,
there is $\tau^q \in K^{(r)}$ for which such an inclusion exists (see exercises 1°, 4°).
The contradiction obtained completes the proof.   □

Exercise 6°. Extend definition 2 and theorem 4 to a compact polyhedron X
homeomorphic to $|L|$.

Exercise 7°. Verify that under the conditions of the Brouwer fixed point
theorem (see Ch. 3, § 4), $\Lambda_f = 1$.

Hint. Construct a homotopy to a constant mapping.

3. The Euler characteristic of a manifold and singular points of a vector field. Now, we dwell on an application of the results obtained to manifold theory.

Theorem 5. *Let M^n be both a smooth, compact manifold and a polyhedron.*[*] *Let $\chi(M^n) \neq 0$. Then for every smooth vector field X on M^n, there exists a point $x^0 \in M^n$ such that $X(x^0) = 0$.*

In other words, there is no vector field without zeros on a manifold with a nonzero Euler characteristic.

Proof. As was mentioned in Ch. 4, § 8, for a vector field X, there exists a one–parameter family of diffeomorphisms $U(x,t)$ such that $(U(x,0) \equiv x$, and the field X is its infinitesimal generator. The orbit $\bigcup_t U(x,t)$ of a point x is called the integral curve of the field X through the point x. It is easy to see that the family of diffeomorphisms $U(x,t)$, $0 \le t \le t_0$, forms a homotopy between the diffeomorphisms

$$U_0 = 1_{M^n} \quad U_{t_0} : M^n \to M^n, \quad \text{where} \quad U_{t_0}(x) \overset{\text{def}}{=} U(x,t_0).$$

Consequently, $\Lambda_{1_{M^n}} = \Lambda_{U_{t_0}}$, but $\Lambda_{1_{M^n}} = \chi(M^n)$; thus, for any t_0, we obtain $\Lambda_{U_{t_0}} = \chi(M^n) \neq 0$. Therefore, (see theorem 4) the diffeomorphism U_{t_0} has a fixed point (for each t_0).

Now, assume that the field X vanishes nowhere on M^n. Then, by the compactness of M^n, there exist positive α and β such that for any $x \in M^n$ in the Riemannian metric, the inequality $\alpha \le \langle X(x), X(x) \rangle \le \beta$ is fulfilled. Hence, each point $x \in M^n$ will necessarily be shifted by the diffeomorphism U_t along the integral curve of the point x for a sufficiently small $t > 0$; this can be verified by considering the integral curve in the chart of the point x. The latter contradicts the existence of a fixed point for the diffeomorphism U_t. □

Corollary. *If n is even, then, there is no vector field without zeros on the sphere S^n.*[**]

---

[*]Note that all smooth, compact manifolds are polyhedra.
[**]This result is colloquially known as the hairy ball theorem

Lemma 3. *There exists a smooth vector field on a compact, smooth manifold such that the sum of the indices of the singular points of this field is equal to the Euler characteristic of the manifold.*

Proof. Let M^n be a compact, smooth manifold, $f : M^n \to R$ a Morse function (a smooth function all of whose critical points are nondegenerate). The space M^n has the homotopy type of a cell complex K, of which the number of cells of dimension λ equals the number $m(\lambda)$ of critical points x_i^λ of index λ of the function f (see Ch. 4, § 11). The Euler characteristic $\chi(K)$ of the space K is

$$\sum_{\lambda=0}^{\infty}(-1)^\lambda \dim_G \widetilde{C}_\lambda(K;G) = \sum_{\lambda=0}^{\infty}(-1)^\lambda \dim_G H_\lambda^*(K;G)$$

(compare with definition 2 and theorem 1 of this section). Thus,

$$(16) \qquad\qquad \chi(M^n) = \chi(K) = \sum_{\lambda=0}^{\infty}(-1)^\lambda m(\lambda).$$

On the other hand, by exercise 10°, § 6, the index of the singular point x_i^λ of the gradient field is equal to $(-1)^\lambda$. Therefore, $\sum_{\lambda=0}^{\infty}(-1)^\lambda m(\lambda)$ is the sum of the indices of singular points of the gradient field of the function f. $\qquad\square$

Lemma 4. *The sum of the indices of singular points of a vector field with isolated singular points on a compact, smooth manifold, does not depend on the choice of the vector field.*

Proof. We give a sketch of the proof. Let M^n be a connected manifold imbedded in R^m, $m > n+1$; we select a sufficiently small "tubular" neighbourhood of the manifold M^n in R^m, i.e., a neighbourhood $U(M^n)$ such that it is a space of a locally trivial fibre bundle with base M^n and fibre homeomorphic to the disc D^{m-n}. Moreover, the projection mapping r of the fibre bundle is a smooth retraction, and the manifold M^n is a strong deformation retract of the space $U(M^n)$. The tubular neigbourhood of the manifold M^n can be intuitively imagined to consist of discs $D_z^{m-n}(x)$ over each point $x \in M^n$ lying in $(m-n)$–dimensional planes which are orthogonal to the tangent planes of the manifold M^n. The set $\overline{U(M^n)}$ is a compact polyhedron. It is not difficult to show that $H_{m-1}^*(\partial U(M^n);Z) \simeq Z$; a generator of this group is the cycle bounding $U(M^n)$. Therefore, any mapping $\phi : \partial U(M^n) \to S^{m-1}$ defines an element $\deg\phi \in Z$. Consider a field $\Phi : \overline{U(M^n)} \to R^m$ which does not vanish on $\partial U(M^n)$. Associate to the field Φ the normed mapping

$$\widetilde{\Phi} : \partial U(M^n) \to S^{m-1}, \quad \widetilde{\Phi}x = \Phi x / \|\Phi x\|.$$

The degree $\deg\widetilde{\Phi}$ of the mapping $\widetilde{\Phi}$ equals the sum of indices of singular points of the field Φ. Now, let v be a vector field on the manifold M^n. We define a field $w : U(M^n) \to R^m$ by the formula $w(x) = v(rx) + x - r(x)$. The sum of indices of singular points of the field w coincides with the sum of the indices of the singular points of the tangent field v (by means of the Sard theorem, the general case can be reduced to the investigation of smooth fields with nondegenerate singular points, and the result of exercise 7°, § 6, can be applied). The field w on $\partial U(M^n)$ is homotopic, without singular points, to the vector field $z(x)=x-r(x)$. Hence, for the normed mappings $\bar{w}$, $\bar{z}$, we obtain deg $\bar{w}$=deg $\bar{z}$, and, therefore, deg $\bar{w}$ does not depend on the field v. □

Lemmas 3 and 4 imply the following theorem.

Theorem 6. *The sum of the indices of singular points of a vector field with isolated singular points on a smooth, compact manifold is equal to the Euler characteristic of the manifold.*

Exercise 8°. Let M^n be a smooth, compact manifold, and $\beta_p(M^n) \overset{\text{def}}{=} \dim_G H_p^s(M^n; G) \neq 0$. Show that any Morse function on the manifold M^n has not less than $\beta_p(M^n)$ critical points of index p (Morse inequalities).

4. The Lefschetz number as a sum of indices of fixed points. A remark-able fact in algebraic topology is the Lefschetz–Hopf theorem connecting the Lefschetz number of a continuous mapping of a compact polyhedron with the characteristics (indices) of fixed points of this mapping. We give several important definitions.

Definition 4. A simplicial complex K is said to be *dimensionally uniform* if there exists a number n such that any simplex from K is a face (possibly, not a proper face) of a certain n–dimensional simplex from K; the polyhedron $|K|$ of the dimensionally uniform simplicial complex K is called a dimensionally uniform polyhedron, and the number n the dimension of the polyhedron $|K|$.

Exercise 9°. Show that if a polyhedron that is at the same time a manifold (with or without boundary) then it is a dimensionally uniform polyhedron.

Exercise 10°. Give an example of a dimensionally uniform polyhedron which is not a manifold.

Let $f : X \to X$ be a continuous mapping of a dimensionally uniform polyhedron X (of dimension n), $X \subset \mathbf{R}^M$, into itself ($n \leq M$).

Definition 5. A point $x_0 \in X$ is called a *regular fixed point of a mapping f* if 1) $f(x_0) = x_0$; 2) there exists a neighbourhood $U(x_0)$ of the point x_0 which is homeomorphic to an n–dimensional disc D^n and has no other fixed points (i.e., points such that $f(x) = x$); 3) the homeomorphism mentioned above

$h : U(x_0) \to D^n$ is the restriction to $U(x_0)$ of a mapping $H : \mathbf{R}^M \to \mathbf{R}^n$ which is a composition of two mappings: the projection of $\mathbf{R}^M$ on a certain n–dimensional subspace and an affine mapping that shifts the projection of the point x_0 to the origin of coordinates and expands the image of the neighourhood $U(x_0)$ to D^n.

If, for instance, $x_0 \in \tau^n \setminus \partial \tau^n \subset X$, then as H we may take a composition of the following three mappings: the transformation which transfers the carrier n–plane of the simplex τ^n to a corresponding n–dimensional subspace, the shift of an image of the point x_0 to the origin of the coordinates, and a large enough expansion such that $H\tau^n \supset D^n$. Then $U(x_0)$ can be chosen as $H^{-1}(D^n) \cap \tau^n$.

Exercise 11°. Let X be both a polyhedron (curvilinear) and a smooth sub-manifold in $\mathbf{R}^M$ (possibly with boundary); let $f : X \to X$ be a continuous mapping with a finite number of fixed points which do not belong to the boundary of the manifold X. Show that all fixed points are regular isolated ones.

Hint. Use the projection on a tangent space.

Exercise 12°. Show that if f has only regular isolated, fixed points, and the polyhedron X is compact, then the number of fixed points is finite.

Obviously, if X is both a polyhedron and a manifold with boundary, then the regular isolated points of the mapping f cannot belong to the boundary.

Let x_0 be a regular isolated, fixed point of a mapping f, let $h : U(x_0) \to D^n$ be a homeomorphism as indicated in definition 5. Consider a neighbourhood $V(x_0)$ of the point x_0 homeomorphic to D^n and so small that $\overline{V(x_0)} \subset U(x_0)$ and $f\overline{V(x_0)} \subset U(x_0)$. In this case, one can consider the vector field $\Phi(x) = 1_{\mathbf{R}^n} - hfh^{-1}$ defined on $h\overline{V(x_0)})$.

Definition 6. The *index, $\mathrm{ind}(f, x_0)$, of a regular isolated, fixed point x_0 of a mapping f* is the index $\mathrm{ind}(0, \Phi)$ of the isolated singular point $0 \in \mathbf{R}^n$ of the vector field $\Phi(x) = 1_{\mathbf{R}^n} - hfh^{-1}$.

Exercise 13°. Show that the index, $\mathrm{ind}(f, x_0)$, does not depend on the choice of the homeomorphism h and the neighbourhood $V(x_0)$.

Hint. Select a sufficiently small neighbourhood $W(x_0)$ and use the fact that

$$1_{\mathbb{R}^n} - h_2 f h_2^{-1} = h_2 h_1^{-1}(1_{\mathbb{R}^n} - h_1 f h_1^{-1}) h_1 h_2^{-1}$$

on $h_2(\overline{W(x_0)})$, when $h_2 h_1^{-1}$ is a linear operator.

Now, we can formulate the main theorem.

Theorem 7 (Lefschetz–Hopf). *Let* $f : X \to X$ *be a continuous mapping of a compact dimensionally uniform polyhedron into itself with regular isolated, fixed points* $x_1, x_2, \dots, x_N$; *moreover, f has no other fixed points. Then the inequality*

$$(17) \qquad \Lambda_f = \sum_{i=1}^{N} \mathrm{ind}\,(f, x_i),$$

holds, where Λ_f is a Lefschetz number of the mapping f, and $\mathrm{ind}\,(f, x_i)$ is the index of the fixed point x_i of the mapping f.

The proof of this theorem requires additional definitions.

Definition 7. The set of those points of a usual (closed) simplex, $(a^0, a^1, \dots \dots, a^k)$, for which all barycentric coordinates are strictly positive, is called the *open k–dimensional simplex* with vertices $a^0, a^1, \dots a^k$.

In other words, an open simplex of dimension $k = 0$ coincides with a closed simplex, and that of dimension $k > 0$ is the interior (with respect to the carrier k–plane) of a closed simplex. It is easy to see that the closure of an open simplex is a closed simplex of the same dimension.

Definition 8. A collection of open simplices whose closures form a simplicial complex in terms of definition 4, § 3, Ch. 4, is called a *complete simplicial complex*.

Definition 9. An arbitrary subset (i.e., a collection of open simplices) of a certain complete simplicial complex, which is not a complete simplicial complex itself, is called an *incomplete simplicial complex*.

Thus, both in complete and in incomplete complexes, the way simplices fit together should be correct, but in an incomplete complex, contrary to a complete one, the simplex can be contained in a complex without some of its faces. Both for complete and incomplete complexes, the concept of a subcomplex (complete or incomplete) can be introduced as in definition 11, § 3, Ch. 5, and the concept of a polyhedron can be introduced as in definition 5, § 3, Ch. 5.

Note that the concept of subordination of one simplex to another, i.e., a situation such that one simplex is the face of another one, works as before on open simplices.

Everywhere below, the term "simplex" denotes an open simplex; the term "complex" is a complete or an incomplete simplicial complex (consisting of open simplices); the term "face of a simplex" is an open simplex which is the face (possibly not a proper face) of the given open simplex; and the term "*star of a simplex*" is the simplicial complex (usually incomplete) consisting of all those simplices whose faces include the given simplex. The term "triangulation" will denote a complex (always complete) whose polyhedron coincides with the given space (or is homeomorphic to it). It should be noted that the polyhedron of a complete complex of open simplices coincides with the polyhedron of the complex consisting of its closures.

Definition 10. Let K be a complex (complete or incomplete). Its *combinatorial closure* is the complete simplicial complex $\bar{K}$ consisting of all simplices from K and of all of its faces. Obviously, the combinatorial closure of a complex is a complete complex; if K is a complete complex, then $\bar{K} = K$.

Definition 11. Let K be a complete simplicial complex, L a complex, $f :$ $|K| \rightarrow |\bar{L}|$ a simplicial mapping transforming each simplex from K into a certain simplex from L. In this case, we shall call f a *simplicial mapping of simplicial complexes* and write $f : K \rightarrow L$.

Definition 12. Let L be an incomplete complex, M a complex, $\bar{g} : \bar{L} \rightarrow M$ a simplicial mapping. The restriction $g = \bar{g}|_L : L \rightarrow M$ of the mapping $\bar{g}$ to L is called a *simplicial mapping of the incomplete complex L to the complex M*.

Further on, we shall need the construction of the so-called *central subdivision of a complex with respect to the given subdivision of a subcomplex of it.*

Let K^n be a dimensionally uniform complex of dimension n, K^l its l–dimensional skeleton, i.e., the subcomplex consisting of all simplices from K^n of dimension not higher than l, $l \leq n$. Let $\widehat{K}^l$ be a certain subdivision K^l, i.e., a complex whose polyhedron coincides with the polyhedron $|K^l|$, and each simplex from K^l is the union of several (possibly one) simplices from $\widehat{K}^l$.

We inductively construct the central subdivision $\widehat{K}^n$. Let us describe the step of induction, that is the construction of the subdivision $\widehat{K}^{m+1}$ of its $(m + 1)$–dimensional skeleton K^{m+1} assuming that the subdivision $\widehat{K}^m$ has already been constructed.

Let the simplex τ^{m+1} belong to K^{m+1}; let the point b_0 be the barycentre of the simplex τ^{m+1}. Then the polyhedron $|\{\partial\tau^{m+1}\}|$ of the boundary, $\partial\tau^{m+1}$, can be represented as the union of certain simplices from $\widehat{K}^m$ and, possibly, also of certain faces of the simplex τ^{m+1} which are not included in K^m, and consequently, also not in $\widehat{K}^m$. A set of points of open intervals which connect the barycentre b_0 of the simplex τ^{m+1} with all the points of one simplex of the enumerated boundary simplices, is itself a simplex of dimension one higher than the dimension of this boundary simplex. The collection of all simplices in τ^{m+1} thus constructed with the barycentre b_0, represents a subdivision of the simplex τ^{m+1}. By taking these subdivisions for all simplices of dimension $m+1$ and by joining them to the already exisisting subdivision $\widehat{K}^m$, we obtain a new subdivision $\widehat{K}^{m+1}$ of the skeleton K^{m+1}, which is said to be central with respect to $\widehat{K}^m$. The induction step is complete.

Now, starting from $\widehat{K}^l$, we construct subdivisions $\widehat{K}^{l+1}$, $\widehat{K}^{l+2}, \ldots, \widehat{K}^n$, i.e. as a result, we construct a subdivision $\widehat{K}^n$ of the whole complex K^n. This subdivision $\widehat{K}^n$ is also said to be central with respect to the original subdivision $\widehat{K}^l$ of the skeleton K^l.

Note that the barycentric subdivision of a complete complex is central with respect to its 0–dimensional skeleton.

Definition 13. Let $\widehat{K}$ be a subdivision of a complex K, $\phi : \widehat{K} \to K$ a simplicial mapping. The simplex $\tau^r \in \widehat{K}$ is said to be *fixed with respect to* ϕ, if $\tau^r \subset \phi(\tau^r)$.

Since we consider open simplices, it follows from definition 13, that dimensions of the simplices τ^r and $\phi(\tau^r)$ coincide (are equal).

Definition 14. A collection of all simplices in K among whose faces is the simplex τ^r (including itself) is called the *star* $St(\tau^r)$ *of the simplex* τ^r in the complex K.

We proceed to the proof of the theorem. It splits into several stages some of which we present as lemmas; for these lemmas, the conditions of the theorem are assumed to be valid.

Lemma 5 (on special simplicial approximation). *For any $\epsilon > 0$, there exist a finite sequence $\{K_{(m)}\}$ of subdivisions of a given triangulation K, and a sequence $\{f_{(m)}\}$ of simplicial mappings ($m = 0, 1, 2, \dots , n$), which possess the following properties:*

1) *$K_{(m+1)}$ is a central subdivision of $K_{(m)}$;*
2) *$f_{(m)} : K_{(m)} \to K$ is a simplicial mapping possessing fixed simplices only of dimension not lower than m;*
3) *$\rho(f_{(m)}(x), f_{(m+1)}(x)) < \epsilon/(n + 1)$ for all $x \in X$ and $m = 0, 1, \dots , n$;*
4) *$\rho(f(x), f_{(0)}(x)) < \epsilon/(n + 1)$ for all $x \in X$;*
5) *the stars of any two different m–dimensional simplices fixed by $f_{(m)}$ do not intersect in $K_{(m)}$.*

In particular, each mapping $f_{(m)}$, $m = 0, 1, \dots , n$ is a simplicial ϵ–approximation of the mapping f, and if f_n possesses fixed simplices, then those are only of dimension n.

Proof. According to the simplicial approximation theorem, there exist an r–multiple barycentric subdivision $K_{(0)} = K^{(r)}$ of mesh $d < \epsilon/2(n + 1)$ of the complex K and a simplicial mapping $f_{(0)} : K_{(0)} \to K$ such that for all $x \in X$, $\rho(f(x), f_{(0)}(x)) < \epsilon/(n + 1)$. Obviously, property 5 is satisfied for $K_{(0)} = K^{(r)}$ when $r \geq 1$. Thus, when $r \geq 1$, the first of the mappings required by the lemma is constructed.

Now, we carry out induction on m. Assuming that the subdivision $K_{(m)}$ and the mapping $f_{(m)} : K_{(m)} \to K$ which possess the desired properties are already constructed, we construct a subdivision $K_{(m+1)}$ and a mapping $f_{(m+1)} : K_{(m+1)} \to K$. We shall subdivide only those simplices which are contained in the stars of simplices left fixed by $f_{(m)}$ and of dimension m. Let τ^m be such a fixed simplex, $St(\tau^m)$ its star in $K_{(m)}$. Take the central subdivision of τ^m with respect to its boundary (not subdivided). Then we perform the central subdivision of the complex $\text{St}(\tau^m)$ with respect to the subdivision of τ^m just performed (note that τ^m is the m–dimensional skeleton of the complex $\text{St}(\tau^m)$), and take the combinatorial closure of the subdivided star. Having performed the indicated subdivision for all stars of fixed simplices and not changing the remaining simplices from $K_{(m)}$, we obtain the desired subdivision $K_{(m+1)}$.

Now, we construct the simplicial mapping $f_{(m+1)} : K_{(m+1)} \to K$. Put $f_{(m+1)} = f_{(m)}$ outside of the subdivided stars of the fixed simplices. We construct a new simplicial mapping on every subdivided star of a simplex fixed with respect to $f_{(m)}$. Let the barycentre of the fixed m–dimensional simplex, τ^m, go to an arbitrary chosen vertex of the star $\text{St}(f_{(m)}(\tau^m))$ of the image $f_{(m)}(\tau^m)$, but not into a vertex of the image itself. And, vice versa, the barycentres of simplices of the subdivision of the star $\text{St}(\tau^m)$ which are not fixed, are to go to arbitrary chosen vertices of the image $f_{(m)}(\tau^m)$. By performing the described construction for all stars (non–intersecting!) of fixed m–dimensional simplices, we obtain the desired simplicial mapping $f_{(m+1)} : K_{(m+1)} \to K$.

Exercise 14°. Verify that the simplicial mapping $f_{(m+1)}$ as constructed above is well defined.

Now, we shall show that the mapping $f_{(m+1)}$ has no fixed simplices of dimensions lower than $m + 1$. Note first, that any simplex from $K_{(m)}$ fixed with respect to the simplicial mapping $f_{(m)} : K_{(m)} \to K$, lies in a simplex of the same dimension (its image) in K and also in a simplex of the same dimension of any of the previous subdivisions. Moreover, from the definition of a fixed simplex and from the construction of a central subdivision it follows that if a vertex b is the barycentre of the simplex τ^r of dimension r of the initial subdivision K or of one of the subsequent subdivision $K_{(0)}, \dots, K_{(m-1)}$, and this vertex b is the vertex of the fixed simplex τ^l of dimension l from K, then $l \geq r$. Indeed, otherwise the simplex τ^l would be a subset of an l–dimensional simplex from K; therefore, the barycentre b of the simplex τ^r, $r > l$, from K or one of the subdivisions $K_{(0)}, \dots, K_{(m-1)}$, cannot be a vertex of the fixed simplex τ^l. The required fact will be proved by contradiction. We assume that in $K_{(m+1)}$ there is a simplex τ^s of dimension s, $s < m + 1$ that is fixed with respect to $f_{(m+1)}$. It follows from the construction of $f_{(m+1)}$ that τ^s is contained in one of combinatorial closures of the subdivided, as described above, stars of type $\text{St}(\tau^m)$, where τ^m is a fixed simplex of the mapping $f_{(m)}$. Outside these stars $f_{(m+1)} = f_{(m)}$, and there $f_{(m)}$ has no fixed simplices. Further, simplices from the combinatorial closure of the subdivided $\mathrm{St}(\tau^m)$, which do not lie in τ^m, are either simplices from $K_{(m)}$ and therefore are not fixed, or they possess, as one of their vertices, a barycentre of a simplex of dimension higher than m, and, therefore, cannot be the fixed simplices of dimension not higher than m. Thus, we obtain that $\tau^s \subset \tau^m$, and, consequently, $s = m$. But this contradicts the construction of $f_{(m+1)}$, since one of the vertices of the simplex τ^s certainly is the barycentre of the subdivided fixed simplex τ^m, and this barycentre is transformed by the mapping $f_{(m+1)}$ outside the set of vertices of the simplex $f_{(m)}(\tau^m)$, which contains τ^m and hence τ^s. Thus, we arrived at a contradiction to the assumption $s < m+1$ and, by the same token, proved that the mapping $f_{(m+1)}$ has no fixed simplices of dimensions less than $m+1$.

Now, we show that the stars of two fixed simplices (with respect to $f_{(m+1)}$) τ_1^{m+1} and τ_2^{m+1} from $K_{(m+1)}$ do not intersect. Indeed, if these stars intersect, then both fixed simplices, in $K_{(m+1)}$, would be $(m+1)$–dimensional faces of the same simplex τ^s of higher dimension $s > m+1$. But $K_{(0)}$ is the r–multiple ($r > 0$) barycentric subdivision of the complex K, and $K_{(m+1)}$ is the result of subsequent central subdivisions, beginning with $K_{(0)}$; therefore, the simplex τ^s may contain no more than one face of each dimension, which itself is a fixed simplex; that is a face that lies in a face of an original s–dimensional simplex from K, of which a subdivision gives the simplex τ^s. Therefore, the stars of two fixed simplices do not intersect.

It remains to verify the inequality

(18) $$\rho(f_{(m)}(x), f_{(m+1)}(x)) < \frac{\epsilon}{n+1}.$$

Indeed, $f_{(m)}$ and $f_{(m+1)}$ differ only on the stars of simplices fixed with respect to $f_{(m)}$. Since the mesh of the triangulation $K_{(0)}$ is given: $d < \epsilon/2(n+1)$, the diameter of any star in $K_{(m)}$ does not exceed $\epsilon/(n+1)$. Further, since the images of the points of the fixed simplex τ^m under the mappings $f_{(m)}$ and $f_{(m+1)}$ do not fall outside the limits of the closure of the star of the image $f_{(m)}(\tau^m)$, inequality (18) is valid for any x from the star of the simplex τ^m, and consequently, for any $x \in X = |K|$, as required.

Thus, starting with the subdivision $K_{(m)}$ and with the simplicial mapping $f_{(m)} : K_{(m)} \to K$, we have constructed a subsequent subdivision $K_{(m+1)}$ and a simplicial mapping $f_{(m+1)} : K_{(m+1)} \to K$ which satisfy the requirements of the lemma. The induction step is complete.

By the same token, we obtain a finite sequence of simplicial mappings $f_{(m)} : K_{(m)} \to K$, $m = 0, 1, \ldots, n$, which only possess fixed simplices of dimensions not lower than m for $f_{(m)}$; consequently, the last of these mappings, $f_{(n)}$, has only n–dimensional fixed simplices (if any). We obtain from inequality (18) and by the choice of $f_{(0)}$, that $\rho(f(x), f_{(n)}(x)) \leq$ $\rho(f(x), f_{(0)}(x)) + \rho(f_{(0)}(x), f_1(x)) + \ldots + \rho(f_{(n-1)}(x), f_{(n)}(x)) < (n+1) \cdot \frac{\epsilon}{n+1}$, i.e.,

$$\rho(f(x), f_{(n)}(x)) < \epsilon$$

for all $x \in X = |K|$. Lemma 5 is proved. $\qquad\square$

The special approximation of $f_{(n)}$ constructed in lemma 5 is useful, as will be shown below, for calculating indices of fixed points.

Lemma 6. *Let $f_{(n)} : K_{(n)} \to K$ be a simplicial mapping, and let the simplex τ^n be fixed with respect to $f_{(n)}$, i.e., $\tau^n \subset f_{(n)}(\tau^n) = T^n \in K$, $\tau^n \in K_{(n)}$. Assume that on the boundary of τ^n, with respect to the carrier n–plane, there are no fixed points of the mapping $f_{(n)}$. Then in τ^n, there exists a unique regular isolated, fixed point x^* such that $f_{(n)}(x^*) = x^*$ and*

$$(19) \qquad ind\,(f_{(n)}, x^*) = \pm 1 = sign\,(\det\,(1_{\mathbf{R}^n} - h f_{(n)} h^{-1})_{h(x^*)})),$$

where h is the mapping described in definition 5, and the subscript $h(x^*)$ indicates the point in the neighbourhood of which the mapping is considered.

Proof. Since $f_{(n)}$ is an affine mapping preserving dimension, it is a homeomorphism on τ_n and, therefore, the composition $\bar{T}^n \xrightarrow{(f_{(n)})^{-1}} \bar{\tau}^n \subset \bar{T}^n$ is a mapping of a closed simplex into itself. Then by the Brouwer theorem, there exists at least one fixed point x^* of this mapping. This point is also a fixed point of the mapping $f_{(n)}$. We shall show that it is unique in τ^n. In fact, assume that there are two such points: $y_1 \neq y_2$, $y_1, y_2 \in \tau^n$. Then all the points of the straight line which passes through y_1, y_2, i.e., the points of the form $y = ty_1 + (1 - t)y_2$, $t \in \mathbf{R}$, are also fixed. But then there exist fixed points on the boundary τ^n (for a certain t). A contradiction. Hence, x^* is a regular isolated, fixed point and $(1_{\mathbf{R}^n} - h f_{(n)} h^{-1})_{h(x^*)}$ is a non–singular linear mapping. It remains to note that by the result of exercise 6°, § 6, Ch. 5, and the definition of index of a fixed point, we have

$$ind\,(f_{(n)}, x^*) = ind\,(h(x^*), (1_{\mathbf{R}^n} - h f_{(n)} h^{-1})_{h(x^*)}) =$$

$$= sign\,(\det\,(1_{\mathbf{R}^n} - h f h^{-1})_{h(x^*)}) = \pm 1.$$

The proof of lemma 6 is finished.   □

The simplicial mapping $f_{(n)}$ constructed in lemma 5 possesses a remarkable property: there are no fixed simplices τ_i^n in the subdivision $K_{(n)}$ that have fixed points on its boundary $\partial\tau^n$. Indeed, otherwise there would exist a fixed simplex of dimension less than n, contradicting the construction of $f_{(n)}$. Then, by lemma 6, the simplicial mapping $f_{(n)}$ has another even more remarkable property: on each of its fixed simplices τ_i^n, $i = 1, \ldots, N$, of dimension n, there exists a unique (in this simplex) regular isolated point x_i fixed respectively to $f_{(m)}$, and, moreover, its index is equal to

$$\mathrm{ind}\,(f_{(n)}, x_i) = \mathrm{sign}\,(\det\,(1_{\mathbf{R}^n} - h_i f_{(n)} h_i^{-1})_{h_i(x_i)}) = \pm1,$$

(20) $i = 1, \ldots, N.$

Now, we explain the connection between these indices and the Lefschetz number of the simplicial mapping $f_{(n)}$.

Lemma 7. *For the Lefschetz number $\Lambda_{f_{(n)}}$ of the special simplicial approximation $f_{(n)}$ constructed in lemma 5, the following formula holds*

$$(21) \qquad\qquad \Lambda_{f_{(n)}} = \sum_{i=1}^{N} \mathrm{ind}\,(f_{(n)}, x_i),$$

where $x_1, \ldots, x_N$ are the fixed points of the mapping $f_{(n)}$.

Proof. Taking into account the definition of the Lefschetz number and equality (15), equality (21) transforms into the equality

$$(22) \qquad \sum_{p=0}^{n}(-1)^p \mathrm{Sp}\,(\Xi_p(f_{(n)})_p) = \sum_{i=1}^{N} \mathrm{ind}\,(f_{(n)}, x_i),$$

where $(f_{(n)})_p : C_p(K_{(n)}; \mathbf{R}) \to C_p(K; \mathbf{R})$ is the homomorphism induced by the mapping $f_{(n)}$, and $\Xi_p : C_p(K; \mathbf{R}) \to C_p(K_{(n)}; \mathbf{R})$ is the chain homomorphism associating an oriented p–dimensional simplex $\tau^p \in K$ with the sum of the compatibly oriented p–dimensional simplices $s_i^p \in K_{(n)}$ into which the simplex τ^p is subdivided (here, we are justified to consider closed simplices).

Compatibility of orientations is introduced similarly to that for the barycentric subdivision in Ch. 5, § 3, subsection 4. It is clear that $\mathrm{Sp}\,(\Xi_p(f_{(n)})_p) = 0$ when $p < n$, since when $p < n$, fixed simplices of the mapping $f_{(n)}$ do not exist. It remains to show that

$$(23) \qquad (-1)^n \mathrm{Sp}\,(\Xi_n(f_{(n)})_n) = \sum_{i=1}^{N} \mathrm{ind}\,(f_{(n)}, x_i).$$

The trace of a chain homomorphism is equal to the sum of the coefficients a_{jj} for all n–dimensional simplices τ_j^n (elements of a basis) in the matrix of this chain homomorphism regarded as a linear operator. We shall explain what these coefficients represent. It can be seen from the construction of homomorphisms Ξ_* and $(f_{(n)})_n$, that the coefficient of τ_j^n differs from 0 only if τ_j^n is a fixed simplex of the mapping $f_{(n)}$, i.e., $\tau_j^n \subset T^n = f_{(n)}(\tau_j^n)$. In this case, the coefficient of a fixed simplex τ_i^n can only be equal to ± 1 depending on whether, in the complex K, the orientation of the image $f_{(n)}(\tau_i^n) = T^n$ is compatible with the one of the simplex T^n or not. Now we pass from the affine mapping $f_{(n)}$ of carrier n–plane of the simplex τ_i^n to the linear mapping $h_i f_{(n)} h_i^{-1}$. Identifying the n–plane with a subspace, and the subspace with $\mathbf{R}^n$, we can write $h_i f_{(n)} h_i^{-1}$ in the form of $F : \mathbf{R}^n \to \mathbf{R}^n$, $F(y) = f_{(n)}(x_i + y) - x_i$. Then it is clear that the desired coefficient of τ_i^n is equal to sign det F, because it is the sign of the determinant that determines whether the orientation of a basis of the linear space $\mathbf{R}^n$ is preserved or not, and, thus, whether the orientation of the simplex is preserved or not. It remains to show that

$$(24) \qquad \mathrm{sign\ det}\ F = (-1)^n \mathrm{ind}\,(f_{(n)}, x_i).$$

Note that sign det F is the index of the isolated singular point x_i of the vector field $f_{(n)}(x) - x_i$ defined on a closure $\bar{\tau}^n$ of the simplex τ_i^n. We construct a homotopy connecting this vector field with the field $f_{(n)}(x) - x = (f_{(n)} - 1_{\mathbf{R}^n})(x)$ on $\bar{\tau}_i^n$ without zeroes on $\partial \tau^n$.

To start with, we shall "extend" the vector field $f_{(n)} - 1_{\mathbf{R}^n}$ by "drawing away" the images of $f_{(n)}(x)$ from the boundary of the simplex τ_i^n. This is done by the homotopy

$$(25) \qquad \Phi_t(x) = \frac{f_{(n)}(x) - x}{\|f_{(n)}(x) - x\|}(\|f_{(n)}(x) - x\| + t \cdot d \cdot w(x)),$$

where t is the homotopy parameter, $0 \le t \le 1$, d the diameter of the simplex τ_i^n, $w(x)$ a continuous function on $\bar{\tau}_i^n$ such that $w(x_i) = 0$ and $w(x) = 1$, when

$x \in \partial \tau_i^n$. Note that $\Phi_0(x) = f_{(n)}(x) - x$, $\Phi_1(x)$ is a vector field on $\bar{\tau}_i^n$ whose vectors on the boundary are directed outwards, i.e., their ends lie outside τ_i^n. Therefore, the following two linear homotopies can be carried out. The first of them,

$$G_s(x) = \Phi_1(x) + s(x - x_i), \quad 0 \le s \le 1,$$

connects the field $G_0(x) = \Phi_1(x)$ with the field

$$G_1(x) = \Phi_1(x) + x - x_i =$$

$$= (f_{(n)} - 1_{\mathbf{R}^n})(x) + \frac{d \cdot w(x)}{\|(f_{(n)} - 1_{\mathbf{R}^n})x\|} \cdot (f_{(n)} - 1_{\mathbf{R}^n})(x) + x - x_i.$$

The second homotopy

$$H_\gamma(x) = G_1(x) - \gamma \Big[x - x_i +$$

$$+ \frac{d \cdot w(x)}{\|(f_{(n)} - 1_{\mathbf{R}^n})(x)\|} (f_{(n)} - 1_{\mathbf{R}^n})(x) \Big], \quad 0 \le \gamma \le 1,$$

connects the field $H_0(x) = G_1(x)$ with the field $H_1(x) = f_{(n)}(x) - x_i$.

Thus, the homotopies Φ_t, G_s, and H_γ applied sucessively, connect the vector field $f_{(n)} - x = (f_{(n)} - 1_{\mathbf{R}^n})(x)$ with the vector field $f_{(n)}(x) - x_i$.

Exercise 15°. Verify that all these homotopies have no zeros on the boundary $\partial \tau_i^n$ of the simplex τ_i^n.

We obtain the following chain of equalities:

$$\text{sign det } F = \text{ind}\,(x_i, f_{(n)}(x) - x_i) =$$

$$= \text{ind}\,(x_i, f_{(n)} - 1^{\mathbf{R}^n}) = (-1)^n \cdot \text{ind}\,(x_i, 1_{\mathbf{R}^n} - f_{(n)}) =$$

$$= (-1)^n \cdot \text{ind}\,(f_{(n)}, x_i),$$

which gives equality (24). The first of these equalities is explained above, the second one is given by the homotopies constructed, the third equality arises from the results of the exercise, and the fourth equality follows from the definition of the index of a fixed point. Summing equality (24) over all fixed points x_i, we obtain that

(26)
$$\text{Sp}\,(\Xi_n (f_{(n)})_n) = \sum_{i=1}^{N} (-1)^n \text{ind}\,(f_{(n)}, x_i),$$

whence equalities (23), (22), and (21) follow immediately. Lemma 7 is proved.
□

Now, we clarify how small $\epsilon > 0$ should be when constructing the approximation $f_{(n)}$ of the mapping f, in order that not only both the Lefschetz numbers Λ_f, but also $\Lambda_{f_{(n)}}$ and the sums of indices of their fixed points be equal.

If the mapping f has only regular isolated, fixed points $y_1, \ldots, y_Q$, then the initial triangulation K of a polyhedron X can be chosen at the very beginning so that each point y_j is inside its n–dimensional simplex τ_j^n, $j = 1, \ldots, Q$. We select the subdivision $K_{(0)}$ in such a way that all the points y_j are inside n–dimensional simplices $s_j^n \subset \tau_j^n$ which are so small that $f(s_j^n) \subset \tau_j^n$. This can be achieved by "stirring" slightly a sufficiently small barycentric subdivision of the initial triangulation K. Now, put

$$\alpha_j = \min_{x \in \partial S_j^n} \rho(x, f(x)), \quad \delta = \min_{x \in X \setminus \bigcup_{j=1}^{Q} s_j^n} \rho(x, f(x)).$$

Note that $0 < \delta \leq \min(\alpha_1, \ldots, \alpha_Q)$. Now, take $\epsilon < \delta$, and construct, as in lemma 5, the approximation $f_{(n)}$ with respect to ϵ. Since $\epsilon < \delta$, the mapping $f_{(n)}$ has no fixed points outside $\bigcup_{j=1}^{Q} s_j^n$. Further, since $\epsilon < \alpha_j$, the characteristics of the vector fields $(1_{\mathbf{R}^n} - h_j f h_j^{-1}, 1_{\mathbf{R}^n} - h_j f_{(n)} h_j^{-1})$ coincide on the boundaries ∂s_j^n of simplices s_j^n. This follows from the Reuth theorem which we leave as an exercise to the reader to prove.

Exercise 16°. Prove the following statement known as the Reuth theorem. Let vector fields ϕ and ψ be defined on a set $B \subset \mathbf{R}^n$ be such that $\|\phi(x)\| > 0$ and $\|\phi(x) - \psi(x)\| < \|\phi(x)\|$ for all $x \in B$. Then the vector fields ϕ and ψ are homotopic without zero vectors on B.

By the choice of s_j^n and the construction of $f_{(n)}$, we obtain $\text{ind}(f, y_j) = \text{ind}(f, x_i)$, for $j = 1, \ldots, Q$, and

(27)
$$\sum_{j=1}^{Q} \text{ind}(f, y_j) = \sum_{j=1}^{Q} \text{ind}(f_{(n)}, x_i).$$

As it was indicated above (exercise 5°) sufficiently near mappings are homotopic, therefore, for a sufficiently small ϵ, all ϵ–approximations are homotopic to the initial mapping so that $\Lambda_f = \Lambda_{f_{(n)}}$ for a sufficiently small $\epsilon > 0$.

The statement of the theorem is proved completely for a polyhedron $|K|$ of a simplicial complex. For the case of a "curvilinear" polyhedron X homeomorphic to $|K|$, the proof is carried over in an obvious way. The Lefschetz–Hopf theorem is proved.   □

In conclusion, it remains to note that for an arbitrary continuous mapping f of a compact polyhedron X into itself, which not necessarily has only regular isolated, fixed points, formula (17) can also be applied if by the sum of indices of its fixed points we understand the sum of the indices of regular isolated, fixed points of a special simplicial approximation of the mapping f.

REVIEW OF THE RECOMMENDED LITERATURE

Contemporary monographs providing a systematic presentation of homology theory and its applications are [23, 24, 53, 72, 84, 79, 33, 90, 39].

For special questions the following literature is useful:

The origins and development of the homology theory in [66].

Homology of chain complexes in [52].

Simplicial homology theory in [67].

Singular homology theory in [47, 85, 29].

The axiomatic approach to homology theory in [27].

Alexander–Čech homology theory in [85].

Lefschetz number, degree of a mapping, characteristic of a vector field and index of a singular point based on the simplicial homology theory in [1, 67], see also [23].

Cellular homology theory in [29].

Triangulation of smooth manifolds in [56].

The sum of indices of singular points of a vector field on a manifold in [55, 57].

A problem book on homology theory [59, 63].

REFERENCES

1. Alexandrov P. S., *Combinatorial Topology*, Gostechizdat, Moscow, 1947, 660 pp. (Russian)
2. Alexandrov P. S., *Introduction to Set Theory and General Topology*, Nauka, Moscow, 1977, 368 pp. (Russian)
3. Alexandrov P. S., Pasynkov B. A., *Introduction to Dimension Theory*, Nauka, Moscow, 1973, 576 pp. (Russian)
4. Alexandrov P. S., Uryson P. S., *A Memoir on Compact Topological Spaces*, Nauka, Moscow, 1971, 144 pp. (Russian)
5. Alexandryan R. A., Mirzakhan E. A., *General Topology*, Vysshaya shkola, Moscow, 1979, 336 pp. (Russian)
6. Aminov Yu. A, *Differential geometry and Topology of Curves*, Nauka, Moscow, 1987, 160 pp. (Russian)
7. Archangelsky A. V., Ponomariov V. I., *First Course of General Topology in Problems and Exercises*, Nauka, Moscow, 1974, 424 pp. (Russian); English translation, Kluwer Academic Publishers, 1984.
8. Archangelsky A. V., Fedorchuk V. V., *Basic Concepts and Construction of General Topology*, Sovr. Probl. Matematiki. Fund. Napr. 17 (1987), VINITI AN USSR, Moscow, 3-110. (Russian)
9. Arnold V. I., *Mathematical Methods in Classical Mechanics*, Nauka, Moscow, 1979, 432 pp. (Russian); English translation.
10. Arnold V. I., *Catastrophe Theory*, MGU, Moscow, 1983, 80 pp. (Russian); English translation, Springer, 1984.
11. Arnold V. I., *Ordinary Differential Equations*, Nauka, Moscow, 1984, 272 pp. (Russian)
12. Arnold V. I., Varchenko A. N., and Gusein–Zade S. M., *Singularities of Differentiable Mappings. Classification of critical points, Caustics and Wave Fronts*, Nauka, Moscow, 1982, 304 pp. (Russian); English translation, Birkhäuser, 1985.
13. Arnold V. I., Varchenko A. N., and Gusein–Zade S. M., *Singularities of Differentiable Mappings. Monodromy and Asymptotics of Integrals*, Nauka, Moscow, 1984, 336 pp. (Russian); English translation Birkhäuser, 1991.
14. Boltyansky V. G., Efremovich V. A., *Visual Topology*, Nauka, Moscow, 1983, 160 pp. (Russian)
15. Borisovich Yu. G, Gelman B. D., Myshkis A. D., and Obukhovsky V. V., *Introduction to the Theory of Multivalued Mappings*, Izd. Voronezh Univ., Voronezh, 1989, 104 pp. (Russian)
16. Bourbaki N., *General Topology. Basic structures*, Nauka, Moscow, 1968, 272 pp. (Russian)
17. Chernavsky A. V., Matveyev S. V., *Outline of Topology of Manifolds*, MGU, Krasnodar, 1974, 176 pp. (Russian)
18. Courant R., Robbins H., *What is Mathematics?*, Oxford Univ. Press, New York, 1941.
19. Coxeter H. S. M., *Introduction to Geometry*, John Wiley and Sons, New York–London, 1965.
20. Crowell R. H., Fox R. H., *Introduction to Knot Theory*, Springer, New York, 1977.

21. Dao Chong Thi, Fomenko A. T., *Minimal Surfaces and the Plateau problem*, Nauka, Moscow, 1987, 312 pp. (Russian); English translation, Gordon and Breach.

22. Dieudonné J., *Foundations of Modern Analysis*, Academic Press, New York, 1969.

23. Dold A., *Lectures on Algebraic Topology*, Springer Verlag, Berlin–Heidelberg-New York, 1972.

24. Dubrovin B. A., Novikov S. P., and Fomenko A. T., *Modern Geometry. Methods of Homology Theory*, Nauka, Moscow, 1984, 344 pp. (Russian); English translation, Springer, 1990.

25. Dubrovin B. A., Novikov S. P., and Fomenko A. T., *Modern Geometry: Methods and Applications*, Nauka, Moscow, 1986, 760 pp. (Russian); English translation , Springer, 1984, 1987.

26. Efremovich V. A., *Basic Concepts of Topology*, Encyclopedia of Elementary Mathematics. V. 5. Geometry, Nauka, Moscow, 1966, pp. 476–556. (Russian)

27. Eilenberg S., Steenrod N., *Foundations of Algebraic Topology*, Prinston Univ. Press, 1952.

28. Engelking R., *General Topology*, Heldermann, 1989.

29. Fomenko A. T., *Differential Geometry and Topology. Additional chapters*, MGU, Moscow, 1983, 216 pp. (Russian)

30. Fomenko A. T., *Variational Problems in Topology*, MGU, Moscow, 1984, 216 pp. (Russian); English translation, Kluwer Acad. Publishers, 1990.

31. Förster O., *Riemannsche Flächen*, Springer-Verlag, Heidelberg, 1979.

32. Freed D., Ulenbeck K., *Instantons and Four-dimensional Manifolds*, Springer, 1987.

33. Fuks D. B., Fomenko A. T., and Gutenmacher V. L., *Homotopy Theory*, MGU, Moscow, 1969, 460 pp. (Russian)

34. Gardner M., *New Mathematical diversions from Scientific American*, Simon and Schuster, 1966.

35. ———, *The unexpected hanging and other mathematical doversions*, Simon and Schuster, 1969.

36. Golubitzky M., Guillemin V., *Stable Mappings and Their Singularities*, Springer, New York, 1974.

37. Grosberg A. Yu., Khokhlov A. R., *Polymers and biopolymers: from the point of view of physisists–theoreticians*, Future of Science **18** (1985), Znanie, Moscow, 122–132. (Russian)

38. Hilbert D., Cohn–Vossen S., *Anschauliche Geometrie*, Springer, Berlin, 1932.

39. Hilton P. J., Wylie S., *Homology Theory*, Cambridge Univ. Press, Cambridge, 1960.

40. Hirsch M. W., *Differential Topology*, Springer–Verlag, New York–Berlin–Heidelberg, 1976.

41.   *History of Soviet Mathematics*, vol. 3, Ch. 9, Naukova Dumka, Kiev, 1968. (Russian)

42. Hu S.-T., *Homotopy Theory*, Academic press, New York–London, 1959.

43. Husemoller D., *Fibre Bundles*, MacGraw–Hill, 1975.

44. Kazakov D. I., *Microworld beyond imagination*, Future of Science **20** (1987), Znanie, Moscow, 70–87. (Russian)

45. Kelley J. L., *General Topology*, Nostrand; Springer–Verlag, New York, 1975.

46. Kolmogorov A. N., Fomin S. V., *Elements of Function Theory and Functional Analysis*, Nauka, Moscow, 1981, 544 pp. (Russian)

47. Kosniowski C., *A First Course in Algebraic Topology*, Cambridge Univ. Press, Cambridge, 1980.

48. Krasnosel'skii M. A., Zabreiko P. P., *Geometric Methods of Nonlinear Analysis (Grundlehren der mathematischen Wissenschaften, vol. 263)*, Springer, 1984.

49. Kudryavtsev L. D., *Mathematical Analysis* Vol I, II, Vysshaya Shkola, Moscow, 1981, **1** 687 pp., **2** 584 pp. (Russian)

50. Kuratovski K., *Topology* vol I, II, Mir, Moscow, 1966, **1** 594 pp., 1969, **2** 624 pp. (Russian)

51. Lusternik L. A., Sobolev V. I., *Elements of Functional Analysis*, Frederick Ungar Publishing Company, New York, 1961.

52. MacLane S., *Homology*, Springer–Verlag, Berlin–Göttingen–Heidelberg, 1963.

53. Massey W, *Homology and Cohomology Theory*, Marcel Dekker, New York–Baasel, 1978.

54. Massey W., Stallings J., *Algebraic Topology. Introduction*, Mir, Moscow, 1977, 278 pp. (Russian)

55. Milnor J., *Morse Theory*, Princeton Univ. Press, Princeton, 1963.

56. Milnor J., Stasheff J., *Characteristic Classes*, Princeton Univ. Press, Princeton, 1974.

57. Milnor J., Weaver D. W., *Topology from the differentiable viewpoint*, Univ. Pr. of Virginia, 1969.

58. Mischenko A. S., *Vector bundles and their applications*, Nauka, Moscow, 1984, 208 pp. (Russian)

59. Mischenko A. S., Solovyev Yu. P., and Fomenko A. T., *Problems in Differential Geometry and Topology*, MGU, Moscow, 1981, 183 pp. (Russian)

60. Mischenko A. S., Fomenko A. T., *A course of Differential Geometry and Topology*, MGU, Moscow, 1980, 439 pp. (Russian)

61. Narasimhan R., *Analysis on Real and Complex Manifolds*, Masson & Cie, Editeur. North–Holland Publ. Comp., Paris, Amsterdam, 1968.

62. Novikov C. P., *Topology*, Sovr. Probl. Matematiki. Fund. Napravleniya **12** (1986), VINITI AN USSR, 5–252. (Russian)

63. Novikov C. P., Mishchenko A. S., Solovyev Yu. P., and Fomenko A. T., *Problems in Geometry. Differential Geometry and Topology*, MGU, Moscow, 1978, 164 pp. (Russian)

64. Novikov C. P., Fomenko A. T., *Elements of Differential Geometry and Topology*, Nauka, Moscow, 1987, 432 pp. (Russian); English translation, Kluwer Acad. Publishers, 1990.

65. *Physics Today*, American Institute of Physics, May,1983, p. 26, p. 48..

66. Poincaré A., *Collected works in 3 vol.*, vol. 2, 1972, 999 pp., vol 3 1974, 771 pp. (Russian)

67. Pontryagin L. S., *Outline of Combinatorial Topology*, Nauka, Moscow, 1976. (Russian)

68. Pontryagin L. S., *Continuous Groups*, Nauka, Moscow, 1984, 520 pp. (Russian)

69. Pontryagin L. S., *Smooth Manifolds and Their Application to Homotopy Theory*, Nauka, Moscow, 1985, 174 pp. (Russian)

70. Postnikov M. M., *Intorduction to Morse Theory*, Nauka, Moscow, 1971, 568 pp. (Russian)

71. Postnikov M. M., *Lectures on Algebraic Topology*, Nauka, Moscow, 1984, 416 pp. (Russian)

72. Postnikov M. M., *Lectures on Algebraic Topology. Homotopy Theory of Cell Spaces*, Nauka, Moscow, 1985, 336 pp. (Russian)

73. Postnikov M. M., *Lectures on Geometry. Semester III. Smooth Manifolds*, Nauka, Moscow, 1987, 480 pp. (Russian)

74. Rohlin V. A., Fuks D. B, *First Course of Topology. Geometric Chapters*, Nauka, Moscow, 1977, 488 pp. (Russian)

75. *Quantum Liquids and Crystals*, Mir, Moscow, 1979, p. 9–42. (Russian)

76. Schwartz A. S., *Quantum Theory of a Field and Topology*, Nauka, Moscow, 1989, 400 pp. (Russian)

77. Shabat B. V., *Introduction to Complex Analysis*, Nauka, Moscow, 1976, 320 pp. (Russian)

78. Siniukov H. S., Matveenko T. I., *Topology*, Visha Shkola, Kiev, 1984, 264 pp. (Russian)

79. Spanier E., *Algebraic Topology*, McGraw–Hill, New York, 1966.
80. Springer G., *Introduction to Riemann Surfaces*, Addison–Wesley, Reading, 1957.
81. Steenrod N., *The Topology of Fibre Bundles*, Princeton Univ. Press, Princeton, 1951.
82. Steenrod N., Chinn W. G, *First Concepts in Topology*, Randon House, 1966.
83. Sternberg S, *Lectures on Differential Geometry*, Prentice Hall, Englewood Cliffs, N.J., 1964.
84. Switzer R. M., *Algebraic Topology. Homotopy and Homology*, Springer.
85. Teleman c., *Elemente de Topologie si Variei Differentiable*, Bucharest, 1964. (Romanian)
86. Volovik G. E., Mineev V. P, *Physics and Topology*, Mir, Moscow, 1972, 496 pp. (Russian)
87. Warner F., *Foundations of Differentiable Manifolds and Lie Groups*, Springer–Verlag, 1983.
88. Seifert H., Trelfall V., *Lehrbruch der Topologie*, Shelsea reprint, 1968; Originally: Teubor, 1934.

Additional references

89. Fomenko A. T., *Visual Geometry and Topology*, Springer, 1994.
90. Fomenko A. T., Fuks D. B., *Course of Homotopic Topology*, KAP (to appear).

SUBJECT INDEX

About the authors

Yu. G. Borisovich. Dr. Sci. (Phys. and Math.), Professor, head of the Department of Algebra and Topological Methods of Analysis of the Voronezh University. His special interests are nonlinear functional analysis, fixed point theory and topology of set–valued mappings. His publications include about two hundred scientific articles and books developing topological methods in nonlinear problems.

Prof. Borisovich is the editor of the series "New Ideas in Global Analysis", which is as a subseries published in "Lecture Notes in Mathematics".

N. M. Blizniakov. Cand. Sci. (Phys. and Math.), is a lecturer at the department of Algebra and Topological Methods of Analysis of the Voronezh University. He has published around forty scientific papers and obtained important results on the problem of calculation of the topological index in the theory of singularities.

T. N. Fomenko. Cand. Sci. (Phys. and Math.), is a lecturer at the Department of Mathematics of the Moscow Institute of Steel and Alloys (Technical University). T. Fomenko has published over thirty scientific works. These include results on Smith theory, on the existence problem of equivariant mappings, and some other problems of algebraic topology.

Ya. A. Izrailevich. Cand. Sci. (Phys. and Math.), is a lecturer at the Department of Mathematical Analysis of the Voronezh University. Ya. Izrailevich has published around forty papers on mathematics and computer science, and obtained interesting results on Smith theory and equivariant mappings of spheres in algebraic topology.

A. T. Fomenko Dr. Sci. (Phys. and Math.), member of the Russian Academy of Science, is the head of the Department of Differential Geometry and Applications of Moscow University. A. Fomenko is a distinguished specialist in topology, differential geometry, minimal surface theory, Hamiltonian system theory. Among other things, he has a passion for painting and drawing. At the request of the authors of this book, A. Fomenko made illustrations in his own abstract style for it. (More of his drawings can be found in his album "Mathematical Impressions", published by Amer. Math. Society, 1990, and in the book "Visual Geometry", Springer, 1994, and also in a number of mathematical publications, where his drawings serve as illustrations.)

About the book

This textbook is mostly a manual of differential topology and homology theory. It contains the basic concepts and theorems of general topology and homotopy theory, the classificatiopn of two–dimensional surfaces, an outline of smooth manifold theory, and of the theory of mappings of smooth manifolds. The elements of Morse theory and homology theory with their applications to fixed points are also included in the book. Finally, the role of topology in mathematical analysis, geometry, mechanics, and differential equation theory is demonstrated. Although originally de-signed for university students, this textbook will also be most useful to specialists in other fields of mathematics, teachers of mathematics, and all others who are interested in becoming conversant with the elements of topology.

Kluwer Texts in the Mathematical Sciences

1. A.A. Harms and D.R. Wyman: *Mathematics and Physics of Neutron Radiography.* 1986   ISBN 90-277-2191-2
2. H.A. Mavromatis: *Exercises in Quantum Mechanics.* A Collection of Illustrative Problems and Their Solutions. 1987   ISBN 90-277-2288-9
3. V.I. Kukulin, V.M. Krasnopol'sky and J. Horácek: *Theory of Resonances.* Principles and Applications. 1989   ISBN 90-277-2364-8
4. M. Anderson and Todd Feil: *Lattice-Ordered Groups.* An Introduction. 1988
                                                                                        ISBN 90-277-2643-4
5. J. Avery: *Hyperspherical Harmonics.* Applications in Quantum Theory. 1989
                                                                                        ISBN 0-7923-0165-X
6. H.A. Mavromatis: *Exercises in Quantum Mechanics.* A Collection of Illustrative Problems and Their Solutions. Second Revised Edition. 1992   ISBN 0-7923-1557-X
7. G. Micula and P. Pavel: *Differential and Integral Equations through Practical Problems and Exercises.* 1992   ISBN 0-7923-1890-0
8. W.S. Anglin: *The Queen of Mathematics.* An Introduction to Number Theory. 1995
                                                                                        ISBN 0-7923-3287-3
9. Y.G. Borisovich, N.M. Bliznyakov, T.N. Fomenko and Y.A. Izrailevich: *Introduction to Differential and Algebraic Topology.* 1995   ISBN 0-7923-3499-X

KLUWER ACADEMIC PUBLISHERS – DORDRECHT / BOSTON / LONDON